HANDBOOK OF
SEALANT TECHNOLOGY

HANDBOOK OF
SEALANT
TECHNOLOGY

EDITED BY
K.L. MITTAL
A. PIZZI

CRC Press
Taylor & Francis Group
Boca Raton London New York

CRC Press is an imprint of the
Taylor & Francis Group, an **informa** business

CRC Press
Taylor & Francis Group
6000 Broken Sound Parkway NW, Suite 300
Boca Raton, FL 33487-2742

First issued in paperback 2020

© 2009 by Taylor and Francis Group, LLC
CRC Press is an imprint of Taylor & Francis Group, an Informa business

No claim to original U.S. Government works

ISBN-13: 978-0-367-57728-5 (pbk)
ISBN-13: 978-0-8493-9162-0 (hbk)
ISBN-13: 978-0-429-19143-5 (ebk)

DOI: 10.1201/9781420008630

Library of Congress Cataloging-in-Publication Data

Handbook of sealant technology / [edited by] K.L. Mittal and A. Pizzi.
    p. cm.
   Includes bibliographical references and index.
   ISBN 978-0-8493-9162-0 (hard back : alk. paper)
   1. Sealing compounds. I. Mittal, K. L., 1945- II. Pizzi, A. (Antonio), 1946- III. Title.

TP988.H36 2010
620.1'99--dc22

2009022075

Visit the Taylor & Francis Web site at
http://www.taylorandfrancis.com

and the CRC Press Web site at
http://www.crcpress.com

Contents

SECTION I   General

SECTION II   Testing and Durability of Sealants and Sealed Joints

SECTION III   Various Types of Sealants and Their Applications

Preface

Sealants have been used for many millennia, as prehistoric people used natural materials, such as earth, loam, grass and reeds, to seal and protect the interior of their homes against the weather.

Sealants are used to seal joints and openings (gaps) between two or more materials. The primary purpose of a sealant is to prevent air, water and other substances from entering or exiting a structure or assembly while permitting a certain amount of movement of mating partners. Although adhesives and sealants are different materials with different characteristics and are used for different purposes, sometimes the distinction between the two is not clear. Generally speaking, the primary purpose of a sealant is to seal a joint, i.e., to prevent leaks, with its adhesion and movement capability being important and requisite properties. Adhesives, on the other hand, are employed to hold similar or dissimilar materials together often as alternatives to other fastening systems, i.e., riveting, nailing, welding, brazing, etc., as adhesive bonding offers many advantages, *inter alia*, more uniform stress distribution over mechanical approaches to hold materials together. Concomitantly, there are some common and some distinctly different requirements for these two classes of materials. However, some materials can fulfill requirements of both sealants and adhesives, and these are termed adhesive-sealants.

Sealants are used in a host of applications in myriad industries and other areas ranging from construction to aerospace to marine to electronics to surgery. For example, sealants are used to seal joints in kitchen and bathrooms. Polysulfide sealants, which are highly resistant to solvents and chemicals, are used to line fuel tanks of aircraft and to seal different riveted joints in aircraft. Sealants which have good water resistance are extensively used in ship building. At the other extreme, fibrin sealants are used as tissue sealants and are very important in surgical procedures to seal wounds instead of using sutures. These few eclectic examples illustrate that sealants, in essence, are used for mundane to sophisticated applications.

Considering the importance and economic impact of sealants, there has been tremendous R&D activity in designing new and improved sealant materials or in ameliorating the existing ones to meet current and future performance requirements and environmental regulations. Also these days the mantra is "Go Green, Think Green." So currently there is keen interest in "green" materials in all sectors, and sealants are no exception. Also it should be underscored that for long-term functioning of sealants their durability against detrimental environmental factors (high humidity, temperature fluctuations, radiation) is of paramount importance, and much effort has been expended in enhancing durability of sealants and sealed joints.

In light of the above, we decided to consolidate recent developments and current activity in the arena of sealants in a single source—this book. This book, containing 17 chapters, is divided into three Sections. The first Section starts with a fascinating history of sealants and takes the reader through a journey how sealants have evolved from natural materials (used by prehistoric people) to synthetic polymers and modern sealant technology. The second chapter deals with the use of silanes in promoting adhesion of sealants. Adequate adhesion of a sealant is *sine qua non* for the functioning of a sealant, and thus adhesion constitutes one of the core requirements of a sealant. Section II encompasses six chapters delving into the testing and durability aspects of sealants and sealed joints. Both durability and testing are of cardinal importance. The topics covered in this Section include: Electrochemical Impedance Spectroscopy (EIS) in sealant testing; chemorheological investigations on the environmental susceptibility of sealants; natural and artificial weathering of sealants; sealant durability and service life of sealed joints; adhesion testing of sealants; and answering the LEED™ challenge to sealant and waterproofing products. Section III expatiates upon many different types of sealants and their very wide ranging applications. Sealants discussed include hot melt sealants, intumescent sealants, waterproofing membrane systems, foam back-up materials, construction sealants, self-adhering flashing products in building openings, and sealants for bridge expansion joints. Section III also covers the application of sealants in automotive electronics packaging and concludes with the successful use of fibrin sealants as tissue sealants in terms of tissue compatibility, biodegradation and clinical utility.

In essence, this book, covering the various ramifications of sealants, reflects the cumulative wisdom of many scientists and technologists actively engaged in the wonderful world of sealants. Also this book provides a commentary on the current state of knowledge related to sealant materials. Furthermore, this book, containing bountiful information, represents a single, easily accessible source. The book is profusely illustrated and copiously referenced. We sincerely hope anyone interested/involved in the broad domain of sealants will find this book of great value and interest.

In closing, we would like to express our thanks to the contributors for their interest, enthusiasm and cooperation without which this book would not have seen the light of day. Also we extend our appreciation to Barbara Glunn for her interest in this book project and members of the production staff at Taylor & Francis for incarnating this book.

K.L. Mittal

A. Pizzi

The Editors

Kashmiri Lal Mittal received his Ph.D. degree from the University of Southern California in 1970 and was associated with the IBM Corp. from 1972 to 1993. He is currently teaching and consulting worldwide in the areas of adhesion and surface cleaning. He is the editor of 97 published books, as well as others that are in the process of publication, within the realms of surface and colloid science and of adhesion. He has received many awards and honors and is listed in many biographical reference works. Mittal is a founding editor of the international *Journal of Adhesion Science and Technology* and has served on the editorial boards of a number of scientific and technical journals. Mittal was recognized for his contributions and accomplishments by the worldwide adhesion community by organizing on his 50th birthday the First International Congress on Adhesion Science and Technology in Amsterdam in 1995. In 2002 he was honored by the global surfactant community, which inaugurated the Kash Mittal Award in the surfactant field in his honor. In 2003 he was honored by the Maria Curie-Sklodowska University, Lublin, Poland, which awarded him the title of doctor *honoris causa*.

A. Pizzi is Professor of Industrial Chemistry, Ecole Nationale Supérieure des Technologies et Industries du Bois (ENSTIB), Nancy University, Epinal, France. A well-known industrial consultant on adhesives and sealants, Dr. Pizzi is the author or coauthor of more than 500 articles, contract reports, patents, and international conference papers and has received many international awards and honors. He serves on the editorial board of many international scientific journals. He is also the author or coeditor of several books. Dr. Pizzi received the Doctor of Chemistry degree (1969) from the University of Rome, a Ph.D. degree in chemistry (1978) from the University of the Orange Free State, South Africa, and a D.Sc. degree (1978) in wood science and technology from the University of Stellenbosch, South Africa.

Contributors

Robert E. Belke, Jr.
Visteon Corporation
Van Buren Township, Michigan

D. J. Benatti
National Institute of Standards and
Technology
Gaithersburg, Maryland

Havazelet Bianco-Peled
Department of Chemical Engineering
Inter-Departmental Program for
Biotechnology
Technion-Israel Institute of Technology
Haifa, Israel

Ronit Bitton
Inter-Departmental Program for
Biotechnology
Technion-Israel Institute of Technology
Haifa, Israel

A. P. Cerra
CSIRO Materials Science and
Engineering
Interface Engineering and Intelligent
Materials Surfaces Team
Melbourne, Australia

Guy D. Davis
DACCO SCI, Inc.
Columbia, Maryland

James Dunaway
Tremco, Inc.
Beachwood, Ohio

W. (Voytek) S. Gutowski
CSIRO Materials Science and
Engineering
Interface Engineering and Intelligent
Materials Surfaces Team
Melbourne, Australia

Pamela K. Hernandez
Tremco, Inc.
Beachwood, Ohio

Misty Huang
Momentive Performance Materials
Tarrytown, New York

Nyeleti S. Hudson
DuPont Building Innovations
Richmond, Virginia

Ju-Ming Hung
National Starch and Chemical Company
Bridgewater, New Jersey

D. L. Hunston
Hunston Scientific
Gaithersburg, Maryland

A. R. Hutchinson
Joining Technology Research Centre
School of Technology
Oxford Brookes University
Oxford, United Kingdom

James D. Katsaros
DuPont Building Innovations
Richmond, Virginia

Jerome M. Klosowski
Klosowski Scientific
Bay City, Michigan

Kent Larson
Dow Corning Corporation
Auburn, Michigan

Eric R. Pohl
Momentive Performance Materials
Tarrytown, New York

Norma D. Searle
Consultant
Deerfield Beach, Florida

Michael Schmeida
Manager of Sustainable Technologies
Tremco Global Sealants
Beachwood, Ohio

D. Stanley
National Institute of Standards and
   Technology
Gaithersburg, Maryland

K. T. Tan
National Institute of Standards and
   Technology
Gaithersburg, Maryland

C. C. White
National Institute of Standards and
   Technology
Gaithersburg, Maryland

Andreas T. Wolf
Dow Corning GmbH
Wiesbaden, Germany

Section I

General

General

1 The History of Sealants

Jerome M. Klosowski and Andreas T. Wolf

CONTENTS

1.1 INTRODUCTION

Sealing has been defined as "the art and science of preventing leaks" [1]. Sealants are used to seal joints and openings (gaps) between two or more substrates. The main purpose of a sealant is to prevent air, water, and other substances from entering or exiting a structure while permitting a certain amount of movement of the substrates. The distinction between sealants and adhesives is not always very clear. The primary function of sealants is to seal a joint, with adhesion and movement capability being important properties. Adhesives, on the other hand, are designed to hold materials together by surface attachment, often as alternatives to mechanical fastening systems. Thus, the primary function of adhesives is to transfer loads between adjacent surfaces, with adhesion and structural strength being important properties. Some materials are truly adhesive-sealants and fulfill the dual role of bonding as well as sealing a joint. These materials often are also called *structural sealants*. All sealants must fulfill three basic functions:

1. Sufficiently fill the joint or gap to create an efficient seal
2. Form a barrier to gas or liquid flow
3. Maintain the seal in the operating environment (while permitting a certain amount of movement of sealed parts)

It is known that people have used sealants and adhesive-sealants throughout history.

Fay, in his *History of Adhesive Bonding* [2], has stated that "the hardest part of writing about the history of adhesives is deciding where to start, since it is impossible to ascertain when or where adhesives were first used. Although there is substantial archaeological and written evidence to suggest that humans have used adhesives for thousands of years, there does not appear to have been a single 'Eureka' moment, when their usefulness was first discovered." The same statement holds true for the history of sealants.

1.2   FROM PREHISTORIC BEGINNINGS TO EARLY INDUSTRIAL USES

Sealing is an age-old problem that dates back to our earliest attempts to create a more comfortable living environment. Prehistoric people used natural "sealants" such as earth, loam, grass, and reeds to protect the interior of their homes against the weather. The use of such natural sealants remained the state of the art for tens of thousands of years.

Neanderthals had used processed birch pitch as an adhesive in tool manufacture as early as 80,000 years ago [3]. Tools dating from around 40,000 BC found at Umm el Tlel in Syria used bitumen, which had been processed at high temperature, as an adhesive joining the tools to their handles [4]. Pitch and bitumen appear to have been used throughout antiquity both as sealants and adhesives [5]. The best surviving examples are those found in the ruins of Babylon, dating from around 1,500 BC, which demonstrate the use of filler materials in bitumen used for bonding and sealing red clay bricks [6].

There are several references to successful uses of sealants in the Bible. When God was giving instructions regarding the building of the ark, he told Noah to "cover it inside and out with pitch" (Genesis 6:14). We are also told that the builders of the Tower of Babel used "bitumen for mortar" (Genesis 11:3) and that the bulrush basket, which carried the baby Moses down the river, was sealed with bitumen and pitch (Exodus 2:3). In these applications, the bitumen and pitch materials served the dual purpose of holding the materials together (an adhesive) and keeping out the water (a sealant).

The early ships of the Mediterranean used tar and resin to help hold back the water. This could be considered an early application of sealants in transportation. Pliny the Elder, writing in around 50 BC, mentioned several other natural sealing materials and their applications [7]. Examples include different varieties of "mastich," gum from the Egyptian thorn, and the resin of the pitch pine used for coating wine casks. These natural thermoplastics based on rosin and other resins, waxes, and bitumens represent early hot melts.

Very few written records exist regarding the use of these relatively sophisticated manufactured sealants and adhesives in the period immediately following the decline of the Greek and Roman empires and it is likely that, similar to so many other technologies, they fell out of common use in Europe for several hundred years. However, by the mid-14th century, the Aztecs were using blood albumin compounded with cement as mortar in their buildings. This and other applications of "curable" binders show that crosslinking has ancient beginnings—the drying of blood, the coagulation of egg albumin by heat and light, the treatment of casein with lime, or the air oxidation of drying oils practiced by painters during the Renaissance [8].

By the 17th century, scientists were beginning to give consideration to the nature of adhesion itself. Francis Bacon in his *Novum Organum* suggested that "there is in all bodies a tendency to avoid breaking up" [9]. He further reports that this tendency is weak in homogeneous substances, but more powerful in bodies made up of heterogeneous substances, reasoning that the "addition of heterogeneity unites bodies." In his arguments, he introduced the concepts of "bonding" ("by which bodies refuse to be torn from contact with other bodies") and "cohesion" ("by which bodies, to differing degrees, abhor their own dissolution"). With the use of adhesives in the furniture industry restarting in the 16th century and developing to full industrial use during the 17th and 18th centuries, the science of adhesion and bonding became increasingly important. In 1717, Newton conjectured that "there are agents in Nature able to make the particles of bodies stick together by very strong attractions" [10].

The first glazing putties—made of linseed oil, drying natural resins, and fillers— were used in the 17th century to seal leaded church windows. This puttying technique survived almost unchanged until the 1960s. Linseed oil putty, consisting in its basic form of 85%–87% chalk and 13%–15% linseed oil [11], was originally prepared laboriously by hand by the glaziers themselves. The powdered chalk was first sieved into a wooden or iron box; the required amount of oil was then poured into a channel in the powder, and the two ingredients kneaded by hand until a crumbly, moist consistency was obtained. The mixture was then beaten with an iron hammer or roller-shaped stick, turned over, and the process was continued until the putty became ductile. Next, lumps of putty weighing some 2–3 kg were placed on a stone, where they were further worked, being either vigorously beaten or flung against a stone. This process was repeated until the putty held together and could be pulled into thin strands, an indication that it had reached a consistency sufficiently malleable for later working.

At the beginning of the 20th century, a small number of companies started the production of linseed oil putties on an industrial scale. Developments in building techniques, and in particular the increasing use of glass, caused a rapid growth in the demand for putty. Linseed oil putty was therefore manufactured in large batches of several hundred kilograms at a time, using motor-driven kneading machines (Figure 1.1).

The finished products were delivered by horse cart to small, specialized retailers, who sold the putty to glaziers (Figure 1.2). They would then install the glass on-site into the window frame using the glazing putty.

The limited availability of linseed oil during World War I led to the identification of other drying oils as binder in the newly emerging oil-based putties. Oil-based putties, which are also often termed *caulks*, are made primarily with calcium carbonate (chalk) and a drying oil, such as linseed, fish, soybean, tung, or castor oil. These putties had a relatively limited movement capability of ±2% to ±5%. Despite their low movement capability, the oil-based putties generally proved to be adequate for sealing joints in building construction of the first half of the 20th century, particularly when painted regularly with oil-based coatings. The coating protected the putty from exposure to sunlight and reduced the rate of oxidation induced by atmospheric oxygen. The drying oil contained in the paint also replenished some of the binder in the putty that was lost as a consequence of the weathering processes. Still, most oil-based putties seldom had more than 5–10 year's life expectancy.

FIGURE 1.1 Production of linseed oil putty on an industrial scale at the beginning of the 20th century. (Courtesy: Dow Corning GmbH.)

FIGURE 1.2 Delivery of glazing putties and accessory materials by horse cart at the beginning of the 20th century. (Courtesy: Dow Corning GmbH.)

1.3   THE ADVENT OF SYNTHETIC POLYMERS AND MODERN SEALANT TECHNOLOGY

Prior to the 1920s, all sealants were derived from vegetable, animal, or mineral substances. The development of modern polymeric sealants coincided with the development of the polymer industry itself, starting sometime in the mid-1920s. Within a short, highly creative period of about 20 years, the key synthetic polymers used in modern sealants were developed and produced on an industrial scale.

Acrylic polymers, both thermoplastic and thermosetting, are among the oldest of synthetic polymers. Acrylic esters were known in the 1890s and reported in 1901 [12]. Commercial production of acrylate ester polymers in dispersion form was begun in 1929 [8]. Butyl rubber, a popular polymer for solvent-borne caulks, was the invention of Thomas and Sparks [13], who, in the 1930s, copolymerized isobutylene with small proportions of diene to provide crosslinking sites. In 1937, Otto Bayer patented the polyaddition reaction of polyisocyanates with polyols that led to the production of high-molecular-weight polyurethanes [14]. In the early 1940s, Patrick and Ferguson laid the foundation for the industrial production of polysulfide polymers with the discovery of the wet reduction process [15]. At about the same time, Rochow and Müller independently developed the direct silane synthesis process, which, when combined with the subsequent hydrolysis or methanolysis process, enabled the large-scale manufacture of silicone polymers [16,17].

The war effort accelerated the introduction of synthetic polymers because of the scarcity of natural rubber supply. The use of these synthetic polymers in adhesives and sealants was frequently discovered shortly after the initiation of their industrial production; however, the commercialization process often was hampered by technical problems. Consequently, synthetic-polymer-based sealants became widely available only by the 1950s and 1960s. Table 1.1 provides an overview of the approximate introduction periods for various modern sealant types. In the following subsections, discussion of the history of modern sealants will be structured by material type.

1.3.1   BUTYL SEALANTS

The generic name *butyl sealant* is used for caulks, mastics, or tapes derived from either polyisobutylene (PIB) homopolymer or butyl rubber, which is a copolymer of isobutylene and 1–3 mole% of isoprene. Most butyl sealant formulations are prepared by mixing butyl polymer with various levels of both low- and high-molecular-weight polyisobutylene to modify properties. Butyl-rubber-based sealants are available as pumpable solvent-release sealants, as preformed tapes, and as hot-applied sealants. Polyisobutylene sealants are commercialized as hot-applied sealants or preformed tapes [18].

Standard Oil Company was the first company to commercialize butyl polymers in the 1930s [19]. Their manufacturing process represents the most successful commercial application of low-temperature cationic polymerization using Friedel-Craft halide catalysts. Butyl rubber became commercially available during World War II. At that time, natural rubber supply was cut off from Indonesia, Burma, and

TABLE 1.1

Approximate Introduction Periods for Various Modern Sealant Types

Introduction	Sealant	Base Polymer	Behavior	Primary Applications
Mid-1940s	Butyl	Polyisobutylene; isobutylene/isoprene copolymer	Thermoplastic hot melt; Elastomeric	Construction, DIY, automotive, electric applications
Late 1940s	Polysulfide	Mercapto-endblocked poly(bis(ethyleneoxy) methanedisulfide) backbone	Elastomeric	Construction, civil engineering, aerospace
Late 1950s	Solvent-borne acrylics	Copolymers of acrylate esters and other comonomers	Thermoplastic	Construction
Early 1960s	Silicone	Poly(dimethylsiloxane)	Elastomeric	Construction, DIY, aerospace, electric and electronic applications, automotive
Mid-1960s	Waterborne acrylics	Copolymers of acrylate esters and other comonomers	Thermoplastic to elastomeric	Construction, DIY
Late 1960s	Polyurethane	Polyether (typically polyoxypropylene); Polybutadiene	Elastomeric	Construction, DIY, civil engineering, automotive
Mid-1980s	Waterborne acrylics (plasticized, "siliconized")	Copolymers of acrylate esters and other comonomers	Elastomeric	Construction, DIY
Early/ Mid-1980s	Silicon-curable polyether	Dimethoxymethylsilyl endblocked polyoxypropylene	Elastomeric	Construction, DIY, automotive, aerospace, electric and electronic applications
Early 1990s	Silicon-curable polyurethane	Di- or trimethoxy- methylsilyl endblocked polymer (typically polyoxypropylene)	Elastomeric	Construction, DIY, automotive, aerospace, electric and electronic applications
Mid-1990s	Silicon-curable polyisobutylene	Dimethoxysilyl- endblocked polyisobutylene	Elastomeric	Electronics, construction
Mid-2000s	Silicon-curable acrylics	Dimethoxysilyl-endblocked acrylic copolymer	Elastomeric	Construction

Note: DIY: Do-it-yourself.

the Malay States, and the need for replacement products to support the war effort became critical. This urgency caused the butyl technology to move ahead at a rapid pace. Because of its low air and moisture permeability, one of the first uses of butyl rubber was in the manufacture of inner tubes for automobile car tires.

By the mid-1940s, the first butyl sealants became commercially available. Early uses included seals in electric wiring boxes, gas pipes, automobiles, and airplanes. Because of their good unprimed adhesion to a large variety of substrates, by the mid-1950s, butyl sealants also became popular in the construction industry. Butyl sealants then were used for glazing joints and splice seals in window units and especially in roofing, where they are still used today. With the introduction of organically sealed insulating glass units in the late 1960s, butyl hot melt sealants gained popularity as primary edge seals, since they have a lower moisture vapor transmission rate than secondary seals based on polysulfide, polyurethane, or silicone rubbers.

Present-day construction uses include insulating glass primary seals; heat, ventilation, and air-conditioning (HVAC) duct sealing (tapes); roof glazing and greenhouse glazing (mastics and tapes); window flashing (tape); roofing membrane splice sealant (tape or mastic); roof and wall penetration seals; and corrugated metal and fiber-reinforced cement roofing seals. Butyl sealants are also used in transportation applications, such as automotive glazing and penetration seals, noise control seals, marine porthole, and container seals. Furthermore, butyl sealants are also still used in electrical wire (splice) seals.

1.3.2 POLYSULFIDE SEALANTS

In 1927, J. C. Patrick patented the discovery of a new synthetic rubber derived from the reaction of alkyl halides and inorganic polysulfides, such as ethylene dichloride and sodium tetrasulfide [20]. This was the first synthetic rubber to be discovered outside Europe. Patrick named the material Thiokol, based on the Greek words for *sulfur* and *glue*. Its special property, resistance to fuels and solvents, was recognized immediately. The early products were flexible, preformed seals based on high-molecular-weight polysulfide rubber. During World War II, with its increased demand for fuel-resistant products, the search for moldable sealing materials intensified. In 1941, Patrick and Ferguson made a discovery that would transform the nascent company that Patrick had founded. By reductively cleaving the high-molecular-weight polysulfide polymer with a mixture of sodium hydrogen sulfide and sodium sulfite, they were able to produce low-molecular-weight liquid polysulfides with mercaptan end groups [15]. These liquid polysulfides formed the basis for the first chemically curing elastomeric sealants. They soon found application in sealing fuel tanks on military and civilian aircraft [21].

Building sealants based on liquid polysulfide polymers came to the market in the late 1940s. They were the first high-performance elastic sealants to be commercialized. Their market introduction occurred at the beginning of curtain wall construction in the United States, which is often credited with creating the need for high-performance sealants [22]. However, one could also suggest that the novel polysulfide sealant, which allowed a moving joint to be sealed successfully, enabled this new construction technique and led to the popularity of the curtain wall. The answer to the question of driver or driven is not certain, but surely one could not exist without the other. In either case, in the 1950s, polysulfides became the sealant of choice in the new curtain wall construction, as well as in cars, trains, and planes.

The early polysulfide sealants were two-component products. The two separate components were mixed together just prior to application. One of the components consisted of liquid polysulfide polymer, filler, and plasticizer, while the other component was formulated from lead dioxide paste as curative and plasticizer. In the 1950s, the cost of polysulfide sealant was approximately ten times that of the more traditionally used oil-based caulks [23]. However, as labor costs steadily increased relative to material costs over the next 20 years, two-component polysulfide sealants enjoyed a tremendous rise in popularity during the 1950s and 1960s. Elastomeric polysulfide sealants were simply the right material available at the right time.

In 1961, one-component polysulfide sealants that did not require mixing became available for use. The popularity of polysulfide sealants continued to increase so that, by the early 1970s, there were many polysulfide sealant manufacturers competing for market share. Furthermore, competition from silicone and polyurethane sealant manufacturers became noticeable. With increasing competition came price erosion. In order to retain profitability, polysulfide manufacturers chose to reformulate their products with lower polymer content. For a while this trend was slowed down somewhat when the principal manufacturer of polysulfide polymers decided to place a seal of quality approval on products that maintained respectable performance. However, this polymer manufacturer eventually ceased exercising control and, as the performance of polysulfide building sealants became more erratic and competition from alternative polymer types continued to increase, the use of polysulfide sealants in building construction diminished substantially. Today, certain high-performance market segments are still served with polysulfide sealants. Polysulfides are still widely used to seal fuel tanks and fasteners in aircraft, in insulating glass manufacture, as well as in various high-value civil engineering applications.

Over the years, considerable effort went into the development of new, environmentally more benign curatives for both one- and two-component polysulfide sealants. As mentioned before, lead peroxide was used as a curative in the early two-component polysulfide sealants. The use of lead peroxide has become less common nowadays because of environmental concerns; its use has been phased out entirely in the United States. and in Western Europe (during the 1990s) and Japan (by the end of 2002). Until the mid-1990s, chromate-cured polysulfide sealants had also been applied to the faying (overlapping) surfaces of joints in aircraft in order to eliminate the potential for joint corrosion. However, products containing hexavalent chromium, a known carcinogen, are being phased out and replaced by alternative systems utilizing other corrosion inhibitors. The majority of two-part polysulfide sealants today are based on manganese dioxide curative.

During the 1990s, protection of groundwater and other environmental concerns created an opportunity for the use of polysulfide sealants in sealing pavement joints at gasoline stations around Europe. These products were first specified in the Netherlands, where the water table is very near to the surface. Dutch, and in the meantime German also, legislation requires that all gasoline service stations be sealed to prevent fuel, particularly diesel, from seeping into the groundwater. Furthermore, the technology has been extended to all applications where the water table is under threat from chemicals, such as chemical storage tanks, fuel spill secondary containment areas in airports and tank farms, chemical plants, refineries, etc.

Because of their excellent water resistance, polysulfides have been widely used since the 1960s in the construction of canals, reservoirs, dams, and other water-retaining civil engineering structures. Furthermore, the excellent resistance of some polysulfide sealant formulations to water and microbial attack has been the key factor in their past and present use in wastewater treatment plants.

Insulating glass manufacture today is still by far the largest application for polysulfide sealants in construction, despite the turmoil that was created in 2001 by the sole U.S. manufacturer and key global supplier of polysulfide polymers exiting the market because of concerns over environmental cleanup costs and the poor economic performance of these products. The scarcity of polysulfide polymers thus provided an opportunity for other sealants, such as polyurethanes and silicones, to enter the supply-critical insulating glass market [24].

1.3.3 ACRYLIC SEALANTS

Historically, the development of acrylates and methacrylates, the key building blocks of acrylic sealants, proceeded slowly. Acrylic (propenoic) acid was first prepared by the oxidation of acrolein with silver oxide in 1843 [25], while methacrylic acid was first obtained in 1865 by dehydration and hydrolysis of ethyl-α-hydroxyisobutyrate [26]. Methyl and ethyl acrylates were prepared in 1873, but were not reported to polymerize at that time [27]. In 1880, G. W. A. Kahlbaum reported the successful preparation of poly(methyl acrylate) [28]. Acrylates and methacrylates received serious attention by Otto Röhm during his thesis work in preparation for his doctoral dissertation in 1901 [29]. A quarter of a century elapsed, however, before he was able to translate his research achievements into commercial reality. Based on the continuing work in Röhm's laboratory, the first small-scale production of acrylates began in 1927 by the Röhm and Haas Company in Darmstadt, Germany. In 1931, BASF and Röhm and Haas produced the first stable dispersions based on acrylic acid esters [30]. Since the initial focus in the commercialization of acrylates was on the clear, monolithic, or laminated, safety glasses that had been invented in 1934 [31], not much effort went into developing other applications.

The first solvent-borne acrylic caulks appeared in the markets in the late 1950s. The mid-1960s then saw the commercialization of the first acrylic latex polymer for waterborne caulks. Both solvent- and waterborne acrylic polymer binders were air-drying, thermoplastic resins, and sealants, compounded with these resins, dried by evaporation of solvent or water, respectively. Those early products were low-performance plasticized pigmented caulks used primarily as gap-fillers in interior building applications. The waterborne acrylic latex products were valued both by professionals and do-it-yourselfers for their low price, low toxicity, and easy application. During the 1970s, sales of latex acrylics grew rapidly, while solvent-borne acrylics experienced only modest sales growth. Over the same time period, the performance of acrylic latex sealants, especially their water resistance, continued to improve. This allowed the use of these sealants in certain exterior applications that placed low demands on their movement capability. "Internally plasticized" caulks, based on acrylic polymers with low glass transition temperature (T_g of 40°C), as well as higher solids polymers (with 65% compared to the previous 55% polymer solids content) became available during the same decade. The internally plasticized acrylic latex polymer yielded sealants of substantially improved adhesion when tested after standard "cure" conditions (1 week at 23°C) without resorting to elevated temperature curing (88°C) as was recommended for testing commercial latex sealants [32]. Furthermore, the low glass transition temperature polymer improved the field performance of sealants by retaining flexibility on long-term exposure. Despite the substantial previous technological advances, by the early 1980s, the performance of acrylic latex sealants had reached a plateau and some people already felt that the end of their life cycle was in sight.

A number of key innovations occurred during the 1980s that drastically changed this perception. First, highly plasticized products were introduced in the market in the mid-1980s. These so-called "plasticized" acrylic latex sealants enjoyed a dramatic rise in sales. The term *plasticized acrylic sealant* was somewhat misleading, since the first-generation acrylic latex sealants were already plasticized products. The key difference between the old and the new generation of acrylic latex sealants was the level of plasticizer used. While the old generation had a polymer-to-plasticizer ratio from 3:1 to 4:1 based on solid acrylic polymer content, the new generation was formulated with a much higher plasticizer ratio, close to 2:1 [33]. This resulted in very flexible sealants with movement capability of ±25% and the ability to meet most of the performance requirements of high-performance sealants specifications, such as ASTM C 920. The second important innovation was based on the use of silane adhesion promoters and/or small amounts of unreactive liquid silicones to impart better water resistance to the adhesion of these sealants. The silane or silicone portion typically represented far less than 2% of the total formulation; however, in a clever marketing ploy, these sealants were advertised as "siliconized" acrylics, projecting the image of silicone performance combined with the ease-of-use attributes of latex acrylics. Finally, transparent (clear) and translucent acrylic latex products were introduced, first based on polymer blends and later on single-polymer formulations, which became an important part of the market. With the increased exterior use of acrylic sealants, resistance of the uncured sealant to rainfall became an important consideration. In the 1980s, the use of acrylic latex sealants was still limited because of the possibility of sealant washout, if the sealant was subjected to a moderate-to-heavy rainfall shortly after placement of the sealant in exposed joints (up to 24 h after application). In order to meet this market need, by the end of the 1980s, the first washout resistant formulations were launched for exterior use.

Acrylic-polymer-based sealants generally offer excellent durability in terms of ultraviolet light resistance and chemical resistance. Waterborne acrylic latex sealants further offer the very desirable application properties of ease of gunning, ease of troweling or smoothing, ease of cleanup, elimination of solvent odors and flammability problems, as well as the ability to accept paint top-coats shortly after joint placement of the latex sealant. Furthermore, because the water-dispersed polymer yields low-viscosity vehicles, low-cost formulations with acceptable rheology (extrusion rate, tooling) can be achieved by adding high levels of low-cost extenders (fillers).

Today, cutting-edge formulations of clear, transparent, and pigmented acrylic latex sealants are capable of meeting the stringent performance requirements of architectural sealant specifications. "Siliconized" acrylic latex formulations now represent the vast majority of acrylic sealants sold. Internally plasticized acrylic sealant formulations are available that meet or exceed the performance requirements of ASTM C 920 without the need for a high level of external plasticizer that could cause staining. Solvent-borne acrylics are still used in a limited number of niche applications; however, because of concerns over their flammability and toxicity, other product types are increasingly replacing them.

The majority of acrylic sealants today are sold for building applications. Because of their ease of application, waterborne acrylics are predominantly sold to the Do-It-Yourself (DIY) market. For the same reason, professional applicators also widely use these products in interior restoration work. Solvent-based acrylics are particularly useful in exterior restoration work owing to their excellent unprimed adhesion to a wide range of substrates. However, they are difficult to apply and have a rather unpleasant odor; therefore, their use remains restricted to professional applicators and will diminish over time, as alternative sealant materials with similar performance and a more benign health profile become available.

1.3.4 SILICONE SEALANTS

In 1863, inspired by the syntheses of other organometallic compounds in previous years, Friedel and Crafts prepared the first compounds with silicon-carbon bonds by the reaction of diethyl zinc with silicon tetrachloride [34]. Ladenburg, an organic chemist who joined with Friedel to continue this work, concluded that "the so-called inorganic elements are capable of forming compounds which are analogous to those of carbon" [35]. He later showed that the hydrolysis of $(C_2H_5)_2Si(OC_2H_5)_2$ gave a stable oil instead of a simple volatile compound analogous to diethyl ketone formed from $(C_2H_5)_2C(OC_2H_5)_2$ [36]. Ladenburg chose to call the stable oil with the empirical formula $(C_2H_5)SiO$ *silicon diethyloxide*.

A great impetus to silicone chemistry came from the 54 papers published by Frederick S. Kipping [37] during the period 1899 to 1944 at the University of Nottingham, England. Kipping made use of the Grignard reagents, which were first reported on in 1900 [38], to prepare organosilicon compounds. He was an organic chemist interested in pure compounds that could be isolated by distillation or crystallization. The oils and glues that he often obtained seemed uninviting to him, but he correctly described them as macromolecules [39]. Kipping, however, had a bias toward forcing similarities between organosilicon compounds and the properties and reactions of the corresponding carbon-based organic compounds, which actually proved to be more of a handicap in his research. It was also Kipping who coined the term *silicone* for compounds of the empirical formula RR′SiO in analogy with ketones RR′CO. The term *silicone* has been retained for all siloxanes even though there is no monomeric ketone analog R_2SiO and already Ladenburg's studies had shown the behavior of siloxanes to be very different from that of ketones.

American chemists were the ones who carried out the research that led to the first commercial production capability of polymers and the development of silicone elastomers. In the United States, Corning Glass Works in the 1930s pioneered work on organosilicone polymers, both in their own laboratory (by J. F. Hyde) and at the Mellon Institute of Industrial Research (by R. R. McGregor). Their objective was to develop silicone resins for use in varnishes and as coating for glass fibers in high-temperature electrical insulation. In the same period, E. G. Rochow and W. I. Patnode at General Electric Company had similar interests, but first chose to work with silicate esters. In 1942, the work at Corning Glass Works had progressed to the point where commercial production was considered. Since Corning Glass Works was in the glass business, the Dow Chemical Company was approached for assistance in manufacturing. As a result, in 1943, the Dow Chemical Company and Corning Glass Works formed Dow Corning Corporation, which focused solely on the commercial development of silicone chemistry [40].

Rochow, in his famous laboratory experiment on May 10, 1940, achieved the direct synthesis of methylchlorosilanes, and by doing so laid the foundation for commercial silicone polymer manufacture [16,41]. Unknown to researchers in the United States, Müller in Germany had also discovered the direct synthesis, although probably some 12–18 months after Rochow [17]. The respective patent applications were not published owing to secrecy restriction and the final patents were only published after the war. The first step in the Rochow–Müller direct process is the reaction between silicon and methylchloride, resulting in the formation of various chlorosilanes. After a complicated distillation process, the hydrolysis of the dimethyldichlorosilane gives a mixture of oligomeric cyclic siloxanes and linear hydroxy-terminated siloxanes. These feedstocks can then be polymerized to form the silicone polymers used in silicone sealants.

The first silicone sealants, introduced in the late 1950s, were two-component products and had good unprimed adhesion only to a limited number of substrates [42]. Therefore, the initial research focused on improving the adhesion of these sealants by use of suitable adhesion promoters. Two-component products were well accepted for industrial production. However, it became obvious that numerous markets, especially the building and construction industry, would prefer one-component products in order to eliminate the need for mixing prior to application.

In the early 1960s, the first one-component silicone sealants were introduced, which were based on the acetoxy-cure system (giving off acetic acid, known for its "vinegar" odor) [43]. Engineers and architects were intrigued by this new sealant because it did not deteriorate in sunlight and did not harden in cold weather or soften in hot summers. This sealant had excellent adhesion to glass and fairly good unprimed adhesion to metals; however, it was corrosive and would not adhere to masonry. A film-forming primer was developed, but the acetoxy silicone sealant could not tolerate large joint movements and would tear if used in an expansion joint. Furthermore, the sealant could not be used in contact with electric or electronic components.

In the mid-1960s, silicone sealants based on oxime [44] and amine [45] cure chemistries were introduced to the market. Both cure chemistries were compatible with masonry, but good unprimed adhesion and joint movement were still lacking.

The amine system was still corrosive, which prevented its use in many markets, and both the amine and the oxime-cure by-products had a pungent odor.

In the early 1970s, a true low-modulus, one-component silicone sealant was developed. The cure chemistry of this sealant released N-methyl acetamide and diethyl hydroxylamine as by-products [46]. The sealant had excellent unprimed adhesion to most building substrates and unsurpassed joint movement capability; however, the shelf life of the sealant in the package was only about 9 months. The mid-1970s saw the launch of a low-modulus silicone sealant that released N-methyl benzamide as by-product [47]. This system had good unprimed adhesion and excellent joint movement, but cure rate was rather slow and the product changed color upon prolonged storage.

In the 1970s, a medium-modulus alkoxy-cure system was introduced that can be considered the first neutral cure sealant to combine good structural strength with sufficient movement capability [48]. The alkoxy-cure combined low odor and relatively fast cure with neutral (noncorrosive) cure and good unprimed adhesion to most substrates. Improvements in shelf stability of this system were achieved during the 1980s [49]. Medium-modulus two-part alkoxy-cure sealants with improved unprimed adhesion were also launched during the 1980s, opening up opportunities in insulating glass, structural glazing, as well as industrial and automotive assembly operations [50].

During the 1980s and 1990s, most development work was directed at improvements in unprimed adhesion to a broad variety of substrates, a fast rate of cure, improvements in shelf life and rheology of the wet sealant, as well as in nonstaining properties.

The construction industry represents the largest market segment for silicones. Silicone sealants are widely used by the construction industry for applications such as sealing building and highway expansion joints, general weatherproofing joints in porous and nonporous substrates, sanitary joints around bathroom and kitchen fixtures, as well as fire-rated joints around pipes, electrical conduits, ducts, and electrical wiring within building walls and ceilings. In a variety of applications, silicone sealants also perform the functions of an adhesive; that is, they act as structural sealants. For example, silicones are used in structural glazing, where the cured sealant becomes part of the overall load-bearing design, or in insulating glass secondary seals, which structurally bond two panes of glass together.

Structural glazing dates back to the 1960s, when a method of bonding glass panes to glass mullions using a medium-modulus silicone sealant was developed in the United States. This system, named *total vision system,* or, TVS, became very popular and is still used today. Figure 1.3 shows, as an example, an original TVS system installed in the United States in 1965.

By the late 1960s, medium-modulus silicone sealants were also used to bond glass panes to aluminum support mullions. The first two-sided structural silicone glazing (SSG) project with mechanical fixation on two opposite sides was completed in 1970 and followed shortly afterward, in 1971, by the first-ever four-sided SSG project (see Figure 1.4) with silicone bonding on all four sides of the glass panels.

During the 1970s and 1980s, the SSG technique rapidly became the fastest-growing form of curtain wall construction in the United States. because it allowed for broader architectural flexibility in achieving dramatic design accents in new

FIGURE 1.3 Early implementation of structural glazing: Total Vision System (TVS) glass wall constructed in 1965. (Courtesy of Dow Corning Corporation.)

FIGURE 1.4 First ever four-sided SSG project (Note: the retention spiders were removed later as the safety of this construction technique had been demonstrated). (Courtesy of Dow Corning Corporation.)

construction or in the renovation of old buildings. Today, SSG is one of the most versatile forms of curtain wall construction in the commercial façade business. SSG has helped architects and designers transform the skylines and downtown landscapes of cities around the world.

Silicone sealants have also found growing applications in the automotive industry based on the trend toward higher under-the-hood temperatures and higher quality and reliability requirements. Furthermore, the increase in automobile production has benefited the silicone sealant demand. Growth in consumption of silicone sealants in electrical or electronic markets has been moderate.

1.3.5 POLYURETHANE SEALANTS

The key reaction on which polyurethane chemistry is based was discovered more than 150 years ago. In 1849, Wurtz reacted isocyanates with alcohols into esters of carbamic acid, which were named *urethanes* [51]. Then, in 1937, Otto Bayer patented the polyaddition reaction of polyisocyanates with polyols that led to the production of high-molecular-weight polyurethanes, which found applications in coatings, paints, foams, elastomers, moldings, and many other forms. The rapid popularity that polyurethanes had attained throughout Europe and the United States was stimulated by shortages of natural rubber materials during World War II.

Polyurethanes have high potential as adhesives due to their self-reinforcing nature, which is due to the phase separation of their soft and hard polymer segments. This polymer structure gives strength, toughness, and tear resistance to the adhesives. The polar nature of the urethane bond further results in strong adhesion to a variety of substrates. Because of these properties, the potential of polyurethanes as adhesives was discovered as early as 1940, and, since then, a wide range of applications for polyurethane adhesives have emerged.

Medium- to low-modulus polyurethane sealants are more difficult to formulate and manufacture than high-modulus materials. Early polyurethane sealant products, introduced in the markets in the mid-1960s, therefore were still medium- to high-modulus products. These products were two-component formulations and, because of their overall property profile, were most suitable as industrial structural sealants. As early as 1968, the use of these structural sealants in bonding and sealing of automotive front and back windscreens was investigated. This method of installing windscreens, termed *direct glazing*, provides increased stiffness to the automotive body, for instance during rollover accidents [52]. However, because the adhesion of polyurethane sealants to glass lacks sufficient resistance to ultraviolet light, a primer and shading coat had to be applied to the perimeter of the windscreen. Designers soon found that this enamel coating, applied by screen-printing an inorganic frit on the windscreen, could be used as a design element to improve the appearance of the transition between glass and automotive body. Today, direct glazing represents the state of the art in automotive glazing and polyurethane structural sealants enjoy nearly 95% market share in this application.

Medium-modulus polyurethane sealants, sold primarily for construction applications, saw a rather slow market introduction hampered by technical problems in the late 1960s. During the early 1970s, the first moisture-cure, one-component formulations were developed. The early one-component polyurethane sealants tended to develop bubbles upon storage and application because of the carbon dioxide released by the curing reaction or by side reactions between the isocyanate crosslinker and water adsorbed on the filler or diffused into the sealant from the substrate. During the early 1980s, one-component formulations containing chemical driers were developed that substantially reduced this tendency toward bubbling. A further problem affecting early one-component polyurethane sealants was slow cure. Hence, movement of building components often caused failure while the sealant was still in a sensitive, uncured, or semicured, state. In order to enhance the cure rate of moisture-cure one-component polyurethane sealant formulations, a variety of catalysts and cure accelerators were patented in the 1980s [53,54]. With most of the early problems resolved during the 1970s and 1980s, the use of polyurethane sealants showed a dramatic increase from the mid-1980s onward.

Polyurethane sealants today are among the most widely used sealant types, second only to silicone sealants. The versatility of urethane chemistry allows for the formulation of a wide range of products, which are available as flowable and nonslump materials in both single-component and multicomponent versions. Furthermore, reactive hot melt and waterborne (dispersion) products exist that utilize similar cure mechanisms as the room-temperature curing polyurethane materials. Reactive polyurethane hot melts were introduced to the market in the early 1990s, while waterborne systems, the newest type of polyurethane sealants, became available in the early 2000s.

Automotive and construction industries today still represent the main applications of polyurethane sealants. In the automotive industry, polyurethane sealants are used in direct glazing of windscreens (both original-equipment-manufacture (OEM) glazing and glass replacement application), in panel bonding/sealing, for flow-in-place gasketing, and as headlamp adhesive-sealant (for lower-temperature headlights). Important applications in the construction industry are horizontal traffic (pavement) joints in highways, airport run- and taxiways, pedestrian walkways, and plazas; penetration seals in walls; civil engineering applications such as seals in wastewater treatment plants and sewage pipe systems; marine applications (teak deck sealer); and seals in some wire and cable applications.

1.3.6 SILICON-CURABLE ORGANIC SEALANTS (HYBRID SEALANTS)

Organic polymers with silicon-curable functionalities, prepared by copolymerizing, grafting, or otherwise reacting silanes with or onto the organic polymer, were discovered and patented as early as the 1950s and 1960s. In 1955, J. Speier, with his discovery of the hydrosilylation reaction in the presence of chloroplatinic acid catalyst, laid the foundation for the preparation of silyl-modified organic polymers [55]. The hydrosilylation method today represents the commercially most widely used method for preparing alkoxysilyl-functional organic polymers for silicon-curable sealants, because it allows controlled introduction of silyl groups with high yields. In order to obtain suitable polymers for the production of low- to medium-modulus silicon-curable sealants, the silyl functionality is typically introduced at the polymer ends, resulting in telechelic functional polymers. Today, silicon-curable sealant polymers are commercially available based on polyether, polyurethane, polyisobutylene, and polyacrylate backbones. The key end-user benefits for this class of sealants are derived from the environment-friendly and worker-health-friendly silicon-curability, the better control of the cure reaction, and the higher durability, when compared to conventional organic sealants.

Most of the silicon-curable organic sealants types were first commercialized in Japan [56]. In 1979, the production of telechelic alkoxysilylated polyether polymers was started in Takasago, Japan. These polymers were prepared by hydrosilylation of allyl-endblocked poly(propylene oxide) polymer with dimethoxy methylsilane. The first-generation sealant products, sold in Japan in the early 1980s and introduced in the European and U.S. markets by the mid- and late-1980s, respectively, were based solely on this polymeric binder [57]. The second-generation products, introduced globally during the mid-1990s, contained a blend of a polyacrylate and a silylated polyether to improve the durability of the cured sealant [58]. Silicon-curable polyether sealants are available both as one- and two-part formulations; however, the majority of commercial silicon-curable polyether sealants sold globally are one-part products. The main applications for silicon-curable polyether sealants are in the construction and transportation markets.

Silicon-curable polyurethane sealants build on the strengths resulting from standard urethane chemistry, such as adhesion and toughness, as well as on the versatility of silicon-cure chemistry. Silylated polyurethane polymers can be obtained either by reaction of an isocyanate-terminated prepolymer with an organofunctional alkoxysilane having a Zerewitinoff-active hydrogen atom, for example, secondary-amino, ureido- or mercapto-organoalkoxysilanes [59,60] or by reaction of a hydroxyl-terminated prepolymer [61] with an isocyanatoorgano alkoxysilane. Thus, silylated polyurethanes are the only type of commercially available silicon-curable organic sealant polymers that are not obtained by hydrosilylation reaction.

Silicon-curable polyurethane sealants were first patented in the 1970s [62]. During the early 1980s, silicon-curable polyurethane sealants were generally high-viscosity, high-modulus, and low-elongation materials, and were used primarily in transportation and industrial applications. During the mid-1980s, efforts were increasingly focused on developing silicon-curable polyurethane sealants for construction applications. Today, various approaches to obtain low-modulus, silicon-curable polyurethane sealants exist that are based on the specific properties of the silane endblocker and the isocyanate prepolymer or polyetherdiol polymer. For instance, low-modulus sealants can also be obtained by reaction of a high-molecular-weight hydroxyl-endblocked urethane prepolymer with an isocyanate-functional alkoxysilane [63]. These polymers have fewer polyurethane hard segments, which reduces the possibility for hydrogen bonding, leading to lower polymer viscosities at equivalent molecular weight and, therefore, lower-modulus sealants.

Silicon-curable polyurethanes, especially the one-component products, have a better balance between reactivity and storage stability when compared to conventional polyurethane sealants. Furthermore, since the cure reaction does not involve free isocyanate, there is no bubble formation due to carbon dioxide, and therefore, no health concerns for the applicator.

The latest generation of silicon-curable polyurethane sealant products is expected to rapidly gain share in construction markets at the expense of traditional polyurethane and polysulfide as well as silicon-curable polyether sealants. Growth of this sealant type will be enhanced by the fact that any polyurethane formulator can manufacture silylated polyurethane polymers without additional equipment needs—a key difference, when compared to other silicon-curable polymer technologies. The main applications for silicon-curable polyurethane sealants are in the construction and transportation markets.

Alkoxysilyl-endblocked polyisobutylene constitutes another class of silicon-curable organic polymers. Although this polymer type was launched in the mid-1990s, formulated sealants based on this polymer are currently only commercially available in Japan. Because of the extremely low vapor permeability of polyisobutylene, sealants based on silylated polyisobutylene can only be formulated as two-part sealants. Due to the high price of the silylated polyisobutylene polymer, most of the sealants commercially available in Japan are being sold for high-value, nonconstruction applications, such as electronics.

Silylated isobutylene polymers are obtained by hydrosilylation of telechelic allyl-terminated polyisobutylene [64]. Allyl-functional telechelic polyisobutylene is accessible via "inifer" polymerization [65]. In this method, a multifunctional compound capable of simultaneously initiating polymerization and acting as a transfer agent is employed. For this compound, the term *inifer* was chosen, derived from the words *initiator* and *transfer*. The tertiary chloro-terminations, generated by the inifer polymerization, can be converted to iso-propenyl ends by selective elimination of hydrochloric acid with a strong base, such as potassium *t*-butanoxide. The alkoxysilyl functionality can then be introduced by hydrosilylation of the iso-propenyl groups with methyldichlorosilane followed by alcoholysis [66].

Silylated acrylates were among the first silicon-curable polymers to be invented during the late 1960s. The initial polymers were prepared by radical polymerization of alkylacrylate or alkylmethacrylate monomers using a radical initiator, for example, azobisisobutyronitrile (AIBN), and a chain transfer agent, typically a mercaptan, for controlling the molecular weight of the final polymer. The alkoxysilyl functionality was introduced by use of one or more of the following three methods: (1) copolymerization with an unsaturated group containing silane, for example, γ-methacryloxypropyltrimethoxysilane; (2) use of an alkoxysilyl-functional radical initiator, for example, α,α′-azobis-5-trimethoxysilyl-2-methyl-valeronitrile; and (3) use of an alkoxysilyl-functional chain transfer agent, for example, γ-mercaptopropyltrimethoxysilane. The alkoxysilyl functionality on the initial silylated acrylate polymers were either pendant or pendant and end-capped on the main chain, depending on the nature of the chain transfer agent used [67–69]. Sealants formulated with these early polymers were either very brittle, when low-molecular-weight polymers were used, or required large amounts of solvents to allow use of high-molecular-weight polymers. They also suffered from a substantial increase in viscosity during their shelf life.

During the mid-1990s, new methods of living radical polymerization for telechelic acrylates were developed. Kusakabe and Kitano realized that atom transfer radical polymerization (ATRP) could be utilized to manufacture polyacrylate polymers with narrow molecular weight distribution having alkenyl groups or curable silyl groups at the chain ends in a high functionality ratio [70]. Telechelic polymers, having silicon-curable groups at the polymer chain ends, impart to the crosslinked sealant better mechanical properties than are obtainable from polymers with pendant functional groups. Telechelic silylated acrylate polymers became commercially available in 2003, and the first sealants based on this polymer type have recently been launched.

1.4 OUTLOOK

The future will probably imitate the past somewhat. Many of the materials will get better, but some will be formulated for lower cost and their properties will not be as good as they are today. There will be more hybrid sealant types, and some will be better than those presently available. For the foreseeable future, industrial research and development activities will continue to focus on the following goals:

1. Higher-performance sealants combining performance aspects of current dissimilar sealant types
2. Increased ease of use, especially for the do-it-yourself market segment, and more consistent unprimed adhesion
3. Simpler installation, and thus reduced overall cost
4. Increased consideration of life-cycle cost and durability of the sealed joint (considering the total cost of the finished component, including mainte-nance, over the lifetime of the component and not just the investment cost of the sealant itself)
5. Increased consideration of environmental impact (e.g., volatile organic con-tent) and worker's health
6. Increased consideration of sustainability (embodied energy, i.e., the energy consumed during the manufacture and distribution of the product, life-cycle analysis, ability to recycle or reuse).

The authors hope that future tests will provide a better prediction of a sealant's long-term performance and suitability for a given sealing task than the current standard-ized test methods. Furthermore, the authors expect that more emphasis will be placed on the quality of the sealant installation, which is so fundamental to success.

In the past, improvements in the quality of sealant installation have not always kept pace with the improvements in the performance of the sealant products. Especially in building construction, sealants are often installed by untrained workers to unclean surfaces in inclement weather. In such situations, even the perfect sealant will fail. The most important advance in the area of construction joint sealing is that training programs are now readily available and there are techniques for continuous inspec-tion of the sealed joints. These job site inspections allow quality assurance of the seal-ing operation and thus catch many of the flaws in material and workmanship before major damage occurs to the building. The authors' wish is that the future will bring a more integrated quality management approach to sealing operations in construction [71] and an increase in nondestructive test methods for all sealant applications.

REFERENCES

1. G.S. Haviland, *Machinery Adhesives for Locking, Retaining, and Sealing*, Marcel Dekker, New York (1986).

2. P.A. Fay, History of adhesive bonding, in *Adhesive Bonding—Science, Technology and Applications*, R.D. Adams (Ed.), pp. 3–22, Woodhead Publishing Limited, Cambridge (2005).

3. J. Koller, U. Baumer, and D. Mania, High tech in the middle Palaeolithic: Neanderthal manufactured pitch identified, *Eur. J. Archaeology*, **4**, 385–397 (2001).

4. E. Boëda, J. Connan, D. Dessort, S. Muhesen, N. Mercier, H. Valladas, and N. Tisnérat, Bitumen as a hafting material on Middle Palaeolithic artefacts, *Nature*, **380**, 336–338 (1996).

5. R.J. Forbes, *Studies in Ancient Technology*, Brill, London (1964).

6. H.S. Alsalim, Construction adhesives used in the buildings of Babylon (in German), *Adhäsion*, **5**, 151–156 (1981).

7. Pliny the Elder (Caius Plinius Secundus), *The Natural History* (original text in Latin), http://penelope.uchicago.edu/Thayer/E/Roman/Texts/Pliny_the_Elder/home.html, for the cited examples see: Liber XII: xxxvi, Liber XIII: xx, and Liber XIV: xxv.

8. I. Skeist and J. Miron, History of adhesives, *J. Macromol. Sci.—Chem.*, **A15**, 1151–1163 (1981).

9. F. Bacon, *Novum Organum* (1620) available in translation by L. Jardine and M. Silverthorne, Cambridge University Press, Cambridge, pp. 140–141 and pp. 191–194 (2000).

10. I. Newton, *Opticks: Or, a Treatise of the Reflections, Refractions, Inflections and Colours of Light*, 2nd edition, with Additions, printed for W. and J. Innys, Printers to the Royal Society, at the Prince's Arms in St. Paul's Church-Yard (1718).

11. Anonymous, RAL 849 B2—Quality Requirements and Designation Rules for Pure Linseed Oil Putty (in German), RAL Gütegemeinschaft, Frankfurt/Main, Germany (1960).

12. H. von Pechmann and O. Röhm, Polymerisation of unsaturated acids. III. α-Methyleneglutaric acid, a product of the polymerisation of acrylic acid, *Ber. Deut. Chem. Ges.*, **34**, 427–429 (1901).

13. R.M. Thomas and W.J. Sparks, Mixed olefinic polymerization process and product, Jasco Inc., U.S. Patent 2,356,128 (1944) and Olefin diolefin copolymers, U.S. Patent 2,356,130 (1944).

14. O. Bayer, W. Siefken, H. Rinke, L. Orthner, and H. Schild, A process for the production of polyurethanes and polyureas, I.G. Farbenindustrie AG, German Patent DRP 728,981 (1937).

15. J.C. Patrick and H.R. Ferguson, Polysulfide polymer, Thiokol Corporation, U.S. Patent 2,466,963 (1949).

16. E.G. Rochow, Preparation of organosilicon halides, General Electric Company, U.S. Patent 2,380,995 (1945) (the patent application was filed on 26 September 1941).

17. R. Müller, Chemisch-Pharmazeutisches Werk von Heyden (VEB Silikonchemie), German Patent DD 5348 (1953) (the patent application D.R.P. C 57411 was filed on 6 June 1942).

18. M.V. Newton, S.D. Halbe, and G.D. Krysiak, Butyl sealants: Formulating, developing, processing, in *ASC Caulks and Sealants Short Course Notes—15–18 January 1990*, The Adhesive and Sealant Council, Washington, DC, U.S.A. (1990), available at www.ascouncil.org/.

19. R.P. Russell, Process for producing polymers of olefines, Standard Oil Development Company, U.S. Patent 2,139,038 (1938).

20. J.C. Patrick, Method of making plastic substances and product obtained thereby, U.S. Patent 1,996,486 (1935) (patent application filed on April 25, 1927).

21. T.C.P. Lee, *Properties and Applications of Elastomeric Polysulfides*, Rapra Review Reports, Report 106, Rapra Technology Limited, Shawbury, Shrewsbury, Shropshire, U.K. (1999).

22. B.S. Kaskel, The metal and glass curtain wall, *CRM—Cultural Resource Management*, **18**, 23–27 (1995), available at http://crm.cr.nps.gov/archive/18-8/18-8-7.pdf.

23. M.J. Scheffler and J.D. Connolly, History of building joint sealants, in *Science and Technology of Buildings Seals, Sealants, Glazing and Waterproofing*, Vol. 5, ASTM STP 1271, M.A. Lacasse (Ed.), pp. 85–94, ASTM International, West Conshohocken, PA (1996).

24. P. Cognard, Sealants for Construction Part III-2—Elastomeric, High Performance Sealants (2004), available at www.SpecialChem4Adhesives.com/.

25. J. Redtenbacher, On the degradation products of glyceryloxide by dry distillation (in German), *Ann. Chem. Pharm.*, **47**, 113–148 (1843).

26. E. Frankland and B.F. Duppa, Investigations of acids of the acrylic acid type (in German), *Ann. Chem. Pharm.*, **136**, 1–31 (1865).

27. W. Caspary and B. Tollens, On acrylic acid ester and acrylic acid (in German), *Nachrichten von der Königlichen Gesellschaft der Wissenschaften und der Georg-Augusts-Universität zu Göttingen*, **17**, 335–336 (1872).

28. G.W.A. Kahlbaum, On polymeric acrylic acid methyl ester (in German), *Ber. Deut. Chem. Ges.*, **13**, 2348–2351 (1880).

29. O. Röhm, On the Polymerization Products of Acrylic Acid, Ph.D. Thesis, University of Tübingen, Germany (1901).

30. Anonymous, *Educational Materials: Bonding/Adhesives Textbook*, Association of European Adhesives Manufacturers (FEICA), Düsseldorf, Germany (2004).

31. O. Röhm and W. Bauer, Process for the manufacture of polymerization products, Rohm & Haas Company, U.S. Patent 2,091,615 (1937), Process for the manufacture of polymerization products, U.S. Patent 2,154,639 (1939), and Glass substitutes and process for preparing, U.S. Patent 2,193,742 (1940).

32. H.C. Young, Acrylic sealants, in *Building Seals and Sealants*, ASTM STP 606, J.R. Panek (Ed.), pp. 62–77, ASTM International, West Conshohocken, PA (1976).

33. J.R. Panek and J.P. Cook, *Construction Sealants and Adhesives*, 3rd edition, John Wiley & Sons, New York (1992).

34. C. Friedel and J. Crafts, On certain new organo silicon combinations (in French), *C.R. Hebd. Acad. Sci.*, **56**, 590–593 (1863).

35. C. Friedel and A. Ladenburg, On silicon chloroform and its derivatives (in German), *Liebig's Ann. Chem. Pharm.*, **143**, 118–128 (1867).

36. A. Ladenburg, On the reduction products of silicic acid ethers and their derivatives (in German), *Liebig's Ann. Chem. Pharm.*, **163/164**, 300–332 (1872).

37. F.S. Kipping and J.T. Abrams, Organic derivatives of silicon. LI. Bis(dihydroxytetraphenylethane) orthosilicate, *J. Chem. Soc.*, **147** 81–84 (1944) (and literature cited therein).

38. V. Grignard, Concerning certain novel organometallic compounds of magnesium and their application in the syntheses of alcohols and hydrocarbons, *C.R. Acad. Sci.*, **130**, 1322–1324 (1900).

39. F.S. Kipping, Organic derivatives of silicon. XXXII. The carbon–silicon binding, *J. Chem. Soc.*, **130**, 104–107 (1927).

40. For a comprehensive review of early silicone chemistry and the events that led to the invention of the "direct process" see: D. Seyferth, Dimethyldichlorosilane and the direct synthesis of methylchlorosilanes—the key to the silicones industry, *Organometallics*, **20**, 4978–4992 (2001).

41. E. Rochow, The direct synthesis of organosilicon compounds, *J. Amer. Chem. Soc.*, **67**, 1772–1774 (1945).

42. C.A. Berridge, Organopolysiloxanes which cure at room temperature, General Electric Company, U.S. Patent 2,843,555 (1958).

43. L.B. Bruner, Acyloxy-endblocked diorganopolysiloxanes, Dow Corning Corporation, U.S. Patent 3,032,532 (1962) and Acyloxysiloxanes and method of using them, U.S. Patent 3,035,016 (1962).
44. E. Sweet, Silicone intermediates and rubbers, Rhone-Poulenc, U.S. Patent 3,189,576 (1965).
45. S. Nitzsche and M. Wick, Curing organopolysiloxanes, Wacker-Chemie, U.S. Patent 3,032,528 (1962).
46. L. Toporcer and I. Crossan, Low modulus room temperature vulcanizable silicone elastomer, Dow Corning Corporation, U.S. Patent 3,817,909 (1974), and J.M. Klosowski, Low modulus room temperature vulcanizable silicone elastomer with improved slump characteristics, Dow Corning Corporation, U.S. Patent 3,996,184 (1976).
47. H. Sattlegger, W. Noll, K. Damm, and H.D. Goelitz, Organopolysiloxane moulding compositions, stable in the absence of water, Bayer AG, U.S. Patent 3,378,520 (1968).
48. D.R. Weyenberg, Method of making one-component room temperature curing siloxane rubbers, Dow Corning Corporation, U.S. Patent 3,334,067 (1967).
49. J.M. Klosowski and M.D. Meddaugh, Methods of improving shelf life of silicone elastomeric sealant, Dow Corning Corporation, U.S. Patent 4,772,675 (1988).
50. E. Joseph and B. Trego, Silicone elastomers and adhesion promoter therefore, Dow Corning S.A., U.S. Patent 4,602,078 (1986).
51. A. Wurtz, On the reactions of cyanuric acid and cyanic acid with ethyl oxide, methyl oxide, amyl oxide, and the resulting products (in German), *Liebig's Ann. Chem. Pharm.*, **71**, 326–342 (1849).
52. K. Dilger, Automobiles, in *Adhesive Bonding—Science, Technology and Applications*, R.D. Adams (Ed.), pp. 357–385, Woodhead Publishing Limited, Cambridge (2005).
53. G.W. De Santis, Polyurethane sealant—primer system, Essex Chemical Corporation, U.S. Patent 3,779,794 (1973).
54. R.N. Coyner and P. Skujins, Moisture curable polyurethane systems, ConTech Inc., U.S. Patent 4,038,239 (1977).
55. J.L. Speier, Process for the production of organosilicon compounds, Dow Corning Corporation, U.S. Patent 2,823,218 (1958).
56. M. Kusakabe, Y. Masaoka, and C. Phanopoulos, Review of innovative developments of silyl-modified polymers for sealant, adhesive and coating applications, *Pitt. Vern.— Europ. Coat.*, **81**, 43–50 (2005).
57. See, for example: T. Mita, H. Nakanishi, J. Takase, K. Isayama, and N. Tani, Curable composition comprising polyether having silyl group, Kaneka Corporation, U.S. Patent 4,507,469 (1985), and T. Hirose, S. Yukimoto, and K. Isayama, Curable polymer composition comprising organic polymer having silicon-containing reactive group, Kaneka Corporation, U.S. Patent 4,837,401 (1989).
58. See, for example: T. Hirose and K. Isayama, Curing composition containing polyether having reactive silicon-containing group and a (meth)acrylate polymer, Kaneka Corporation, U.S. Patent 4,593,068 (1986), and S. Kohmitsu, H. Wakabayashi, T. Hirose, and K. Isayama, Curable polymer composition, Kaneka Corporation, European Patent 0,339,666 B1 (1995).
59. G.M. Seiter, Polyurethane sealant containing trialkyloxysilane end groups, 3M Company, U.S. Patent 3,627,722 (1971).
60. G.L. Brode and L.B. Conte, Jr., Vulcanizable silicon terminated polyurethane polymers, Union Carbide Corporation, U.S. Patent 3,632,557 (1972).
61. S.D. Rizk, H.W.S. Hsieh, and J.J. Prendergast, Silicon-terminated polyurethane polymer, Essex Chemical Corporation, U.S. Patent 4,345,053 (1982).
62. E.R. Bryant and J.A. Weis, Vulcanizable silicon terminated polyurethane polymer composition having improved cure speed, Inmont Corporation, U.S. Patent 3,979,344 (1976).

63. R.R. Johnston and P. Lehmann, Process for producing prepolymers which cure to improved sealants, and products formed thereby, Witco Corporation, U.S. Patent 5,990,257 (1999).
64. J.P. Kennedy, D.R. Weyenberg, L. Wilczek, and A.P. Wright, Method of preparing allyl-terminated polyisobutylene, Dow Corning Corporation, U.S. Patent 4,758,631 (1988).
65. J.P. Kennedy, R.A. Smith, and L.R. Ross, Jr., Novel telechelic polymers, block copolymers, and processes for the preparation thereof, University of Akron, U.S. Patent 4,276,394 (1981).
66. T. Iwahara, K. Noda, and K. Isayama, Curable isobutylene polymer, Kaneka Corporation, U.S. Patent 4,904,732 (1990).
67. B.J. Sauntson, Ceramic tile adhesives, Scott Bader Company Ltd., U.S. Patent 4,333,867 (1982).
68. E.P. Plueddemann, Bonding thermoplastic resins to inorganic materials, Dow Corning Corporation, U.S. Patent 3,306,800 (1967) and Room temperature curable acrylate rubbers, U.S. Patent 3,453,230 (1969).
69. K. Kohno, S. Nishikawa, Y. Hattori, and K. Kitao, Room temperature curing elastic composition, Sunstar Giken K.K., U.S. Patent 4,478,990 (1984).
70. M. Kusakabe and K. Kitano, Processes for preparing (meth)acrylic polymers having functional groups at the chain ends, Kaneka Corporation, European Patent Application 0,789,036 A2 (1997).
71. A.T. Wolf, Improving the service life of sealed cladding joints through a total quality management approach, in *Durability of Building and Construction Sealants*, A.T. Wolf (Ed.), pp. 45–59, RILEM Publications, Cachan, France (1999).

2 Organofunctional Silanes for Sealants

Misty Huang and Eric R. Pohl

CONTENTS

2.1 INTRODUCTION

A sealant is a viscous material that when applied changes state to become a solid barrier. The change of state is driven by evaporation of solvents, cooling of heated resinous or polymeric materials, or chemical reactions. It is used to prevent the transmission of liquids, gases, noise, dust, or heat from one area into another.

Sealants have a long history. Stone-age humans used bitumen in constructing their dwellings. The walls of the ancient city of Jericho, the Tower of Babel, the brick baths of India, and Egyptian ships dating from antiquity all used sealants made of bitumen, tar, or pitch in their construction. Prior to the 20th century, most sealants were derived from natural sources such as vegetable extracts, bitumen, tar, or pitch. During the 1930s, organic polymers began to replace these older materials because they were more insoluble in organic solvents and fuels, had longer use-life, were resistant to environmental exposure and corrosion, and had better adhesion. These features made them suitable for many applications such as construction, insulating glass, transportation, aerospace, marine, and industrial. Concurrent with the development of organic polymer-based sealants, the discovery and use of organofunctional silanes in sealants has led to improved adhesion, durability, and stability.

Organofunctional silanes are silicon-containing organic compounds with two reactive centers: an organic functional group and a hydrolyzable silyl group. These centers are covalently bonded to each other through a hydrocarbon linking group. A general chemical formula illustrating the bifunctional nature of these materials is

$$Y\text{-}R\text{-}Si(CH_3)_a(OR^1)_{3-a}$$

where R is an alkylene group, typically methylene or propylene; R^1 is an alkyl, acyl, alkenyl, or alkylidene amino group, typically CH_3^- Y is an organofunctional group; and a is 0, 1, or 2. The dual reactivity is well suited for bonding dissimilar materials together, crosslinking organic polymers, or functioning as a desiccant.

The most common use of organofunctional silanes in sealants is as an adhesion promoter. The dual-reactive centers of these compounds are designed to interact with both the organic polymer of the sealant and surfaces to which the sealant must adhere to perform its barrier function. The interactions of the organofunctional silanes with the polymer and surfaces result in higher adhesion strength, especially under hydrothermal environments. The organofunctional group is chosen to interact with organic polymer or participate in the curing reactions of the sealant. Over 20 different polymers are used as the base for sealants. Typical organic polymers are silicones, polyurethanes, polysulfides, silyl-terminated polyethers, poly(vinyl chloride) and solvent or waterborne acrylics as well as the older materials, bitumen, and oleoresins. The surfaces are generally inorganic siliceous minerals, metals, and metal oxides, although these silanes have also been used to improve adhesion to various organic substrates, such as plastics, paints, and natural materials.

A closely related role for these silanes is that of coupling agent [1,2]. Sealants often contain inorganic reinforcing fillers, extenders, and pigments. The silane couples these fillers or pigments to the organic polymers, resulting in improved strength and durability of the sealants.

Organofunctional silanes have more recently been used as crosslinkers for organic polymers [3,4]. The organofunctional group of the silane reacts with reactive end groups or reactive pendent groups on the polymer backbone or is copolymerized with other monomers during the preparation of polymers. The silane is covalently bonded to the polymer. The cure is initiated when the sealant is exposed to moisture. The silyl group reacts with water to hydrolyze to silanols and condenses to form siloxanes. The siloxane bond, Si-O-Si, is very stable, leading to sealants with good aging and physical properties. Controlling the moisture content of the sealant is necessary to achieve a long shelf life of the formulated product. Although various desiccants can be used, organofunctional silanes have been found to be effective materials for this function. In this use, the electron-withdrawing nature of the organofunctional group, such as the vinyl group, accelerates the hydrolysis rate of the silane relative to the hydrolysis rate of the silylated polymer, so that the vinyl silane can chemically remove the water from the sealant before it reacts with the silylated polymer.

2.2 MECHANISM OF ADHESION

The mechanism by which organofunctional silanes increase adhesion of sealants to substrates is complex and may involve one or more facets. Generally, the silane adhesion promoter is formulated into the sealant. The silane migrates to the interface region on applications [5,6]. The chemistry and physical processes that occur once the sealant is applied have been extensively studied [7,8,9]. Substrate wetting, chemical bonding, interpenetrating polymer networks, and acid–base mechanisms may all contribute to the adhesion strength.

2.2.1 SUBSTRATE WETTING

A minimum requirement for adhesion is that the organic polymer adequately wet the substrate [10]. The wetting of the substrate occurs when its surface tension (energy) is greater than the polymer liquid surface tension (energy) and when the viscosity of the polymer is sufficiently low to flow under the forces created by the differences in surface tensions (energies). Many inorganic surfaces, such as metal oxides, ceramics, and minerals, have higher surface tensions (energies) than the polymers typically used in sealants [11,12]. The adsorption of these polymers onto the high-surface-energy surfaces may produce adhesion strengths in excess of the cohesive strength of the polymer.

High adhesion strength of sealants adhering to mineral surfaces is often observed when joints are made with sealants that do not contain organofunctional silanes and are kept in dry or arid environments. When exposed to moist or humid environments, water vapor diffuses through the polymer matrix of these sealants and accumulates at the interface between the sealant and substrate. The accumulation of water was demonstrated in a glass-fiber-filled composite in which the amount of water absorbed was in large excess of the water solubility limit of the resin [13]. A thin film of liquid water formed between the polymer matrix and substrate. When liquid water accumulates at the glass fiber–matrix interface, a loss of adhesion occurs because the cohesive strength of the water film is low, and failure occurs in the water layer.

In sealants, silanes diffuse to and interact with the substrate after application. When the silane contains an alkyl group bonded to the silicon, it modifies the surface but is nonreactive with the polymer. The surface tension (energy) of the substrate is reduced and may become similar to or lower than the surface tension (energy) of the polymers [14]. Poorer wetting and loss of adhesion strength occur [1].

2.2.2 INTERPENETRATING POLYMER NETWORK (IPN)

Alkoxysilanes hydrolyze when exposed to moisture to form silanols [15]. These silanols are reactive. They interact with the surface and condense with themselves to form siloxane oligomers and polymers. The silanols, siloxane oligomers, and polymers adsorb onto the surface of the inorganic (metal) substrate and form covalent bonds with the metal hydroxyls as well as secondary bonding interactions such as van der Waals, dipole–dipole interactions, hydrogen bonds, and ionic bonds. The siloxane oligomers are dispersed within the polymer matrix to form a separate phase or are soluble in the polymer matrix. These siloxane oligomers and polymers generate a network that is anchored to the surface of the substrate. The entanglement between the organic polymer and siloxane network enhances the adhesion strength [16].

Harding and Berg investigated adhesion using single particles embedded into a matrix [14]. The glass particles were treated by a series of silanes and then embedded in a polar poly(vinyl butyral) matrix to form the composite. They found that alkoxysilanes with nonpolar groups, such as octyl, reduced interfacial strength, whereas silanes with polar functional groups, such as aminoalkyl, increased interfacial strength. The improvement increased as the chain length increased from monoaminosilane to diaminosilane to triaminosilane. Their study suggests that the adhesion is improved when the surface tension of the substrate is higher than the surface tension of the organic polymer matrix, and the silane can interpenetrate into this organic polymer.

2.2.3 CHEMICAL BONDING

The organofunctional silane is bifunctional and capable of forming covalent bonds with both the organic polymer of the sealant and the inorganic surface of the substrate [17]. Covalent bonds are stronger than the van der Waals, dipole–dipole, and hydrogen-bonding interactions that may be involved in surface wetting and interpenetrating polymer networks [18]. The formation of these chemical bonds involves various processes. The organofunctional silane diffuses to the substrate surface. The bound water on the surface of the substrate reacts with the alkoxysilyl group to form silanols, which then react with the hydroxyl groups that are present on nonmetal or metal oxide substrate to form covalent bonds and with themselves to form siloxane oligomers and polymers [19,20]. The formation of the siloxane network affects adhesion strength. Parker and MacLachlan found that adhesion was related to the degree of hydrolysis and the final state of the siloxane network [21]. The chemical structure of the organofunctional silane and the presence of catalysts affect the hydrolysis and condensation rates of the organofunctional silane and the structure of the resulting siloxane network. Finally, the organofunctional group of the siloxane network reacts with the organic polymer of the sealant during cure. The substrate is then covalently linked to the sealant through the siloxane network. This siloxane network has formed multiple covalent bonds to both the organic polymer and the inorganic surface. Debonding of the organic polymer from the surface would require breaking of numerous covalent bonds to effect adhesion failure.

Formation of a siloxane network is beneficial to the adhesion of the sealant to the surface. Monolayer coverage of the silane on the substrate is insufficient for optimal adhesion [22]. Generally, multilayers with thicknesses of about 10 nm provide better adhesion strength. Belton and Joshi demonstrated that the time necessary to remove a polyimide film from a silicon wafer upon immersion in water increased fourfold as the aminopropyltriethoxysilane film increased from a monolayer to about 10 monolayers [23]. Another factor affecting adhesion is the formation of an interphase region [22,24]. After reaction of the siloxane network with the organic polymer of the sealant, an interphase is generated that forms a gradient of varying modulus. This modulus is intermediate between the high modulus of the substrate and the low modulus of the organic polymer. The gradient dissipates stresses over a larger region

TABLE 2.1

Equilibrium Constants for the Reversible Reaction of Water with Silylated Silica Surface

$$\equiv Si\text{-}O\text{-}SiR_n(OH)_{3-n} + H_2O \rightleftharpoons \equiv Si\text{-}OH + HOSiR_n(OH)_{3-n}$$

Species	Equilibrium Constant, K_{eq}
$\equiv Si\text{-}O\text{-}Si(OH)_3$	10^{-5} (estimated)
$\equiv Si\text{-}O\text{-}SiR(OH)_2$	10^{-4} (estimated)
$\equiv Si\text{-}O\text{-}SiR_2(OH)_1$	2.7×10^{-3}
$\equiv Si\text{-}O\text{-}SiR_3$	5×10^{-2}

when the sealant joint is under stress. It prevents concentration of these stresses at a sharp interface between two dissimilar materials of very different moduli.

The chemical bonds between the organofunctional silane and the nonmetal or metal surface arc labile and will readily react with water [13]. In the case of nonmetals, the nonmetal surface should have surface hydroxyl groups. If the nonmetal is a plastic, the surface hydroxyl groups can be created by oxygen plasma [25]. Pohl and Osterholtz demonstrated that the siloxane bond was hydrolyzable in the presence of water and established an equilibrium between the siloxane and silanol in aqueous solution [26]. Ishida and Koenig extended the work to silanes bonded to silica or glass surfaces and demonstrated the reversibility of the siloxane bond using Fourier transform infrared spectroscopy [27]. Plueddemann estimated the equilibrium constants for the hydrolysis reaction of silane coupling agents on silica [16]. Table 2.1 contains the equilibrium constants for the reversible reaction of water with the silylated silica surfaces [16].

From the data in Table 2.1, the organofunctional silanes that contain three hydrolyzable groups will form more stable Si-O-Si bonds with silica. In addition, because the concentration of water is low in the interphase region, the equilibrium lie far to the left in the preceding equation. The interphase has primarily covalent bonds between the silane and substrate and only a small percentage of the silanols. The hydrolysis rate of these covalent bonds between the silane and substrate depends on the composition of silicon-oxygen-nonmetal or silicon-oxygen-metal bonds that form between the silane adhesion promoter and surface. Certain bonds, such as Ti-O-Si, hydrolyze more rapidly than the siloxane bond, and therefore may not be as resistant to moist conditions. The inorganic surface composition affects adhesion strength, especially under moist conditions.

2.2.4 ACID–BASE INTERACTIONS

An alternative bonding mechanism is predicated on the acidity or basicity of the surface [28,29,30]. The acidity and basicity of the surface depend on the metal hydroxides or oxides. The interactions of polar compounds or polymers with oxides depend on the isoelectric point of the surface and the logarithm of the dissociation constant of the acid functional groups on the polymer, referred to as the pK_a of the acid. The isoelectric point of a surface (IEPS) is the pH value at which the number of positive and negative charges is equal and the zeta potential is zero. Qualitative measures of the surface interactions with an acid and base are given by Equations 2.1 and 2.2, respectively:

$$\Delta A = IEPS - pK_a \tag{2.1}$$

$$\Delta B = pK_a - IEPS \tag{2.2}$$

Large positive values of ΔA or ΔB indicate strong ionic bonds. Large negative values of ΔA or ΔB indicate weak dipole interactions. The ability of water to interrupt these interactions depends on the pH of the water. The bonds are most stable when the pH value of the water is between the IEPS and the acid or base dissociation constant.

Schreiber and colleagues studied a variety of silanes as adhesion promoters in a nonreactive polyurethane adhesive [31]. Their results confirmed that an acid–base interaction of the organofunctional group with the polymers and the substrates could enhance the adhesion strength at the interface. In this study, the organofunctional group of the alkoxysilane did not form covalent bonds with the polyurethane polymer. The acid–base interactions were calculated from the results of inverse gas-phase chromatography. The results indicate that the higher the interactions between polymer and substrate, the stronger the adhesion. Epoxysilanes were found to be the most effective adhesion promoter to different substrates because these raise the potential for both acid and base interactions. Aminosilane is a good choice for acidic substrates such as poly(vinyl chloride).

2.2.5 Adhesion to Organic Substrates

Organofunctional silanes have been traditionally used at the interface between organic polymers and inorganic substrates. However, Landon and coworkers have demonstrated enhanced adhesion to plastic substrates when organofunctional silanes are added to silylated polyurethane sealants [32]. In this study, polyurethane pre-polymers were terminated with aminosilanes. These aminosilanes function mainly as crosslinkers in this hybrid polymer and not as adhesion promoters. The mobility of the silanes that are bonded to the polyurethane polymer is restricted and, therefore, the silyl end group of the polymer does not migrate effectively to the substrate. Thus, strong interactions between the silylated polyurethane and substrate do not occur. Additional monomeric organofunctional silanes are needed to promote the adhesion. The monomeric silanes not only enhance the adhesion to inorganic substrates, such as glass and aluminum, but also to selected plastics, such as poly(vinyl chloride), polystyrene, and polyacrylic substrates.

The mechanism for improved adhesion may involve the transport of the monomeric organofunctional silane across the interface to the plastic substrate surface. Because organofunctional silanes are soluble in many plastics, the silane may diffuse into the plastic substrate and form an interpenetrating network. The organofunctional group may also react with the plastic polymer to further enhance its adhesion to the plastic substrate. Eklund et al. studied the adsorption of silanes onto inorganic

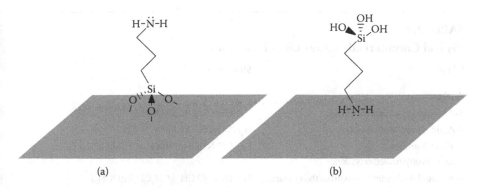

FIGURE 2.1 Schematic representations showing two different orientations of the aminopropyltrimethoxysilane molecule on a glass surface. The hydrolyzed silane reacts with the surface and the amino group is pointed toward the air side (a). The unhydrolyzed silane is oriented in the opposite direction and the trimethoxysilyl group is oriented toward the air side (b). (From S.J. Landon, N.B. Dawkins, and B.A. Waldman, *Proc. Eurocoat '97*, Lyon, France, Vol. 1, p. 171, 1997.)

substrates [33]. The bifunctional silane, aminopropyltrimethoxysilane, was oriented with both the amino and trimethoxysilyl groups pointing to the air side, as shown in Figure 2.1. Because the orientation depends on minimizing the surface tensions (energies) of the silane, substrate, and organic polymer, and because the substrate does not have reactive surface hydroxyls, the orientation of aminosilane on plastics would tend to position the silyl group upward toward the sealant and away from the plastic substrate. Hydrolysis of the aminosilane in orientation B would create a silanol-rich surface to which silanes that are still dissolved in the sealant and have not migrated across the sealant–plastic interface can react to form siloxane bonds.

Other mechanisms have been proposed that may also be useful in explaining the adhesion promotion benefits of organofunctional silanes. These mechanisms are oxide reinforcement [34,35], compatibility theory [1], deformable layer theory [22], and restrained layer theory [24].

2.3 SILANES AS ADHESION PROMOTERS

Numerous commercial organofunctional silanes are used as adhesion promoters in sealants. The type of organic polymer, the chemistry used to cure the sealant, and to a lesser extent the substrate usually dictate the choice of adhesion promoter. A listing of the more commonly used commercial silanes is given in Table 2.2.

Global suppliers of these commercial silanes include Momentive Performance Materials, Dow Corning, Evonik, Shin Etsu, and Wacker. Their web sites are given in Table 2.3.

2.3.1 SILICONE SEALANTS

Room-temperature vulcanizing silicone sealants are based on a hydroxyl terminated poly(dimethylsiloxane) or poly(methylphenylsiloxane) that are reacted with

TABLE 2.2
Typical Commercial Silanes Used in Sealants

Silane	Structure

Aminosilanes

3-Aminopropyltrimethoxysilane	$H_2NCH_2CH_2CH_2Si(OCH_3)_3$
3-Aminopropyltriethoxysilane	$H_2NCII_2CH_2CH_2Si(OCH_2CH_3)_3$
N-Ethyl-3-amino-2-methylpropyltrimethoxysilane	$CH_3CH_2NHCH_2CH(CH_3)CH_2Si(OCH_3)_3$
4-Amino-3,3-dimethylpropyltrimethoxysilane	$H_2NHCH_2C(CH_3)_2CH_2CH_2Si(OCH_3)_3$
bis-(3-Trimethoxysilylpropyl) amine	$HN[CH_2CH_2CH_2Si(OCH_3)_3]_2$

Epoxysilanes

3-Glycidoxypropyltrimethoxysilane

$CH_2OCH_2CH_2CH_2Si(OCH_3)$

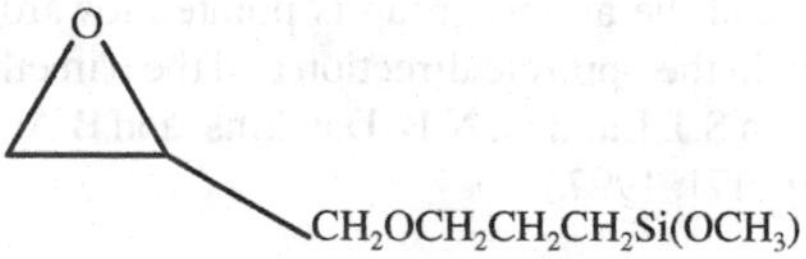

3-Glycidoxypropylmethyldiethoxysilane

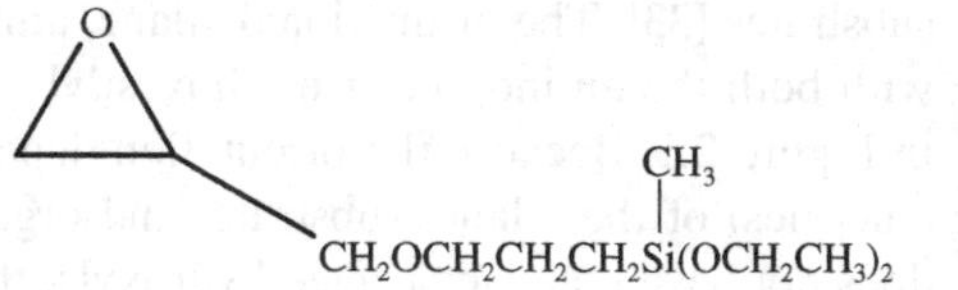

CH_3

$CH_2OCH_2CH_2CH_2Si(OCH_2CH_3)_2$

3,4-Epoxycyclohexylethyltriethoxysilane

$CH_2CH_2Si(OCH_2CH_3)_3$

Isocyanurate silane

Tris-(3-trimethoxysilylpropyl) isocyanurate

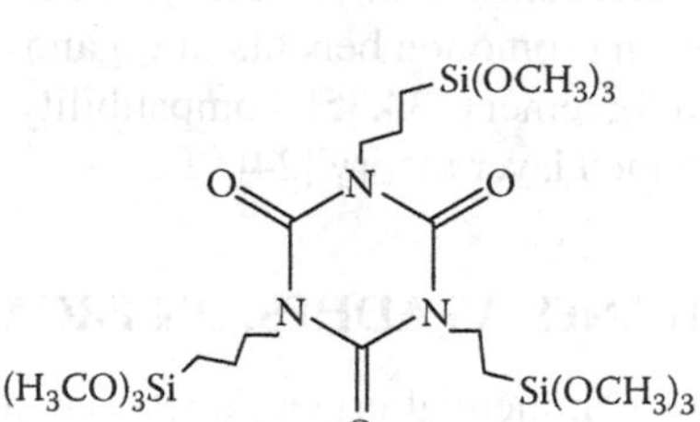

$Si(OCH_3)_3$

$(H_3CO)_3Si$ ... $Si(OCH_3)_3$

Isocyanatosilanes

Isocyanatomethyltrimethoxysilane	$O=C=NCH_2Si(OCH_3)_3$
3-Isocyanatopropyltrimethoxysilane	$O=C=NCH_2CH_2CH_2Si(OCH_3)_3$
3-Isocyanatopropyltriethoxysilane	$O=C=NCH_2CH_2CH_2Si(OCH_2CH_3)_3$

Mercaptosilanes

| 3-Mercaptopropyltrimethoxysilane | $HSCH_2CH_2CH_2Si(OCH_3)_3$ |
| 3-Mercaptopropyltriethoxysilane | $HSCH_2CH_2CH_2Si(OCH_2CH_3)_3$ |

— continued

TABLE 2.2 (continued)
Typical Commercial Silanes Used in Sealants

Ureidosilane	Structure
3-Ureidopropyltrimethoxysilane	$H_2NC(=O)NHCH_2CH_2CH_2Si(OCH_3)_3$
Vinyl silanes	
Vinyltrimethoxysilane	$CH_2=CHSi(OCH_3)_3$
Vinyltriethoxysilane	$CH_2=CHSi(OCH_2CH_3)_3$
Alkylsilanes	
Methyltrimethoxysilane	$CH_3Si(OCH_3)_3$

TABLE 2.3
Global Suppliers of Organofunctional Silanes

Supplier	Web Site
Momentive Performance Materials Wilton, Connecticut	http://www.momentive.com/geam/en/HomePage/Home/home.html
Dow Corning Corporation Midland, Michigan	http://www.dowcorning.com/
Evonik GmbH Düsseldorf, Germany	http://www.Degussa.com/Degussa/en/
Shin Etsu Silicones Tokyo, Japan	http://www.silicone.jp/e/
Wacker Chemie AG Munich, Germany	http://www.wacker.com/cms/en/home/index.jsp

organosilicon compounds. Poly(methylphenylsiloxane) is used when high temperature performance is required. The curing reaction involves the reaction of moisture with the organosilicon end groups to generate silanols, which condense to siloxanes. Crosslinking occurs as a result of the condensation reactions.

The general structure of these orthosilicates or organosilicon compounds is

$$R^2_b Si\text{-}X_{4-b}$$

where R^2 is a hydrocarbon, usually methyl, ethyl, or vinyl; X is a hydrolyzable group; and b is an integer from 0 to 2. Common structures for X are acetoxy, oximato, amino, benzamido, alkenyloxy, and alkoxy. These organosilicon compounds are used in large excess of the amount needed to react with the hydroxyl groups on the silicone polymer. The excess organosilicon compound is needed to control the cure rate, physical properties, and shelf stability of the organosilicon-endblocked silicone polymer. In addition to its crosslinking role, the organosilicon compound also functions as a desiccant as well as an adhesion promoter. However, further improvements in adhesion can be attained if organofunctional silanes are added to the sealant formulation. The choice of the silane adhesion promoter depends on the organosilicon compound used to endblock the poly(dimethylsiloxane) or poly(methylphenylsiloxane) polymer.

Organoacetoxysilicon-endblocked silicone polymers generally have good-to-excellent adhesion to a large variety of substrates. The acetic acid that forms conditions the surface and removes impurities such as calcium or magnesium salts. However, the pungent odor of the acetic acid by-products makes them undesirable for use in consumer applications. Di-tert-butoxydiacetoxysilane has been used to modify the adhesion strength of the silicone sealant [36]. It may function as a chain extender to decrease the modulus and ultimate tensile strength of the cured sealant and change the failure mode from interfacial to cohesive.

Similar to the acetoxy-endblocked silicone polymers, organoaminosilicon-endblocked silicone polymers also have good-to-excellent adhesion to many substrates [37]. On reaction with moisture, these polymers generate amines, which have an unpleasant fishy odor. These amino-endblocked polymers, therefore, have very limited use, even in original equipment manufacturers (OEMs) or industrial applications. When additional adhesion strength is needed, aminosilanes are often selected as adhesion promoters.

Organoalkoxybenzamidosilicon, organooximatosilane, and organoalkenyloxy-silicon-endblocked silicone polymers generally require an organofunctional silane adhesion promoter. 3-Aminopropyltriethoxysilane and N-(2-aminoethyl)-3-aminopropyltrimethoxysilane are often used for this purpose [38,39]. Organoalkoxysilicon-endblocked silicone polymers have often used tris-(3-trimethoxysilylpropyl) isocyanurate adhesion promoter [40,41].

2.3.2 POLYURETHANE SEALANTS

Polyurethane sealants are based on organic polymers that are endcapped with a di- or polyisocyanate group [42]. In one-part isocyanate-endcapped polyurethane-based sealants, 3-isocyanatopropyltrimethoxysilanes, 3-isocyanatopropyltriethoxysilanes, or 3-glycidoxypropyltrimethoxysilanes are used as adhesion promoters.

In two-part isocyanate-endcapped polyurethane-based sealants, or one-part silane-terminated polyurethane-based sealants, aminosilanes such as N-(2-aminoethyl)-3-aminopropyltrimethoxysilane and 3-aminopropyltriethoxysilane are used [38]. The isocyanate end group in some cases is blocked to improve the stability of the polymer [43]. Epoxysilanes can be used in these systems to improve adhesion to substrates.

Partial endcapping of the isocyanate-capped polymer with bis-(3-trimethoxysilylpropyl) amine was shown to improve the adhesion of sealants, especially to normally difficult-to-adhere-to substrates, such as stainless steel or concrete [44].

2.3.3 POLYSULFIDE SEALANTS

Polysulfide sealants are based on organic polymers that contain a thioether group in the backbone and are terminated with a mercapto functional group. The thioether group makes these sealants resistant to solvents and hydrocarbons used as fuels. The cure generally involves an active oxidizing agent for the mercaptan-capped polysulfide polymers to form disulfide crosslinks. The organofunctional silanes often used with these polymers are epoxy, polysulfide, mercapto, and isocyanato functional silanes such as 3-glycidoxypropyltrimethoxysilane, bis-(3-triethoxysi-lylpropyl) polysulfide, 3-mercaptopropyltriethoxysilanes, and 3-isocyanatopropyl-triethoxysilanes [45,46].

2.3.4 Silyl-Terminated Polymers

Silyl-terminated polymers are divided into two groups: silyl-terminated polyethers and silyl-terminated polyurethanes. Although these polymer systems contain reactive silyl end groups, additional organofunctional adhesion promoters are often required to further enhance the adhesion to substrates.

Landon and coworkers have demonstrated enhanced adhesion of sealants based on silyl-terminated polyurethanes to various metal oxide, glass, and plastic substrates [32]. The adhesion was achieved by using various organofunctional silanes. In their study, the isocyanate-terminated polyurethane prepolymers were reacted with aminosilanes. These silanes function mainly as crosslinkers in this hybrid polymer. The mobility of the silanes is limited after the reaction with the polymer (due to the constraints that the polymer places onto the ability of the silane to react with the surface). Monomeric organofunctional silanes that were added to the sealant formulation were found to improve adhesion to various substrates, as shown in Table 2.4. Surprisingly, it was observed that the silanes not only enhanced the adhesion to inorganic substrates, such as glass and aluminum, but also to selected plastic substrates, such as poly(vinyl chloride), acrylonitrile-butadiene-stryrene (ABS) copolymer, polystyrene, and acrylics [32].

Aminosilanes are effective as adhesion promoters in various silyl-terminated polymers. However, the amine group is susceptible to air oxidation, which makes the product slowly turn yellow over time. This phenomenon is known as the *yellowing effect*. This effect becomes more noticeable as the amine content of the silane is increased, such as the case for diamino and triaminosilanes. New aminosilanes have demonstrated their ability to reduce or eliminate this yellowing while still maintaining good adhesion [47]. The key structural feature of these new silanes is a branched alkylene group that links the amino group with the silicon atom. The structures of these silanes are illustrated in Figure 2.2.

The branching of the alkylene bridging group in beta position to the amino functional group can retard the oxidation of the amine. The improvement has been demonstrated in an accelerated yellowing test for a model formulation consisting of

Aminosilane	2 wt%
Precipitated calcium carbonate coated with stearic acid	33 wt%
Ground calcium carbonate coated with stearic acid	36 wt%
Di-isododecyl phthalate	30 wt%
p-Toluenesulfonic isocyanate	1 wt%

TABLE 2.4

Adhesion of Silylated Polyurethane-Based Sealants to Selected Substrates after 7-Day Water Soak

Organofunctional Silane	1	2	3	4	5	6	7
	Adhesion Strength (N/m)						
Aluminum	3.7 (C)	3.5 (C)	4.2 (C)	1.9 (C)	2.8 (C)	3.0 (C)	3.7 (C)
Glass	4.0 (C)	4.0 (C)	4.9 (C)	1.9 (C)	3.5 (C)	3.5 (C)	4.2 (C)
Poly(vinyl chloride)	4.0 (C)	3.7 (C)	4.9 (C)	2.3 (C)	3.7 (C)	2.8 (C)	1.1 (C/A)
ABS	3.5 (C/A)	4.4 (C)	3.3 (C)	1.2 (A)	1.7 (C)	2.5 (C)	0.4 (C/A)
Polystyrene	4.0 (C)	0.5 (A)	4.2 (C)	0.2 (A)	0.5 (C/A)	0.4 (A)	1.1 (C/A)
Acrylic	0.5 (A)	1.1 (A)	1.2 (A)	1.2 (C/A)	0.42 (A)	1.2 (A)	1.1 (A)

Note: Silanes evaluated were as follows: 1 = N-2-(aminoethyl)-3-aminopropyltrimethoxysilane; 2 = N-2-(aminoethyl)-3-aminopropyldimethoxysilane; 3 = 3-aminopropyltrimethoxysilane; 4 = *bis*-(3-trimethoxysilylpropyl)amine; 5 = 3-ureidopropyltrimethoxysilane; 6 = 3-glycidoxypropyl-trimethoxysilane; 7 = 3-glycidoxypropylmethyldiethoxysilane. The adhesion tests were conducted in accordance with the ASTM C794 after 7 days' immersion in water. The definition of the terms C, C/A, and A are as follows: C is greater than 80% cohesive failure; C/A is between 50 and 80% cohesive failure; A is less than 50% cohesive failure.

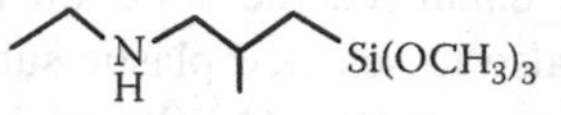

FIGURE 2.2 The structures of aminosilanes having lower propensity to yellow when used as adhesion promoters. The aminosilanes 8, 9, and 10 are supplied by Momentive Performance Materials under the registered trademarks Silquest* A-Link 15 silane, Silquest* A-1639 silane, and Silquest* A-2637 silane, respectively.

The prepared samples were then placed in an oven at 80°C for 5 days. Color of the initial sample and the sample after the heat treatment was measured by a Minolta colorimeter and recorded in Table 2.5. The change in b value indicates the effect of yellowing. The higher the Δb value, the more yellow the sample has become.

In addition to less yellowing, it was also observed that branched aminosilanes (8 and 9) improved the flexibility of moisture-curable sealants. The evaluation involved doping a sample of the silylated polyurethane-based sealant with 0.5 wt% (2 phr) of each silane. Tensile strengths of the sealants remained essentially the same. However, the sealants containing silanes with a branched bridging group, such as N-ethyl-3-amino-2-methylpropyltrimethoxysilane (8) and

TABLE 2.5
The Effect of Silane Structure on Yellowing in Model Sealant Formulation, after 5 Days in an Oven at 80°C

	Initial			After 5 Days in an Oven at 80°C			Change
Silane	L	a	b	L	a	b	Δb
1	89.83	0.05	6.52	86.96	−1.55	20.83	14.31
3	91.1	0.16	4.8	91.45	−1.52	14.76	9.96
8	91.58	0.14	4.6	89.71	−0.1	12.07	7.47
9	91.2	0.22	4.52	91.57	−0.25	6.3	1.78

Note: Silanes evaluated were as follows: 1 = N-(2-aminoethyl)-3-aminopropyltrimethoxysilane; 3 = 3-aminopropyltrimethoxysilane; 8 = N-ethyl-3-amino-2-methylpropyltrimethoxysilane; 9 = 4-amino-3,3-dimethylbutyltrimethoxysilane. L is a measure of the lightness of a sample, and ranges from 0 (black) to 100 (white). a defines the degree of redness (positive a) to greenness (negative a). b defines the degree of yellowness (positive b) to blueness (negative b).

4-amino-3,3-dimethylbutyltrimethoxysilane (9), exhibited about 25% reduction in 100% modulus, and about a twofold increase in elongation, as shown in Table 2.6. These features are all favorable to ASTM C-920 type of construction sealants.

The change in sealant properties may be due to the reactivity of the aminosilane with moisture. Sealants containing N-(2-aminoethyl)-3-aminopropyltrimethoxysilane or 3-aminopropyltrimethoxysilane cured rapidly. Sealants containing N-ethyl-3-amino-2-methylpropyltrimethoxysilane or 4-amino-3,3-dimethylbutyltrimethoxysilane cured more slowly. The reactivity of the aminosilane containing a branched alkylene linking group with water is lower owing to the steric hindrance of the branched alkylene [15]. The stability of the resulting silanols is also greater. The aminosilanes containing the branched alkylene group participate less in the curing of the

TABLE 2.6
Impact of Different Aminosilanes on Silylated Polyurethane Resin Sealant Properties

Silane	1	3	8	9
Tensile strength (MPa)	1.11	1.21	1.11	1.26
Elongation (%)	651	626	1150	1073
100% Modulus (MPa)	0.41	0.39	0.29	0.27
Tear resistance (N/m)	8.1	7.0	7.9	7.9
Hardness shore A	20	22	13	13

Note: Silanes evaluated were as follows: 1 = N-(2-aminoethyl)-3-amino-propyltrimethoxysilane; 3 = 3-aminopropyltrimethoxysilane; 8 = N-ethyl-3-amino-2-methylpropyltrimethoxysilane; 9 = 4-amino-3,3-dimethylbutyl trimethoxysilane.

TABLE 2.7
Adhesion of Silylated Polyurethane Sealants with Different Aminosilanes to Various Substrates

Silane	1	3	8	9
	Wet Adhesion (N/m / failure)			
Glass	3.7/A	4.6/C	3.0/C	3.3/C
Aluminum	3.3/C	3.2/C	3.0/C	2.8/C
ABS	3.9/C	3.5/C	3.0/C	3.3/C
Poly(vinyl chloride)	2.8/C	3.0/C	3.5/C	2.6/C
Polystyrene	3.7/C	4.0/C	3.0/C	3.2/C
Glass-filled nylon	4.6/A	4.2/C	2.8/A	2.8/C

Note: Silanes evaluated were as follows: 1 = *N*-(2-aminoethyl)-3-aminopropyltrimethoxysilane;
3 = 3-aminopropyltrimethoxysilane; 8 = *N*-ethyl-3-amino-2-methylpropyltrimethoxysilane;
9 = 4-amino-3,3-dimethylbutyltrimethoxysilane.
The definitions of the terms C and A are as follows: C = greater than 80% cohesive failure; A = less than 50% cohesive failure.

silyl-terminated polyurethane. The average crosslink density of the cured sealant is reduced. The lower reactivity of the aminosilanes containing a branched alkylene group did not decrease the adhesion to various substrates, as shown in Table 2.7. The test results for the peel strengths were obtained after 7-day water immersion.

2.3.5 ACRYLIC SEALANTS

Waterborne acrylic polymers are particularly well suited for use in consumer sealants. These sealants are easy to apply, easy to clean with soap and water, and eliminate exposure of the user to organic solvents and chemicals. The inherent hydrolytic reactivity of the commonly used organofunctional silanes makes their use a challenge in waterborne acrylic-based sealants. Performance of the commonly used silanes as adhesion promoter diminishes over time. Once the silanes are added to the sealant, they react and form oligomeric and polymeric siloxane resins that are insoluble in both water and oil phases of the waterborne acrylic. The oligomeric and polymeric siloxane resins phase-separate and become inactive. Much effort has been devoted to developing silanes for use in waterborne sealants. The efforts are focused on three approaches: stabilizing silanols, retarding hydrolysis, and emulsifying the silane.

Aminosilanes are commonly used in waterborne acrylic sealants with mixed results [49]. The performance of these aminosilanes may be related to the stability of the silanols in the sealant. Chiang et al. proposed that 3-aminopropylsilonetriol is stable because it can form an intramolecular hydrogen bond with itself [50].

Hydrogen bonding between the amine and silanol may reduce the reactivity of the silane. The silanes may also be capable of forming salts with themselves. The silanol is deprotonated by the amino group to form an intramolecular ammonium silanoate salt. These structures are able to better maintain their solubility in water and reactivity with the substrates.

In mildly alkaline solution, the silanol groups are much less stable and condense to polysiloxane [1]. Zazyczny and Steinmetz [51] studied the hydrolysis and condensation of amino-containing silane in an aqueous solution using a method based on gel permeation chromatography. They observed a cloud point indicating phase separation due to the formation of siloxane tetramer. Condensation of the silanols to higher-molecular-weight species was avoided under highly alkaline conditions.

A series of waterborne organofunctional silanes were evaluated as adhesion promoters in acrylic latex sealants [51]. These waterborne silanes were alkaline; their pH values were in a range of 10 to 12, except for the glycol functional silane, whose pH value was in a range of 3 to 5. Conventional silanes and waterborne silanes were added to the sealant formulation at 1 weight percent. In order to test the shelf life of the sealants containing organofunctional silanes, peel tests were performed after storing the sealants for 30 and 90 days. The adhesion of the sealant to glass or aluminum was different when different silanes were used in the formulation, as shown in Table 2.8. The waterborne silanes containing amino functional groups (11, 12, 13), as well as the conventional aminosilanes (10 and 15), provided better wet adhesion than the other conventional and waterborne silanes screened. The adhesion was poorer after 90 days. The loss in adhesion may be due to the alkalinity of the sealant, which could hydrolyze the covalent bonds between the silane and substrate. The acrylic latex has a pH value in the range of 7–9, which has been shown to contribute to the instability of the silanols [51].

The efficacy of silane adhesion promoters can be improved by controlling the hydrolysis rate of the silane. 3-Glycidoxypropyltrimethoxysilane is used in many acrylic latex sealant formulations because the glycidyl group is reactive with the carboxylic acid and hydroxyl groups that are pendent or terminus on the polymer chains. The products have good adhesion initially, but over time, these sealants containing the 3-glycidoxypropyltrimethoxysilane lose their adhesion strength. The hydrolysis and condensation of the epoxysilane are thought to contribute to the poor aging characteristics of these sealants.

Huang and Waldman explored the use of epoxysilanes that are slower to hydrolyze and more soluble in the oil phase of the acrylic latex sealant [52,53]. One material that provided better aging properties is 3-glycidoxypropylmethyldiethoxysilane. Based on extrapolations from the work done by Pohl et al. [54] and Smith [55], this silane hydrolyzes significantly slower than 3-glycidoxypropyltrimethoxysilane in acrylic latexes. Pohl et al. found that the acetate-catalyzed hydrolysis of vinyl-trimethoxysilane was 20 times faster in aqueous acetone than that of vinyltriethoxysilane [54]. Smith found that under alkaline conditions, dimethyldimethoxysilane hydrolyzed slower than methyltrimethoxysilane [55]. The silanols generated on hydrolysis are also more stable in water. The equilibrium constant for the reaction of the 3-glycidoxypropylmethylsilanediol to form siloxanes and water is 210 [26]. The equilibrium constant for the reaction of silanetriols to form siloxanes and water is

TABLE 2.8

Comparison of Conventional Silanes 6, 10, 15, and 17 with Waterborne Silanes 11, 12, 13, 14, 16, and 18 in Acrylic Latex Sealants

	Glass				Aluminum			
	Dry Adhesion (N/m)		Wet Adhesion (N/m)		Dry Adhesion (N/m)		Wet Adhesion (N/m)	
Silane	30 Days	90 Days	30 Days	90 Days	30 Days	90 Days	30 Days	90 Days
No silane	8.8	1.8	< 0.2	< 0.2	4..4	3.5	< 0.2	< 0.2
10	7.9	3.7	2.8	1.9	5.3	2.6	2.5	1.8
11	7.5	3.9	4.4	2.1	6.8	1.8	2.8	< 0.2
12	7.5	2.6	3.7	1.6	3.7	1.8	2.1	1.2
13	8.8	2.1	3.7	< 0.2	6.7	3.5	1.4	0.4
6	2.6	1.4	0.7	< 0.2	1.8	0.9	< 0.2	< 0.2
14	1.8	0.9	0.9	< 0.2	3.5	1.4	< 0.2	< 0.2
15	8.4	5.6	3.2	< 0.2	4.7	2.6	< 0.2	< 0.2
16	2.6	1.6	< 0.2	< 0.2	5.8	4.4	< 0.2	< 0.2
17	2.6	6.1	1.1	< 0.2	4.4	2.8	< 0.2	< 0.2
18	7.5	2.6	2.6	< 0.2	2.1	1.8	< 0.2	< 0.2

Note: Silanes evaluated were as follows: 10 = 3-aminopropyltriethoxysilane; 11 = aminopropylsilanol solution, supplied by Evonik under the trade name Dynasylan® Hydrosil HS2627siloxane oligomer; 12 = 4, 7, 10-triazadecylsilanol solution, supplied by Evonik under the trade name Dynasylan® Hydrosil HS2776 siloxane oligomer; 13 = N-(2-aminoethyl)-3-aminopropylsilanol solution, supplied by Evonik under the trade name Dynasylan® Hydrosil HS2775 siloxane oligomer; 6 = 3-glycidoxypropyltrimethoxysilane; 14 = 6,7-dihydroxy-4-oxa-heptylsilanol solution, supplied by Evonik under the trade name Dynasylan® Hydrosil HS2759 siloxane oligomer; 15 = 3-methacryloxypropyltrimethoxysilane; 16 = 3-methacryloxypropylsilanol solution, supplied by Evonik under the trade name Dynasylan® Hydrosil HS2789 siloxane oligomer; 17 = vinyltrimethoxysilane; 18 = vinylsilanol solution, supplied by Evonik under the trade name Dynasylan® Hydrosil HS2781 siloxane oligomer.

485. The smaller equilibrium constant for the condensation reaction of 3-glycidoxypropylmethylsilanediol indicates that significant amounts of the silanediol remain in the acrylic latex when used at low levels. The combination of slower hydrolysis rate, higher solubility in the oil phase, and more stable silanols endows acrylic latexes with better aging properties. Huang and Waldman found that the viscosities of sealants without silane and sealants containing 1 phr of 3-glycidoxypropylmethyldiethoxysilane were stable for up to a year, while the sealant containing 3-glycidoxypropyltrimethoxysilane gelled, as shown in Figure 2.3 [53].

The drying and curing process for acrylic latex-based sealants may require as long as 3 weeks at room temperature to complete. A three-stage process has been proposed for the coalescence, as illustrated in Figure 2.4 [56]. Controlled reactivity of the silane with

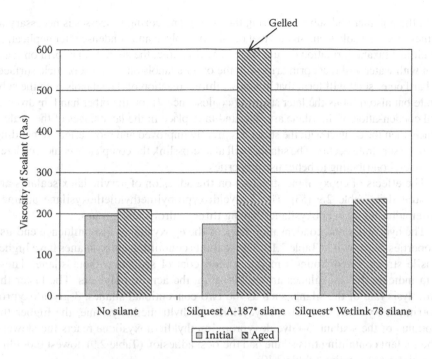

FIGURE 2.3 Viscosity of sealants with and without epoxysilanes. Silquest A-187* silane is 3-glycidoxypropyltrimethoxysilane and Silquest* Wetlink 78 silane is 3-glycidoxypropylmethyldiethoxysilane, supplied by Momentive Performance Materials. Sealants were aged in an oven at 50°C for 4 weeks.

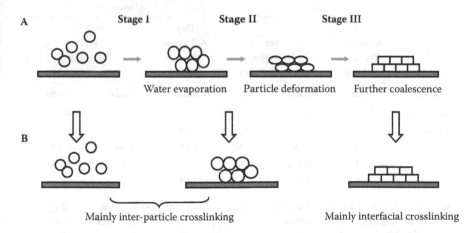

FIGURE 2.4 Schematic showing the three stages of (a) latex coalescence and (b) formation of siloxane bonds.

both the polymer and substrate during the drying and curing processes is necessary to achieve the best adhesion. Silanes that rapidly hydrolyze and condense after application result in siloxane formation during stage I. At this time, the silanols that form on reaction with water will react primarily with the other silanols on the latex particle surface. A hard outer shell will form that not only inhibits reaction of the silanols with the substrate but also inhibits the later stages of coalescence. If, on the other hand, hydrolysis and condensation of the silane are slow and take place in the later stages of the coalescence, then the contact with the surface is greatly improved and more effective bonding with the substrate occurs. The silanols will also crosslink the completely coalesced latex particles, contributing to better use properties.

The effects of epoxysilane structure on the adhesion of acrylic latex sealants are illustrated in Table 2.9 [53]. The 3-glycidoxypropylmethyldiethoxysilane attained better adhesion than epoxysilanes having three hydrolyzable groups.

The hydrolysis and condensation rates of the epoxysilanes also influence end-use properties, as shown in Table 2.10. The sealants containing epoxysilanes have higher tensile strength and Young's modulus than control sealant without silane. These data indicate that the silanes are crosslinking the acrylic polymers. The faster the hydrolysis rate of the silane, such as the two conventional silanes, 3-glycidoxypropyltrimethoxysilane and 3,4-epoxycyclohexylethyltrimethoxysilane, the higher the modulus of the sealant. 3-Glycidoxypropylmethyldiethoxysilane reacts the slowest. The sealant containing this silane had the best adhesion (Table 2.9), lowest modulus, and highest elongation (Table 2.10).

TABLE 2.9
Peel Strength of Acrylic Latex Sealants Containing Epoxysilanes

| | Glass | | | Aluminum | | |
| | Dry Adhesion (N/m / failure) | Wet Adhesion (N/m / failure) | | Dry Adhesion (N/m / failure) | Wet Adhesion (N/m /failure) | |
Silane	Initial	Initial	Aged	Initial	Initial	Aged
No silane	2.2/C	0	0	2.3/C	0.9/A	0
6	NA	0	Gel	NA	1.2/A	Gel
19	2.3/C	2.6/C	2.5/C	2.3/C	2.5/C	1.3/C
20	1.6/C	0.3/A	NA	2.2/C	0.8/A	NA
21	2.1/C	2.1/A	0.1/A	1.8/C	1.9/C	0.6/A

Note: Silanes evaluated were as follows: 6 = 3-glycidoxypropyltrimethoxysilane; 19 = 3-glycidoxypropylmethyldiethoxysilane; 20 = 2-(3,4-epoxycyclohexyl)ethyltrimethoxysilane; 21 = 2-(3,4-epoxycyclohexyl)ethyltriethoxysilane. The definitions of the terms C and A are as follows: C = greater than 80% cohesive failure; A = less than 50% cohesive failure. The sealants were aged in an oven at 50°C for 4 weeks. The adhesion tests were carried out in accordance with ASTM C 794. Wet adhesion was measured after 7 days water immersion.

TABLE 2.10

Mechanical Properties of Acrylic Latex Sealants Containing Different Epoxysilanes

Silane	Tear Strength (N/m)	Elongation (%)	Tensile Strength MPa	Young's Modulus MPa	Hardness Shore A
No silane	6.1	306	0.53	0.41	37
6	3.3	179	1.35	1.05	38
19	8.0	321	1.12	0.31	38
20	10.4	242	1.97	1.18	45
21	9.6	256	1.18	0.83	47

Note: Silanes evaluated were as follows: 6 = 3-glycidoxypropyltrimethoxysilane; 19 = 3-glycidoxypropylmethyldiethoxysilane, supplied by Momentive under the registered trademark of Silquest® Wetlink 78 silane; 20 = 2- (3,4-epoxycyclohexyl) ethyltrimethoxysilane supplied by Momentive under the registered trademark of Silquest® A-186 silane; 21 = 2- (3,4-epoxycyclohexyl)ethyltriethoxysilane, supplied by Momentive under the registered trademark of CoatOSil® 1770 silane.

The modulus and elongation were dependent on the amount of silane used. As the concentration of the 3-glycidoxypropylmethydiethoxysilane was increased, the modulus became higher and the elongation was reduced. The shelf life of the sealant was reduced as more silane was incorporated. The optimum loading of the 3-glycidoxypropylmethyldiethoxysilane (19) was found to be between 0.5–1 phr [52].

Stable emulsion of the silane was found to also improve its performance in waterborne coatings. Stable emulsions of polar and hydrolytically unstable silanes, such as 3-glycidoxypropyltrimethoxysilane, are difficult to produce. However, stable emulsions have been made using nonpolar and less reactive silanes such as 2-(3,4-epoxycyclohexyl)ethyltriethoxysilane (21) [57,58]. This silane emulsion was used successfully as a primer [52] or as an additive for polyurethane dispersions.

2.4 SILANES AS CROSSLINKERS

Organofunctional silanes are used as crosslinkers for organic polymers that are commonly used in sealants. The organofunctional group of the silanes reacts with the functional group on the polymer. These silylated polymers cure when exposed to atmospheric moisture. This technology was developed nearly 30 years ago. It is a hybrid polymer that possesses characteristics of RTV silicone and polyurethane sealant.

Two major categories of silylated polymers available are silylated polyethers and silylated polyurethanes. Silylated polyethers are made by reacting dimethoxysilane with allyl-terminated poly(propylene glycol) [59]. The silylated polyurethanes are made either by reacting an amino or mercapto functional silane with an isocyanate-terminated polymer or by reacting an isocyanatosilane with a polyol [3]. Various organofunctional silanes have been evaluated as crosslinkers. However, the structure of the prepolymer dominates the mechanical properties of the cured elastomer. The structure of the silane crosslinker has only minor impact on these mechanical properties, although it affects the viscosity and stability of the finished sealant. Table 2.11 illustrates the effects of different aminosilane crosslinkers on sealant performance.

The secondary aminosilane, N-phenyl-3-aminopropyltrimethoxysilane, reacts faster with the isocyanate-terminated polymer than either the mercaptosilane (22) or ureidosilane (26), but without the viscosity surge that occurs when primary aminosilanes (3,23) are used. The sealant made of N-phenyl-3-aminopropyltrimethoxysilane-capped polyurethane polymer has shorter tack-free time and better-balanced mechanical properties when compared with other silane-endcapped polyurethanes.

Aminosilanes are most often used for endcapping the isocyanate-terminated polymer. In order to avoid unfavorable crosslinking, secondary aminosilanes are preferred. The Michael adducts resulting from the reaction of dialkyl maleate with primary aminosilanes have been used [60,61]. Mercaptosilanes have been also used as endcappers in commercial products, mainly for automotive applications.

Isocyanatosilanes such as isocyanatomethylmethyldimethoxysilane and 3-isocyanatopropyltrimethoxysilane have been used to endblock polyols. Isocyanatomethylmethyldimethoxysilane is supplied by Wacker under the trade name Geniosil® XL42 silane, and 3-isocyanatopropyltrimethoxysilane is supplied by Momentive Performance Materials under the registered trademark Silquest* A-Link 35 silane.

These two isocyanatosilanes form carbamate linking groups when they react with the polyols, rather than the urea linking group that forms when an aminosilane reacts with an isocyanate-terminated polymer. Bectause these isocyanatosilane-terminated polymers do not form a urea group that bonds the silyl group to the polymer, they have lower viscosity than polymers resulting from the reaction of aminosilane with isocyanate-terminated polymers. When faster cure is required, isocyanatomethylmethyldimethoxysilane is used. The strong electron-withdrawing group alpha to the dialkoxysilyl group significantly increases its reactivity with moisture.

TABLE 2.11
Physical Properties of Silylated Polyurethane Sealants

Silane Crosslinker	Tensile Strength (MPa)	Elongation (%)	100% Modulus (MPa)	Tear Strength (N/m)	Hardness Shore A	Tack-free Time (h)
22	1.38	580	0.21	7.0	20	4
23	2.00	585	0.28	7.9	25	4
24	2.41	590	0.34	7.5	30	0.5
25	1.24	350	0.34	5.3	20	3
3	1.79	390	0.41	8.2	25	2

Note: Silanes evaluated were as follows: 22 = 3-mercaptopropyltrimethoxysilane; 23 = 3-aminopropylmethyldimethoxysilane; 24 = N-phenyl-3-aminopropyltrimethoxysilane; 25 = 3-ureidopropyltrimethoxysilane; 3 = 3-aminopropyltrimethoxysilane.

Silanes have also been used to crosslink acrylic latex polymers. Silanes containing a reactive free-radical group, such as vinyl or methacryl, are copolymerized with the other monomers during the preparation of the latex. However, vinyltrimethoxysilane and 3-methacryloxypropyltrimethoxysilane are difficult to use and result in the formation of gel particles during preparation. Pohl et al. evaluated silanes containing sterically hindered leaving alkoxy groups [54]. Conventional methoxy- and ethoxy-leaving groups were replaced by *n*-propoxy and iso-propoxy, respectively. The reactivity decreased by a factor of 650. These sterically hindered silane-modified latex coatings have excellent adhesion and chemical resistance [62].

2.5 CONCLUSION

Organofunctional silanes are widely used in sealant technology to improve adhesion, crosslink polymers, couple fillers, and act as desiccants. The use of these additives lengthens the useful life of sealants, especially when exposed to harsh conditions such as high humidity, exposure to water, or other chemicals. Organofunctional silanes achieve these benefits because they form strong interactions with many substrates and fillers or form stable siloxane bonds that make them suitable as crosslinkers for organic polymers. These characteristics are key to improving the durability of these sealants when used in a vast array of applications, which range from construction to automotive to aerospace.

ENDNOTES

* Silquest and CoatOSil are trademarks of Momentive Performance Materials, Inc.

REFERENCES

1. E.P. Plueddemann, *Silane Coupling Agents*, Plenum Press, New York (1982).
2. J.G. Marsden, in *Handbook of Adhesives*, I Skeist (Ed.), p. 536, Van Nostrand Reinhold, New York (1990).
3. T.M. Feng and B.A. Waldman, *Adhesives Age,* **38**(4), 30 (1995).
4. M. Huang, *Adhesives Age,* **44**(7), 37 (2001).
5. J. Comyn, F. de Buyl, N.E. Shephard, and C. Subramaniam, *Intl. J. Adhesion Adhesives,* **22**, 385 (2002).
6. J. Comyn, F. de Buyl, and T.P. Comyn, *Intl. J. Adhesion Adhesives,* **23**, 495 (2003).
7. F.A. Bergman, Moisture Crosslinkable Polymers: Studies on the Synthesis, Crosslinking and Rheology of Methoxysilane Functional Poly(Vinyl Esters). Thesis (doctoral)— Technische Universiteit Eindhoven (2001).
8. V.V. Severnyi, Ye. Minsker, and N.A. Ovechkina, *Polym. Sci. U.S.S.R.,* **19**(3), 42 (1977).
9. S.H. Ding, D.Z. Liu, and L.l. Duan, *Polym. Degradation Stability,* **91**, 1010 (2006).
10. W.A. Zisman, *Ind. Eng. Chem.* **35**(10), 19 (1963).
11. W.A. Zisman, *Ind. Eng. Chem., Prod. Res. Devel.* **8**(2), 98 (1969).
12. W. Riegler and K. Rhodes, *Med. Device Diagnostic Ind.,* **28**(9), 80 (2006).
13. A. Lekatou, *Mater. Res. Soc. Sym. Proc.,* **304**, 27 (1993).
14. P.H. Harding and J.C. Berg, *J. Appl. Polym. Sci.,* **67**, 1025 (1998).
15. F.D. Osterholtz and E.R. Pohl, *J. Adhesion Sci. Technol.,* **6**, 127 (1992).

16. E.P. Plueddemann, in *Silanes and Other Coupling Agents,* K. L. Mittal (Ed.), p. 3, VPS, Utrecht (1992).
17. M. Rosen, *J. Coatings Technol.*, **50**(644), 70 (1978).
18. R.J. Good, in: *Treatise on Adhesion and Adhesives*, R.L. Patrick (Ed.), Vol. 1 p. 15, Marcel Dekker, New York (1967).
19. M.A. Lutz, T.E. Gentle, S.V. Perz, and M.J. Owen, in *Silanes and Other Coupling Agents,* Vol. 2, K.L. Mittal (Ed.), p. 3, VSP, Utrecht (2000).
20. J.C. Bolger and A.S. Michaels, in *Interface Conversion*, P. Weiss and D. Cheevers (Eds.), Chap. 1, Elsevier, New York (1969).
21. A.A. Parker and J. MacLachlan, in: *Silanes and Other Coupling Agents,* Vol. 2, K.L. Mittal (Ed.), p. 27, VSP, Utrecht (2000).
22. E.P. Plueddemann, *Silane Coupling Agents,* Chap. 1, 2nd edition, Plenum Press, New York (1991).
23. D.J. Belton and A. Joshi, in *Molecular Characterization of Composite Interfaces*, H. Ishida and G. Kumar (Eds.), p. 187, Plenum Press, New York (1985).
24. C.A. Kumins and J. Roteman, *J. Polym. Sci. Part A*, **1**, 527 (1963).
25. M. Strobel, C.S. Lyons, and K.L. Mittal (Eds.) *Plasma Surface Modification of Polymers: Relevance to Adhesion,* VSP, Utrecht (1994).
26. E.R. Pohl and F.D. Osterholtz, in S*ilanes, Surfaces and Interfaces*, D. E. Leyden (Ed.), p. 481, Gordon & Breach Science Publishers, New York (1986).
27. H. Ishida and J.L. Koenig, *J. Polym. Sci. Polym. Phys. Ed.* **18**, 233 (1980).
28. F.M. Fowkes, *J. Adhesion Sci. Technol.*, **1**, 7 (1987).
29. K.L. Mittal (Ed.), *Acid-Base Interactions: Relevance to Adhesion Science and Technology*, Vol. 2, VSP, Utrecht (2000).
30. K.L. Mittal and H.R. Anderson, Jr. (Eds.), *Acid-Base Interactions: Relevance to Adhesion Science and Technology*, VSP, Utrecht (1991).
31. H.P. Schreiber, R. Qin, and A. Sengupta, *Proc. of the 21st Annual Meeting of the Adhesion Society*, p. 71 (1998).
32. S.J. Landon, N.B. Dawkins, and B.A. Waldman, *Proc. Eurocoat '97*, Lyon, France, Vol. 1, p.171 (1997).
33. T. Eklund, J. Bäckman, P. Idman, A.E.E. Norström, and J.B. Rosenholm, in *Silanes and Other Coupling Agents*, K.L. Mittal (Ed.), Vol. 2, p. 55, VSP, Utrecht (2000).
34. P. Walker, *J. Adhesion Sci. Technol.*, **5**, 279 (1991).
35. L.J. Broutman and B.D. Agarwal, *Society of Plastics Industry, 28th Ann. Tech. Conf. Reinf. Plast.*, Section 5-B (1973).
36. T.A. Kulpa, *U.S. Patent 3,296,161* (1967).
37. C.E. Creamer, *U.S. Patent 3,451,964* (1969).
38. R.Johnston and P.Lehmann, *U.S. Patent* 6310170 (2001).
39. J.A. Knepper, D.R. Flackett, and E.T. Asirvatham, *U.S. Patent 5,569,750* (1996).
40. A.H. Smith and M.D. Beers, *U.S. Patent 3,708,467* (1973).
41. J.H. Wengrouvius and G.M. Lucas, *U.S. Patent 5,175,057* (1992).
42. M.N. Sathyanarayana and M. Yaseen, *Prog. Org. Coat.*, **26**, 275 (1995).
43. D.A. Wicks and Z.W. Wicks, Jr., *Prog. Org. Coat.*, **41**, 1 (2001).
44. M.H. Berger, W.P. Mayer, and R.J. Ward, *U.S. Patent 4,374,237* (1983).
45. G.B. Lowe, T.C. Lee, J. Comyn, and K. Huddersman, *Intl. J. Adhesion Adhesives*, **14**, 85 (1994).
46. M.W. Raney and R.J. Pickwell, *U.S. Patent 4,000,347* (1976).
47. R. Gauthier and C. Lacroix, *U.S. Patent 2007/0066768A1* (2007).
48. R. Johnston and P. Lehmann, *U.S. Patent 6,310,170B1* (2001).
49. A.R. Bullman, *U.S. Patent 4,340,524* (1982).
50. C.H. Chiang, H. Ishida, and J. Koenig, *J. Colloid Interface Sci.*, **74**, 396 (1980).
51. J.M. Zazyczny and J.R. Steinmetz, *Pitture e Vernici Europe*, **72**(18), 38 (1996).

52. M.W. Huang, *U.S Patent 6,001,907* (1999).
53. M.W. Huang and B.A. Waldman, *Proc. of the 22nd Annual Meeting of the Adhesion Society*, p. 357 (1999).
54. E.R. Pohl, A. Chaves, C.T. Danehey, A. Sussman, and V. Bennett, in *Silanes and Other Coupling Agents*, Vol. 2, K. L. Mittal (Ed.), p. 15, VSP, Utrecht (2000).
55. K.A. Smith, *J. Org. Chem.*, **51**, 3827 (1986).
56. L.A. Cannon and R.A. Pethrick, *Macromolecules*, **32**, 7617 (1999).
57. M.J. Chen, F.D. Osterholtz, A. Chaves, P.E. Ramdatt, and B.A. Waldman, *J. Coat. Technol.*, **69**, 43 (1997).
58. F.D. Osterholtz, E.R. Pohl, M.J. Chen, and A. Chaves, *U.S. Patent 5,714,532* (1998).
59. K. Hashimoto, *Adhesives Age*, **41**(8), 18 (1998).
60. M.S. Shaffer, R.R. Roesler, and L. Schmalstieg, *U.S. Patent 6,005,047* (1999).
61. J.T. Staurt, *U.S. Patent 6,265,517* (2001).
62. M.J. Chen, A. Chaves, F.D. Osterholtz, E.R. Pohl, and W.B. Herdle, *Surface Coatings Intl.*, **79**(12), 539 (1996).

Section II

Testing and Durability of
Sealants and Sealed Joints

3 Electrochemical Impedance Spectroscopy in Sealant Testing

Guy D. Davis

CONTENTS

3.1 INTRODUCTION

A good sealant provides a durable barrier against whatever environment is being contained or excluded, such as water or fuel. The barrier requires good wetting of and bonding to the substrate and low permeability of the environment into the sealant. Depending on the application, other desirable properties could include corrosion inhibition, adhesion strength, and electrical or thermal conductivity.

Laboratory-based prognostic tests that can evaluate the performance of sealants are needed to differentiate between sealants in order to qualify materials and select an appropriate sealant that will meet specific requirements (e.g., sealing performance, lifetime, corrosion protection, appearance, conductivity, or cost). This is especially true because environmental concerns have encouraged or required sealants with low volatile organic compound (VOC) emission or nonchromate corrosion inhibitors. Failure to differentiate between sealants and to predict performance could result in placing sealants in service that are not suitable for a given application. Inadequate screening could lead to leaks or significant increases in corrosion or required maintenance to prevent premature corrosion. Alternatively, an expensive sealant could be chosen when a less expensive material would be satisfactory.

3.2 ELECTROCHEMICAL IMPEDANCE SPECTROSCOPY

Electrochemical impedance spectroscopy (EIS, also known as *ac impedance spectroscopy*) is an established laboratory tool used to inspect coatings and corrosion of metals [1–9] and has recently been applied to the testing of sealants [10,11]. EIS involves using a potentiostat to apply a small ac voltage (10–50 mV) between the working electrode (specimen) and a counterelectrode, and measuring the current that is induced between the specimen and a reference electrode. (In some cases, the reference and counter electrodes can be combined for a two-electrode measurement.) Corrosion and degradation of both bare and coated metals can be investigated using EIS, but the technique is especially suited for coated metals because the coating acts as a capacitor and is amenable to ac techniques, but not to dc techniques such as linear polarization or cyclic voltammetry, which are commonly used to study corrosion and passivation of metal surfaces. Traditionally, EIS is performed on immersed coated specimens using separate counter and reference electrodes immersed in the same electrolyte bath. Both the magnitude and phase of the current with respect to the voltage are measured, and a complex impedance is calculated. This measurement is repeated over a wide range of frequencies (e.g., 0.01–10,000 Hz).

It is common to represent the impedance spectrum of a coating as an equivalent circuit. The equivalent circuit shown in Figure 3.1 is very popular in describing a coating with defects. The leftmost capacitance and resistance represent the coating, while the rightmost parameters represent the interface between the coating and the metal. Initially, the coating is a barrier to the electrolyte and the resistance is very high, so that the impedance spectrum is dominated by the coating capacitance. This behavior is illustrated by the "good" spectrum of Figure 3.2. In the Bode magnitude representation (log impedance versus log frequency), the impedance spectrum has a slope of −1 and a high value, typically 10^8 to 10^{10} Ω, at low frequencies. The phase angle is close to −90° over most of the frequency range. As a coating degrades, moisture penetrates and allows electrical conductance via pores and holidays or voids,

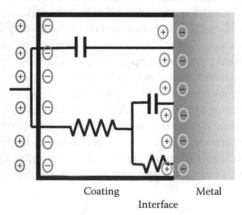

Coating   Metal

Interface

FIGURE 3.1   Equivalent circuit for a defective coating: The leftmost capacitor and resistor represent the coating. The rightmost components represent the coating–metal interface.

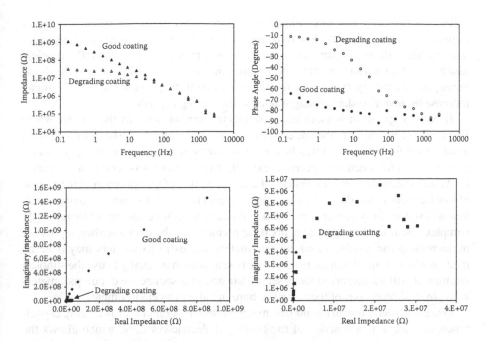

FIGURE 3.2 Bode magnitude (upper left) and phase (upper right) plots of "good" (solid symbols) and "degrading" (open symbols) coatings: Nyquist plots (lower graphs) show the same data. The scale of the right plot has been expanded to show the degrading coating spectrum more clearly.

so the coating resistance decreases. The impedance at low frequencies decreases by one or more orders of magnitude and becomes independent of frequency. In the phase angle plots, the phase angle is close to zero (resistive) at low frequencies and decreases to near −90° (capacitive) at high frequencies. The "degrading coating" spectrum of Figure 3.2 is an example of such a spectrum. Once enough moisture reaches the interface, the metal starts to corrode. The interface circuit parameters will reflect this corrosion, and a second time constant can often be detected in the spectrum, especially in the phase angle spectrum. An alternative display of the data is a Nyquist plot (imaginary impedance versus real impedance). In this representation, a good coating follows along a segment of a semicircle. As the coating degrades, the semicircle becomes smaller in size and the spectrum completes more of the semicircle. Once corrosion begins at the interface, a second semicircle is formed that represents these reactions.

The changes are detected by EIS well in advance of any visual signs of corrosion (e.g., rust or blistering) or coating failure, so that EIS can be used to predict coating performance. Good correlation has been reported between EIS measurements and long-term coating performance for immersion in different electrolytes [3,6,12–16].

As shown in Figure 3.2, the most obvious EIS parameter that monitors coating health is low-frequency impedance. It varies by several orders of magnitude as the coating degrades and, hence, is very sensitive to coating performance. It can be defined as the impedance magnitude at a given (low) frequency or an average of impedance values at several low frequencies. Once the resistive portion of the spectrum includes all the frequencies chosen, the two approaches give virtually identical results. Another parameter that is less commonly used to monitor coating health is breakpoint frequency, that is, the frequency at which the phase angle is −45°. It can increase by several orders of magnitude as the coating degrades.

If an EIS spectrum is modeled as an electric circuit, such as the one shown in Figure 3.1, or is similar to the one used for modeling the sealants to be discussed here, additional information can often be obtained. Equivalent circuit modeling (ECM) is especially useful when the spectrum exhibits two or more time constants, because ECM uses the entire spectrum instead of a small portion of the spectrum as in the case of low-frequency impedance measurements. It must be noted that there is no one circuit that will successfully model a spectrum, but rather an infinite number of increasingly complex circuits. Although better fits or agreements to the data are always possible by increasing the number of circuit parameters, the extra parameters may have no relationship with specimen properties. The best approach is usually to use the simplest circuit that will adequately model a set of data and to associate the circuit parameters with physical properties of the coated or bare metal and its environment.

One particularly useful circuit parameter as shown in Figure 3.1 is coating capacitance, because it is a function of the coating dielectric constant, which allows the moisture content of the coating or sealant to be calculated from it. Analytically, for a coating or sealant, the capacitance C of a parallel-plate configuration can be expressed as

$$C = \varepsilon_0 \varepsilon_S A / d \tag{3.1}$$

where ε_0 is the permittivity of free space, ε_S is the dielectric constant of the coating or sealant, d is the thickness of the sealant, and A is the area. In the presence of moisture, this becomes

$$C = \varepsilon_0 \left(\varepsilon_S + M \left(\varepsilon_W - \varepsilon_S \right) \right) A / d \tag{3.2}$$

where M is the moisture concentration (volume fraction) and ε_W is the dielectric constant of water (80). The significantly larger dielectric constant of water relative to that of the sealant or coating (typically, 3.0–4.0) [17] allows low concentrations of moisture to be readily detected from increases in the capacitance. Equation 3.2 can be rewritten as

$$M = \left(C / C_0 - 1 \right) / \left(\varepsilon_W / \varepsilon_S - 1 \right) \tag{3.3}$$

or

$$M = 0.046 \left(C / C_0 - 1 \right) \tag{3.4}$$

where C_0 is the initial capacitance before exposure to water. Thus, this approach allows the moisture absorption by a sealant to be determined by comparing the capacitance before and after moisture exposure. Because the impedance Z of a capacitor is inversely related to the capacitance

$$Z = j/\omega C \qquad (3.5)$$

where j is the square root of -1 and ω is the angular frequency, as the sealant absorbs moisture, the impedance of the capacitive region of the Bode magnitude plot decreases. However, this decrease is much smaller than the decrease observed in the resistive low-frequency region of the spectrum and may be subtle on the log–log representation of the Bode plot.

The sum of the two resistances in the circuit reflects the low-frequency imped-ance if the impedance is independent of frequency (i.e., resistive). If the frequency range of the spectrum is not low enough for the low-frequency impedance to become resistive (at low enough frequencies, the impedance of the capacitance becomes high enough that the impedance of the resistance dominates the spec-trum), the sum of the resistances reflects the impedance where it is independent of frequency.

The traditional remote-electrode EIS approach can be used for sealants by treating the sealant as a coating. The sealant would be used to encapsulate the metal coupon (or at least the area that was exposed to the test solution), and EIS would be used to detect moisture absorption by the sealant or any porosity or defects in the sealant.

An alternative approach, results of which will be discussed later, treats the seal-ant as an adhesive and uses corrosion sensors [18] that have been used to study the durability of adhesively bonded joints in lap-shear or wedge test configurations [19–23]. These sensors allow EIS measurements to be taken under ambient conditions instead of during immersion. Removal from the immersion environment is critical in the case of adhesive bonds so that the EIS signal is obtained from the adhesive and adhesive–adherend interface and not from the electrolyte path between the two adherends (electrodes) in immersion. The same requirements also hold for bonded sealant specimens.

The specimen for this approach is a lap-shear specimen using a ring of sealant as the adhesive (Figure 3.3) [10,11]. EIS measurements are taken across the sealant by applying corrosion sensors to each of the two adherends. By using titanium and aluminum adherends, a strong galvanic couple is created. Because titanium is near the noble end of the galvanic series in seawater and aluminum is near the active end [24], the aluminum acts as a sacrificial anode and corrosion of this adherend is pro-moted. The specimen tests the ability of the sealant to prevent corrosion by both the exclusion of electrolyte from inside the sealant ring and the inhibition of corrosion by leaching of a corrosion inhibitor. The specimens are immersed in an appropri-ate solution, such as salt water. Elevated temperatures can be used to exacerbate the exposure conditions by accelerating diffusion and chemical reactions. Periodically, the specimens are removed from the solution, and EIS measurements between the two adherends are taken. In addition to the EIS measurements, lap-shear adhesion strength measurements and inspection for corrosion of the aluminum in the sealant

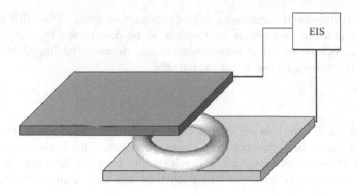

FIGURE 3.3 Schematic representation of sealant test specimen: A ring of sealant is used to make a lap-shear specimen with two different metals.

ring and just outside of the sealant ring help to judge sealant performance. The latter two measurements are destructive and, hence, can only be performed once on a given specimen unless mold release is applied to one adherend to prevent adhesion to that adherend and the assembly is held together by nonconductive screws [10,11]. In this case, visual inspection for corrosion can be done at the same time as taking the EIS measurements.

3.3 TESTING OF SEALANTS

3.3.1 NONCONDUCTIVE SEALANTS

Nonconductive sealants exhibit EIS spectra similar to those for most coatings. Results from two nonconductive sealants, labeled N1 and N2, will serve to demonstrate this approach. Sealant N1 is a two-part, manganese-dioxide-cured polysulfide compound with a chromate inhibitor; sealant N2 is a two-part, epoxy-cured polythioether compound without a corrosion inhibitor.

Upon exposing the N1 specimens to hot salt water, a clear delineation of the panels into two sets was apparent (Figure 3.4). Slightly less than half showed a large decrease (three orders of magnitude) in low-frequency impedance following only 3 days of immersion. The remainder showed only little change even with longer exposures (60 days of immersion). One specimen exhibited intermediate behavior—initially little change followed by a gradual decrease. Inspection of the specimens showing the rapid initial drop in low-frequency impedance indicated defects in the sealant at or near the interface (primarily voids or dewetted areas) that allowed moisture intrusion into the interior and corrosion inside the ring (Figure 3.5). All defective specimens exhibited aluminum corrosion inside the sealant ring. By comparison, the good specimen exhibited no corrosion in the ring and 100% cohesive failure in the sealant after 91 days of immersion. EIS was readily able to detect defects, regardless of whether they were present initially or developed over time. The dichotomy between the two sets suggests that either the sealant had narrow processing windows or that operator error contributed to defective specimens.

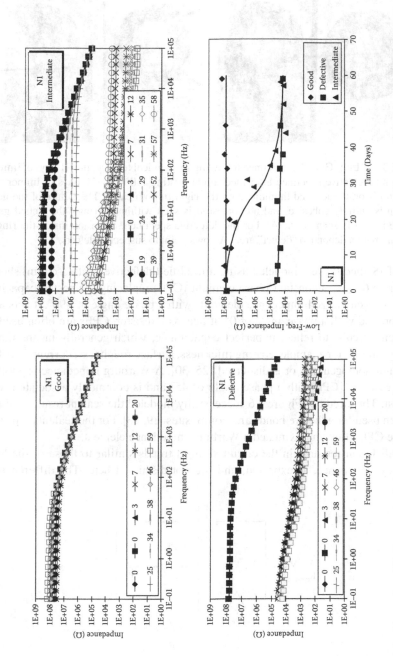

FIGURE 3.4 Examples of Bode magnitude impedance spectra of N1 sealant specimens as functions of time: good, intermediate, and defective. The legend gives the exposure time in days. Also shown is the low-frequency impedance for these three specimens as a function of exposure time (lower right).

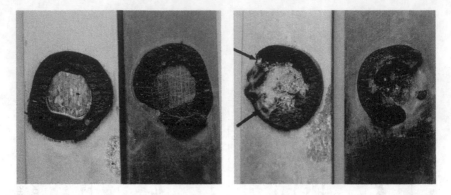

FIGURE 3.5 Left: Good N1 specimen following mechanical testing after 91 days of immersion. Right: Defective specimen after mechanical testing following 30 days of immersion. Aluminum is on the left and titanium is on the right in both cases. The failure of the good specimen is entirely cohesive, and no corrosion is seen within the ring—the etched grain structure is clearly seen on the Al. For the defective specimen, corrosion inside the ring is seen along with a debonded "defect" area. Arrows point to suspected defects.

The EIS spectra were modeled using a modified defective coating circuit shown in Figure 3.6, with a constant phase element (CPE) replacing the nested capacitor. A CPE is a generalized impedance element; with a phase angle of −90°, it acts as a capacitor, and with a phase angle of 0°, it acts as a resistor. CPEs are often used in equivalent circuits to reflect imperfect capacitance, which generally means that a system property, for example, coating thickness, surface roughness, current distribution, is not homogeneous or is dispersed [25–30]. A Warburg impedance is another special case of a CPE with a phase angle of 45° and is commonly associated with diffusion. The circuit of Figure 3.6 successfully modeled the sealant results and has also been used for adhesive bonds and composites [19–23]. For the sealants reported here, the CPE often approximated a Warburg impedance element.

The different resistors in the circuit showed trends similar to those of the low-frequency impedance, as expected, and are not discussed here. The difference in

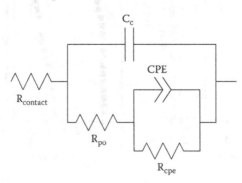

FIGURE 3.6 Equivalent circuit for a defective coating circuit with a constant phase element (CPE).

capacitance C between a "good" sealant and a "defective" sealant was three orders of magnitude (Figure 3.7), which was surprising as most changes of capacitance with absorption of moisture are by a factor of only up to two [19–23], caused by increases in the dielectric constant ε of the coating or sealant as the moisture content increases according to Equation 3.1. In this case, the large increase in capacitance cannot be solely explained by changes in the dielectric constant. Rather, the ingress of salt water effectively reduced the separation of the two metal plates (d in Equation 3.1) and allowed the large increase in capacitance.

The good specimens exhibited a decrease in pull strength of approximately one-third following 90 days of immersion. The defective specimens exhibited pull strengths of approximately half of those of the good specimens, indicating that the moisture ingress was caused by either poor interfacial bonding or reduced sealant strength. Failure of most specimens was cohesive in the sealant, with a few defective specimens failing at the Al interface.

The pull strength and the amount of corrosion in the sealant ring are given in Figure 3.8 as functions of low-frequency impedance. Although there was scatter in the pull strength results, higher low-frequency impedances correlated with higher pull strengths. The correlation between corrosion and low-frequency impedance is striking. No specimen with low-frequency impedance above $10^6\ \Omega$ showed any corrosion in the sealant ring regardless of time of exposure. All specimens with low-frequency impedance below $10^6\ \Omega$ showed corrosion, with most of them having more than 50% of the interior area corroded even at the first mechanical testing time (as soon as 7 days).

The N2 specimens, in general, behaved similarly to the good N1 specimens. Only one N2 specimen exhibited the defective behavior. The low-frequency impedance of the good N2 specimens decreased by approximately an order of magnitude early in the exposure and was comparable to that of N1 (Figure 3.9). The N2 resistive components were similar to those of N1 and are not shown here. The capacitances of

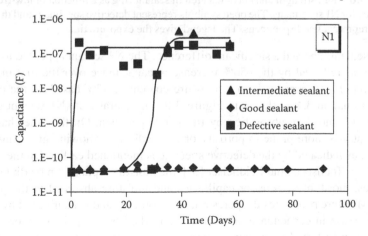

FIGURE 3.7 Capacitance of the N1 specimens: good, defective, and intermediate. The good sealant and defective sealant results are averages of multiple specimens.

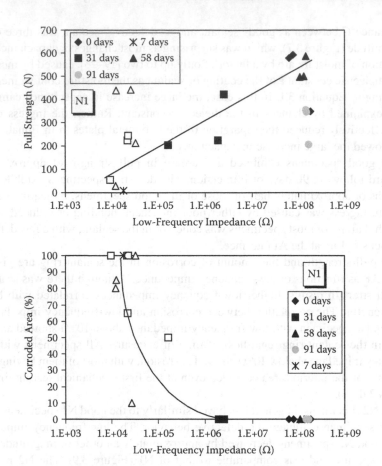

FIGURE 3.8 Pull strength and corroded area in sealant ring as a function of low-frequency impedance for N1 specimens. The open symbols represent defective specimens and the closed symbols represent good specimens. The legend gives the exposure time.

the two sealants showed a significant difference. The N2 sealant appeared to absorb moisture, as reflected by the ~50% increase in capacitance over the first month of immersion. This corresponds to a moisture content of ~3% based on the calculations of Equation 3.4 as shown in Figure 3.10. In contrast, the N1 sealant showed less than 1% moisture absorption by the good specimen. On the other hand, the N1 sealant was more prone to porosity or poor adhesion allowing gross ingress of moisture, as indicated by the defective specimens mentioned earlier. On the basis of the work by Touhsaent and Leidheiser [31], the moisture uptake can be divided into two types: moisture in pores or capillaries and moisture absorbed in the polymer matrix. Moisture penetrates the pores relatively quickly and is manifested by a rapid initial increase in capacitance. Moisture is absorbed by the polymer more slowly, and this is manifested by a slower increase in capacitance at longer periods. The amount of moisture in the pores can be calculated by extrapolating the slower rise in capacitance to zero time; this value is denoted by C_1 in Figure 3.9. For N2, the pore

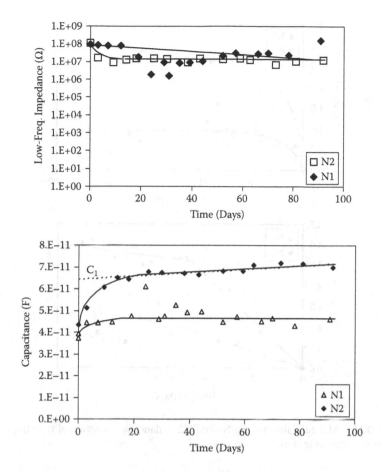

FIGURE 3.9 Low-frequency impedance and capacitance for the two nonconductive sealants as a function of immersion time. The significance of C_1 is explained in the text.

moisture is 2.3%, with the remainder being absorbed into the sealant resin. For N1, the pore moisture is 0.9%, with very little being absorbed into the sealant resin, suggesting a less permeable polymer.

The plot of moisture content as a function of the square root of time (Figure 3.10) suggests that the moisture diffusion is Fickian (class I), with different diffusion constants for the two types of water uptake (pores and absorption into the polymer) for N2 and the pore water uptake for N1. Such behavior is expected for polymers above their glass transition temperature as is the case with these sealants.

The N2 specimens exhibited the highest initial bond strength of all the sealants tested, but the bond strength decreased rapidly upon immersion in hot salt water (Figure 3.11). Failure was within the sealant in all cases, indicating poorer cohesive strength when exposed to hot salt water. This may be a result of moisture absorption by the sealant (as opposed to moisture ingress via defects as seen in the N1 specimens).

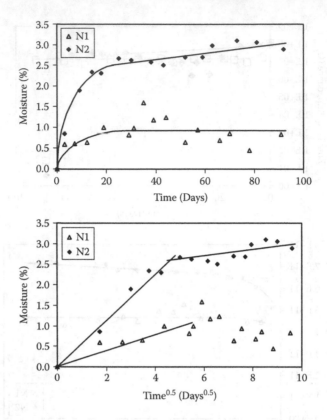

FIGURE 3.10 Moisture absorbed by N1 and N2 sealants as a function of time (top) and the square root of time (bottom).

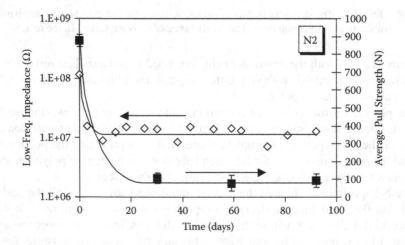

FIGURE 3.11 Low-frequency impedance and pull strength of N2 sealant specimens following salt water immersion.

3.3.2 CONDUCTIVE SEALANTS

Nickel, or other metal, particles/flakes are added to a conductive sealant to provide a low-impedance path through the sealant. Consequently, the EIS spectra of conductive sealants are very different from those of nonconductive sealants. This will be demonstrated using results from two conductive sealants, labeled C1 and C2. Sealant C1 is a two-part, nickel-filled, manganese-dioxide-cured polythioether compound; sealant C2 is a two-part, nickel-filled, epoxy-cured polythioether compound. Both of these sealants have a chromate inhibitor.

Upon exposure to hot salt water, the impedance of the C1 specimens varied greatly (Figure 3.12). The geometric average low-frequency impedance increased by three orders of magnitude in 3 days of immersion, but the measurements of the individual specimens ranged from no increase in impedance to more than five orders of magnitude increase. After several weeks of exposure, the impedance of all specimens returned to near-baseline values. No change in impedance was seen for a specimen that was kept dry in the oven. The conductivity of these filled sealants relies on the physical contact of the network of Ni particles or flakes to provide an electrical path, and this could be a strong function of the density of Ni flakes and their distribution, depending on the loading amount of the flakes. This suggests that immersion in salt water caused a disruption in the electrical network of Ni. One possibility would be severe corrosion of Ni, but this is unlikely. Ni is corrosion resistant, and a 3-day immersion would not be sufficient to form extensive corrosion products that could break the electrical contact, and there is no mechanism to restore conductivity. The second possibility, which is more likely, is that a swelling of the polymer component of the sealant forced the Ni filler flakes apart. The effect of swelling on conductivity would depend on the exact distribution of the Ni filler and can explain the large variability in the results. Open-faced specimens were prepared to confirm this hypothesis, and the results are presented as follows. In this scenario, salt can also diffuse in and restore conductivity. The variability of impedance would be expected to be the greatest for small sealing areas such as the ones in these specimens. For large aircraft panels or other structures, it is likely that there would always be an area of low impedance that would provide electrical contact to the airframe or mating component.

For the C1 specimens, the maximum low-frequency impedance correlated with the amount of corrosion noted on the aluminum inside the ring of sealant once the specimens were mechanically tested (Figures 3.13 and 3.14). If the low-frequency impedance never rose above 10^3 Ω, no corrosion was observed inside the sealant ring even after 87 days of immersion. In contrast, if the low-frequency impedance rose above 10^3 Ω (even if it subsequently decreased to less than 10^3 Ω), corrosion was found inside the sealant ring. This is supportive of the mechanism in which moisture is absorbed by the sealant and causes the Ni flakes to lose interconnectivity and then subsequently induces corrosion inside the sealant ring.

The pull strength of these specimens decreased by approximately a factor of two to three during immersion (Figure 3.15). This decrease was correlated with the maximum low-frequency impedance seen in the early stages of immersion even though the impedance returned to near its original value with longer exposure; if the maxi-

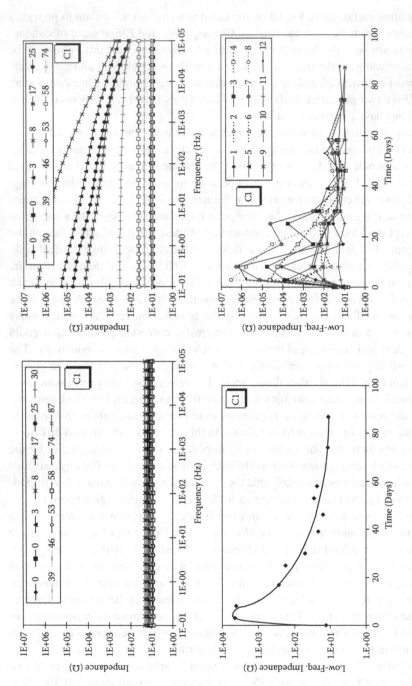

FIGURE 3.12 Upper panels: Examples of no change in impedance, and large increases and then decreases in impedance for the C1 specimens. Lower left: Geometric mean of the low-frequency impedance of all C1 specimens as a function of exposure time to hot salt water. Lower right: Low-frequency impedances of individual C1 specimens. The legend gives the exposure time in days.

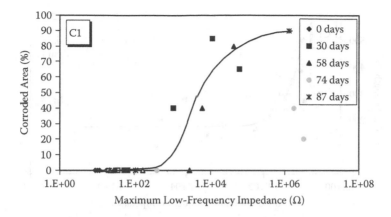

FIGURE 3.13   Corroded area inside sealant ring as a function of maximum low-frequency impedance for C1 specimens. The legend gives the exposure time.

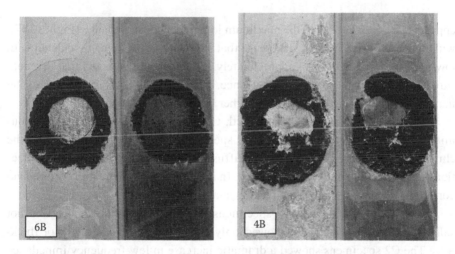

FIGURE 3.14   C1 specimens after pulling following 58 days of immersion. Specimen 6B (left pair) showed no interior corrosion, while specimen 4B (right pair) showed extensive aluminum corrosion. Specimen 4B had one of the highest low-frequency impedances; specimen 6B had one of the lowest impedances. For each pair, the Al is on the left and Ti is on the right.

mum low-frequency impedance was greater than ~30 Ω, the pull strength was below ~300 N.

Many of the C2 specimens experienced a rapid failure (deadhesion along the titanium interface) within just 3 days of immersion. Some of these early-failing specimens exhibited a large increase in the electrochemical impedance (up to 3×10^4 Ω) upon exposure; others became noisy and poorly behaved in the last measurement before failure, when the interface was presumably breached (Figure 3.16). This behavior may serve as a prognostic tool. As with the C1 specimens, pull strength

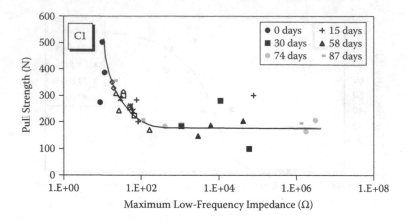

FIGURE 3.15 Pull strength as a function of maximum low-frequency impedance for C1 specimens. The legend gives the exposure time.

appears to be correlated with the maximum low-frequency impedance (Figure 3.17), with higher pull strengths (at least those that were measured) being associated with low-frequency impedances of approximately 10 Ω. It should be noted that the measured maximum low-frequency impedance in all cases is a minimum value; the maximum impedance may have been higher between measurements.

Unlike any of the other sealants tested, C2 exhibited a ring of uncorroded aluminum outside the sealant ring on some specimens (Figure 3.18). In this case, the chromate inhibitor appears to be able to diffuse out of the sealant and provide protection outside the sealant for a limited time. In all but one specimen (after 59 days), no corrosion was seen in the interior of the sealant ring.

A few open-faced C1 and C2 specimens were prepared and immersed in hot salt water. The C1 specimens showed only little change in impedance or thickness. The C2 specimens showed a dramatic increase in low-frequency impedance and a corresponding swelling or increase in film thickness (Figure 3.19). These results support the hypothesis that the reduced conductivity exhibited by some of the conductive sealant specimens was caused by moisture ingress that disrupted the network of Ni flakes.

Simple electron microscope (SEM) x-ray images were taken of the open-faced C1 and C2 specimens to explain the high impedance of the open-faced C2 specimens compared to the low impedance of the C1 specimens (Figure 3.20). After exposure, the C2 specimen exhibited a lower density of Ni filler compared to the specimen before exposure, while the C1 specimen exhibited no decrease in Ni-filler density compared to the baseline. That is, the Ni-filler network of the C2 specimens was disrupted, leading to decreased conductivity and increased low-frequency impedance.

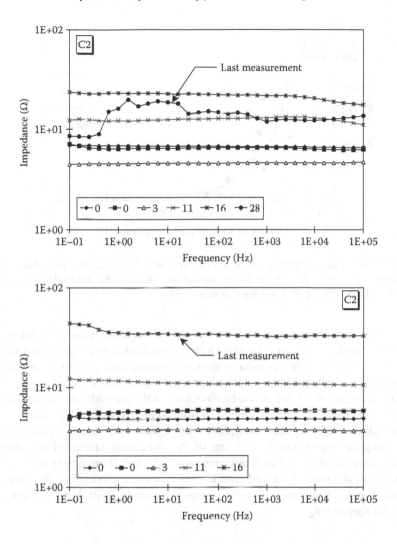

FIGURE 3.16 Impedance spectra of two C2 specimens showing either an erratic last spectrum or increased impedance spectrum shortly before failure. The anomalous behavior may be an indicator of pending failure. The legend gives the immersion time in days.

3.4 SUMMARY AND CONCLUSIONS

EIS is an established laboratory technique to study degradation of coatings and corrosion of metals. By modeling the impedance spectra with an equivalent circuit, the amount of moisture absorbed by a coating can be calculated from the coating capacitance. Traditionally, EIS results could only be obtained by immersing the specimen or part of the specimen surface (using a bottomless beaker arrangement). Such a procedure allows the barrier properties of a sealant to be investigated by applying the sealant as a coating.

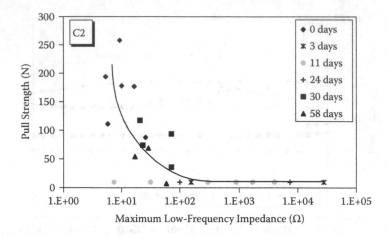

FIGURE 3.17 Pull strength as a function of maximum low-frequency impedance for C2 specimens: Specimens that failed before mechanical testing are arbitrarily given a pull strength of 10 N. The legend gives the exposure time.

Recently developed corrosion sensors allow EIS measurements to be taken from adhesively bonded specimens. By using a ring of sealant as an adhesive between two dissimilar metals, the effectiveness of the sealant in excluding moisture or other fluids from the interior and in preventing corrosion can be evaluated. Adhesion (interfacial) or cohesive strength can also be obtained. Both nonconductive and conductive sealants can be tested in this manner. For nonconductive sealants, sealant failure is readily detected, and moisture absorption via pores and resin can be separately quantified using equivalent circuit modeling of the impedance spectra. For conductive sealants, absorption of moisture can cause swelling and disruption of the electrical continuity of the metal filler and lead to a loss of conductivity. For the specimens presented here, intrusion of salt water restored the conductivity, but led to corrosion inside the sealant ring.

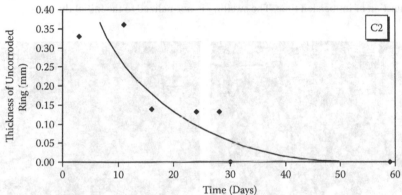

FIGURE 3.18 Top: C2 specimen after 11 days of immersion showing a ring of uncorroded aluminum surrounding the sealant. Arrows point to the uncorroded area. Al is on the left, and Ti is on the right. Bottom: Average thickness of this uncorroded ring as a function of immersion time.

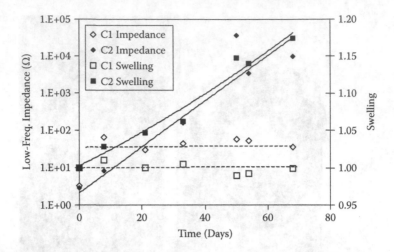

FIGURE 3.19 Low-frequency impedance and swelling of open-faced C1 and C2 specimens as a function of immersion time. The swelling is normalized to the initial thickness.

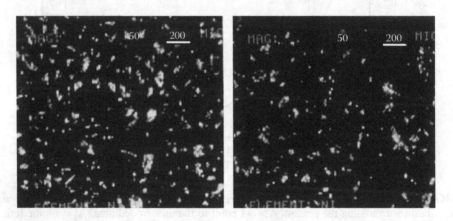

FIGURE 3.20 X-ray images of Ni for open-faced C2 specimens before (left) and after (right) 50 days of immersion. The bar is 200 μm.

REFERENCES

1. A.J. Bard and L.R. Faulkner, *Electrochemical Methods: Fundamentals and Applications*, Wiley, New York (1980).
2. F. Mansfeld, *J. Appl. Electrochem.* **25**, 187 (1995).
3. F. Mansfeld, *Corrosion* **37**, 301 (1981).
4. G.W. Walter, *Corrosion Sci.* **26**, 681 (1986).
5. F. Mansfeld, M.W. Kendig, and S. Tsai, *Corrosion* **38**, 570 (1982).
6. J.R. Scully, *J. Electrochem. Soc.* **136**, 979 (1989).
7. J.R. Scully, in *Corrosion Testing and Evaluation: Silver Anniversary Issue*, ASTM STP 1000, R. Baboian and S.W. Dean, (Eds.), p. 351, ASTM, Philadelphia (1990).
8. U. Rammelt and G. Reinhard, *Prog. Org. Coatings* **21**, 205 (1992).

9. J.N. Murray, *Prog. Org. Coatings* **31**, 375 (1997).
10. G.D. Davis and J.G. Dillard, *J. Adhesion Sci. Technol.* **20**, 1215 (2006).
11. G.D. Davis and J.G. Dillard, *J. Adhesion Sci. Technol.* **20**, 1235 (2006).
12. M. Kendig, F. Mansfeld, and S. Tsai, *Corrosion Sci.* **23**, 317 (1983).
13. M. Kendig and J. Scully, presented at Corrosion89, Paper 32, NACE, Houston (1989).
14. W.S. Tait, *J. Coatings Technol.* **61**(768), 57 (1989).
15. J.N. Murray and H.P. Hack, presented at *Corrosion90*, Paper 140, NACE, Houston (1990).
16. J.A. Grandle and S.R. Taylor, *Corrosion* **50**, 792 (1994).
17. D.R. Lide, (Ed.), *CRC Handbook of Chemistry and Physics*, 71st Edition, pp. 13–12, CRC Press, Boca Raton, FL (1990).
18. G.D. Davis and C.M. Dacres, U.S. Patent 5,859,537 (1999); G.D. Davis and C.M. Dacres, U.S. Patent 6,054,038 (2000); G.D. Davis and C.M. Dacres, U.S. Patent 6,313,646 (2001); G.D. Davis, C.M. Dacres, and L.A. Krebs, U.S. Patent 6,328,878 (2001).
19. G.D. Davis, P.L. Whisnant, and J.P. Wolff, Jr., *Proc. 41st Intl SAMPE Symp.*, p. 544, SAMPE, Covina, CA (1996).
20. G.D. Davis, L.A. Krebs, L.T. Drzal, M.J. Rich, and P. Askeland, *J. Adhesion* **72**, 335 (2000).
21. G.D. Davis, K. Thayer, M.J. Rich, and L.T. Drzal, *J. Adhesion Sci. Technol.* **16**, 1307 (2002).
22. G.D. Davis and B.J. Harkless, *Intl. J. Adhesion Adhesives* **22**, 323 (2002).
23. G.D. Davis, *Proc. 34th Intl. SAMPE Tech. Conf.*, p. 1086, SAMPE, Covina, CA (2002).
24. R. Baboian, in *Corrosion: ASM Handbook 13A*, S.D. Crammer and B.S. Covino, Jr. (Eds.), p. 210, ASM International, Metals Park, OH, (2003).
25. J.R. Macdonald and W.R. Kenan, *Impedance Spectroscopy*, John Wiley & Sons, New York (1987).
26. C.A. Schiller and W. Strunz, *Electrochimica Acta* **46**, 3619 (2001).
27. C.S. Hsu, and F. Mansfeld, *Corrosion* **57**, 747 (2001).
28. M.E. Orazem, P. Shukla, and M.A. Membrino, *Electrochimica Acta* **47**, 2027 (2002).
29. J.-B. Jorcin, M.E. Orazem, N. Pebere, and B. Tribollet, *Electrochimica Acta* **51**, 1473 (2006).
30. M.E. Orazem, N. Pébère, and B. Tribollet, *J. Electrochem. Soc.* **153**, B129 (2006).
31. R.E. Touhsaent and H. Leidheiser, Jr., *Corrosion* **28**, 435 (1972).

4 Chemorheological Investigation on Environmental Susceptibility of Sealants

K.T. Tan, C.C. White, D.J. Benatti,
D. Stanley, and D.L. Hunston

CONTENTS

4.1 INTRODUCTION

Sealants are filled elastomers that are used in structures to prevent moisture penetration through gaps, joints, and other openings. These structures span a wide range of diverse applications such as transportation vehicles and medical equipment, but the greatest use of sealants is in construction. Although modern commercial sealants are durable, eventually they fail. Studies in the construction industry have shown a 50% failure rate in less than 10 years and a 95% failure rate within 20 years after installation [1–3]. What makes these failures particularly detrimental is that sealants are often used in areas where moisture-induced degradation is difficult to monitor and expensive to repair. Consequently, sealant failure is frequently detected only after considerable damage has already occurred. In the housing market alone, premature failure of sealants and subsequent moisture intrusion damage contributes significantly to the $65 billion to $80 billion spent annually on home repair [4]. The environmental susceptibility issue, therefore, is the most demanding requirement of a sealant since it is the property that ultimately determines long-term service life. Over the past few decades, extensive research has been undertaken to evaluate the effects of environment on the long-term durability of such materials, and to develop a better understanding of the underlying mechanisms of environmental attack [5–7].

When considering the durability of sealants, it is essential to recognize that there are three potential weak links in a sealant system: (1) the bulk sealant itself, (2) the interface between the substrate and sealant, and (3) the substrate. Failure occurs whenever the weakest link in this system fails. This chapter focuses on the environmental susceptibility of bulk sealants. Depending on their applications, these materials are exposed to environments having different combinations of aging factors. Temperature, oxygen, pollutants, humidity, solar ultraviolet (UV)-visible radiation, and cyclic loading have been highlighted as primary factors [4–9]. The interaction of these aging factors with sealants inevitably largely determines their long-term durability.

As an example, the work of Paroli et al. [10] documents the deleterious effects of water and UV radiation on a polyurethane sealant with a UV irradiance of 0.37 W/m^2·nm. The chemical changes were measured using photoacoustic Fourier transform infrared (FTIR) spectroscopy. After only 1000 h, there were substantial changes in the chemical composition involving cleavage of urethane linkage, as evidenced by the decrease in the absorption band at 1539 cm^{-1}. In addition, the sealant exhibited a high degree of oxidation, as can be seen from the formation of broad carbonyl and hydroxyl absorption bands. In another study that used direct pyrolysis mass spectroscopy, Montaudo et al. [11] showed that an aliphatic polysulfide sealant based on polythiohexamethylene, when heated from 30°C to 400°C, decomposed via a β-CH hydrogen transfer process with formation of SH and olefin end groups. There is also evidence showing that the olefin groups were subsequently hydrogenated to form methyl end groups. Chemical deterioration in other types of sealants upon exposure to environments, including silicones [8,12–14] and acrylics [15,16], has also been extensively reported.

As with chemical composition, mechanical-rheological properties are also altered by environmental attack. Boettger and Bolte [17], for instance, found that elongation to break for a room-temperature-curing two-part polysulfide sealant decreased by 40% from its initial reference value after 6000 h of exposure at 80°C, but a more rapid decrease of 100% was seen for exposure at 90°C. In the case of silicone sealants, despite their excellent environmental resistance, prolonged exposure can deteriorate their polymeric network structures. An example of this can be found in the work of Virlogeux et al. [18], who clearly showed an increase in crosslink density of a silicone sealant based on poly(dimethylsiloxane) when exposed to UV-visible radiation. The embrittlement was indicated by an increase in indentation hardness of the exposed sealant. Demarest et al. [19] found that significant surface crazing and chalking occurred in two commercial polyurethane sealants after 15 years of natural exposure in moving joints of low-rise industrial buildings. Since both accelerated and natural outdoor aging lead to significant modifications in both chemical properties and mechanical-rheological behavior over time, it is important to characterize both to gain a better understanding of the durability of sealants. In other words, a holistic approach must be adopted.

Since there are various generic types of sealants, generalized statements on the sealant durability issue are often too broad to arrive at a judgment regarding a specific product. Indeed, within each sealant family there is a large number of different formulations, and the durability of a sealant is strongly affected by its constituents. Different cure conditions used for a particular sealant type can produce different polymeric network structures, and can accrue different curative or catalyst residues, which can lead to varying long-term durability. Furthermore, the interpretation of environmental effects is handicapped by the fact that the formulation of the sealant is usually unknown [20]. To set the scene for a discussion on environmental susceptibility, a model sealant system based on styrene-butadiene-styrene (SBS) triblock copolymer will be used as an example in the present chapter. Particular emphasis will be placed on (1) elucidating the underlying failure mechanism of the model system using chemorheological tools when the specimens are exposed to a combination of three major environmental factors, that is, UV-visible radiation, temperature, and moisture; and (2) discussing the effects of temperature and moisture on the photo-degradation, both in terms of chemical and mechanical-rheological properties.

4.2   EXPERIMENTAL CONSIDERATIONS

4.2.1   MATERIALS AND SPECIMEN PREPARATION

The SBS triblock copolymer used in the experiment was a commercially available product. Specimens were prepared by solution casting using a mixture of 80% mass fraction of toluene and 20% mass fraction of the copolymer. Two types of specimens were prepared. For spectroscopic studies, specimens were made by spin-casting the solution mixture on calcium fluoride disks at 104.7 rad s^{-1} for 20 s in a virtually CO_2-free dry glove box. The calcium fluoride disks having a diameter of 19 mm and a thickness of 4 mm were selected as the substrate due to their excellent resistance to heat and moisture and their transparency to both UV and infrared radiations [21]. The coated disks were cured at 50°C for an hour in a nitrogen gas blanket. The thickness of the resulting specimen coating was estimated to be in the range of 15 μm to 20 μm. Specimens for mechanical-rheological characterization were prepared by a drawdown technique. The solution was spread on a release paper, and the film thickness was controlled by layering strips of masking tape along the length of the paper. The excess liquid was then removed by firmly drawing a metal bar across the paper. Free-standing film specimens with a nominal thickness ranging from 400 to 500 μm were obtained by evaporating the solvent gradually at 50°C for an hour in a nitrogen gas blanket. The cure conditions for the FTIR spectroscopy and dynamic mechanical thermal analysis (DMTA) specimens were kept the same so that a direct comparison between the chemical and mechanical-rheological changes could be made.

4.2.2 EXPOSURE CONDITIONS

Specimens were exposed to UV-visible radiation with wavelengths between 290 and 600 nm using an integrating sphere-based weathering chamber [22]. The intensity of the radiation is depicted in Figure 4.1. During exposure, the specimens were subjected to one of four environments involving combinations of temperature and relative humidity (RH), that is, (1) 30°C at 0% RH, (2) 30°C at 80 % RH, (3) 55°C at 0% RH, and (4) 55°C at 80% RH.

4.2.3 FTIR AND UV-VISIBLE SPECTROSCOPIES

Specimens were periodically removed from the weathering chamber for FTIR and UV-visible spectroscopic analyses. Precise positioning of specimens was achieved by fitting them into a dismountable 150-mm-diameter ring of an automated sampler as described previously [23]. This demountable ring enabled the spectroscopic measurements to be recorded precisely at the same location for each measurement. Five replicates were exposed to each exposure environment. The transmission-absorption FTIR was carried out using a Nexus 670 Spectrometer* equipped with a liquid-nitrogen-cooled mercury cadmium telluride detector. All spectra were collected over a range of 650 to 4000 cm^{-1}, and were averaged over 132 scans at a nominal

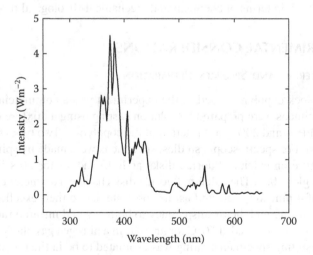

FIGURE 4.1 Intensity of output UV-visible radiation from the weathering device.

---

* Certain commercial products or equipments are described in this paper to specify the experimental procedure adequately. In no case does such identification imply recommendation or endorsement by the National Institute of Standards and Technology, nor does it imply that it is necessarily the best available for the purpose.

resolution of 4 cm^{-1}. UV-visible absorption spectra were recorded over a wavelength range of 280 to 900 nm using a Perkin–Elmer Lambda 900 spectrometer.

4.2.4 Dynamic Mechanical Thermal Analysis (DMTA)

Dynamic mechanical temperature sweep measurements were made using a Rheometrics Solids Analyzer (RSA) III (Rheometrics Scientific, Inc., Piscataway, New Jersey) in the tension mode at a frequency of 1 Hz. The experiments were conducted from −100°C to 120°C at a heating rate of 2°C min^{-1}. The specimens were rectangular with a width of 5 mm and a gauge length of 10 mm. Measurements were made on two independent specimens.

4.2.5 Calculation of Absorbed Radiation Dose

The absorbed dose is defined as the radiation that is incident on and absorbed by the specimen, and it can be obtained by integrating the product of spectral irradiance, $I_0(\lambda, t)$, and the spectral absorption of the material, $1-e^{-A(\lambda,t)}$, over the wavelengths of the UV-visible radiation impinging on the specimen for the exposure duration. It has the form

$$\text{Absorbed dose} = \int_0^t \int_{290}^{600} I_0(\lambda,t)\left(1-e^{-A(\lambda,t)}\right) d\lambda dt \qquad (4.1)$$

where $A(\lambda,t)$ is the absorbance of the specimen at a specified UV-visible wavelength, λ, and at exposure time t. A detailed experimental description is given elsewhere [24].

4.3 MECHANISM OF FAILURE

4.3.1 Chemical Property Changes upon Exposure

Before considering the evolution of FTIR spectra in detail, it should be pointed out that the various combinations of temperature and humidity over a wide exposure range did not change the primary mechanism of failure of the specimens, which was evident from the fact that the general trends in the spectra as a function of exposure time were very similar for all environments (not shown here for brevity). However, the environment had considerable effect on the kinetics of failure, which will be further discussed in the next section. Since the mechanism of failure was the same for all environments, only the data obtained at 55°C and 80% RH will be presented here.

Subtraction of the initial spectrum before exposure from the spectra recorded after various exposure times reveals several interesting features, as shown in Figure 4.2. In this figure, the bands below and above the zero baseline indicate the loss and the formation, respectively, of certain functional groups. As can be seen from Figure 4.2, broad absorption bands corresponding to oxidation products, that

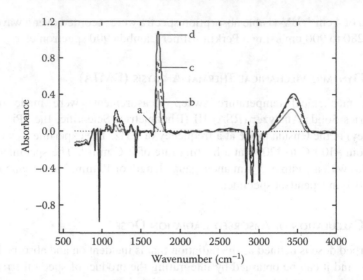

FIGURE 4.2 Transmission FTIR difference spectra obtained by subtracting the initial unexposed specimen spectrum from the spectra of the exposed specimens for different exposure times (spectra for only four exposure times are shown, for clarity): The specimens were exposed to 55°C at 80% RH for (a) 0, (b) 3, (c) 6, and (d) 12 days.

is, hydroxyl functional groups (O–H) in the region 3200 cm^{-1} to 3600 cm^{-1} and carbonyl functional groups (C=O) in the region 1660 cm^{-1} to 1800 cm^{-1}, increase progressively with increasing exposure time. It has been established that the mechanism of oxidation of 1,4-butadiene involves primary radical attack on the methylene group in the α-position relative to the double bond [25,26], while the mechanism of oxidation of 1,2-vinylbutadiene occurs at the tertiary site [27]. Subsequent oxygen addition and hydrogen abstraction results in the formation of unsaturated hydroperoxides. The hydroperoxides are unstable to thermal oxidation and photooxidation, so they rapidly decompose into alkoxy and hydroxyl radicals [27]. The resulting alkoxy radicals undergo further oxidation to form secondary hydroxyl and carbonyl functional groups, which are indeed detected in the FTIR spectra in the regions 3200 to 3600 cm^{-1} and 1600 to 1800 cm^{-1}, respectively.

The intensities of the absorption bands associated with the 1,2-vinyl unsaturation of butadiene (911, 995, 1640, 2988, and 3075 cm^{-1}) and 1,4-unsaturation of butadiene (967, 1438, 1652, and 3004 cm^{-1}) decrease with exposure time. In addition, there is a linear relationship between the intensity loss of absorption bands at 911 and 967 cm^{-1}, which are assigned to the out-of-plane bending of CH$_2$ in the 1,2-vinyl unsaturation and the out-of-plane bending of CH in the trans-1,4 unsaturation, respectively. Also, after 10 days of exposure, these absorption bands are no longer apparent. The loss of the CH and C=C absorptions may be attributed to crosslinking and saturation reactions occurring simultaneously [27] during the early stage of exposure.

As can be seen from Figure 4.2, there were no significant changes in absorption bands related to phenyl functional groups (1493 and 1609 cm^{-1}) and skeletal stretching of styrene (1190 cm^{-1}). Therefore, the styrene units did not change within the

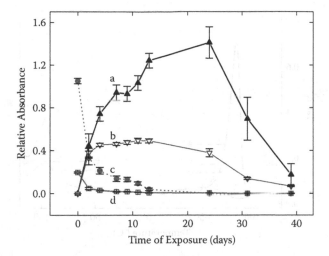

FIGURE 4.3   Kinetic evolution of the oxidation products and the consumption of unsaturations during the course of experiment for specimens exposed to 55°C at 80% RH. Different lines represent different absorption bands: (a) relative absorbance of 1724 to 1030 cm^{-1}, (b) 3440 to 1030 cm^{-1}, (c) 967 to 1030 cm^{-1}, and (d) 911 to 1030 cm^{-1}. Lines are added to aid the reader, and error bars are the standard deviations of the data.

timescale of the experiments. Some earlier studies [28] have reported that the radical species formed during oxidation of the butadiene unit can thermally destabilize the styrene unit and induce oxidation. It is likely that the timescale of the present experiments may not be long enough or the exposure conditions may not be severe enough for the styrene unit to degrade significantly.

It is noteworthy that the absorbances of the hydroxyl and carbonyl functional groups reach asymptotic values after 10 days of exposure (see Figure 4.3). The presence of these limiting intensities is presumably associated with crosslinking reactions, which may result in the creation of a relatively high-modulus outer "skin" in the specimens. Since segmental molecular mobility is restricted in the outer crosslinked structure, the rate of penetration of oxygen into the surface region of the specimen is slower. Hence, the probability of reaction of the butadiene with oxygen is reduced significantly, resulting in the limiting intensities of hydroxyl and carbonyl functional groups for longer times of exposure. The alternative explanation is that the crosslinked structure may shield the underlying material from incident radiation, causing the degradation to concentrate in a thin outer surface layer [29]. The effect of the crosslinking will be further examined in the next section.

4.3.2   Mechanical-Rheological Property Changes upon Exposure

From the FTIR analysis, it was observed that both thermal degradation and photodegradation led to the formation of hydroxyl and carbonyl functional groups, and consumption of unsaturations. Subsequent recombination of the radicals led to the occurrence of crosslinking reactions [28]. The chemical analysis based on FTIR spectroscopy is unable

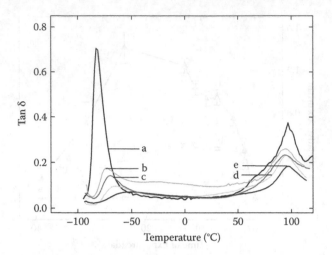

FIGURE 4.4 Evolution of tan δ as a function of temperature for specimens exposed to 55°C at 80% RH. Different lines represent measurements at different times of exposure: (a) 0, (b) 6, (c) 12, (d) 21, and (e) 24 days.

to directly establish the formation of a network by the crosslinking reactions. Therefore, DMTA was employed to support the suggestion of crosslinking and to establish the relationship between the chemical and rheological changes that occurred in the samples.

The variation of tan δ (i.e., dimensionless ratio of loss modulus, E'', to storage modulus, E') as a function of temperature for different exposure times is shown in Figure 4.4. It is apparent that the value of tan δ generally decreases with increasing exposure time. It should be recalled that E'' is associated with energy loss due to intermolecular friction, while E' is related to the elastic energy stored in the system. For a highly crosslinked structure, the molecular motion of chain segments is limited, and hence the material behaves similar to a perfect elastic spring with little or no heat dissipation, resulting in a rigid material with low damping and high storage modulus. Conversely, if a chain segment within a polymer structure is free to move as a result of a chain scission reaction, the extent of dissipation of mechanical energy as heat is enhanced, giving rise to a material with high damping and low storage modulus. Therefore, the decrease in tan δ is consistent with the occurrence of crosslinking in the material.

Each tan δ curve shows two transition peaks, corresponding to the glass transition temperatures, T_g, for the butadiene unit at the lower temperature and for the styrene unit at the higher temperature. From Figures 4.4 and 4.5, T_g for the styrene units remains unchanged at 95°C over the timescale of the experiment. This is consistent with the earlier observations from FTIR spectroscopy, where it was concluded that the styrene units did not play a major role in the degradation process. However, Figure 4.4 shows a significant change in T_g for butadiene segments, which shifted from −82°C to −47°C after about 25 days of exposure (see Figure 4.5). Such an increase in T_g indicates a restriction in molecular mobility in the butadiene units as the network tightens due to crosslinking reactions.

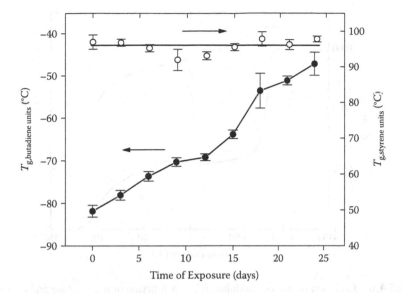

FIGURE 4.5 Evolution of the glass transition temperature of the butadiene and styrene units for the specimens exposed to 55°C at 80% RH. Lines are added to aid the reader, and error bars are the standard deviations of the data.

Figure 4.6 gives the storage modulus plots corresponding to the tan δ curves in Figure 4.4. The results show that the exposure produced an increase in E' in the region between the T_g's. This increase can be attributed only to crosslinking in the butadiene units since the styrene units are glassy at these temperatures. Moreover, Figure 4.7 shows that there is a good correlation between T_g of butadiene units and the value of E' in this region.

The crosslinking in the butadiene units can be further analyzed by examining the modulus data between the two T_g's. The value of E' contains contributions from both the styrene and butadiene units, so an expression is needed to express E' in term of contributions from the two constituents. The expression chosen was the modified semiempirical Halpin–Tsai equation [30–32]:

$$E'_{PB} = 0.54 \exp \left[\frac{\log E' - 0.18 \log \left(0.20 E'_{PS} \right)}{0.82} \right] \tag{4.2}$$

where E'_{PS} and E'_{PB} are the storage moduli of the styrene and butadiene units, respectively. The magnitude of E'_{PB} in the butadiene rubbery plateau region indicates the extent of crosslink density in the butadiene segments. The degree of crosslink density can indeed be estimated using the kinetic theory of elasticity [33], in which the relationship between E'_{PB} and the concentration of effective chains per cubic meter, v_e, is given by

$$v_e = \frac{E'_{PB}}{2RT(1 + v_{PB})} \tag{4.3}$$

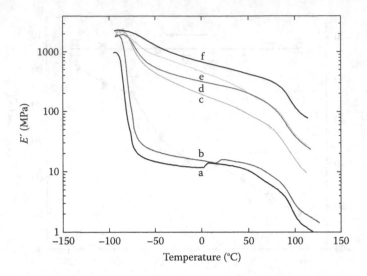

FIGURE 4.6 Evolution of storage modulus, E', as a function of temperature and exposure time for specimens exposed to 55°C at 80% RH. Different lines represent measurements at different times of exposure: (a) 0, (b) 3, (c) 6, (d) 12, (e) 21, and (f) 24 days.

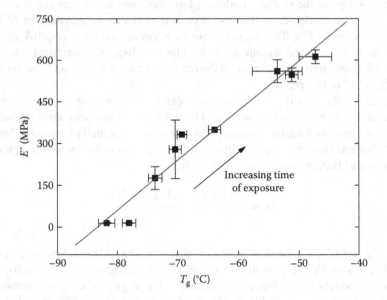

FIGURE 4.7 Relationship between the storage modulus, E', in the rubbery region at 20°C and the glass transition temperature of the butadiene unit, T_g, for the specimens exposed to 55°C at 80% RH. The error bars are the standard deviations of the data. The straight line is the linear regression fit to the data with a regression coefficient of 0.99.

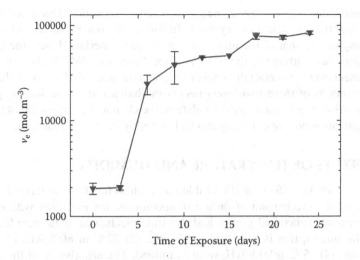

FIGURE 4.8 Evolution of the concentration of effective chains per cubic meter, v_e, for the specimens exposed to 55°C at 80% RH. A line is added to aid the reader, and error bars are the standard deviations of the data.

where R is the gas constant, T is the temperature in kelvin, and v_{PS} is Poisson's ratio of the butadiene unit. Combining Equations 4.2 and 4.3 yields

$$v_e = \frac{0.54}{2RT(1+v_{PB})}\exp\left[\frac{\log E' - 0.18\log(0.20E'_{PS})}{0.82}\right] \tag{4.4}$$

Using E'_{PB} at 20°C for the calculation, the value of v_e in the specimen increased by a factor of 35 after 25 days of exposure as shown in Figure 4.8, indicating the occurrence of extensive crosslinking reactions in the material. The mechanism of crosslinking reactions may be represented in its simplest and most general form by the following equations:

$$\text{Initiation: ROOR} \rightarrow \text{RO}^{\bullet} + {}^{\bullet}\text{OR} \tag{4.5}$$

$$\text{Propagation: RO}^{\bullet} + \text{PH} \rightarrow \text{ROH} + \text{P}^{\bullet} \tag{4.6}$$

$$\text{Termination: } 2\text{P}^{\bullet} \rightarrow \text{P–P} \tag{4.7}$$

where PH represents a hydrocarbon, P• a polymeric radical, and RO• an alkoxy radical.

During the early stage of exposure, the presence of two distinct peaks in the tan δ plot implies that the butadiene and styrene units were phase-separated (see Figure 4.4). However, as the exposure time increases, the peak for butadiene broadens as well as shifts to a higher temperature. This suggests an increase in structural heterogeneities in the butadiene regions at the size scale of the molecular motion associated with the butadiene glass-to-rubber transition. One possible explanation is that the coefficient of oxygen diffusion for the outer crosslinked surface is lower compared to that of the underlying protected material. Hence, the chemical degradation was confined to the outer surface. Since the DMTA gives an average measurement over the overall thickness of the specimen, the difference in the chemical compositions of these two layers results in changes as a function of specimen thickness. Also, as a result of the degradation, the chemical composition of the butadiene segments is no longer homogeneous but highly heterogeneous [27].

4.4 EFFECTS OF TEMPERATURE AND HUMIDITY

In order to study the effects of UV-visible radiation as functions of temperature and humidity on the mechanism of failure in specimens, the evolution with exposure time of carbonyl functional groups and dynamic mechanical data recorded for the four environments, that is, (1) 30°C at 0% RH, (2) 30°C at 80% RH, (3) 55°C at 0% RH, and (4) 55°C at 80% RH, were examined. The absorbance of the carbonyl functional band (1724 cm^{-1}) was normalized with the absorbance at 1030 cm^{-1}. The absorbance of the 1030 cm^{-1} band did not change over the course of the degradation, so normalization accounted for any variation in the thickness of specimens since the absorbance ratio of the two bands is independent of specimen thickness according to Beer's law [34].

4.4.1 EFFECT OF TEMPERATURE

The formation of carbonyl functional groups as a function of absorbed radiation dose is depicted in Figure 4.9. It can be seen that the absorbance generally rose with increasing exposure time and then leveled off. Further examination of the figure reveals that the formation of groups depends considerably on the exposure temperature. The specimens exposed at 55°C exhibited a higher absorbance than those exposed at 30°C. Therefore, the UV-visible-radiation-induced degradation was more severe at higher temperatures. This observation may be attributed to the fact that degradation in a solid polymer is generally a diffusion-controlled process; therefore, the diffusion rates of oxygen and radicals in the system are greater at higher temperatures [35,36]. In addition, a higher temperature may enhance oxidative degradation by increasing the molecular mobility of polymeric chains, further promoting intermolecular reactions via an increase in the accessibility of oxygen for the degradation process [37].

Turning now to the effect of temperature on mechanical-rheological properties, as can be seen from Figures 4.10 and 4.11, the exposure to a higher temperature led to a greater increase in both E' and T_g of the specimens. This is a direct indicator of the significant detrimental effect of temperature in increasing the extent of crosslinking reactions. As remarked earlier, crosslinking reactions occur via recombination of radicals formed over the course of degradation. The mobility of radicals would be greater at higher temperatures, facilitating the occurrence of crosslinking reactions. This was indeed supported by the substantial increase in crosslink density for the specimens exposed to a higher temperature, as shown in Figure 4.12.

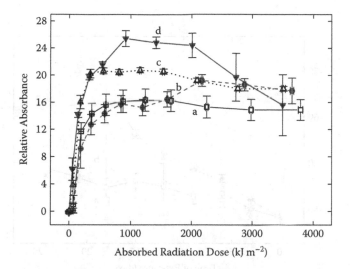

FIGURE 4.9   Evolution of the relative absorbance of 1724 cm⁻¹ to 1030 cm⁻¹ as a function of absorbed radiation dose. Different lines represent data collected from various environments: (a) 30°C at 0% RH, (b) 30°C at 80% RH, (c) 55°C at 0% RH, and (d) 55°C at 80 % RH. Error bars are the standard deviations of the data.

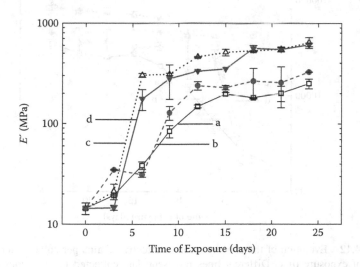

FIGURE 4.10   Evolution of the storage modulus, E', in the rubbery region at 20°C as a function of exposure time. Different lines represent data collected from various environments: (a) 30°C at 0% RH, (b) 30°C at 80% RH, (c) 55°C at 0% RH, and (d) 55°C at 80% RH. Error bars are the standard deviations of the data.

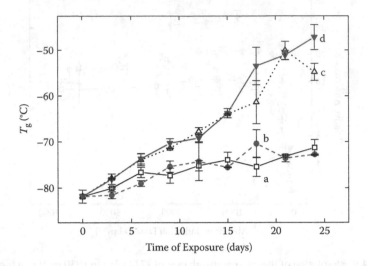

FIGURE 4.11 Evolution of the glass transition temperature, T_g, for the butadiene unit as a function of exposure time. Different lines represent data collected from various environments: (a) 30°C at 0% RH, (b) 30°C at 80% RH, (c) 55°C at 0% RH, and (d) 55°C at 80% RH. Error bars are the standard deviations of the data.

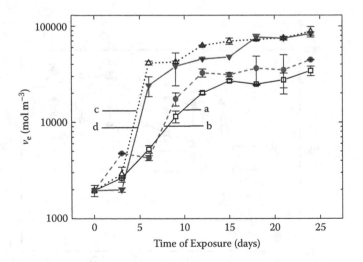

FIGURE 4.12 Evolution of the concentration of effective chains per cubic meter, v_e, as a function of exposure time. Different lines represent data collected from various environments: (a) 30°C at 0% RH, (b) 30°C at 80% RH, (c) 55°C at 0% RH, and (d) 55°C at 80% RH. Error bars are the standard deviations of the data.

4.4.2 EFFECT OF HUMIDITY

It is noteworthy that, within experimental scatter, the initial evolution of carbonyl functional groups was independent of the humidity level during the first few days of exposure, as depicted in Figure 4.9. However, the effect of humidity level became apparent when the total absorbed dose exceeded 500 kJ m^{-2} (corresponding to 10 days) for exposure at 30°C and 1500 kJ m^{-2} (corresponding to 20 days) for exposure at 55°C, where specimens exposed to a higher humidity level of 80% RH exhibited greater oxidative degradation. It is possible that there was some plasticization by absorbed water molecules occurring in the specimens, and this enhanced the rate of diffusion of oxygen and radicals into the specimens [36]. Higher diffusion rates would increase the formation of carbonyl functional groups. The level of humidity, however, had no apparent effect on either E' or T_g within the timescale of the experiment.

4.4.3 SYNERGISTIC EFFECT OF TEMPERATURE AND HUMIDITY

The synergistic effect of temperature and relative humidity on the extent of UV-visible degradation can be appreciated by comparing the absorbances of carbonyl functional groups formed in the environment of 55°C and 80% RH with the environment of 30°C and 0% RH. The absorbance in the former environment was found to be more than 50% greater than that in the latter. This clearly indicates the deleterious effect of combining temperature and relative humidity. This result is in agreement with the general findings reported in the literature [38].

4.5 CONCLUSIONS

One of the major problems with any sealant system is its susceptibility to hostile environmental conditions, such as temperature, oxygen, pollutants, humidity, UV-visible radiation, and cyclic loading. Prolonged exposure to such environments leads to substantial changes in physical, chemical, and mechanical-rheological properties over time. From the discussion in this chapter, it is clear that chemorheological investigations provide a useful tool to characterize the mechanisms of environmental attack on sealants. This ability to investigate both chemical and mechanical-rheological properties is important, as sealants are polymeric in nature; so any chemical modification can be directly related to the corresponding mechanical-rheological modifications.

The usefulness of such a tool was borne out by the study of the UV-visible degradation for a model sealant system based on styrene-butadiene-styrene triblock copolymer using FTIR spectroscopy and dynamic mechanical thermal analysis. The model system was subjected to one of four environments involving combinations of temperature and RH, that is, (1) 30°C at 0% RH, (2) 30°C at 80% RH, (3) 55°C at 0% RH, and (4) 55°C at 80% RH. It was found that the degradation proceeded by end-chain scission and crosslinking reactions in the butadiene units. The latter was the predominant process; it rendered specimens brittle, causing them to lose their desired mechanical properties as shown by substantial increases in both storage modulus and glass transition temperature with increasing exposure time. Higher temperature led to greater degradation as manifested by greater formation of oxidation products and greater extent of brittleness in the model system. Increasing humidity level, however, produced a rather small increase in degradation rate at a given temperature.

REFERENCES

1. R. Woolman and A. Hutchinson, *Resealing of Buildings: A Guide to Good Practice*, Butterworth-Heinemann, Oxford (1994).
2. E. Grunau, *Service Life of Sealants in Building Construction* (in German), Research Report, Federal Ministry for Regional Planning, Building and Urban Planning, Bonn, Germany (1976).
3. R. Chiba, H. Wakimoto, M. Kadono, F. Kawakubo, H. Koji, M. Karimori, E. Hirano, T. Amaya, S. Sasatani and K. Hosokawa, *Proceedings of International Conference*, Japan Sealant Industry Association, Kyoto, Japan, pp. 175–199 (1992).
4. Expenditures for Residential Improvements and Repairs—1st Quarter, Current Construction Reports, U.S. Census Bureau, Department of Commerce, Washington, DC (2002).
5. A.T. Wolf, in: *Durability of Building Sealants*, J.C. Beech and A.T. Wolf (Eds.), pp. 63–89, E & FN Spon, London, UK (1996).
6. K.K. Karpati, *J. Coat. Technol. 56 (719)*, 57–60 (1984).
7. J.M. Cahill, *Building Res. 38*, 17–24 (1966).
8. M.J. Owen and J.M. Klosowski, in *Adhesives, Sealants and Coatings for Space and Harsh Environments*, L.H. Lee (Ed.), pp. 281–291, Plenum Publishing Corp., New York (1988).
9. A. Wolf, *Polym. Degrad. Stab. 23*, 135–163 (1989).
10. R.M. Paroli, K.C. Cole and A.H. Delgado, *Polym. Mater. Sci. Eng. 71*, 435–436 (1994).
11. G. Montaudo, E. Scamporrino, C. Puglisi and D. Vitalini, *J. Polym. Sci., Part A: Polym. Chem. 25*, 475–487 (1987).
12. N. Grassie and I.G. Macfarlane, *Eur. Polym. J. 14*, 875–884 (1978).
13. N. Grassie, I.G. Macfarlane and K.F. Francey, *Eur. Polym. J. 15*, 415–422 (1979).
14. G. Camino, S.M. Lomakin and M. Lazzari, *Polymer 42*, 2395–2402 (2001).
15. O. Chiantore, L. Trossarelli and M. Lazzari, *Polymer 41*, 1657–1668 (2000).
16. A. Torikai, M. Ohno and K. Fueki, *J. Appl. Polym. Sci. 41*, 1023–1032 (1990).
17. T. Boettger and H. Bolte, in *Durability of Building and Construction Sealants*, A.T. Wolf (Ed.), pp. 125–149, RILEM Publications S.A.R.L., Cachan Cedex, France (1999).
18. F. Virlogeux, D. Bianchini, F. Delor-Jestin, M. Baba and J. Lacoste, *Polym. Intl. 53*, 163–168 (2003).
19. V.A. Demarest, J.A. Dionne, M. Lertora and J.R. Magnotta, in *Science and Technology of Building Seals, Sealants, Glazing and Waterproofing*, J.M. Klosowski (Ed.), Vol. 7, pp. 22–42, American Society for Testing and Materials, Philadelphia (1998).
20. P. Vandereecken, A.T. Wolf, H. Bolte and T. Boettger, in *Durability of Building and Construction Sealants*, A.T. Wolf (Ed.), pp. 243–258, RILEM Publications S.A.R.L., Cachan Cedex, France (1999).
21. J.D. Rancourt, *Optical Thin Films User's Handbook*, McGraw-Hill, New York (1987).
22. J.W. Martin, J.W. Chin, W.E. Byrd, E.J. Embree and K.M. Kraft, *Polym. Degrad. Stab. 63*, 297–304 (1999).
23. T. Nguyen, J.W. Martin, W.E. Byrd and E.J. Embree, *J. Coat. Technol. 74 (932)*, 31–45 (2002).
24. J.W. Martin, T. Nguyen, E. Byrd, B. Dickens and N. Embree, *Polym. Degrad. Stab. 75*, 193–210 (2002).

25. M. Piton and A. Rivaton, *Polym. Degrad. Stab. 53*, 343–359 (1996).
26. C. Adam, J. Lacoste and J. Lemaire, *Polym. Degrad. Stab. 24*, 185–200 (1989).
27. S.W. Beavan and D. Phillips, *Eur. Polym. J. 10*, 593–603 (1974).
28. N.S. Allen, M. Edge, A. Wilkinson, C.M. Liauw, D. Mourelatou, J. Barrio, M.A. Martinez-Zaporta, *Polym. Degrad. Stab. 71*, 113–122 (2001).
29. R.L. Clough and K.T. Gillen, *Polym. Degrad. Stab. 38*, 47–56 (1992).
30. J.C. Halpin and S.W. Tsai, Air Force Materials Lab Technical Report, 67–423 (1969).
31. L.E. Nielsen, *J. Appl. Phys. 41*, 4626–4627 (1970).
32. K.T. Tan, C.C. White, D.J. Benatti and D.L. Hunston, *Polym. Degrad. Stab. 93*, 648–656 (2008).
33. L.E. Nielsen and R.F. Landel, *Mechanical Properties of Polymers and Composites*, Marcel Dekker, New York (1994).
34. N.B. Colthup, L.H. Daly and S.E. Wiberley, *Introduction to Infrared and Raman Spectroscopy*, Academic Press, Boston (1990).
35. M.R. Kamal and Bing Huang, in *Handbook of Polymer Degradation*, S.H. Hamid, M.B. Amin and A.G. Maadhah (Eds.), pp. 127–168, Marcel Dekker, New York (1992).
36. J. Crank, *The Mathematics of Diffusion*, Clarendon Press, Oxford (1979).
37. A.C. Somersall, E. Dan and J.E. Guillet, *Macromolecules 7*, 233–244 (1974).
38. G.B. Lowe, in *Durability of Building Sealants*, A.T. Wolf (Ed.), pp. 225–234, RILEM Publications S.A.R.L., Cachan Cedex, France (1999).

5 Natural and Artificial Weathering of Sealants

Norma D. Searle

CONTENTS

5.1   INTRODUCTION

The resistance of sealants to weathering is a major factor in their suitability for use in outdoor applications that depend on their durability. Therefore, determining the weatherability of sealants is an essential step in development of new and improved products. *Weathering* may be defined as irreversible changes in the chemical and physical properties in a direction that usually shortens the useful life of the sealant. Changes in appearance and mechanical properties result from modification of the chemical structures of the sealants due to complex interactions with environmental elements, primarily solar radiation, heat and cold, moisture (solid, liquid, and vapor), oxygen, and atmospheric contaminants, such as acid rain. Oxygen, because of its high chemical reactivity, is by far the most important atmospheric constituent responsible for material degradation. Photooxidation, which results from the effect of absorbed solar actinic radiation by the sealants in the presence of oxygen, accounts for most failures that occur during outdoor exposure [1]. In many of the applications of sealants, cyclic movements are imposed by climatic and construction factors, which, in combination with environmental stress factors, synergistically accelerate their degradation. Although all weather factors, particularly in combination with one another, play a very important role in the deteriorating effect of the environment on materials, the absorbed actinic radiation of the sun, that is, the spectral region responsible for the damage, is the critical factor. It has sufficient energy to break many types of chemical bonds, thus initiating the reactions that result in chemical and physical changes. The other weather factors promote degradation through their influence on the secondary reactions that follow the breaking of bonds. The rate of the secondary reactions is strongly influenced by temperature, being more rapid at higher temperatures. The significance and effect of each of the weather factors on the degradation of sealants is described in this chapter and in Chapter 6.

The resistance of sealants to weathering depends on many factors, including the type of sealant, its formulation, and the level of weather factors in the geographic location in which it is used. Most outdoor exposure tests are carried out in geographic locations that have relatively high levels of solar radiation, temperature, and moisture in order to test the effects of the most severe environmental conditions on the sealants. Because of the improvement over the years in the durability of sealants for outdoor applications, the life expectancy of many well-formulated sealant products exposed to the environment is at least 25–30 years. Therefore, it is impractical in many cases to screen potential new formulations by standard outdoor weathering tests, even those that provide severe environmental conditions. Accelerated outdoor weathering techniques and laboratory accelerated tests simulating the effects of natural weathering are required for the development of weatherable formulations. Maximum acceleration of degradation during outdoor weathering is obtained using the Fresnel-reflector technique for intensified solar radiation. Several types of laboratory accelerated artificial weathering techniques have been developed over the years. These have the advantage of allowing control of the three main weathering factors, radiation, heat, and moisture, for more consistent evaluations of weathering resistance and for research studies on the effect of these parameters. However, when conducting exposures in devices that use laboratory light sources, it is important to consider how well the artificial test conditions will reproduce property changes and failure modes caused by end-use environments on the sealant being tested as well as the stability ranking of sealants. Development of valid laboratory accelerated tests requires measurement of the critical weather factors in the end-use environment and determination of their effects on the sealant materials.

This chapter describes the methods currently used to evaluate weather resistance of sealants and, based on reported studies, suggests modifications to the current methods to reliably predict the long-term performance of sealants. At the current state of the art, short-term natural outdoor exposure tests as well as accelerated outdoor and laboratory weathering tests can provide an assessment of the relative weathering resistance of sealants to the test conditions, but cannot predict the service life of sealants under end-use conditions.

5.2 ENVIRONMENTAL WEATHERING TESTS

Outdoor exposure testing under natural conditions exposes test specimens to weather factors and their patterns that cannot be readily duplicated by laboratory artificial weathering techniques. In the United States, tests are generally carried out in either south Florida or central Arizona, or both, where materials typically fail fastest due to intensification of the weather elements responsible for degradation. These are international "benchmark" environments for natural weathering tests. South Florida has a subtropical climate with higher levels of all three critical weathering factors, solar radiation, temperature, and moisture, than are present in most end-use environments. The central Arizona desert is a worldwide recognized standard exposure environment for testing in a climate typical of the hot and dry conditions of the desert. In the summer, solar irradiance and temperatures are higher than those in south Florida. The large daily and summer-to-winter temperature swings also differentiate central Arizona from south Florida. Other test sites are in tropical and temperate climates, industrial areas, and salt air locations.

The destructive effect of the weather varies with location, season, time of day, cloud cover, atmospheric contaminants, and exposure orientation as well as with type of sealant material because of differences in the critical weather factors and in the varying sensitivities of materials to the latter. Outdoor tests in one location cannot be expected to provide information on durabilities of sealants in a different location because of differences in solar radiation, temperature, rainfall, and atmospheric conditions. Exposures in several locations with different climates that represent a broad range of anticipated service conditions are recommended. Because of seasonal and year-to-year climatological variability of outdoor conditions in any one location, results from a single exposure test cannot be used to predict the absolute rate at which a sealant degrades. Several years of repeat exposures are needed to obtain an "average" test result for a given location. Because of the variability of weather, it is strongly recommended that a control sealant material with known out-door performance and having similar composition and construction as well as failure modes be included with each exposure test for evaluation of the performance of sealants relative to that of the control. It is preferable to use two controls, one having good durability and one having poor durability. At least three replicate specimens for each test and control material are recommended.

During their entire service life, joint seals are exposed to frequent cyclic move-ments, which impose cyclic mechanical strain on the seals. It is considered to be one of the major factors contributing to sealed joint failure by environmental exposure [2,3]. Studies have shown [4,5] that joint movements can be large during the spring and summer months and that diurnal cyclic movements are superimposed onto the seasonal cycle of movements. The amount, type, and frequency of movement that a sealant experiences during its lifetime strongly depends on the materials used in con-struction, on the orientation of the sealant toward sunlight, and many other factors. When combined with appropriate mechanical movement, natural weathering tests can provide information on the ability of a sealant to withstand climate influences along with the stresses of cyclic movement.

5.2.1 WEATHERING FACTORS

5.2.1.1 Solar Energy

Solar radiation on the earth's surface (daylight) consists of energy received both directly from the solid angle of the sun's disk (sunlight) and diffusely reflected by the atmosphere (skylight). Figure 5.1 [6] shows the direct normal solar spectral irra-diance and hemispherical (direct plus diffuse) solar spectral irradiance incident on a 37°-tilted surface facing the equator at an air mass of 1.5, representative of the aver-age latitude of the 48 contiguous states of the United States under reasonably cloud-less atmospheric conditions, referred to as *U.S. Standard Atmosphere*. The spectral curves are plotted from the tabular data in ASTM G173 [6], which are standard reference spectral irradiances applicable to weathering and durability exposure stud-ies. The spectral energy ranges from about 298 nm in the ultraviolet (UV) region to about 2500 nm in the near-infrared (NIR) region. The diffuse component consists of radiation that has been scattered in the atmosphere either from the solar beam on its travel downward or reflected by the earth's surface into the atmosphere. It represents a substantial proportion of the total incident solar radiation, particularly in the ultra-violet region, because of greater scattering of the shorter wavelengths. Therefore, the spectral power distribution (SPD) of the diffuse component differs from that of the solar beam in that it contains more short-wavelength radiation. In an equatorial location on a cloudless day, 50% of the total radiation at 325 nm is due to the diffuse radiation from the sky [7]. The diffuse component is a much smaller percentage of the total radiation in the visible and infrared spectral regions. Therefore, a building wall facing away from the sun receives proportionately a larger percentage of short-wavelength than of long-wavelength solar radiation.

At all latitudes, the altitude of the sun, and thus total solar irradiance, is highest at solar noon when the path length that solar radiation traverses through the atmosphere

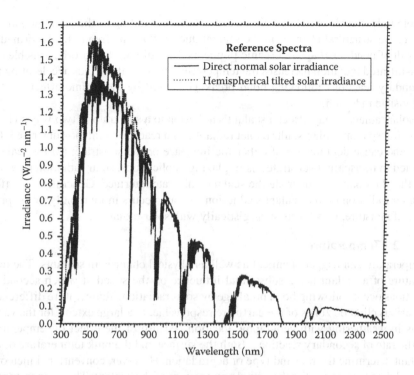

FIGURE 5.1   Solar spectral irradiance: direct normal and hemispherical (direct plus diffuse) on a 37°-tilted surface facing the equator (air mass 1.5). (Taken from ASTM G173-03, Standard Tables for Reference Solar Spectra Irradiances: Direct Normal and Hemispherical on 37° Tilted Surface, Annual Book of ASTM Standards, Vol. 14.04. With permission of ASTM International.)

is at a minimum. Both the intensity and spectral power distribution of solar radiation vary with time of day, season, altitude, geographical location, and atmospheric conditions. The shorter the wavelength, the more sensitive it is to these variables. The UV portion (298–400 nm), the portion that has the largest actinic effect on most materials, varies from less than 1% to about 6% of total solar radiation. At 41° north latitude, the short wavelength cut-on (i.e., the 5% transmittance wavelength) shifts from below 300 nm in the summer to about 310 nm in the winter.

According to the first law of photochemistry, light must be absorbed by a material to have any effect on it. For most materials, their capacity to absorb radiation increases with decrease in wavelength. Since the energy of radiation is inversely proportional to its wavelength, ultraviolet radiation is not only more strongly absorbed, but has higher-energy photons than those of visible and infrared radiation. Therefore, the UV portion of solar radiation has greater potential to break chemical bonds and thus is primarily responsible for initiating the weathering of sealants. However, the visible and infrared portions of solar radiation also contribute to degradative processes. Short-wavelength visible radiation can have a significant actinic effect on sealants, such as aromatic-type polyurethanes, that absorb these wavelengths. The visible radiation absorbed by colored sealants causes fading of the pigment and can also cause structural changes in the sealant due to attack by free radicals formed as a result of bond breakage in organic pigments. In addition, the absorbed visible and near-infrared radiation increases the temperature of the sealant, thus accelerating the secondary reactions following bond breakage caused by the actinic effects of UV and visible radiation.

Solar radiation may affect a sealant's adhesion to a substrate in two ways: (1) for some transparent sealants, ultraviolet radiation can reach the sealant–substrate interface and cause degradation of either the interface or the substrate; (2) for sealants applied to transparent substrates as in glazing applications, solar radiation penetrating the glass pane can degrade the sealant–substrate interface. Generally, the effect of solar radiation on the sealant's adhesion to glass occurs more rapidly in the presence of moisture, which acts synergistically with solar radiation.

5.2.1.2 Temperature

Temperature can trigger chemical as well as physical changes in sealants. The temperature of a sealant has a substantial influence on the speed at which secondary reactions occur following bond breakage by solar radiation. Temperature differences in various climatic zones of the earth are responsible, to a large extent, for the variations in weathering with geographic location because of the effect of temperature on the rate of secondary reactions. Both the surface and the bulk temperature of the sealant determine the rate and type of degradation. However, conventional meteorological data cannot predict the actual temperature of the sealant. The temperature of the sealant's surface depends on the ambient temperature; the sealant's absorptivity of solar energy, particularly of the visible and infrared portions; its emissivity; thermal conductivity and heat capacity; and the wind speed and convection rate. Since sealants generally have low thermal conductivity, exposure of pigmented sealants to solar radiation results in heating of the sealant surface by the absorbed radiation. Because of their selective absorption of the visible and near-infrared portions of solar radiation, the surface temperature of pigmented sealants is closely related to color, being low for white surfaces and high for black surfaces and varying with color and its spectral absorption properties in relation to the spectral power distribution of solar radiation. Thus, the maximum surface temperatures vary, depending on the color of the pigments, but generally fall in the range of 40°C–70°C.

The temperature gradient of sealants heated by solar radiation causes differences in the surface and bulk temperatures, and thus in stresses, which, under the influence of the diurnal and seasonal temperature variations, can cause degradation of the sealant. Temperature variations will also cause stresses at the sealant–substrate interface because of differences in thermal expansion coefficients of the sealant and substrate, leading to adhesion failure. Daily temperature cycling and changes in solar irradiance causing the building façade components to expand and contract impose cyclic movement on the sealed joints, and thus mechanical stress on composite systems. Joint movement is highest in dark-colored facades facing the sun. Under extreme weather conditions, freeze/thaw cycling of absorbed water in the sealant matrix or thermal shocks due to cool rain hitting hot, dry surfaces will induce mechanical stresses that can cause structural failures in some systems. Temperature and its cycles are also closely linked to water in all of its forms. Increase in temperature accelerates hydrolysis reactions, while reduction in temperature results in condensation as dew on the sealant.

5.2.1.3 Moisture

Moisture may be delivered to the surface of a sealant and sealed joint in the form of humidity, dew, rain, snow, frost, or hail, depending on the atmospheric conditions and ambient temperature. Dew (condensed moisture) occurs primarily at night, when the sealant temperature drops below the dew point. The cooling effect, and thus the formation of dew, is most prevalent for materials that have low heat capacity and thermal conductivity, such as most sealants. Moisture contributes to the weathering of sealants and sealed joints both by reacting chemically in hydrolytic processes and by imposing mechanical stresses when it is absorbed or desorbed. The action of water is largely responsible for lowering the adhesion of the sealant to the substrate in a sealed joint. The stresses exerted by repeated swelling and drying can cause bond separation at the sealant–substrate interface. In test methods for glazing sealants, the sealant–glass interface is exposed either to water spray, water immersion, or high humidity along with exposure to the source of radiation, which acts synergistically with moisture to cause loss of the sealant's adhesion to glass. The mechanical stresses caused by repeated swelling and drying can also result in cracking of the sealant surface, especially if the surface has been embrittled by photochemical aging processes. Moisture can also act as a solvent or carrier, for example, in transporting dissolved oxygen or in leaching away plasticizers, thus altering the physical properties of the sealant. Thus, moisture contributes to the degradation of the appearance and physical properties of a sealant and its adhesion to a substrate.

The penetration depth of moisture into the material, and thus the influence on weathering, is substantially greater when the total amount of precipitation is distributed over a longer time period. Therefore, the time the sample surface is exposed to wetness is more important in weathering of materials than the total amount of precipitation. All commercial exposure sites provide data on time-of-wetness (TOW), which is the length of time liquid water is present on the surface of the sealant as condensation and precipitation. Exposed specimens are wet more than 50% of the time in humid environments, such as in Miami, Florida, and only 5% or less of the exposure time in arid locations, such as in Phoenix, Arizona [8].

5.2.1.4 Atmospheric Contaminants

Ozone is present in the earth's atmosphere both as a result of UV photolysis of oxygen in the upper atmosphere and reaction between terrestrial solar radiation and atmospheric pollutants such as nitrogen oxides and hydrocarbons from automobile exhausts. It is a powerful oxidant that can react rapidly with elastomers and other unsaturated polymeric materials to cause stiffening and cracking, particularly in materials under mechanical stress.

Common air pollutants in industrial environments include sulfur oxides, hydrocarbons, nitrogen oxides, and particulate matter such as sand, dust, dirt, and soot. Some of these may react directly with organic materials, but have a much more severe effect in combination with other weather factors. For example, dilute sulfuric acid is formed only when sulfur dioxide (SO_2) and water on the surface of materials are exposed to solar radiation. Acid rain is an important consequence of pollutants generated by modern industrial societies and generally acts synergistically with radiation to accelerate the effects of weathering. When it reacts with pigments, it causes rapid discoloration of the pigment as well as crosslinking [9] and embrittlement [10] of the polymeric binder. Acid rain enhances hydrolytic degradation and thus is an important factor in weathering of sealants that involves hydrolysis.

5.2.2 ENVIRONMENTAL EXPOSURE METHODS

5.2.2.1 Natural Environmental Exposure Tests

Outdoor weathering tests are commonly characterized as *natural* or *accelerated*. The term *natural* is generally used for outdoor exposure on fixed-angle racks in locations and orientations that maximize the effects of weathering components, particularly solar radiation. Temperature and wetting are also intensified by various techniques of natural outdoor weathering tests to produce higher degradation rates than materials experience under normal end-use conditions. These weathering tests are of necessity shorter than the service life of sealants. Since degradation of most polymeric materials is not a linear function of exposure time or radiant exposure (energy received in joules per square meter, J/m^2), short-term tests cannot predict the absolute long-term performance of sealants. Therefore, prediction of service life based on short-term exposures may not be representative of the actual service life of a sealant. Durabilities of sealants are often evaluated in terms of performance in comparison to that of sealant specimens of known durabilities (controls) [11] exposed at the same time.

Procedures for natural outdoor exposures of sealants are described in ASTM Practice C1589 [12] and in several publications on sealants [13,14]. General procedures for outdoor exposures of materials and effects of variables in environmental parameters are described in other publications and in ASTM and ISO practices [1,15–18]. The most widely used method is direct exposure to the environment, in which the front surface of the test specimen is exposed to unfiltered solar radiation and to all elements of the environment [16,18]. Indirect exposures, in which specimens are placed behind glass, are used for sealants that are exposed to solar radiation through glass in end-use applications [17,18]. Specimens of sealants exposed to solar radiation through glass, for example, structural silicone sealants, are prepared using clear uncoated float glass as the substrate. They are exposed directly to the sun with the substrate as the front surface. The glass-filtered specimens are protected from the shortest UV wavelengths of solar radiation. Different test results may be obtained compared with direct exposure because of differences in SPD and temperature. Figure 5.2 [19] shows the effect of window glass on the SPD of sunlight incident on specimens. The UV absorption of the glass, and thus its filtering effect, increases with decreasing wavelength. The UV transmission of the glass is inversely related to its thickness and the iron and sulfur content. In a study on the UV transmittance of window glass [20], it was found that the 310 nm transmittance of glass in thicknesses between 2.3 and 2.7 mm from various sources ranged from 0.5% to 6.3%. The UV

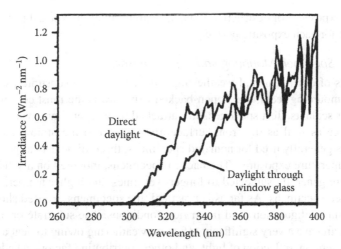

FIGURE 5.2 Effect of window glass on the UV spectral irradiance of daylight. (Taken from ASTM G154-06, Standard Practice for Operating Fluorescent Light Apparatus for UV Exposure of Nonmetallic Materials, Annual Book of ASTM Standards, Vol. 14.04. with permission of ASTM International.)

transmission at 310 nm of the different glasses was reduced by 30 to 60% during the first 3 months of exposure to solar radiation and did not change significantly on further exposure. At these thicknesses, wavelengths of 300 nm and shorter are completely filtered out. Increasing amounts of radiation are transmitted between 310 and 370 nm, and all radiation longer than 370 nm is transmitted except for the small amount specularly reflected by the glass. When in use outdoors, the glass should be cleaned periodically to eliminate the dirt and mildew that may accumulate and reduce its transmittance of solar radiation.

5.2.2.1.1 Test Specimens
Preparation of test specimens of sealants for outdoor exposure is described in ASTM Practice C1589 [12]. As far as is practical, test specimens simulate those used in service conditions, consisting of seal or sealant material as well as suitable substrate or installation materials, which can have a significant effect on the weathering of the sealants. Test specimens may be of any size or shape that can be mounted in a fixture or a holder or applied directly to the racks. The size and shape may be determined by the specific test method used to evaluate the effect of weathering or may be a large specimen from which smaller ones are cut after weathering. However, the cutting operation on exposed specimens may affect the property measured, particularly for materials that embrittle upon exposure. The specimens should be large enough so that mounting edges that would affect test results can be removed. In order to allow for variability of weathered materials, at least three replicate specimens are required of each material being evaluated by nondestructive tests; for destructive tests of mechanical properties, twice the number required by the evaluation method

should be exposed. For destructive tests, a separate set of unexposed test specimens is required for each exposure period.

5.2.2.1.2 Specimen Mounting and Exposure Angle

Test results of environmental weathering can vary with the substrate used and the specimen-mounting technique. Open-backed exposure is the most commonly used method for sealants. Test specimens are attached to an open framework so that the back surface as well as the front surface are exposed to the atmosphere. Backed exposure is generally used for nonrigid specimens, three-dimensional parts, and for higher-temperature exposures. The backed specimens, mounted on a solid plywood backing, are generally subjected to longer wet times and higher temperatures than open-backed specimens. As far as is practical, the specimens exposed should simulate the joint configuration used in service conditions. The substrate or installation materials can have a very significant effect on weathering owing to heat absorption, moisture retention, reflection of light, and other contributing factors. The location of the specimen on the exposure rack, the spacing between specimens, and the color of the specimens can affect specimen temperature and time of wetness. In order to minimize variability within a single study caused by the location of the specimens on the rack, it is recommended that they be placed on a single test panel or on test panels adjacent to each other.

The exposure angle, that is, the tilt with respect to the horizontal of the exposed surface of specimens, has a significant effect on the total dosage of solar radiation, the amount of UV the specimens receive, the specimen temperature, and the amount of moisture and time of wetness. Samples positioned for maximum direct beam exposure receive less short-wavelength radiation than those positioned for maximum exposure to sky radiation, that is, to diffusely scattered radiation. The typical outdoor exposure angle and the angle of choice for sealants is 45° relative to the horizontal, facing the equator. The alternate option for sealants is the latitude angle, that is, the angle from the horizontal that is equal to the geographic latitude of the exposure site. It provides the highest annual solar radiation dosage of any fixed angle exposure. Some protocols require exposure at latitude angle minus 10°, at which the annual ultraviolet radiant exposure is highest [21], particularly in the summer months, because of the larger amount of diffuse radiation on the specimens. A 90° exposure angle would be most realistic for building sealants, but is not widely used because of the lower solar radiant exposure and temperature and fewer hours of wetness. In a temperate climate, this type of exposure results in about half the amount of ultraviolet radiant energy as that obtained at 45° [21] or at horizontal [22] exposure. Vertical surfaces are less prone than horizontal surfaces to moisture condensation during the night, and do not experience the accumulation of water on the surface after rainfall that occurs on horizontal surfaces. At 0° (horizontal) and 5° angles, specimens are exposed essentially to the entire sky dome and at exposure sites close to the equator they receive more annual solar radiation than at 45°, especially during the summer months. The seasonal effect of the angle of the sun on the spectral power distribution of direct normal solar radiation in the ultraviolet region is shown in Figure 5.3 [23].

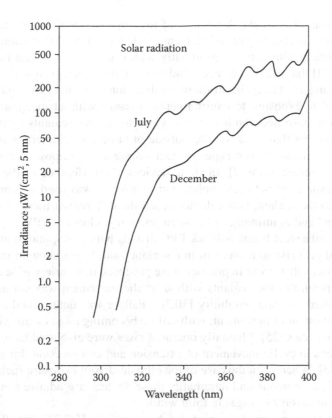

FIGURE 5.3 Seasonal variation of direct normal solar UV spectral irradiance at 41° N latitude at noon under clear sky conditions. (Reprinted from R. C. Hirt, R. G. Schmidt, N. D. Searle and A. P. Sullivan, *J. Opt. Soc. Am.* 50, 704–713, 1960. With permission of the Optical Society of America.)

5.2.2.1.3 Cyclic Movement

It was recognized at an early stage in research on the durability of sealants that cyclic joint movement is one of the major factors leading to premature failure of sealants in sealed joints [24–27]. It is generally believed that natural outdoor weathering of sealants under forced mechanical cycling yields results that show better correlation with actual in-service performance than any other test. Without simultaneous mechanical cycling, natural outdoor weathering can reproduce most of the surface degradation observed in actual service joints, but does not provide good prediction of changes in the bulk properties of the sealants. Initial results obtained using forced strain cycling suggest that mechanical cycling during exposure accelerates both the surface and bulk deterioration of the sealant. Therefore, the researchers recommended that in order for test methods to provide a reliable assessment of the performance of sealants, they must subject the sealant to cyclic movement in conjunction with exposure to the relevant weather factors [24,28,29]. However, techniques for incorporating movement, such as expansion and compression, during exposure have not yet been introduced into routine outdoor testing of sealants. A number of investigative studies have reported on acceleration of weathering resulting from cyclic movement either simultaneously with exposure or alternating periodically with exposure [3]. It has been shown [5,25,27,30,31] that the occurrence of adhesion failure increased when the specimens were subjected to cyclic strain in combination with environmental exposure as compared to exposure to environmental stresses without mechanical strain. Many devices have been built over the years to subject sealants to the type of cyclic movements that occur on the outside of buildings. Numerous researchers have utilized climate-driven exposure racks designed to impose cyclic strain as a function of temperature [2]. In some devices, the difference in thermal coefficients of expansion between timber and aluminum was used to impart cyclic movement to the sealant. Other devices used the differential thermal coefficient between steel and aluminum, or between poly(vinyl chloride) (PVC) pipes filled with water and a steel frame or black PVC driving bars to respond to temperature changes and give rise to movement in a sealant joint. It was shown that the same types of failures that occur in practice were produced in a variety of sealant types and that exposure of the sealants without cyclic movement was not adequate to reliably evaluate sealant durability [30,32]. Failure was determined to be influenced by the extent of movement, with failure becoming more extensive at higher yearly strain cycles [25]. Manually operated vises were also used to subject sealant specimens to cyclic movement of extension and compression during outdoor exposure [33]. Whereas in the case of the climate-driven exposure racks the joint movement is continuous and automatic, these devices are adjusted manually to provide the necessary changes in joint width.

A unique design of sample supports has been described [13,34] that allows simultaneous extension at one end of the test specimen and compression at the other end during exposure. Mechanical cycling is achieved by manually extending one end of the test specimen and holding it in extension for 1 month and then extending and holding the other end in extension for 1 month. This is repeated six times in 1 year. Initial results show that movement appears to increase the degree of degradation by only twofold or less for most of the sealants and to have a negligible effect on some types of sealants tested. However, the effect of compression/expansion could only be compared with the portions of the specimens that were subjected to shear movement during exposure, rather than with no movement of specimens.

Lacher et al. [35] described a new computer-controlled cyclic fatigue test apparatus (CFTA) designed for use during outdoor exposure of sealants. The computer control allows the CFTA to subject specimens to consistent strain/time profiles or replicate input strain data in real time from thermal and/or moisture-driven cyclic movement of materials such as metals, glass, ceramics, and wood. Previously, mechanical strain was imposed on sealant specimens by dimensional changes in these materials due to thermal changes and to swelling and shrinking caused by absorption and desorption of moisture [36,37]. The computer-controlled CFTA can be programmed to execute a specific strain versus time profile or to mimic mechanical strain caused by any type of moisture/thermal material response.

5.2.2.2   Accelerated Environmental Exposure Tests

The maximum acceleration of aging processes in outdoor weathering is obtained with the use of a Fresnel-reflector panel rack for exposure. Descriptions of the device and guidelines for its use are given in ISO 877 [18], ASTM G90 [38], and ASTM D4364 [39]. It provides high-intensity solar radiation by following the sun and reflecting the sun's rays from an array of 10 flat mirrors onto a single target area in which the test material is placed. A blower forces air across the target area to cool the specimens and generally limits the specimen surface temperatures to about 15°C above the maximum temperature of equatorially mounted samples exposed to unconcentrated normal incidence radiation. Testing of samples exceeding 13 mm in thickness is not recommended, because of ineffective cooling. Some Fresnel-reflector accelerated test devices use a patented dynamic temperature control system that controls a set temperature and replicates variations in ambient daytime temperatures of materials. Periodic spraying with deionized water is used to simulate and accelerate the results of conventional testing in semihumid subtropical and temperate regions. Nighttime spray cycles simulate rain and dew. Tests carried out in the absence of a programmed moisture cycle are intended to simulate conventional exposure testing in desert, arid, and semiarid regions.

The Fresnel-reflector device can be used to intensify either direct exposures or exposures behind glass. Since the device only concentrates the direct rays of the sun and not the diffuse radiation, it requires clear atmospheric conditions with little moisture and thus less scattering of the direct rays, such as is prevalent in Arizona. The solar flux is approximately eight times as high as on an equatorial mount without mirrors on which the surface of samples is maintained perpendicular to the direct beam of the sun throughout the day. In the ultraviolet portion of the solar spectrum, approximately 1400 MJ/m² of ultraviolet radiant exposure (295 to 385 nm) is received over a typical 1-year period when these devices are operated in a central Arizona climate. This compares with approximately 333 MJ/m² of ultraviolet radiant exposure from a central Arizona at-latitude exposure. However, the radiation contains a lower percentage of short-wavelength ultraviolet than global daylight because it only includes direct beam radiation, which is deficient in short-wavelength ultraviolet since it is more easily scattered by the atmosphere. Also, mirrors are typically less efficient at shorter ultraviolet wavelengths.

Ultraviolet radiant exposure levels should not be used to compute acceleration factors since acceleration is material dependent, varying with type of material and its formulation due to differences in the effect on materials of increase in irradiance and temperature. Acceleration factors, based on comparison with conventional outdoor tests and elapsed time to a predetermined change in property, have been reported to vary from 2 to 11 for various polymer types and compositions [40]. The acceleration factor is small for materials that are degraded primarily by the short UV solar wavelengths because of the lower percentage of these wavelengths in the solar concentrator. Thus, the stability rankings may differ from that of natural exposure. Good correlations as well as discrepancies have been reported between Fresnel-reflector exposures and standard outdoor tests based on stability ranking of polymeric materials [41–46].

5.2.3 Exposure Stages and Timing of Exposures

Methods of "timing" outdoor weathering tests include the following: (1) calendar basis—exposure for a specified number of days, months, or years; (2) radiant exposure basis—exposure to a specified level of solar radiant energy in a specified wavelength region using hemispherical radiation measurements at the same tilt and azimuth angle as the test specimens; (3) deterioration basis—exposure until a specified change has occurred; and (4) property change in a weathering reference material (WRM)—exposure until a specified property change has occurred in a weathering reference material exposed at the same time. Timing on a calendar basis has been the most widely used method of evaluating stabilities because of its simplicity. However, results of exposures conducted for less than 12 months will often depend on the particular season of the year in which the exposure was carried out. Although variabilities due to seasonal differences are reduced by exposures lasting several years, the results may still depend on the particular season in which the exposure was started. For example, exposures begun in the spring may exhibit more degradation than exposures begun in the autumn.

Since solar radiation is one of the most important factors in the deterioration of seals and sealants exposed to the environment, testing to specific levels of solar radiant exposure is preferable to tests based on elapsed time. Timing based on radiant exposure can reduce seasonal and year-to-year variations in weathering caused by inconsistent conditions of total solar and solar UV irradiance. Total solar UV radiant exposure is the recommended measurement for exposure stages because ultraviolet has a much greater effect than visible and near-infrared radiation on most sealants and its variations are not readily detected in measurements of full-spectrum solar radiation because of the small percentage of UV in the full spectrum. Studies have shown that better correlations are obtained among exposures made at different times when timing is based on incident solar UV radiation rather than on incident total solar radiation [42,47–52].

Correlations can potentially be improved further by timing on the basis of UV radiant exposure in the spectral region responsible for the damage by the light source to which the sealant is exposed, that is, the actinic region identified by the activation spectrum of the material [53,54]. However, since all actinic wavelengths are not equally destructive, the optimum method of timing on the basis of radiant exposure would be in terms of "effective" radiant exposure (i.e., effective dosage, D_{Eff}). The "effective" irradiance (E_{Eff}) is determined by assigning to each wavelength in the spectral region its relative effect based on the activation spectrum in accordance with Equation 5.1.

$$E_{Eff} = \sum_{\lambda_1}^{\lambda_2} E_0(\lambda) \times \frac{[\Delta(\lambda)]}{[\Delta(\lambda_{max})]}$$

(5.1)

$E_0(\lambda)$ = Incident irradiance at specified wavelength
$\Delta(\lambda_{max})$ = Property change at wavelength of activation spectrum peak
$\Delta(\lambda)$ = Property change at specified wavelength

The irradiance at the wavelength corresponding to the peak of the activation spectrum ($\Delta(\lambda_{max})$) is normalized to 1.0 and irradiances at other wavelengths are multiplied by the fractional sensitivities obtained from the activation spectrum. The sum is the "effective" irradiance. The effective dosage (D_{Eff}) is obtained by multiplying the total dosage (total radiant exposure D_{Total}) by the ratio of the effective irradiance divided by the total irradiance (Equation 5.2).

$$D_{Eff} = \frac{E_{Eff}}{E_{Total}} \times D_{Total}$$

(5.2)

Although timing exposures in terms of radiant exposure or effective dosage can potentially improve correlations between outdoor exposures at different times and locations compared with timing on a calendar basis, it does not take into consideration the variations in other weather factors, particularly temperature, moisture, and atmospheric contaminants, which, in conjunction with solar radiation, have a significant influence on weathering. Differences in the other factors can cause large differences in the extent of degradation produced by the same radiant exposure. Timing based on property change in a WRM that has the same sensitivity as the test material to all weather factors would be the optimum way to time exposures. However, identifying such a material is an impractical expectation.

5.2.4 MEASUREMENT OF WEATHERING FACTORS

Characterizing and monitoring the environmental factors during exposure is an important part of environmental exposure tests. It is necessary for correlating the damage produced during outdoor exposures with the stress factors in order to develop more environmentally resistant products and for development of accelerated tests that are capable of reproducing property changes and failure modes caused by environmental exposure. Defining the damage solely in terms of exposure time ignores the effects of variations in intensity and spectral irradiance of solar radiation and moisture and temperature differences between different exposure tests.

Instruments for recording climatological data during the exposure period should be operated within 1000 m of the exposure racks [16]. Instruments for measurement of daily maximum and minimum ambient temperatures are mounted inside a white ventilated enclosure and must have a calibration traceable to a national standards calibration body. In the United States, this is the National Institute of Standards and Technology (NIST). A specially designed insulated black-panel thermometer (black standard) or uninsulated black-panel thermometer (black panel) that complies with the requirements in ISO 4892, Part 1 [55] or ASTM G151 [56] is mounted on the specimen rack so that the surface of the black panel is in the same relative position and subject to the same influences as the surface of the test specimens. The temperature sensor and metal panel are coated with a nonselective, light-stable, black finish that absorbs most of the solar radiation reaching the earth's surface. The temperature measured represents the surface temperature of a dark-colored sealant, which is the maximum surface temperature attained on exposure. The uninsulated sensor generally registers temperatures that are between 3°C and 12°C lower than that of the insulated sensor, the difference being larger at higher irradiance levels. The amount of rain, snow, or hail falling on a specified surface area is measured by weight or volume with the help of various devices in which the precipitation is caught in an open container. The duration of rainfall is also measured. Exposure laboratories typically measure two main types of moisture, relative humidity and time of wetness (TOW). Measurements of relative humidity by a shaded device such as a wet bulb/dry bulb or solid-state sensor enclosed in a ventilated shelter indicate the amount of water vapor in the air mass relative to its maximum capacity at the ambient temperature. TOW is the length of time liquid water is present on the surface of the specimen because of condensation and precipitation. It is measured either by increase in electrical conductivity of a cotton wick when it is wet or by the electrical potential developed in a moisture-sensing galvanic cell. One example of a TOW sensor of this type is described in ASTM G84 [57]. Other instruments for measuring TOW on surfaces are also available.

Commercial outdoor testing facilities measure and monitor the irradiance and radiant energy of full-spectrum solar radiation (nominally 300 to 2500 nm) as well as the ultraviolet portion alone, both total UV and narrow spectral bands of the UV. Full-spectrum hemispherical (direct plus diffuse) solar radiation is measured by means of a pyranometer calibrated at least annually either against a reference pyranometer traceable to the World Radiometric Reference (WRR) [58–61] or with the use of a pyrheliometer in accordance with ASTM G 167 [62] or ISO 9846 [63]. The pyranometer is mounted within ±2° of the angle of the rack on which the samples to which it relates are exposed. It is connected to an integrating device to record the total energy received over a given period. The pyrheliometer is used to measure the direct beam of full-spectrum solar radiation, a measurement required for exposures using Fresnel-reflecting concentrators. The calibration is traceable to the WRR [58,64,65], and recalibration is carried out annually by either the manufacturer, a qualified calibration laboratory, or internally using a laboratory-standard radiometer in accordance with ASTM E824 [59], ISO 9847 [61], or ASTM E816 [64] unless more frequent recalibration is indicated by the required checks.

Total (hemispherical) solar UV radiation (TUVR) and narrow spectral bands of solar UV are measured by means of suitably filtered radiometers that are cosine-corrected to include diffuse ultraviolet radiation. They are calibrated at least annually against a calibrated reference radiometer with transfer of calibration from reference to field radiometer according to ASTM E 824 [59] or ISO 9847 [61]. The reference radiometer is calibrated according to ASTM G130 [66]. The spectral response of the narrow-band radiometer should be as flat as possible throughout the spectral region measured. Total solar UV radiation (TUVR) has traditionally been measured with radiometers that have a spectral response between 295 and 385 nm. The average 12-month TUVR radiant exposure is reported to be 308 MJ/m^2 in a subtropical climate and 333 MJ/m^2 in a desert climate at latitude angle (tilt angle equal to the latitude of the exposure site) [67]. Use of a radiometer that responds to UV wavelengths as long as 400 nm can result in recorded UV radiant exposures that are 25% to 30% higher.

The highest irradiance of natural daylight measured at normal incidence in Miami at solar noon during the vernal equinox is about 0.68 W/m^{-2} per nm at 340 nm. Measurement of the direct component of solar UV radiation is required for exposures using Fresnel-reflecting concentrators. It is measured according to ASTM G90 [38] using two ultraviolet radiometers, one of which is fitted with a shading disk to block the direct solar ultraviolet radiation. The direct component is determined as the difference between the readings from the two radiometers. Irradiance is reported in watts/square meter (W/m^2) and radiant exposure, that is, irradiance integrated over time, is reported in megajoules/square meter (MJ/m^2). The radiometers are positioned at the same tilt and azimuth angle as the test specimens. Use of pyranometers, pyrheliometers, and UV radiometers is described in ASTM G 183 [68] and ISO 9370 [69].

5.3   LABORATORY ACCELERATED WEATHERING TESTS

The advantage of laboratory accelerated weathering tests lies in their ability to accelerate the weathering processes under controlled and reproducible conditions. These tests are particularly important for research and development of new sealant formulations and are also useful for specification testing. Laboratory weathering tests lend themselves to evaluations by certain chemical analysis techniques that are not applicable to outdoor weathering tests, in which surface contamination often interferes. However, adequate correlation with weathering under environmental conditions requires that the chemical and physical changes induced in the sealant duplicate those produced by outdoor exposure. Because of the complex nature of the weathering process, this can only be realized if the balance of weathering factors in the environment is correctly simulated and intensification of the weathering factors is not excessive.

The main weathering factors, that is, solar simulated radiation, heat and moisture in the form of humidity and either water spray, condensation, or immersion, are included in the basic design of most artificial weathering devices. Intensification of one weathering factor alone over that normally experienced by the sealants can distort the results. Acceleration of degradation that is much greater than a factor of ten may also distort the results in terms of the relative weather resistance of sealant products. The acceleration factor, that is, the exposure time in an outdoor weathering test divided by exposure time in a laboratory accelerated test required to produce equivalent degradation, varies with the type of sealant as well as with its formulation. Therefore, it is not possible to establish an acceleration factor that applies to all sealants or even to a sealant material class. For this and other reasons, there is no direct way to predict service life under use conditions from the results of laboratory accelerated weathering tests. Techniques for predicting service life of sealants are still under development (see Section 5.6).

5.3.1 SPECTRAL POWER DISTRIBUTIONS (SPDs) OF LABORATORY
ACCELERATED WEATHERING SOURCES

The single most important consideration when conducting laboratory accelerated weathering tests is the SPD of the radiation source, that is, the incident wavelengths and their relative intensities. The relation between the absorption properties of the sealant and the SPD of the light source determines the fraction of the incident radiation at each wavelength absorbed by the sealant. The wavelengths responsible for the degradation depend on the quantum efficiencies of the absorbed wavelengths. The wavelength sensitivity of the sealant to the source of radiation to which it is exposed is graphically represented by its activation spectrum [53,54]. Activation spectra of various types of sealants exposed to filtered xenon arc radiation showed that the 360–380 nm spectral region had the major effect on loss of adhesion [70]. Wavelengths shorter than 340 nm had only a minor effect on adhesion in most of the sealants tested. The activation spectrum based on exposure to the filtered xenon arc is representative of the activation spectrum based on exposure to terrestrial solar radiation. The wavelengths responsible for degradation vary with the SPD of the radiation source and often determine the mechanism of degradation. Both absorption of light by the sealant, which is a prerequisite to degradation, and bond breakage, the primary photochemical step following absorption of light, are wavelength dependent. Therefore, simulation of the SPD of terrestrial solar radiation is critical to simulation of the degradation caused by exposure to natural weather conditions.

Studies [71] of several aromatic types of polymeric materials have shown that short UV wavelengths absorbed by the structural components cause photolytic reactions that lead to photo-Fries rearrangements, while longer UV wavelengths, absorbed by impurities such as hydroperoxides and unsaturated carbonyl-containing structures, initiate photooxidation reactions involving free radicals. The relative amount of surface versus bulk degradation, bond scission versus crosslinking, chalking versus cracking and crazing, and other diverse types of degradation will differ if the relative intensities of short versus long actinic wavelength radiation differ. If the predominant mechanism of degradation varies with the test source, the effectiveness of stabilizers that act by interfering with the mechanism of degradation will differ. In addition, the stability ranking of sealants will often vary with the type of light source because of differences in the absorption properties of sealants and, thus, differences in the relative amounts of radiation absorbed by the sealants from each type of light source.

The better the simulation of the SPD of the natural source by the laboratory light source and the better control of irradiance, temperature, and moisture conditions that the laboratory accelerated device can achieve, the more suitable it is for the purposes for which it is used. The devices commonly used for sealants employ as radiation sources either filtered xenon arcs, fluorescent ultraviolet lamps, or filtered open-flame carbon arcs. An appropriately filtered xenon arc provides the closest simulation to the spectral power distribution of terrestrial solar radiation. The fluorescent UV lamps lack the long-wavelength UV radiation as well as the visible and near-infrared radiation of terrestrial solar radiation, and the SPD of the open-flame carbon arc also differs from the SPD of terrestrial solar radiation. Because of differences in the spectral power distributions of the exposure sources as well as differences in temperature, type, and amount of moisture, and test cycles used in the three types of laboratory weathering devices, tests in these devices may not result in the same performance ranking or types of failure modes of sealants. Thus, they cannot be used interchangeably and comparisons should not be made of the relative stability of sealants exposed in the different types of apparatus. The exposure duration required by each of the three types of exposure devices for testing the weathering performance of sealants may differ.

5.3.1.1 Xenon Arc Devices

The xenon arc has become established worldwide as the radiation source for optimum simulation of terrestrial solar radiation because, when properly filtered, it provides a closer match to the spectral irradiance of terrestrial solar radiation than any other artificial light source. It has the full-spectrum radiation of terrestrial solar radiation, that is, ultraviolet, visible, and near-infrared. Simulation of the SPD of the shortest UV of terrestrial solar radiation is important in the case of sealants primarily sensitive to this radiation. "Daylight" filters are used to screen out wavelengths shorter than about 295 nm to simulate terrestrial daylight in the UV region. *Daylight*, in the context of weathering, refers to the direct beam of solar radiation plus diffuse sky radiation in the UV and visible regions. *Sunlight* refers only to the direct beam of solar radiation, but it is often incorrectly used when referring to daylight. Figure 5.4 compares the SPD of UV and visible radiation of the xenon arc through a daylight-type filter with peak daylight in Miami, Florida, measured at a 26° tilt angle during

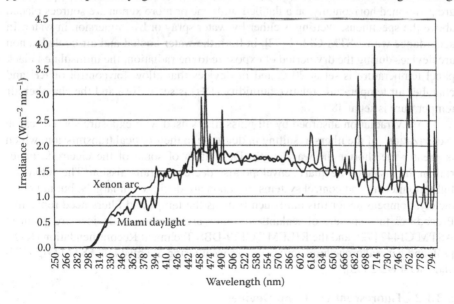

FIGURE 5.4 Spectral irradiance of xenon arc with daylight filters compared with peak daylight in Miami, Florida normalized at 420 nm. (Courtesy of Atlas Material Testing Technology LLC.)

the spring equinox. Filters that screen out wavelengths shorter than 310 nm are used to simulate exposure to daylight through window glass.

Unlike carbon arc exposure devices, which have minimal control over radiation intensity, modern xenon arc devices control irradiance through electrical power management. The irradiance setting historically used in ASTM standards is 0.35 W/$(m^{-2}.$ nm) at 340 nm. The specification was based on simulation of the average solar irradiance at this wavelength received over a full day in South Florida or Arizona under optimum sky conditions. This is only about 50% of the solar irradiance at 340 nm at noon time in these locations and is less than specimens receive between about 10 AM to 2 PM, which is about 0.5 W/$(m^{-2}.$ nm) at 340 nm. Therefore, the irradiance setting that is increasingly being used is 0.51 W/$(m^{-2}.$ nm) at 340 nm, equivalent to the irradiance setting used in ISO standards (550 W/m^2 at 300–800 nm). Higher irradiance levels are also used in some studies.

Two related xenon arc systems have been developed: air-cooled and water-cooled. The type of cooling has some influence on the overall design and the design of the optical filtering system. In the water-cooled xenon arc devices, the light source is positioned vertically in the center of the device and the specimens are rotated on a rack around the axis of the light source. At predefined intervals, water is sprayed on the specimens. The cycle historically used is 102 min light only followed by 18 min light plus water spray on the front surface. However, for increased exposure to moisture, the alternate cycle specified in ASTM C1442 [72], the general weathering standard for sealants, is 2 h light only followed by 2 h light plus wetting. It is more representative of the amount of moisture that sealants are exposed to in humid climates, such as in Miami, Florida. In the air-cooled xenon arc devices, specimens are positioned horizontally on a flat bed with one or more xenon arc sources placed above the specimens. Wetting is either by water spray or by immersion in water. In accordance with ASTM C1442 [72], in both the water-cooled and air-cooled xenon arc devices during the dry period of exposure to the radiation, the uninsulated black panel temperature is set at 70°C and in devices that allow for control of RH and chamber air temperature, relative humidity (RH) is set at 50% and the chamber air temperature is set at 48°C.

The UV radiation absorbed by all glass filters used with exposure sources cause "solarization" of the filters, resulting in decrease in their optical transmission, which is greater the shorter the wavelength. Deposition of some of the electrode material on the inside of the quartz envelope also decreases transmission. The automatic light monitoring and control systems of xenon arc devices can largely, but not completely, compensate for this until such time as the lamps and filters need replacing. Procedures for exposure of sealants in xenon arc weathering devices are given in ASTM C1442 [72], and the RILEM TC139-DBS Technical Recommendation (RTR) [73], and general procedures for xenon arc weathering are given in ASTM G155 [74] and ISO 4892-2 [75].

5.3.1.2 Fluorescent UV Lamp Devices

The emission properties of two types of fluorescent UV lamps, UVB-313 and UVA-340, are shown in Figure 5.5 in comparison with the daylight spectrum. The emission of the UVB lamp has no relation to the spectral power distribution (SPD) of solar

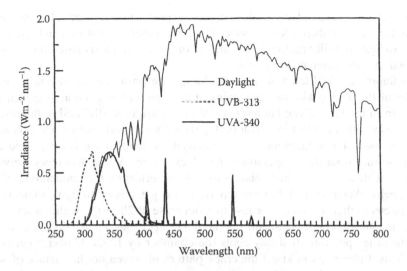

FIGURE 5.5 Spectral irradiance of fluorescent UVB-313 lamps, fluorescent UVA-340 lamps and noon daylight in Miami, Florida, at 26°S exposure during the spring equinox. (Courtesy of Atlas Material Testing Technology LLC.)

radiation on the earth's surface. A large portion of its energy is at shorter wavelengths than are present in terrestrial solar radiation and the emission intensity decreases sharply at wavelengths longer than 313 nm. Most of the energy is in the spectral region below 340 nm. This type of lamp had been used because it could rapidly test the sensitivity of many polymeric materials to UV radiation, owing to the high photon energy of its short wavelengths. However, these highly energetic wavelengths can initiate different mechanisms and types of degradation than the wavelengths present in solar radiation. Because of the presence of shorter-wavelength radiation than is present in terrestrial solar radiation, exposures using these lamps have produced reversals in stability rankings of polymers and errors in the performance of stabilizers compared with outdoor tests.

The UVA-340 lamps have largely replaced UVB lamps in tests on sealants because of the good match of the short UV spectral power distribution to that of solar UV radiation. However, they only provide good simulation of solar radiation in the spectral region from 300 to about 340 nm and, in common with all fluorescent UV lamps, lack the long-wavelength UV, visible and near-infrared radiation present in daylight and xenon arc radiation. Therefore, they lack the actinic effects of full-spectrum radiation on materials sensitive to long-wavelength UV and visible radiation. For example, while short-wavelength UV generally has a greater effect in causing color change in pigments, longer-wavelength UV and visible wavelengths can also have a significant effect. Thus, it can be expected that chrome yellow pigments, which are very sensitive to blue light, would have very different light fastness based on fluorescent UVA-340 exposure than on exposure to solar radiation. Because of the lack of visible and near-infrared radiation, in contrast to the temperature variations of differently colored materials exposed to solar and xenon arc radiation, all materials exposed to fluorescent UV radiation attain the same (ambient) temperature. Thus, for materials that vary in their sensitivity to heat, stability rankings may differ compared with tests based on exposure to solar or solar-simulated radiation.

In fluorescent UV lamp devices, the lamps are positioned horizontally in the center of the device with the specimens in a horizontal position at a defined distance from the lamps. The fluorescent UV lamps are typically used in fluorescent UV/condensation devices in which dark periods with condensation are alternated with periods of UV radiation. The condensation temperature is controlled at a lower setting than the temperature of UV exposure. A key difference between wetting with water spray and condensation is the effect on the temperature of the specimens. Water spray, following a period of dry exposure to light, subjects the specimens to thermal shock because of a decrease of 20–40°C within a few minutes. In contrast, condensation during the dark period has a much milder effect on specimen temperature. It slowly cools the chamber by 10–20°C over a period of 1 h. This difference can affect the crack pattern observed on the surface of sealants. Thermal shock generally results in more numerous and deeper cracks. The fluorescent UV devices may also incorporate short periods of water spray on the front surface of the samples during exposure to radiation for thermal shock and mechanical erosion.

Modern fluorescent UV lamp devices have capabilities for irradiance setting and control. In accordance with ASTM C1442 [72], when used with the UVA-340 lamps, the irradiance is set at 0.77 W/(m^2. nm) at 340 nm in apparatus with irradiance control and the exposure cycle is 8 h UV at an uninsulated black panel temperature of 60°C followed by a dark period of 4 h wetting by condensation at an uninsulated black panel temperature of 50°C. For specimens that are too thick for adequate heat transfer (more than 20 mm including support dimensions) so that wetting by condensation is not applicable, the exposure cycle is 5 h UV only followed by 1 h UV plus wetting by water spray on the front surface. These devices do not control relative humidity. Procedures for exposing sealants in devices with fluorescent UV lamps are given in ASTM Standard Practice C1442 [72] and the RILEM TC139-DBS Technical Recommendation (RTR) [73]. Other standards that describe the use of the fluorescent UV devices for laboratory accelerated weathering are ASTM G154 [19] and ISO 4892-3 [76].

5.3.1.3 Carbon Arc Devices

Figure 5.6 shows representative UV/visible spectral emissions of two types of carbon arc sources compared with the daylight spectrum. The enclosed carbon arc Weather-Ometer™ (Atlas Material Testing Technology LLC, Chicago, Illinois), which had been used for weathering tests of sealants in the past, is not currently used in any ASTM or ISO sealant standards. The SPD differs significantly from that of daylight in both the UV and visible regions. In the UV region, short wavelengths have weaker intensity and long wavelengths have much greater intensity than in daylight. Thus, this device would have a much weaker effect than solar radiation on materials that absorb only short-wavelength UV radiation, but a stronger effect on materials that also absorb long-wavelength UV radiation. Consequently,

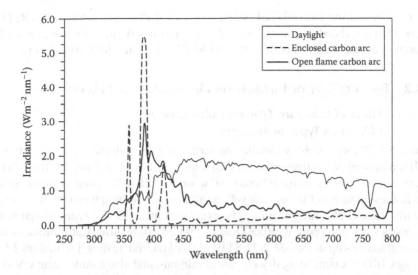

FIGURE 5.6 Spectral irradiance of the enclosed carbon arc, the open-flame carbon arc with Corex D^R filters and noon daylight in Miami, Florida, at 26°S exposure during the spring equinox. (Courtesy of Atlas Material Testing Technology LLC.)

the stability ranking of materials that differ in ultraviolet absorption characteristics in these regions can be expected to be distorted by this source compared with exposure to daylight.

The filtered open-flame carbon arc, also referred to as the Sunshine Carbon Arc™, is one of the artificial light sources used for laboratory accelerated weathering of sealants. It provides a better match to daylight in the 300–360 nm region, and deviates less at wavelengths longer than 360 nm than the enclosed carbon arc. However, it has more intense long-wavelength UV radiation and weaker intensity of visible radiation than daylight. Figure 5.6 shows the spectral emission through Corex D^R type borosilicate glass filters (Atlas Material Testing Technology LLC, Chicago, Illinois). These filters are commonly used to reduce the excessive short-wavelength UV radiation of the unfiltered arc, but they transmit shorter wavelengths than are present in solar radiation. Other types of sharp-cut glass filters can be used to more closely match the solar cut-on or to simulate daylight through window glass. Use of the unfiltered open-flame carbon arc for faster testing has often produced reversals in stability ranking compared to outdoor exposure because of the unnatural short-wavelength UV radiation present.

The carbon arc device has the capability for periodic water spray on the samples and condensation during a dark period as well as humidity and temperature control. Irradiance is not monitored or controlled as in xenon arc devices, but exposure is maintained constant by the requirement of daily replacement of the carbon rods. The open-flame carbon arc uses three pairs of cored carbon rods and operates in a free flow of air. The arc is rotated among the pairs to provide approximately a day's operation per set of rods. The filters are arrayed around the arc. Procedures for using filtered open-flame carbon arc devices for exposures of sealants can be found in ASTM

C1442 [72], and the RILEM TC139-DBS Technical Recommendation (RTR) [73]. Other standards that describe the use of the filtered open-flame carbon arc device for laboratory accelerated weathering are ASTM G152 [77] and ISO 4892-4 [78].

5.3.2 Effect of Type of Exposure on Degradation of Sealants

5.3.2.1 Effect of Exposure Type on Color Change of Various Types of Sealants

Figure 5.7 [79] shows the yellowing, in terms of CIE (International Commission on Illumination) Δb* values, produced in various types of sealants by three types of exposures. Increase in the b* color value of the CIE 1976 color scale measures lightfastness in terms of increase in yellow color, which for many materials occurs at an early stage of photodegradation. The exposures included two types of xenon arc devices, a fully equipped Weather-Ometer (WOM) simulating wet climate conditions and a table-top device, SUNTEST™ (Atlas Material Testing Technology LLC, Chicago, Illinois), simulating dry climate conditions, and also a fluorescent UV condensation device, UVCON™ (Atlas Material Testing Technology LLC, Chicago, Illinois). In the WOM, specimens were subjected to water spray for 18 min in each

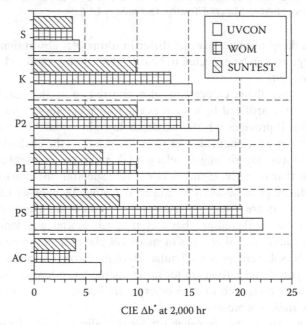

FIGURE 5.7 Dependence of lightfastness (yellowing in terms of CIE Δb*) of sealants (S = silicone; K = kraton-based thermoplastic; P_2 and P_1 = one-component polyurethanes [P_2 softer than P_1]); PS = polysulfide; AC = siliconized acrylic) on type of exposure (UVCON = fluorescent UVA-340; WOM = xenon arc with moisture; SUNTEST = xenon arc without moisture). (Reprinted from G. Wypych, F. Lee and B. Pourdeyhimi, in *Durability of Building and Construction Sealants*, A. T. Wolf (Ed.), pp. 173–197, RILEM Publications S.A.R.L., Cachan, France (2000). With permission of RILEM Publications S.A.R.L.)

2 h exposure period. In the SUNTEST the specimens were exposed to radiation only without added moisture. In both xenon arc devices the irradiance level was 0.35 W/ (m^{-2}. nm) at 340 nm and the uninsulated black panel temperature was 63°C. The UVCON contained fluorescent UVA-340 lamps and was operated using the cycle of 8 h light at 60°C followed by 4 h condensation at 50°C. The irradiance was uncontrolled, but estimated to be 0.7 W/(m^{-2}. nm) at 340 nm. The sealants were applied to solvent-cleaned aluminum panels and cured for 28 days at ambient conditions prior to exposure, which was 2000 h in all three devices.

Regardless of the sealant, CIE Δb* is largest for specimens tested in the UVCON. The more intense yellowing might be attributed to the higher irradiance of short wavelength UV radiation on specimens exposed in this device. However, the specimens were exposed to light only two-thirds of the total exposure time compared with exposure for 100% of the time in the other devices. The larger CIE Δb* can be explained by the difference in the spectral emission properties between the fluorescent UV-340 lamps and the xenon arc source. The fluorescent lamps are deficient in UV wavelengths longer than 350 nm as well as in visible radiation. It is well known that long-wavelength UV and short-wavelength visible radiation have the effect of destroying (photobleaching) the yellow species produced by shorter-wavelength UV radiation. Thus, photobleaching concurrent with the formation of yellow species on exposure in the xenon arc devices is absent during exposure in the fluorescent UV device.

The differences in rates of yellowing by the three types of exposures varied with the type of sealant. Of particular interest is the fact that the two xenon arc tests exhibited different rates of yellowing of the Kraton™-based thermoplastic (K) (Kraton Polymers, Houston, Texas), the two polyurethanes (P$_1$ being a harder polymer than P$_2$), and polysulfide (PS). For all of these polymers, the WOM produced more severe yellowing than the SUNTEST. The difference was largest in the case of polysulfide, the polymer most sensitive to moisture, the weather factor that differed between the two types of xenon arc tests. The presence of moisture during exposure in the WOM also had a significant effect on the stability ranking of these sealants. Tests in the WOM ranked polysulfide as less stable than K and P$_2$, but tests in the SUNTEST, which lacked moisture, ranked polysulfide as more stable than K and P$_2$. The data demonstrate that radiation is not the only important environmental stress factor causing degradation. Other stress factors, such as moisture, can have a very significant influence, depending on the sensitivity of the sealant to moisture.

5.3.2.2 Effect of Type of Exposure on Surface Degradation of Acrylic Sealant

Figure 5.8 [79] shows the differences in surface morphology of an acrylic white sealant exposed for 2000 h in the three types of weathering devices. The UVCON had a much more severe effect on the acrylic surface than either of the xenon arc devices, which caused very little change. The UVCON produced substantial surface erosion and melting of the material at the edges as well as rigid surface layers. It is an abnormal effect compared with outdoor exposure of the sealant. The most logical explanation is that the temperature of the acrylic specimen was higher in the UVCON device. Although the black panel temperatures were approximately the

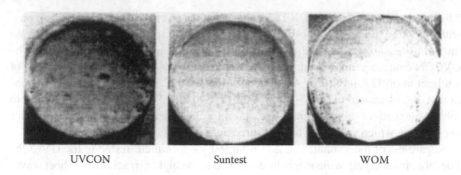

UVCON   Suntest   WOM

FIGURE 5.8 Effect of type of exposure (UVCON = fluorescent UVA-340; Suntest = xenon arc without moisture; WOM = xenon arc with moisture) on surface morphology of acrylic sealant. (Reprinted from G. Wypych, F. Lee and B. Pourdeyhimi, in *Durability of Building and Construction Sealants*, A. T. Wolf (Ed.), pp. 173–197, RILEM Publications S.A.R.L., Cachan, France (2000). With permission of RILEM Publications S.A.R.L.)

same in all the devices, the temperature of a white specimen is lower than that of the black panel in the xenon arc device, but is the same as that of the black panel in the fluorescent UV device. The explanation is based on the spectral differences between the exposure sources. In the xenon arc devices, the temperature of the black panel is largely due to the heat received from the visible and near-infrared radiation absorbed from the source by the black coating on the panel. Only a small portion of the heat is supplied by the air temperature in the chamber, which is lower than that of the black panel. A white specimen in the xenon arc device is at approximately the same temperature as the air in the chamber, since it reflects rather than absorbs the visible and some of the near-infrared radiation. In contrast, the black panel in the fluorescent UV device is heated to the specified temperature primarily by hot air because the source lacks visible and near- infrared radiation. Thus, all specimens, regardless of color, are heated only by the air in the chamber and, therefore, white specimens are at the same temperature as the black panel.

5.3.3 LABORATORY ACCELERATED WEATHERING TEST STANDARDS

5.3.3.1 ASTM Sealant Weathering Standards

ASTM standards for sealants are developed by Committee C 24 on "Building Seals and Sealants." General procedures for exposures of sealants in laboratory acceler- ated weathering devices are given in ASTM C1442 [72], which is referenced by all ASTM C 24 standards that include weathering tests. The standards include test methods, practices, and specifications for determining the effect of laboratory accel- erated weathering on sealants. The use of any one of the three types of weathering devices described in Sections 5.3.1.1 through 5.3.1.3 is allowed, with the caveat that they cannot be used interchangeably, because test results may differ owing to dif- ferences in the exposure sources as well as in the test conditions. The choice of the type of weathering test is by mutual agreement between the interested parties. The type of sealants include building construction sealants (C1501 [80] and C1519 [81]), elastomeric joint sealants, (C510 [82], C793 [83], C794 [84], and C1248 [85]), structural silicone sealants (C1184 [86]), precured elastomeric joint sealants (C1518 [87] and C1523 [88]), EIFS (exterior insulation and finish system) sealants (C1382 [89]), latex sealants (C732 [90] and C734 [91]), and solvent-release-type sealants (C1257 [92]).

5.3.3.1.1   Test Specimen Preparation, Cure, and Exposure

The test sealant may be applied to a variety of types of substrates or tested as a free film. In many of the test methods, the default substrate is anodized aluminum. Some common specimen configurations may include slab, tensile bar, H-block samples, patties, sheets, drawdowns, preformed joint sealants, prevulcanized elastomeric joint materials, beads, channels, and others. For tests on sealants to be used for application in joints, the sealant is adhered to two parallel contact surfaces (substrates). The size and configuration of the specimens are determined by the specifications of the test method used to evaluate the change in properties following exposure. Where practical, it is recommended that specimens be sized to fit the sample holders supplied with the apparatus. Specimens configured for movement during exposure to artificial weathering conditions may also be used. At least four replicates of each sealant being tested are generally prepared. Three replicates are exposed, and one is kept as an unexposed file specimen for comparison with the exposed specimens. When destructive tests are used to evaluate weathering stability, sufficient unexposed file specimens are retained so that the property of interest can be determined on unexposed file specimens each time exposed materials are evaluated.

Prior to exposure, the sealant specimens are cured statically (no movement) in accordance with all of the ASTM C24 weathering standards, but some of the standards offer the option for dynamic cure (exposure to cyclic movement). In most cases, the specimens are cured under standard conditions of $23 \pm 2°C$ and $50 \pm 5\%$ relative humidity for 21 days or less. For some types of configurations, the cure period is as short as 1 or 2 days. In a few standards, the option is given of either 21 days at standard conditions or 21 days consisting of the first and last 7 days at standard conditions and days 8–14 at $38 \pm 2°C$ and 95% relative humidity. After curing, three replicates are exposed to the radiation source, heat, and moisture in one of the three types of exposure devices. At the end of the exposure period, the test sealant is examined for property change in comparison with the unexposed file specimen. It is recommended that a similar material of known performance under use conditions (a control) be exposed simultaneously with the test specimens for evaluation of the performance of the test materials relative to that of the control under the same laboratory exposure conditions. It is preferable to use two control materials, one with relatively poor durability and the other with good durability. Evaluation of the durabilities of test specimens in relation to the durability of a control rather than in terms of absolute change greatly improves the agreement in test results among different laboratories.

5.3.3.1.2   Cyclic Movement in ASTM C1519

Cyclic movements alternating with periods of weathering are included in ASTM C1519 [81]. Because joint sealants are subjected to cyclic mechanical strain in addition to environmental degradation factors during their lifetime, the addition of movement to the weathering tests is considered to be important in assessing the durability of a sealant under simulated use conditions as well as to increase the acceleration of degradation. In the ASTM C1519 test, after specimens are exposed to artificial weathering for 4 weeks, they are placed in a cyclic movement machine and subjected to 6 cyclic movements of extension and compression at room temperature at a rate of 12.7 mm/h. Any degree of extension and compression can be used, but the most common test movements are ± 12.5% and ± 25%. After the movement cycles, the specimens are extended to the prescribed extension, blocked open with appropriate spacers, and examined for any cohesive or interfacial failure and its extent. The depths of any cracks or breaks are measured and recorded and any pertinent observations, such as sealant deformation and bubble formation, are noted. Other measurements such as hardness, tensile strength, elongation, and modulus can be made, if specified. The 4 weeks of weathering followed by cyclic movement and examination are repeated as often as specified.

5.3.3.1.3 Duration of Exposures

Exposure tests are often used to compare sealants for their relative resistance to weathering. The most reliable method of ranking the stability of a group of sealants is the determination of the exposure time or radiant exposure necessary to produce the same defined change in the measured property in each of the sealants. Thus, periodic evaluation of the effect of exposure is necessary to monitor the property change as a function of exposure time or radiant exposure. If a single exposure period is used for the test, the exposure time or radiant exposure that produces the largest performance difference between test materials should be selected.

The validity of a laboratory accelerated weathering test depends on whether the sealants that pass the test will perform adequately in the field and not fail prematurely. The minimum exposure duration should be of sufficient length to produce a statistically significant change of the property evaluated in a sealant known to give poor performance when used in the application of interest. If the exposure duration is not of sufficient length to identify an unacceptable sealant, the test is invalid. Since the exposure duration required for one type of sealant cannot be assumed to be applicable to other types of sealants, specification of exposure duration should be based on experimental determination for each type of sealant. In the past, exposure times specified in many of the standards were inadequate because they were not based on experimental verification of the duration required to ensure that the sealant will not fail prematurely in the field. As new specifications are written, consideration is given to the importance of adequate exposure duration. The ASTM sealant test methods specify minimum exposure times or radiant exposures, and the ASTM specification standard provides the specific exposure duration for the type of sealant.

The minimum exposure times in the weathering test methods range from 100 h to 5000 h. Assuming an acceleration factor (see Section 5.4.2) of 5 for a specific sealant product based on the exposure time required for a property change by the laboratory accelerated test compared with outdoor exposure in South Florida, 4 years of laboratory accelerated weathering exposure would be required to ensure adequate performance for 20 years to South Florida conditions. Since the atmospheric conditions that sealants will be exposed to in most other locations are not as severe as in South Florida, the acceleration factor for the laboratory accelerated test will be somewhat larger and the duration necessary to ensure 20 years of adequate performance under in-service weather conditions will be less than 4 years. However, it is still considerably longer than the exposure durations specified in the current standards. Improvements to the current accelerated test methods are required to provide acceleration factors large enough for practical test times to predict long-term performance of sealants based on accelerated tests. Although addition of cyclic movement can increase the acceleration of degradation, the acceleration factors based on test data in which cyclic movement is included in both laboratory and outdoor tests may not differ significantly from those that have been based on static exposures under both laboratory and natural conditions.

5.3.3.2   ISO 11431 and JIS A1439 Sealant Weathering Standards

The sealant weathering standard of the International Standards Organization, ISO 11431 [93], was developed by Technical Committee ISO/TC59 on Building Construction, Subcommittee SC8 on Jointing Products. The standard is specifically for glazing applications. Specimens are prepared in which the sealant to be tested adheres to two parallel glass surfaces and is generally statically cured for 28 days at standard conditions, $23 \pm 2°C$ and $50 \pm 5\%$ relative humidity. These cure conditions are referred to as Conditioning Method A. As an alternative option for slower curing sealants, Conditioning Method B is also allowed. This method exposes the specimen after completion of Conditioning Method A to an additional 21 days of alternating exposure to dry heat in an oven at $70 \pm 2°C$ and to immersion in water at $23 \pm 2°C$.

The test in a device with automatic cycling consists of a 500 h cyclic exposure of the test specimens to heat and artificial light through the glass substrate for 102 min followed by an 18 min period of exposure to water, either by water spray or immersion, at a black standard thermometer (see Section 5.2.4) temperature of 65 $\pm 5°C$. The wet period can be either with or without exposure to light. The reference (default) method uses a light source that represents the spectral distribution of daylight in the ultraviolet and visible portions as defined in CIE publication No. 85 (1989) [94] with an irradiance of 550 W/m^2 between 290 and 800 nm at the surface of the test specimens. The xenon arc and special metal halide lamps with suitable filters are satisfactory. However, fluorescent UV lamps, a source that deviates from natural light, can also be used.

For tests in which the specimens are transferred manually between the dry radiation chamber and the wet phase chamber, the 504 h of exposure is made up of 3 cycles, each being 168 h and consisting of 5 h of immersion in water at $25 \pm 3°C$ and 19 h of exposure to radiation at $65 \pm 5°C$ each day for 5 days, followed by exposure to radiation at $65 \pm 5°C$ for 2 days. After the 504 h of exposure, the sealants are elongated to the percentage agreed on at a rate of 5.5 ± 0.7 mm per minute. The elongation is maintained for 24 h, and the specimens are examined for evidence of loss of adhesion or cohesion. The depths of any cracks or breaks due to loss of adhesion or cohesion are measured, using a suitable measurement device capable of reading up to 0.5 mm.

The sealant standard of the Japanese Standards Association, JIS A1439 [95], was developed by the Japanese Industrial Standards (JIS) Committee. The weathering test method in the standard is similar to the ISO 11431 [93] weathering test except that the only light source specified is the xenon arc and the light is on during the wet period. Also, test conditions are specified only for a device with automatic cycling. Specimens are subjected to either elongation or shearing deformation prior to examining them for adhesion or cohesion failure. Specimens are cured either by Method A or Method B, with a slight modification of Method B in ISO 11431, and the duration of exposure is 500 h.

5.3.3.3   RILEM Technical Recommendations (RTRs)

5.3.3.3.1   RILEM TC139-DBS RTR for Laboratory Accelerated Weathering

RILEM TC139-DBS Durability of Building Sealants developed a technical recommendation (RTR) on a durability test method for sealants [73] that had been initiated by ISO TC59/SC8 "Jointing Products—Sealants" (Work Group 6) and was in direct support of the ISO group. It describes a laboratory procedure for assessing the effects of cyclic movement and artificial weathering on laboratory-cured elastomeric waterproofing joint sealants for use in high-movement building façade applications. Test specimens are prepared in which the sealant to be tested adheres to two parallel contact surfaces (substrates). The default substrate is anodized aluminum. Although any substrate can be used, the results may not be predictive of adhesion to other substrates. The specimens are conditioned either statically (no movement) for 28 days at standard conditions (Method A) or for an additional 21 days of alternating heat and water immersion (Method B) or dynamically (exposed to cyclic movement) for 2 weeks followed by 2 weeks of static conditioning at standard conditions (Method C).

The conditioned specimens are exposed to repetitive cycles of 8 weeks of artificial weathering to xenon arc radiation as the default or fluorescent UVA-340/water spray as an option followed by two thermomechanical cycles as defined in ISO 9047 [96] using the full amplitude suggested as the movement range of the sealant under test. As an option, prior to thermomechanical cycling, the specimens are subjected to rapid mechanical fatigue cycling at their rated movement capability (default 200 cycles) at a rate of 5 cycles per minute. A thermomechanical cycle consists of 1 week of treatment as follows: Day 1—Refrigeration of the test specimens at −20°C ± 2°C for 24 h with extension of the specimens in the test machine to the required amplitude after the first 3 h; Day 2—At the end of the 24 h, the extension is released and the test specimens are placed in the oven at 70°C ± 2°C for 24 h with compression in the test machine to the required amplitude after the first 3 h; Day 3—The compression is released and the procedure of Day 1 is followed; Day 4—The procedure of Day 2 is followed; and Days 5 to 7—The compression is released and the specimens are stored at 23°C ± 2°C and 50°C ± 5% relative humidity without applying any mechanical force. After the weathering and cyclic movement, the sealants are extended to their full rated extension and examined for any failures as in the ASTM C1519 [81] test or in accordance with ISO/DIS 11600 [97]. The degradation cycle, which consists of the weathering exposure, the cyclic movement, and the examination for failures, is repeated as often as required. The default number of cycles is three, but it should be as many as required to induce a substantial (visible) degradation in a sealant that is known to have poor performance for the application. The repeated exposure of sealant specimens to cycles of artificial weathering and cyclic movement is meant to simulate the effects of weathering by the natural environment on sealants installed in curtain-wall joints exposed to high joint movement. Artificial weathering is not intended to simulate the deterioration caused by localized environmental conditions, such as atmospheric pollution, biological attack, or exposure to salt water.

The RILEM RTR was evaluated and validated by two studies: the first one was carried out by Oxford Brookes University [98] and the second one was carried out by the Tokyo Institute of Technology in Yokohama under the auspices of the Japanese Sealant Industry Association (JSIA) [99]. Both studies included various types of sealants, but the products were different for the two studies. The first study used a fluorescent UV/condensation device for the weathering phase, and the second one used a xenon arc device. The results of the two studies are reported in Chapter 6. The type of failures and changes in surface appearance were similar to those observed in actual service and both studies found that the RILEM test method was able to differentiate between products with regard to their resistance to laboratory accelerated weathering and mechanical cycling. Faster degradation appeared to be induced by the Japanese test conditions using xenon arc exposure. In the British study, after the fourth degradation cycle it became apparent that degradation of a number of sealants was not accelerated as rapidly as had been expected. Therefore, 1000 fatigue cycles with an amplitude of ± 25% at a rate of 5 cycles/minute were added following the fourth durability cycle. In the Japanese study, durability cycles were carried out with and without inclusion of a 200-cycle fatigue aging. Both studies found that fatigue cycling substantially accelerates sealant degradation. It was also shown that the dynamic cure procedure could be used to simulate movement during cure. The RTR method is being considered by ISO TC59/SC8 for the development of a future durability test standard.

5.3.3.3.2   RILEM TC190-SBJ RTRs for Laboratory Accelerated and Natural Weathering

RILEM TC190-SBJ was chartered to continue the work of TC139-DBS on the development of durability test methods for wet-applied (gun-grade) sealant products. Two draft RTR methods based on work initiated by the Architectural Institute of Japan are in the process of being published. One is a laboratory accelerated artificial weathering test [100] and the other is a natural outdoor weathering test [13], both of which include cyclic movement. Both are based on a novel test specimen configuration that allows simultaneous exposure of the sealant to compression and extension during movement cycles. The design differs from any specimen configuration previously used in that the support substrates to which the sealant adheres are riveted to the base plate with pivoted hinges. By extending one end of the joint test specimen, the other end gets compressed so that the test specimen experiences varying amounts of extension/compression along its length in a single movement. For instance, changes in the surface appearance and interfacial or cohesive failure may be reported for 0%,

±10%, ±20%, and ±30% tensile compression/extension movements. Thus, within a single test specimen, sealant durability can be assessed for a range of movement amplitudes. Therefore, it allows reduction in the total number of specimens required to evaluate the different amplitudes of movement and also simplifies the handling operations needed to induce the cyclic movement, which makes simultaneous exposure to weathering and movement cycles more feasible. The latter is most representative of the interaction of the environment with joint sealants under use conditions. The novel design can be used for both laboratory accelerated artificial and outdoor weathering exposures.

In the RTR for laboratory accelerated weathering [100], cyclic movement with sequential exposure to artificial weathering is specified as the default procedure. Weathering is carried out for 6 weeks (default) followed by 6 weeks of movement cycles (default). Alternatively, the option is cyclic movement simultaneous with weathering for 6 weeks. The default exposure source is the xenon arc with daylight filters at an irradiance level of 550 W/m² at 300–800 nm. The exposure consists of alternate periods of 102 min dry at 65°C ± 3°C and 18 min wet either by water spray or immersion in water. Optionally, fluorescent UVA-340 lamps or the filtered open-flame carbon arc may be used. The weekly mechanical cycle, which is repeated for 6 weeks, consists of extension of one end of the specimen for 2 days followed by extension of the other end for 2 days; for the remainder of the week (3 days) the specimen is kept in a relaxed (unstressed) state. The combination of artificial weathering and mechanical movement cycles, carried out either sequentially or simultaneously, constitutes an exposure cycle. After completion of each exposure cycle, the specimens are extended to the full rated movement capability of the sealant tested and visually examined for changes in appearance and for cohesion and/or adhesion failures. The depths of any flaws due to loss of cohesion or adhesion are determined according to the rules in ISO 11600 [97], and the general condition of the sealant is reported. The default value of the total number of degradation cycles is four. However, the minimum number of cycles should be as many as required to induce a substantial (visible) degradation in a sealant that is known to have poor performance for the application.

In the RTR for natural outdoor exposure [13], mechanical movement is carried out simultaneously with the exposure. The extension and compression positions are switched monthly, and the specimens are visually examined every 3 months in the same way as they are examined using the laboratory accelerated weathering test. The default exposure angle for the specimens is 45° facing the equator. Optionally, the specimens can be exposed at the latitude angle. The default exposure duration is 3 years, and the alternative option is 1 year. Because test results of outdoor exposure vary with geographic location as well as with seasonal and annual climate changes, results from a single exposure cannot be used to determine the absolute rate at which a sealant degrades under natural exposure conditions. Therefore, unless the stabilities of a series of sealants exposed at the same time are being compared, the stability of a sealant is determined relative to one or more controls having known performance and similar composition and construction to the test material, and exposed at the same time. It is advisable to use a control that has relatively good durability as well as one that has relatively poor durability.

The objective of the RILEM TC190-SBJ program suggested by the Japanese researchers was to develop an accelerated artificial weathering test method and to compare the results against the results of long-term outdoor weathering. To this end, the investigation undertaken by the Japanese researchers under the auspices of RILEM 190-SBJ included 24 commercially available sealants of seven chemical types exposed in the three laboratory accelerated weathering devices described in the RTR and at three outdoor exposure sites at southern, central, and northern locations in Japan over a period of 20 years. The test parameters in the study [34] deviated in some respects from the RILEM 190-SBJ RTRs. The compression/extension switching interval for the laboratory accelerated weathering test was 500 h instead of the weekly cycle specified in the RTR, and the fluorescent UV exposure consisted of light/dark periods rather than constant light. In addition, the specimens were exposed to less moisture in the fluorescent UV exposure because the condensation procedure was used during the dark period instead of the recommended water spray for wetting during exposure to light.

The results of the initial evaluations [101] indicate that the test methods are capable of differentiating between products with regard to their resistance to accelerated or outdoor weathering and mechanical cycling. The type of failure and the changes in surface appearance observed during the test regimes were similar to those observed in actual service conditions. However, movement did not have as large an effect on the degradation as anticipated. For all sealants, comparisons of the effect of the ±15% versus the ±30% amplitudes based on the scheme used for determining degree of degradation (DOD), that is, multiplication of the quantity by the size of the cracks, showed very little difference, if any. Since the center sections of the specimens experienced some shear movements and were not devoid of movement, it was only possible to compare the effect of tensile cyclic movement on degradation with the effect of shear movement on degradation. However, comparison of the DOD in the section that experienced ±15% movement with the DOD in the center section showed that the increase was only twofold or less for both outdoor exposure in southern Japan and laboratory artificial weathering. The effect of movement varied with the type of sealant. For some sealants ±15% cyclic movement had no effect on the DOD by comparison with the DOD in the central section. Investigations reported earlier in the chapter, including the studies based on the RILEM 139-DBS RTR, indicated that fatigue cycling has a much greater effect than tensile cycling movement on acceleration of degradation. Therefore, future studies should include fatigue cycling after determining the optimum rate of cycling and degree of cyclic movement required to adequately accelerate degradation without distorting the effects produced by natural fatigue cycling in installed sealant joints.

The acceleration factor for xenon arc artificial weathering versus weathering in southern Japan, both without movement, based on the exposure time to produce the same DOD, ranged from less than 1.8 (more than 5000 h in xenon arc weathering versus 1 year outdoors) to ~29 (~300 h versus 1 year) for seven different sealants. The acceleration factor with movement ranged from 4.4 (2000 h versus 1 year) to 22 (400 h versus 1 year). The acceleration factor was largest for the two-part polyurethane and smallest for the one-part polysulfide. The acceleration factor without movement was between 9 and 11 for the other sealants, except for the one-part water-based acrylic, for which it was 18. Cyclic movement had the effect of increasing the acceleration factor of the two-part silicone-modified polyether sealants and the one-part polysulfide by about twofold. Thus, it appears that for these sealants, the cyclic movement had a greater effect on artificial weathering than on natural exposure.

Another RILEM research program on natural and artificial weathering of sealants, which was initiated by RILEM TC139-DBS in 1996 and subsequently carried out under the auspices of RILEM TC190-SBJ, is managed by the Commonwealth Scientific and Industrial Research Organization (CSIRO) of Australia. The study involves 12 gun-grade sealants tested on 10 different substrates and exposed at 8 exposure sites in 6 countries (the United States, Canada, United Kingdom, Singapore, Australia, and New Zealand) under outdoor natural weathering over a period of 10 years and to three different accelerated artificial weathering conditions. All exposures of the sealants are conducted with and without enforced cyclic movement to study the effect of cyclic strain on the degradation rates.

5.4 LABORATORY ACCELERATED VERSUS NATURAL WEATHERING

Appropriate laboratory-accelerated tests have the potential of providing information on the resistance of sealants to natural weathering, and they have played an important role in the development of sealants with highly improved weatherability. However, since all stresses present in an outdoor exposure cannot be simulated in a laboratory accelerated test, the latter cannot replace natural exposure. Therefore, before drawing any final conclusions regarding the ability of a product to withstand the outdoor environment, it is necessary to validate the results of artificial weathering tests by conducting outdoor exposure tests for a reasonable length of time. Two fundamental issues must be considered in sequence when selecting a laboratory accelerated weathering test: correlation and acceleration. Laboratory accelerated weathering is a complementary technique, the usefulness of which largely depends on how closely it reproduces the chemistry and weathering effects, that is, the degradation mechanisms, the failure modes, and stability ranking of various sealants with different durabilities, that result from the slower outdoor exposures. Correlation depends on reasonably representing the critical weather factors, that is, radiation, heat, and moisture, present in the service environment. Any accelerated weathering test, whether it is based on an artificial or natural light source, can only approximate the in-service exposure conditions since it is inherently impossible to both intensify and precisely simulate the critical weather factors and their effects. Because of the synergistic effects of the weather factors and the complex nature of the weathering processes, acceleration by intensification of one factor alone can distort failure modes and stability ranking of materials. When the balance of weather factors is altered, the balance between the various reactions that constitute the weathering processes can change. The closer the simulation of the natural balance of all important weathering factors by the accelerated test, the better the correlation that will be achieved with outdoor tests and end-use service results.

5.4.1   CORRELATION

The term *correlation* refers to the ability of the accelerated test to produce results that agree with real-time outdoor exposures. "Results that agree," as applied to weathering tests, has various meanings. Correlation frequently refers to the relative performance of sealants with different durabilities, that is, their stability ranking. It also often refers to the types of degradation produced. Another criterion of correlation between two types of weathering tests is based on the use of linear regression analysis for comparing the tests in terms of property change as a function of exposure. Qualitative comparison is made of the profiles of the graphed data on property change versus exposure time or radiant exposure. Ideally, results obtained using laboratory accelerated tests should satisfy all correlation criteria when compared with results from natural outdoor exposures in order to provide a reliable early evaluation of the weatherability of a sealant. The extent of correlation between two types of tests also depends on the material property measured as the criterion of degradation.

Poor correlation of laboratory accelerated weathering with outdoor test results can be attributed to inadequate simulation of the spectral power distribution of the natural source. However, when using an appropriately filtered xenon arc source, lack of correlation can be attributed to other factors that are not adequately simulated. These may be the irradiance level, lack of a dark period, temperature of specimens, amount and type of moisture and humidity levels, and dry/wet cycles. Light intensities much higher than those in actual use can change the primary mechanism of degradation and alter the stability ranking of the sealants since some types are more sensitive than others to increase in irradiance. For some materials, continuous exposure to light without a dark period can preclude correlation with natural exposures because of elimination of critical dark reactions that occur in outdoor exposures. Specimen temperatures that are abnormally high relative to those under natural exposure can also alter the primary degradation mechanism and stability ranking of sealants since some types are more temperature sensitive than others. Abnormally high frequency of rapid temperature changes can produce mechanically induced cracking or other types of degradation not seen in exposure under environmental conditions. Moisture, in conjunction with other environmental factors, is very important in causing degradation of many sealants. If the amount and type of moisture to which sealants are exposed in a laboratory accelerated weathering device differ from that under natural exposure, the mechanism of degradation and stability ranking of sealants may be very different. Also, just as the season in which the outdoor weathering test is started can affect the test results, whether the samples are exposed initially to a dry radiation period or to a wet dark period in laboratory tests can alter the type of degradation and stability ranking.

Comparisons based on "timing" exposures in terms of the actinic portion only of the radiant exposure should improve correlations between laboratory accelerated and outdoor exposures just as it has been shown to improve correlations among outdoor tests carried out at different times (see Section 5.2.3). Conditions that give good correlation with one exposure site for one type and form of material and formulation do not necessarily ensure good correlation for others, and may not be applicable to other sites. It is important to note that because of differences in exposure conditions at different outdoor sites, good correlation between sites is also often not attained. Some studies have claimed a high correlation between artificial and outdoor weathering, at least for certain performance properties. In a study by Fedor and Brennan [102], xenon arc and fluorescent UV laboratory accelerated tests reproduced the main features of surface degradation, that is, type of cracking, pitting, and color change in a number of different types of sealants, as observed in South Florida outdoor exposure. Beech and Beasley [103–105] showed highly significant correlation for modulus changes in polyurethane, polysulfide, and silicone sealants between 1 year in a temperate climate in England and 8 weeks in a fluorescent UVB/condensation device. In the study referred to in Sections 3.2.1 and 3.2.2 [79], the exposures on a number of types of sealants in two types of xenon arc devices and a fluorescent UV condensation device were compared with exposure outdoors in South Florida and Arizona [106]. Very good correlation was obtained between xenon arc exposure that included water spray and humidity control and exposure in South Florida. Exposures in the xenon arc table-top device without added moisture were useful in predicting color changes, but did not reflect the cracking behavior that occurs in a moist climate, such as in Florida. Its simulation of Arizona exposure could not be determined because of lack of sufficient data from Arizona tests at the time of the report. Exposure of the sealants in the fluorescent UV device did not correlate with the color change or cracking behavior of the sealants in outdoor exposure.

5.4.2 ACCELERATION FACTORS

The term *correlation* is often erroneously used to describe the relation between the exposure times for laboratory and outdoor weathering required to cause the same extent of degradation. The appropriate term is *acceleration factor*, which is the time to failure (or change in property) under natural environmental exposure divided by the time to failure (or change in property) by the accelerated test, both evaluated by the same criterion of degradation. Acceleration over "real-time" weathering can be accomplished in several ways: (1) by continuous exposure to the same level of critical weather factors, uninterrupted by the diurnal cycle and variations in weather conditions; (2) by continuous exposure to irradiance levels that are the highest encountered outdoors during optimum atmospheric conditions; and (3) by setting temperatures, relative humidity, and thermal and moisture cycles to high stress levels, but not more than the maximum under end-use conditions. Determination of acceleration factors is only valid if the laboratory test does not cause unnatural failure modes or distort stability ranking of the test materials. For accelerated test conditions that allow correlation with outdoor exposure, the acceleration factor is generally less than 10. Unfortunately, typically, the greater the acceleration, the poorer the correlation.

Acceleration factors are material dependent and can be significantly different for each sealant type as well as for different formulations of the same sealant. For example, in a study by Fedor and Brennan [102] on surface degradation and color change of a number of different types of sealants, they found that the exposure time to reproduce the changes observed after 1 year of outdoor exposure in South Florida ranged from 1200 to 2600 h. Thus, the acceleration factors ranged from 3 to 7. Significantly, the largest difference in the acceleration factor was exhibited for different formulations of the same type of sealant material. The two-part polyurethane required 1200 h, whereas the one-part polyurethane required 2600 h. This exemplifies the important fact that the acceleration factor is formulation specific and the relation between a specific type of accelerated test and outdoor weathering cannot be generalized over broad generic classes. Therefore, it is erroneous to attempt to establish a single acceleration factor for a specific type of laboratory accelerated test to be used to predict lifetimes under natural weather conditions for various sealants and their formulations.

Because of the complex nature of the interaction of the weather factors with a material, there is presently no simple way to estimate the acceleration factor for a particular material. For most polymeric materials, the rate of degradation is not simply a linear function of the level of irradiance; the relation between increase in irradiance and rate of degradation varies with the material. In addition, the acceleration factor does not depend only on the irradiance level. Temperature, moisture, and other weather factors also have a significant effect on acceleration of degradation. For these reasons, the increase in irradiance cannot be equated with increase in acceleration of degradation. There is no substitute for experimentally determining the acceleration factor for a given material. A practical approach to determining the acceleration factor for sealants that have a long life expectancy is to compare accelerated with natural weathering during the early stages of degradation using a very sensitive analytical technique for detecting chemical changes [107,108]. These are precursors to the physical property changes that occur later in the weathering processes. The acceleration factor can then be used to extrapolate physical property changes by laboratory accelerated exposure to determine the time to reach the same physical property change under natural exposure conditions. The validity of this technique requires that the degradation mechanisms be the same in the laboratory accelerated and natural exposure and that the same relation exist between the chemical and physical property changes in the two types of exposures.

The acceleration factor for a particular sealant is dependent not only on the type of accelerated test, the specific test parameters, and the property change measured, but also on the geographical, seasonal, and environmental conditions of outdoor exposure. For example, Sandberg [109] reported that for significant degradation of surface appearance (cracking) in polyurethane, the required exposure in the weathering device (fluorescent UVB/condensation) was 500 h to simulate the damage caused by 1 year in the snowy, rainy, cool climate of upper Michigan and 2500 h to simulate the damage caused by 1 year in Hilo, Hawaii. Thus, the acceleration factor ranged from 3 for simulation of exposure in Hawaii to 17 for simulation of exposure in upper Michigan.

Since variability in rate of degradation in both laboratory accelerated and natural exposures can have a significant effect on the acceleration factor calculated, the latter should be based on a sufficient number of exterior and laboratory accelerated exposures. At least two, preferably three, replicate specimens are required for each test so that time to failure in each can be analyzed statistically. Standardizing test conditions with a reference material, such as polystyrene, in devices that provide for control of weather factors can improve reproducibility of laboratory tests. Often, when problems related to reproducing the results of weathering occur, they can be traced to differences in the samples, either because of differences in polymer batches or nonuniformity among the test specimens. Even specimens cut from the same sheet often vary significantly. Nonuniformity in response to weathering is largely due to either nonuniformity within the specimen of the UV-absorbing impurities that initiate the degradation or to the nonuniform distribution of stabilizers. Aliphatic-type polymers are particularly prone to this problem because it is only the impurities in these polymers that are capable of absorbing terrestrial solar radiation. In some aromatic-type polymers, degradation by solar radiation is also due, to a large extent, to the ultraviolet-absorbing impurities present. Preparation of the samples and their treatment during storage can also have a substantial effect on the test results.

5.5   EVALUATION OF WEATHERING EFFECTS

The weathering effects on sealants can be manifested in the surface, bulk, or interface to the substrate. The failure of a sealant in an active joint is usually manifested either by cohesive failure in the sealant or adhesion failure between the sealant and the substrate, or both. Visual evidence of cohesive failure of the sealant or loss of adhesion of the sealant to the substrate is one of the main tests for effect of weathering. In some standards, after weathering, the sealant is elongated [81,93,95] or subjected to shearing deformation [95] for 24 h prior to examining it for cohesive or adhesion failure. When either of these failures is observed, the depth of the cracks is measured to determine how far they have propagated into the bulk of the sealant, in accordance with the requirements defined in ISO 11600 [97].

Weathering of all polymeric materials proceeds from the irradiated surface inward. Therefore, the effects of exposure to environmental factors can be detected sooner by surface-oriented techniques than by methods that measure changes in mechanical (bulk) properties. Surface changes can take the form of discoloration, crazing, cracking, chalking (whitening), loss of gloss, and surface erosion. In most of the standard test methods for determining the resistance of sealants to weathering, evaluations are based on visual appearance of changes. In some standards, sealants are examined just for appearance changes such as cracking, crazing, and chalking, whereas other standards consider color changes of sealants along with visual evidence of other effects of weathering [90,92]. A few standards were written specifically for color changes of sealants and staining of substrates [80,82,85]. In one of these standards [80], the color change is determined quantitatively, either by a visual method or instrumentally.

The degradation of sealants or sealed joints is generally quantified by measuring macroscopic performance indicators, such as the sealant's modulus, tensile strength, or elongation at break. In the case of precured sealants, tear propagation and modulus properties are measured to evaluate the effect of laboratory accelerated weathering (C1518 [87] and C1523 [88]), and tensile adhesion (sealant adhesion under applied tensile force) is measured in structural silicone sealants and EIFS joints (C1184 [86] and C1382 [89]). Modulus is one of the most common properties studied as a function of sealant aging. In most cases, the modulus increases as the sealant ages, but it can also decrease. Effect of weathering on hardness and elongation properties can also be determined. The effect of weathering on the flexibility of the sealant is determined in ASTM C734 [91] and ASTM C793 [83] by bending the sealant around a mandrel at a temperature at or below the freezing point of water and examining it for cracks and adhesion failure.

Many sensitive analytical methods are available for characterizing the chemical and physical changes that occur in sealants in the early stages of weathering [110]. Although these techniques have been used for research studies on sealants, they have not been used for routine monitoring of the aging processes in sealants. Thermoanalytical methods are widely used for the characterization of sealant formulations and are ideally suited for studies of the effects of aging. Since they require little sample preparation and only small amounts of sample, they can be used to identify changes in field-installed sealant joints as well as to identify aging mechanisms in laboratory accelerated weathering studies. Techniques such as differential scanning calorimetry (DSC) [111,112], thermogravimetry/differential thermal analysis (TG/DTA) [113], and dynamic mechanical thermal analysis (DMTA) [114,115] have been used to study the effects of accelerated weathering on sealants. Spectroscopic methods allow for the monitoring of material degradation on a molecular level. Infrared spectroscopy (IR) and x-ray photoelectron spectroscopy (XPS) are the most commonly used spectroscopic techniques for the study of chemical changes induced by aging of sealants [116–120]. The IR spectrum provides information on the chemical structure of a material, particularly on the formation of new functional groups as, for example, due to photooxidation, or reduction of existing functional groups caused by the breaking of chemical bonds. Photoacoustic Fourier Transfer Infrared (PAS-FTIR) spectroscopy was reported by Boettger and Bolte [112] to be the technique of choice for determining chemical changes in sealant materials since surface roughness is not a hindrance to this technique. X-ray photoelectron spectroscopy (XPS) provides information on the atomic composition of surfaces and interfaces. It has been found useful in identifying the cause of interfacially failed tensile adhesion joints prepared from polysulfide sealants based on the XPS spectra of the surface of the glass substrate [116].

5.6   SERVICE LIFE PREDICTION

The service life of a sealant is defined as the length of time between its installation and failure. The sealant fails when a critical property that is required for proper functioning of the sealant falls below a predefined level. The prediction of service life by extrapolation of data obtained under accelerated test conditions is not a simple procedure and is still a major unresolved weathering problem. At the present time, there is no mathematical model or reliable technique for predicting long-term performance of sealants (or any other polymeric material) based on short-term tests [121], a requirement for faster product development cycles. One approach is to obtain an "acceleration factor" between exposure times in the accelerated test and natural weathering during the early stages of weathering by a sensitive analytical technique. Since determination of an acceleration factor based on physical property changes would require an impractical length of time for outdoor exposure, particularly for sealants that have long life expectancy, sensitive analytical techniques for detecting chemical changes [107,108] during the early stages of weathering have been explored.

However, the method is based on several assumptions and requires that the profiles of the graphed property changes versus exposure be similar for the two types of exposures. Simms [122] developed a curve-fitting method to determine the "acceleration shift factor" (ASF) to account for changes in the relation between accelerated and natural exposure with progression of weathering. For the reasons given in Section 5.4.2, the acceleration factor must be determined for each type and formulation of a sealant material as well as for the type of property change measured. Therefore, it is not possible to establish an "acceleration factor" valid for all sealants or even for a sealant material class. Thus, obtaining data on life expectancy of sealants based on acceleration factors is a very labor-intensive procedure.

Regression analysis is another technique for deducing lifetimes under natural weathering from lifetimes obtained by laboratory accelerated tests. It is based on measurements of property changes as a function of exposure by both types of tests. It depends on good linear relations over the period measured or finding the mathematical expression that gives a linear relationship between property change and exposure [123]. However, the same limitations exist as those inherent in using acceleration factors.

There have been many other empirical approaches to predicting service life from accelerated test results, some based on relations between changes in polymer properties and exposure variables [124]. A simple empirical method for estimating lifetimes of polymeric materials by comparison with control materials has been proposed by Fischer and Ketola [125]. The method consists of exposing the test materials in the laboratory accelerated test device simultaneously with several control materials having similar composition and construction as the test material and a known range of failure times outdoors. It requires, as do all service life methods, that the accelerated test produce the same failure modes as natural exposure. In addition, the correlation between the two exposures for the stability ranking of sealants should be very high. If these conditions prevail and the service life of the control materials is well defined, the service life of the test material can be bracketed by two of the control materials.

On the basis of studies by Karpati [25,26,27] that demonstrated that cyclic movement of sealant joints during outdoor weathering is the most important factor influencing their performance, she investigated the effect of laboratory fatigue tests by using vises to extend and compress specimens of a two-part polysulfide sealant to different amplitudes and at different cycling rates [126]. It reproduced all three stages of permanent deformation that had been observed in outdoor exposure using the strain cycling exposure rack, and the results could be plotted in such a manner as to provide an indication of the long-term performance of the sealant. For example, extrapolation of the best fit line showed that after 20 years, movements of ±19% could be achieved and the results were consistent with those obtained from sealants subjected to long-term outdoor exposure with cyclical movement. Lacasse [127] did a similar study on two commercially available polyurethane-based sealants that had been cured for 2 months in the laboratory. Using the same technique as Karpati for plotting the results, a 20 year life expectancy was predicted if the cyclic movement does not exceed ±15% joint movement. However, no attempt has been made yet to correlate these results with long-term outdoor exposure tests. It was pointed out by

Lacasse [127] that the use of an automated fatigue testing device capable of subjecting sealants to much higher rates of cyclic deformation has the potential of reducing the time required to assess the performance of sealants.

The "Standard Guide for Statistical Analysis of Accelerated Service Life Data," ASTM G172 [128], developed by Subcommittee G03.08, describes a methodology for service life prediction based on use of the Eyring Model for multiple stress variables. The objective of the guide is establishment, in an accelerated time frame, of the distribution of frequency (or probability) of failure versus time in service. The Weibull distribution has been found to be the most useful for providing a mathematical description of the lifetime distribution as a function of time. The principal criterion for estimating service life is the shape of the distribution and its position along the time scale axis.

The validity of the estimate of service life depends on the validity of the assumption that the failure mechanism at the higher stress levels of the accelerated test is the same as the failure mechanism under usage conditions. Accurate service life prediction also depends on precise control of the variables in the laboratory accelerated test device in order to obtain useful information for service life prediction. Other methods that have been proposed for service life prediction of sealants are described in Chapter 6, Section 6.6.

5.7 CURRENT VERSUS FUTURE WEATHERING TESTS

The major goal of weathering tests on building seals and sealants is to simulate and accelerate the effects of environmental exposure in order to screen new sealant types and formulations for their resistance to long-term weathering. Although the current natural and accelerated weathering tests have been valuable in contributing to the development of improved sealant products, a reliable method for estimating long-term performance of sealants based on short-term tests is not yet available. However, studies using the current methods have provided a considerable body of knowledge toward the development of methods to eventually satisfy the needs of the sealant industry.

The importance of adequately curing specimens of sealants prior to subjecting them to weathering tests was demonstrated in studies by Beech and Beasley [104]. They showed that, for some sealants, the changes observed during the first year of exposure could be attributed almost entirely to effects of continuing cure of the sealant. When stored under standard laboratory conditions, some sealants were still undergoing marked changes in modulus 12 months after preparation. Large differences were found in the rates of cure of various sealants. The study by Beech and Beasley was in the form of a two-phase investigation. In phase I, freshly prepared sealant specimens were used, whereas in phase II, the tests were repeated using a second set of specimens that had been stored for 12 months under standard laboratory conditions. The changes in properties on exposure to natural and artificial weathering were much smaller for the phase II specimens, but were more readily discerned. Thus, the influence of cure processes on the observed changes in properties during the second phase of the study was markedly less important than in the first phase. Due to differences in the effects of natural and artificial weathering conditions on the cure of the sealants, unless well-cured specimens are exposed, correlations between the two types of weathering are unreliable. Thus, in the design of improved weathering tests of sealants, it is important to utilize information on their cure requirements.

After installation of sealants in joints, they are often exposed to continuous cyclic movement during the curing period, which can be much more than 4 weeks for one-component products [2]. Studies have shown that movement during the curing process causes failures of some sealants at the curing stage, and for others it can have a detrimental effect on the properties of the cured sealant. Many systems that were cycled during cure exhibited a reduction in modulus and extension capability. In a movement-during-cure experimental program, Jones et al. [129] measured rates and magnitudes of joint movements on actual service joints in curtain walls on buildings and fed this information into specially designed stepper-motor-driven cycling devices. Immediately after the sealant was gunned into the joint, it was subjected to simulated movements for 2 weeks followed by 1 week at rest. Control joints were cured in a static state under normal laboratory conditions for the same 3-week period. Dynamic mechanical thermoanalytical (DMTA) measurements were made on bulk sealant samples to monitor the development of cure, and the key properties of "25% modulus" and "extension at peak load" were used to assess the mechanical performance of the joint sealants following cure. The strains used were ±7.5% and ±12.5% at fixed rates of both 1 and 10 cycles per day at a constant temperature of 21°C. In addition, temperature cycles subjecting the joints to an elevated temperature during compression and a reduced temperature during tension were applied. Results from the mechanical cycling during cure showed that the performance of tensile adhesion joints prepared from one-part systems was significantly reduced, but the effect on two-part systems was minimal. The DMTA tests showed that movement during cure increases the rate of cure for a one-part sealant, but has no effect on the rate of cure of a two-part sealant. In the joints that were subjected to temperature cycling, larger reduction in mechanical properties resulted than in joints kept at a fixed temperature. In polyurethane and silicone sealants, voids were created in the sealant bead, which were deformed permanently, and thus significant reductions in apparent modulus were measured. Sealant joint properties were significantly affected even at low rates and amplitudes and performance deteriorated with increase in movement rate or amplitude.

Thus, studies have demonstrated that movement during cure can be a significant factor in reducing the performance of joint sealants, and it is a real problem that occurs in high-movement curtain wall systems. It is now becoming evident that joint movement during cure, that is, dynamic cure of sealants, should be used as a conditioning procedure and may be used as a screening process for the selection of sealants used in high-movement applications. However, further studies are required to determine the optimum type and rate of cycling for various types of sealants and the effect of temperature control in conjunction with dynamic cure. Since state-of-the-art automated test equipment can monitor joint movements under in-service conditions, its use should be investigated for subjecting joint sealants to a dynamic cure that uses real-time movement parameters.

It has been pointed out (see Section 5.3.3.1.3) that the exposure durations specified in the current sealant weathering standards were not verified experimentally and many are too short to be able to identify a product that will fail prematurely in the field. The minimum exposure duration should be determined experimentally for each type of sealant since the duration sufficient to identify an unacceptable product of one type of sealant may not be long enough for other types of sealants. In most of the current standards, the same duration is specified for the two or three types of exposures included in the standard. An exposure duration determined to be appropriate for specific test conditions in one type of laboratory accelerated device may not be applicable using other test conditions or another type of laboratory accelerated device. Therefore, for future weathering tests, the duration appropriate for the specific laboratory accelerated test and type of sealant should be determined.

The importance of simulating the main environmental stress factors, that is, the full spectrum of solar radiation, the temperatures and moisture conditions and their variations that sealants are subjected to under use conditions, has been well established. However, studies have shown that intensification of the stress factors by means of laboratory accelerated weathering tests has had only a relatively moderate influence on acceleration of the degradation produced by exposures in outdoor benchmark environments such as in South Florida and in Arizona, where the stress factors are maximized. Acceleration factors based on the current laboratory accelerated weathering tests generally range from about 3 to 10, but can be considerably higher in some cases. Unless the acceleration factor for the material of interest is much larger than 10, the current weathering tests do not provide sufficient acceleration for estimating performance for 20 years or more based on a practical period of exposure in a laboratory accelerated device. Therefore, improvements to the current accelerated test methods are required to reduce test times for sealants designed for long-term performance.

Over the past 20 or more years, it has become increasingly evident that cyclic movement is a major factor in the effect of the environment on sealants, particularly on joint sealants. Both daily and annual cycles of expansion and compression are caused by variations in the temperature of the sealants and in their moisture content. It leads to permanent deformation and eventually to failure of the sealant for the function it is meant to perform. Investigations to determine the effect of added cyclic movement during outdoor exposures have shown its importance for reliably assessing the performance of sealants as well as for acceleration of the test results [2]. In comparison with in-service deterioration, outdoor exposure tests without mechanical cycling accelerate only the surface changes. However, when sealant specimens are subjected to periodic expansion and compression during weathering, bulk deterioration of the sealant is also accelerated. Karpati [130] showed that with the use of a specially designed strain-cycling exposure rack (SCR) for outdoor exposure that could simultaneously subject sealant specimens to cyclic movement and temperature change, failures of various latex and solvent-borne acrylic sealant products started to occur after only a few months of exposure. Tests on one-part chemically cured silicone sealants using this device showed that strain-cycling movement is the predominant factor that causes failure during weathering of these sealants and that outdoor weathering alone has a negligible effect [25]. A procedure was developed for the evaluation of silicone sealants in which unsatisfactory products could be identified within a 2-month period [131]. The types of failures were similar to those that occur during in-service use of sealants. In the Boettger and Bolte studies to determine useful performance test methods [112], model joint sealants were subjected to mechanical strains in both laboratory accelerated and outdoor exposures. The studies showed that artificial aging with extension/compression cycling most closely resembled in-service conditions as opposed to artificial aging without movement. Thus, both outdoor and laboratory accelerated weathering tests of sealants should include mechanical cycling, either sequentially or simultaneously with weathering, but preferably the latter, which more closely simulates in-service conditions.

Although mechanical cycling has been shown to increase the acceleration of degradation provided by the current outdoor and laboratory accelerated exposures alone, the increase is too small to sufficiently reduce the exposure time required to evaluate the performance of sealants expected to be useful for at least 20 years. In the study of laboratory accelerated weathering of sealants in conjunction with sequential cyclic compression/extension movements carried out by Oxford Brookes University [98] in accordance with the protocol of the RILEM TC 139-DBS durability test method [73], four durability cycles (each cycle consisting of 8 weeks of artificial weathering in a fluorescent UV/condensation device followed by 4 days of cyclic compression/extension movements) were not sufficient to determine the weatherability of the sealants. However, addition of 1000 rapid fatigue cycles with amplitude of ±25% at a rate of 5 cycles/minute following the fourth durability cycle did degrade the sealants sufficiently to allow their relative performance to be evaluated. The Tokyo Institute of Technology study in accordance with the protocol of the RILEM TC 139-DBS durability test method [73] was carried out with and without the added option of the 200-cycle fatigue aging after the exposure period and before the thermomechanical cycling [99]. The study clearly showed that sealants are more likely to fail when fatigue cycling is included. In addition, the investigators concluded that it is important to include fatigue cycling in order to determine the performance that the sealants would exhibit in actual service conditions. Based on these two studies, fatigue cycling should be added to natural and laboratory accelerated exposures to significantly accelerate the testing of sealants and to provide the potential for reliably predicting their long-term performance. Studies to determine the appropriate type of fatigue cycling are required. The pattern and rate of movement of service joints can be reliably monitored, and this information can be used to build devices that expose sealant test joints to the same movement patterns as observed in the actual service joints.

Accelerated outdoor weathering using the Fresnel-reflector technique has not yet been investigated to any large extent for application to the aging behavior of sealants for predicting long-term performance. Exposure to intensified solar radiation in conjunction with mechanical and/or fatigue cycling should increase the acceleration factors that are obtained by current techniques as well as simulate the type of surface and bulk changes caused by in-service conditions.

5.8 SUMMARY

The weatherability of sealants is an important consideration in development of new and improved products. The environmental elements mainly responsible for changes in appearance and mechanical properties of sealants, namely, solar radiation, heat/

cold, moisture, oxygen, and atmospheric contaminants, are reviewed and their effects described. The current natural and accelerated environmental test methods are described and the importance of including cyclic movement is discussed. The types of laboratory accelerated tests used to simulate the effects of weather on sealants are reviewed and the importance of simulating the full spectral power distribution of solar radiation and the presence of moisture is demonstrated by examples.

Correlation and acceleration are fundamental issues in selecting a laboratory accelerated test. Good correlation between artificial and natural weathering is based on reproducing the type and mechanism of degradation and stability ranking of sealants resulting from environmental exposure. Factors that contribute to poor correlation are discussed. The acceleration factor, that is, the relation between the exposure times by artificial and natural weathering required to produce equivalent changes in the property measured, varies with the type of material and its formulation. Therefore, it is not possible to assign a general acceleration factor to a laboratory accelerated test. Acceleration factors have ranged from less than 2 to 10 or larger for various types of polymeric formulations. Since each material responds differently to increase in irradiance and temperature as well as to moisture and other exposure parameters and the acceleration factor can vary with the property change measured, it cannot be predicted, but must be determined experimentally for each material and property change.

Methods of evaluating weathering effects on sealants as well as sensitive analytical techniques for determining chemical changes during the early stages of weathering are reviewed. The latter techniques have the potential for use in predicting service life of sealants through determination of acceleration factors. Other methods that have been explored for service life prediction are summarized. Prediction of long-term performance of sealants (or any other polymeric material) is still a major unsolved weathering problem. Suggestions are given for future modifications to natural and artificial weathering tests of sealants to provide the capability of predicting their long-term performance.

REFERENCES

1. N. D. Searle, "Weathering", in *Encyclopedia of Polymer Science and Technology*, 3rd ed., Vol. 4, J.I. Kroschwitz, (Ed.), pp. 629–659, Wiley-Interscience, Hooboken, NJ (2003).
2. T. G. B. Jones and M. A. Lacasse, in *Durability of Building Sealants*, RILEM Report 21, A. T. Wolf (Ed.), pp. 73–105, RILEM Publications, Cachan, France (1999).
3. A. Wolf, *Polym. Degrad. Stab.*, 23, 135–163 (1989).
4. J. C. Beech and C. H. C. Turner, *Building Research and Practice*, 11(5), 287–291 (1983).
5. A. R. Hutchinson, T. G. B. Jones and K. E. Atkinson, in *Proceedings of the International RILEM Symposium on Durability of Building Sealants*, A. T. Wolf (Ed.), pp. 99–116, E & FN Spon, London (1999).
6. ASTM G173-03, Standard Tables for Reference Solar Spectra Irradiances: Direct Normal and Hemispherical on 37° Tilted Surface, Annual Book of ASTM Standards, Vol. 14.04.
7. N. Robinson, *Solar Radiation*, pp. 161–195, Elsevier Science Publishing Co., New York (1966).

8. R. M. Fischer and W. D. Ketola, in *Service Life Prediction: Challenging the Status Quo*, J. W. Martin, R. A. Ryntz and R. A. Dickie (Eds.), pp. 79–92, Federation of Societies for Coatings Technology, Blue Bell, PA (2005).

9. M. E. Nichols and C. A. Darr, *ACS Symposium Series*, 722, 333–353 (1999).

10. D. Patil, R.D. Gilbert and R.E. Fornes, *J. Appl. Polym. Sci.*, 41, 1641–1650 (1990).

11. ASTM G113-06, Standard Terminology Relating to Natural and Artificial Weathering Tests of Nonmetallic Materials, Annual Book of ASTM Standards, Vol. 14.04.

12. ASTM C1589-05, Standard Practice for Outdoor Weathering of Construction Seals and Sealants, Annual Book of ASTM Standards, Vol. 04.07.

13. A. T. Wolf, *Materials and Structures*, 41(9), 1487–1495 (2008).

14. M. J. Crewdson, in *Durability of Building Sealants*, RILEM Report 21, A. T. Wolf (Ed.), pp. 173–180, RILEM Publications, Cachan, France (1999).

15. K. Hardcastle and N. D. Searle, in *Plastics and Coatings, Durability—Stabilization Testing*, R. A. Ryntz (Ed.), pp. 189–240, Hanser Gardner Publishers, Cincinnatti, OH (2000).

16. ASTM G7-05, Standard Practice for Atmospheric Environmental Exposure Testing of Nonmetallic Materials, Annual Book of ASTM Standards, Vol. 14.04.

17. ASTM G24-05, Standard Practice for Conducting Exposures to Daylight Filtered Through Glass, Annual Book of ASTM Standards, Vol. 14.04.

18. ISO 877-94, Plastics—Methods of Exposure to Direct Weathering, to Weathering Using Glass-Filtered Daylight, and to Intensified Weathering by Daylight Using Fresnel Mirrors.

19. ASTM G154-06, Standard Practice for Operating Fluorescent Light Apparatus for UV Exposure of Nonmetallic Materials, Annual Book of ASTM Standards, Vol. 14.04.

20. W. D. Ketola and J. S. Robbins, III, in *Accelerated and Outdoor Durability Testing of Organic Materials*, W. D. Ketola and D. Grossman (Eds.), pp. 133–151, Special Technical Publication (STP) 1202, ASTM International (1994).

21. G. C. Newland, R. M. Schulken, Jr. and J. W. Tamblyn, *Mater. Res. Stand.*, 3, 487–488 (1963).

22. M. M. Quayyum and A. Davis, *Polym. Degrad. Stab.*, 6, 201–209 (1984).

23. R. C. Hirt, R. G. Schmidt, N. D. Searle and A. P. Sullivan, *J. Opt. Soc. Am.* 50, 704–713 (1960).

24. K. K. Karpati, *Adhesives Age*, 16 (11), 27–30 (1973).

25. K. K. Karpati, *Adhesives Age*, 23 (1), 41–47 (1980).

26. K. K. Karpati, *J. Coatings Technol.*, 56 (719), 57–60 (1984).

27. K. K. Karpati, *Materials and Structures*, 22, 60–63 (1989).

28. J. C. Beech, *Materials and Structures*, 18, 473–482 (1985).

29. H. E. Ashton, in *Permanence of Organic Coatings*, G. G. Schurr (Ed.), pp. 67–85, Special Technical Publication (STP) 781, ASTM International (1982).

30. J. I. Fry and R. S. Whitney, BRANZ Technical Paper P26, Building Research Association of New Zealand, Judgeford, NZ, 22–26 (1979).

31. D. H. Nicastro and J. P. Solinski, *The Construction Specifier*, 4, 51–62 (1997).

32. W. R.Sharman, J. I. Fry and R. S. Whitney, *Durability of Building Materials*, 2, 79–90 (1983).

33. K. K. Karpati, *J. Coatings Technol.*, 50 (641), 27–30 (1978).

34. N. Enomoto, A. Ito, I. Shimizu, T. Matsumura, Y. Takane, and K. Tanaka, in *Durability of Building and Construction Sealants and Adhesives*, 2nd edition, A. T. Wolf (Ed.), pp. 82–90. Special Technical Publication (STP) 1488, ASTM International (2007).

35. S. Lacher, R. S. Williams, C. Halpin and C. White, in *Service Life Prediction: Challenging the Status Quo*, J. W. Martin, R. A. Ryntz and R. A. Dickie (Eds.), pp. 207–216, Federation of Societies for Coatings Technology, Blue Bell, PA (2005).

36. R. S. Williams, A. Sanadi, C. Halpin and C. White, in *Proceedings 3rd International Woodcoatings Congress*, pp. 25–30, The Hague, The Netherlands (2002).

37. R. S. Williams, S. Lacher and C. Halpin, in *Proceedings 3rd International Symposium on Service Life Prediction*, Sedona, AZ (2004).
38. ASTM G 90-05, Standard Practice for Performing Accelerated Outdoor Weathering of Nonmetallic Materials Using Concentrated Natural Sunlight, Annual Book of ASTM Standards, Vol. 14.04.
39. ASTM D4364-02, Standard Practice for Performing Outdoor Accelerated Weathering Tests of Plastics Using Concentrated Sunlight, Annual Book of ASTM Standards, Vol. 08.02.
40. B. L. Garner and P. J. Papillo, *Ind. Eng. Chem. Prod. Res. Dev.,* 1(4), 249–253 (1962).
41. M. P. Morse in *Permanence of Organic Materials,* G. G. Schurr (Ed.), pp. 43–66, Special Technical Publication (STP) 781, ASTM International (1982).
42. G. A. Zerlaut and M. L. Ellinger, *J. Oil Colour Chem. Assoc.* 64(10), 387–397 (1981).
43. G. A. Zerlaut in *Testing of Polymers*, Vol. 4, W. E. Brown (Ed.), pp. 10–34, Wiley-Interscience, New York (1969).
44. R. J. Martinovich and G. R. Hill in *Weatherability of Plastic Materials*, M. R. Kamal (Ed.), pp. 141–154, Applied Polymer Symposium No. 4, Wiley-Interscience, New York (1967).
45. J. B. Howard and H. M. Gilroy, *Polym. Eng. Sci.,* 9, 286–294 (1969).
46. C. R. Caryl, in *Testing of Polymers*, Vol. 4, W. E. Brown (Ed.), pp. 379–397, Wiley-Interscience, New York (1969).
47. G. A. Zerlaut, in *Accelerated and Outdoor Durability Testing of Organic Materials,* W. D. Ketola and D. Grossman (Eds.), pp. 3–26, Special Technical Publication (STP) 1202, ASTM International (1994).
48. G. A. Zerlaut, M.W. Rupp, and T.E. Anderson, *Paper 850378, Proceedings of SAE Int'l Congress*, Detroit (1985).
49. R. W. Singleton, R. K. Kunkel, and B. S. Sprague, *Textile Res. J.,* 35, 228–237 (1965).
50. R. W. Singleton and P. A. C. Cook, *Textile Res. J.,* 39, 43–49 (1969).
51. G. A. Zerlaut, in *Permanence of Organic Coatings,* G. G. Schurr (Ed.), pp. 10–34, Special Technical Publication (STP) 781, ASTM International (1982).
52. B. Zahradnik and B. Juriaanse, Preprint, *ANTEC '84, 42nd Annual Technical Conference*, 397–400 (1984).
53. N. D. Searle, in *Handbook of Polymer Degradation*, 2nd edition, S.H. Hamid (Ed.), pp. 605–643, Marcel Dekker, New York (2000).
54. ASTM G178-03, Standard Practice for Determining the Activation Spectrum of a Material (Wavelength Sensitivity to an Exposure Source) Using the Sharp Cut-On Filter or Spectrographic Technique, Annual Book of ASTM Standards, Vol. 14.04.
55. ISO 4892-1: 1999, Plastics—Methods of Exposure to Laboratory Light Sources—Part 1: "General guidance."
56. ASTM G 151-06, Standard Practice for Exposing Nonmetallic Materials in *Accelerated Test Devices that Use Laboratory Light Sources*, Annual Book of ASTM Standards, Vol. 14.04.
57. ASTM G 84-05, Practice for Measurement of Time-of-Wetness on Surfaces Exposed to Wetting Conditions as in Atmospheric Corrosion Testing, Annual Book of ASTM Standards, Vol. 3.02.
58. WMO, Guide to Meteorological Instruments and Methods of Observation, 5th ed. OMM Vol. No. 8, Secretariat of the World Meteorological Organization, Geneva, Switzerland (1983).
59. ASTM E 824-05, Standard Test Method for Transfer of Calibration from Reference to Field Radiometers, Annual Book of ASTM Standards, Vol. 14.04.
60. ISO 9060:1990, Solar Energy—Specification and Classification of Instruments for Measuring Hemispherical Solar and Direct Solar Radiation.
61. ISO 9847:1992, Solar Energy—Calibration of Field Pyranometers by Comparison to a Reference Pyranometer.

62. ASTM G 167-05, Standard Test Method for Calibration of a Pyranometer Using a Pyrheliometer, Annual Book of ASTM Standards, Vol. 14.04.
63. ISO 9846:1993, Solar Energy—Calibration of a Pyranometer Using a Pyrheliometer.
64. ASTM E 816-05, Standard Test Method for Calibration of Pyrheliometers by Comparison to Reference Pyrheliometers, Annual Book of ASTM Standards, Vol. 14.04.
65. ISO 9059:1990, Solar Energy—Calibration of Field Pyrheliometers by Comparison to a Reference Pyrheliometer.
66. ASTM G 130-06, Standard Test Method for Calibration of Narrow- and Broad-Band Ultraviolet Radiometers Using a Spectroradiometer, Annual Book of ASTM Standards, Vol. 14.04.
67. ASTM D 1435-05, Standard Practice for Outdoor Weathering of Plastics, Annual Book of ASTM Standards, Vol. 08.01.
68. ASTM G 183-05, Standard Practice for Field Use of Pyranometers, Pyrheliometers and Radiometers, Annual Book of ASTM Standards, Vol. 14.04.
69. ISO 9370:1997, Plastics—Instrumental Determination of Radiant Exposure in Weathering Tests—General Guidance and Basic Test Method.
70. N. D. Searle, in *Durability of Building and Construction Sealants and Adhesives*, A. T. Wolf (Ed.), pp. 355–371, Special Technical Publication (STP) 1453, ASTM International (2004).
71. N. D. Searle, in *Accelerated and Outdoor Durability Testing of Organic Materials*, W. D. Ketola and D. Grossman (Eds.), pp. 133–151, Special Technical Publication (STP) 1202, ASTM International (1994).
72. ASTM C 1442-06, Standard Practice for Conducting Tests on Sealants Using Artificial Weathering Apparatus, Annual Book of ASTM Standards, Vol. 04.07.
73. A. T. Wolf, *Materials and Structures*, 34, 579–588 (2001).
74. ASTM G 155-05a, Standard Practice for Operating Xenon Arc Light Apparatus for Exposure of Nonmetallic Materials, Annual Book of ASTM Standards, Vol. 14.04.
75. ISO 4892-2:2006, Plastics—Methods of Exposure to Laboratory Light Sources—Part 2: "Xenon Arc Lamps."
76. ISO 4892-3:2006, Plastics—Methods of Exposure to Laboratory Light Sources—Part 3: "Fluorescent UV Lamps."
77. ASTM G 152-06, Standard Practice for Operating Open Flame Carbon Arc Light Apparatus for Exposure of Nonmetallic Materials, Annual Book of ASTM Standards, Vol. 14.04.
78. ISO 4892-4:2004, Plastics—Methods of Exposure to Laboratory Light Sources—Part 4: "Open-flame Carbon Arc Lamps."
79. G. Wypych, F. Lee and B. Pourdeyhimi, in *Durability of Building and Construction Sealants*, A. T. Wolf (Ed.), pp. 173–197, RILEM Publications S.A.R.L., Cachan, France (2000).
80. ASTM C1501-04, Standard Test Method for Color Stability of Building Construction Sealants as Determined by Laboratory Accelerated Weathering Procedures, Annual Book of ASTM Standards, Vol. 04.07.
81. ASTM C1519-04, Standard Practice for Evaluating Durability of Building Construction Sealants by Laboratory Accelerated Weathering Procedures, Annual Book of ASTM Standards, Vol. 04.07.
82. ASTM C510-05a, Standard Test Method for Staining and Color Change of Single- or Multicomponent Joint Sealants, Annual Book of ASTM Standards, Vol. 04.07.
83. ASTM C793-05, Standard Test Method for Effects of Laboratory Accelerated Weathering on Elastomeric Joint Sealants, Annual Book of ASTM Standards, Vol. 04.07.
84. ASTM C794-06, Standard Test Method for Adhesion-In-Peel of Elastomeric Joint Sealants, Annual Book of ASTM Standards, Vol. 04.07.
85. ASTM C1248-05, Standard Test Method for Staining of Porous Substrates by Joint Sealants, Annual Book of ASTM Standards, Vol. 04.07.

86. ASTM C1184-05, Standard Specification for Structural Silicone Sealants, Annual Book of ASTM Standards, Vol. 04.07.
87. ASTM C1518-04, Standard Specification for Precured Elastomeric Silicone Joint Sealants, Annual Book of ASTM Standards, Vol. 04.07.
88. ASTM C1523-04, Standard Test Method for Determining Modulus, Tear and Adhesion Properties of Precured Elastomeric Joint Sealants, Annual Book of ASTM Standards, Vol. 04.07.
89. ASTM C1382-05, Standard Test Method for Determining Tensile Adhesion Properties of Sealants When Used in Exterior Insulation and Finish Systems (EIFS) Joints, Annual Book of ASTM Standards, Vol. 04.07.
90. ASTM C732-06, Standard Test Method for Aging Effects of Artificial Weathering on Latex Sealants, Annual Book of ASTM Standards, Vol. 04.07.
91. ASTM C734-06, Standard Test Method for Low-Temperature Flexibility of Latex Sealants after Artificial Weathering, Annual Book of ASTM Standards, Vol. 04.07.
92. ASTM C1257-06a, Standard Test Method for Accelerated Weathering of Solvent-Release-Type Sealants, Annual Book of ASTM Standards, Vol. 04.07.
93. ISO 11431:2002, Building Construction—Jointing Products—Determination of Adhesion/Cohesion Properties of Sealants after Exposure to Heat, Water and Artificial Light Through Glass.
94. CIE Publication No. 85:1989, Recommendations for the Integrated Irradiance and the Spectral Distribution of Simulated Radiation for Testing Purposes; Solar Spectral Irradiance, ISBN 3 900734 224.
95. JIS A 1439–1997, Test Methods of Sealants for Sealing and Glazing in Buildings.
96. ISO 9047: 2001, Building Construction—Jointing Products—Determination of Adhesion/Cohesion Properties of Sealants at Variable Temperatures.
97. ISO 11600:2002, Building Construction—Jointing Products—Classification and Requirements for Sealants.
98. T. G. B. Jones, A. R. Hutchinson and A. T. Wolf, *Materials and Structures*, 34, 332–341 (2001).
99. H. Miyauchi, N. Enomoto, S. Sugiyama and K. Tanaka, in *Durability of Building and Construction Sealants and Adhesives*, A. T. Wolf (Ed.), pp. 206–212, Special Technical Publication (STP) 1453, ASTM International (2004).
100. A. T. Wolf, *Materials and Structures*, 41(9), 1497–1508 (2008).
101. N. Enomoto, Test Results for Weathering of Sealants, presentation at RILEM TC190-SBJ meeting, Berlin, Germany (Oct. 9, 2006).
102. G. R. Fedor and P. Brennan, *Adhesives Age*, 33 (5), 25–29 (1990).
103. J. C. Beech and J. L. Beasley, in *Science and Technology of Building Seals, Sealants, Glazing, and Waterproofing*, Third Volume, J. C. Myers (Ed.), pp. 33–50, Special Technical Publication (STP) 1254, ASTM International (1994).
104. J. C. Beech and J. L. Beasley, in *Science and Technology of Building Seals, Sealants, Glazing, and Waterproofing*, Fourth Volume, D. H. Nicastro (Ed.), pp. 65–76, Special Technical Publication (STP) 1243, ASTM International (1994).
105. J. L. Beasley, in *Durability of Building Sealants, RILEM Proceedings* 28, J. C. Beech and A. T. Wolf (Eds.), pp. 17–25, E & FN Spon., London (1996).
106. G. Wypych, S. Kuberski and F. Lee, in *Durability of Building and Construction Sealants and Adhesives*, A. T. Wolf (Ed.), pp. 310–321, Special Technical Publication (STP) 1453, ASTM International (2004).
107. D. R. Bauer, J. L. Gerlock, and R. A. Dickie, *Prog. Organic Coatings*, 15, 209–221 (1987).
108. J. L. Gerlock, D. F. Mielewski, and D. R. Bauer, *Polym. Degrad. Stab.*, 20, 123–134 (1988).
109. L. B. Sandberg, *J. Materials in Civil Engineering*, 3 (4), 278–291 (1991).
110. A. T. Wolf and K. Oba, in *Durability of Building Sealants*, RILEM Report 21, A. T. Wolf (Ed.), pp. 323–342, RILEM Publications, Cachan, France (1999).

111. R. Squire, Chemical Changes in Roof Membranes and Building Sealants, Building Research Establishment Occasional Paper, OP 60, Building Research Establishment, Garston, U.K. (1994).

112. T. Boettger and H. Bolte in *Science and Technology of Building Seals, Sealants, Glazing, and Waterproofing*, Seventh Volume, J. M. Klosowski (Ed.), pp. 66–80, Special Technical Publication (STP) 1334, ASTM International (1998).

113. M. A. Lacasse and R. M. Paroli, in *Science and Technology of Building Seals, Sealants, Glazing, and Waterproofing*, Fourth Volume, D. H. Nicastro (Ed.), pp. 65–76, Special Technical Publication (STP) 1243, ASTM International (1994).

114. T. M. Malik and N. E. Leonard, in *Science and Technology of Building Seals, Sealants, Glazing, and Waterproofing*, Third Volume, J. C. Myers (Ed.), pp. 21–32, Special Technical Publication (STP) 1254, ASTM International (1994).

115. T. M. Malik, *Polym. Mater. Sci. Eng.*, 75, 291–292 (1996).

116. G. B. Lowe, T. C. P. Lee, J. Comyn and K. Huddersman, *Intl. J. Adhesion Adhesives*, 14, 85–92 (1994).

117. R. M. Paroli, A. H. Delgado and K. C. Cole, *Canadian J. Appl. Spectroscopy*, 39, 7–14 (1994).

118. R. M. Paroli and A. H. Delgado, *American Chemical Society Symposium Series*, 581, 129–148 (1994).

119. R. M. Paroli, K. C. Cole and A. H. Delgado, *American Chemical Society Symposium Series*, 598, 117–136 (1995).

120. I. Hatsuo, *J. Adhesives and Sealants Council*, 1, 521–533 (1996).

121. A. T. Wolf, in *Durability of Building and Construction Sealants and Adhesives*, A. T. Wolf (Ed.), pp. 385–398, Special Technical Publication (STP) 1453, ASTM International (2004).

122. J.A. Simms, *J. Coatings Technol.* 59 (748), 45–53 (1987).

123. F. Gugumus, Preprint, Symposium on Polymer Stabilization and Degradation: Problems, Techniques and Applications, Manchester, England, September 1985.

124. R. P. Brown, *Polymer Testing*, 14, 403–414 (1995).

125. R. M. Fischer and W. D. Ketola, in *Handbook of Polymer Degradation*, 2nd edition, S.H. Hamid (Ed.), pp. 645–669, Marcel Dekker, New York (2000).

126. K. K. Karpati, *Durability of Building Materials*, 5, 35–51 (1981).

127. M. A. Lacasse, in *Science and Technology of Building Seals, Sealants, Glazing, and Waterproofing*, Third Volume, J. C. Myers (Ed.), pp. 5–20. Special Technical Publication (STP) 1254, ASTM International (1994).

128. ASTM G 172-03, Standard Guide for Statistical Analysis of Accelerated Service Life Data, Annual Book of ASTM Standards, Vol. 14.04.

129. T. G. B. Jones, A. R. Hutchinson and K. E. Atkinson, in *Proceedings of the International RILEM Symposium on Durability of Building Sealants*, A. T. Wolf (Ed.), pp. 117–132, E & FN Spon, London (1999).

130. K. K. Karpati, *Adhesives Age*, 31 (5), 20–23 (1988).

131. K. K. Karpati, *J. Coatings Technol.*, 56 (710), 29–32 (1984).

6 Sealant Durability and Service Life of Sealed Joints

Andreas T. Wolf

CONTENTS

6.1 INTRODUCTION

Sealants play a vital role in many products common in our daily lives. Sealants are generally chosen for their ability to fill gaps, accommodate relative movement of the substrates, and exclude or contain another material. Sealants are available in a variety of forms, from pastelike to flowable materials. Both single-component and multicomponent versions are available, each with different cure chemistry.

Applications of sealants are extremely broad; a partial list includes construction, automotive, aerospace, electronics, packaging, health care, medical, food, textile, and general industries (appliance, assembly, furniture, maintenance, etc.), as well as consumer uses.

In these applications, sealants are exposed to a variety of environmental and service influences such as chemical reagents, hot or cold water, moisture, oils, gasoline, solvents, hydraulic fluids, chemical atmospheres (e.g., ozone, acidic gases, salt spray, etc.), outdoor weathering, biological agents (bacteria, molds, insects, or rodents), radiation (sunlight, x-ray, radioactivity, etc.), high vacuum, and many others. One of the most important characteristics of a sealed joint in any of the aforementioned applications is its endurance to the operating environment. This endurance is also referred to as the *joint's permanence* or *durability*. Testing, assessing, and predicting the durability of sealed joints has become increasingly important across all industries utilizing "engineered" sealant solutions.

Over the past two decades, the sealants industry has undergone rapid technological and structural changes. On the one hand, advancements in technology have enabled the launch of a multitude of new sealant products based on novel polymers, cure chemistries, and formulations. On the other hand, increasing competitive pressure and more demanding customers have required shorter product development cycles. Unlike the well-established sealants, which have been available for more than 20 years based on the same formulations, these new sealant products do not have documented performance histories. At present, generating a reliable performance history for a new sealant product still requires long-term outdoor testing and extensive in-service field evaluations. Attempts at reducing test time by employing various forms of short-term laboratory-based aging tests have had only limited success and are viewed with suspicion by sealant specifiers, mainly because of the lack of an established correlation with the actual in-service performance of sealants. Other than the manufacturers' literature, often only little information is available on the life span of these sealants.

Thus, the sealants industry urgently needs a method for determining long-term performance rapidly and with assured reliability. Accelerated laboratory aging experiments are the most promising method for acquiring durability information in the shortest possible time; however, a methodology for conducting and interpreting these experiments needs to be developed that improves the predictive value of this technique. Various approaches have been proposed recently in developing protocols that incorporate both accelerated and outdoor exposures for predicting the service life of sealants used in construction. Ultimately, these protocols will be used to develop standards, in cooperation between academia, industry, and standard-setting organizations, which in turn will enhance industry confidence in accelerated testing.

6.2 CLASSIFICATION OF DEGRADATION FACTORS

Depending on their application, sealants in service joints are exposed to different combinations of individual degradation (aging) factors. This exposure causes the sealant properties to degrade over time until the sealed joint fails. The occurrence of failure marks the end of the service life of the sealed joint.

Degradation factors can attack a sealed joint on three fronts: the sealant itself, the sealant–substrate interface—or "interphase," if one considers the thickness of the distinct phase between the bulk sealant phase and the substrate phase—and the substrate. Since a great variety of application substrates exists, with each substrate having its own degradation characteristics, this chapter cannot cover the effects stemming from the degradation of the substrates. Rather, the author focuses here on the degradation of the sealant bulk and the sealant–substrate interface.

For any application, the degradation factors can be grouped into two main classes, that is, environmental factors and service factors. Figure 6.1 shows, as an example, the degradation factors that influence the service life of sealed building joints (adapted from References 1 and 2). Outdoor environmental factors can be further segmented into weathering factors and biological factors, while service factors can be further classified as design, installation, and use factors.

Service factors augment and modulate the environmental factors. Mechanical stress, for instance, is a degradation factor by itself; however, it also acts in synergy with different environmental factors. The installation conditions not only determine the initial level of the functioning of the sealed joint but also determine the rate of degradation caused by the environmental factors. In the case of sealed building joints, for example, the installation conditions also determine whether the sealant's first movement will be in compression or in extension. They further influence whether the sealant during its service life will be exposed to compression and extension cycles of about equal magnitude or will be exposed mainly to extension or compression. The building location, the compass direction of the joint, and its inclination angle determine the level of solar irradiance and the time of wetness. For industrial applications, specific service factors may sometimes outweigh environmental factors in their importance for the degradation of

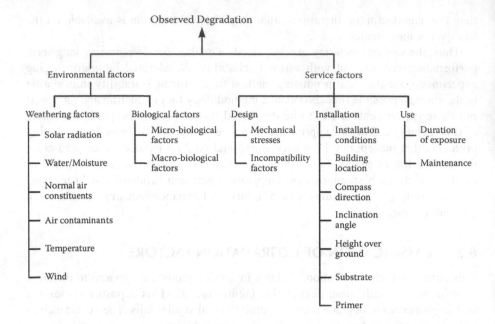

FIGURE 6.1   Degradation factors influencing the service life of sealed building joints.

the sealed joint. For example, the joint design may induce large localized stress peaks in the sealant, or residual stresses may be "frozen" in the cured sealant owing to loss of volume upon cure. However, even in indoor applications, moisture often plays a critical role in the degradation of sealed joints, either as the primary degradation factor or by acting in synergy with other environmental or service factors.

The combination of degradation factors that act on a sealed joint under service conditions is always unique. Therefore, a careful analysis of the degradation factors effective in each individual application should be conducted. However, for applications involving direct or indirect exposure to the outdoor environment, experience has shown that four weathering factors, that is, spectral (solar) radiation, oxygen, temperature, and water or moisture, are responsible for the degradation of most sealed joints. Furthermore, cyclic mechanical stress is the most damaging service factor leading to sealant fatigue in sealed joints exposed to repetitive joint movements (see Reference 3 and literature cited therein). Such repetitive movements may be induced in the joint, for example, by expansion and contraction of adjacent substrates or by vibrations of a component.

6.3   EFFECTS OF DEGRADATION FACTORS ON SEALANTS

Due to the vast variety of sealant applications, this chapter cannot cover all possibly encountered degradation factors, but will focus on those environmental and service factors that are of importance in the majority of commercial sealant applications.

6.3.1 OUTDOOR ENVIRONMENTAL (WEATHERING) FACTORS

Outdoor degradation processes in sealants are primarily caused by the weathering factors of solar radiation (especially the ultraviolet portion), water, oxygen, and temperature and its cycles. Aggressive atmospheric contaminants or components may also contribute to the degradation of sealed joints in highly polluted areas or in industrial environments. In certain applications, specific environmental degradation factors may override the importance of the preceding weathering factors. In this regard, microbiological attack often is the most important environmental (biological) degradation factor. For instance, in sealant applications below grade (underground), in wastewater treatment plants, or in food processing plants, the primary degradation factor often is microbiological attack. The following discussion will focus on the aforementioned primary environmental degradation factors.

6.3.1.1 Solar Radiation

Solar radiation accounts for the most widespread damage to outdoor materials. Its effectiveness as a degradation factor is due to the fact that certain portions of the sun's spectrum, either alone or in the presence of other weathering factors such as moisture or oxygen, are capable of inducing chemical reactions in the materials undergoing irradiation. The chemical and physical changes that result from exposure to sunlight often take the form of discoloration, crazing, loss of gloss, surface erosion, cracking, or loss of tensile strength, extensibility, or adhesion.

According to the first law of photochemistry, light must be absorbed by a material in order to have any effect on it. Since the energy of radiation is inversely proportional to its wavelength, photons of ultraviolet radiation carry a higher energy than those of visible light or infrared radiation. Light of shorter wavelengths is also more frequently absorbed by most materials and, carrying higher photon energy, has a greater potential to break chemical bonds. In order to be capable of cleaving a specific chemical bond, the solar radiation must contain wavelengths whose corresponding photon energies exceed the binding energy of the specific bond. Various chemical bonds (C-N, C-Cl, C-O, C-C, N-H) occurring in organic sealant polymers can be broken by photolysis with solar ultraviolet radiation (300–400 nm). Some bonds can also be broken by visible radiation (see Figure 6.2).

Since natural weathering of sealants occurs in the earth's atmosphere, the effect of sunlight, oxygen, and elevated temperature are inextricably connected. The degradative effect of sunlight and oxygen acting in concert is substantially larger than when both factors are acting individually (synergistic effect). Sunlight, whose photon energies are insufficient to cause photolytic bond cleavage, is, however, capable of initiating photooxidation reactions in the presence of oxygen. Conversely, absorption of oxygen by polymers occurs faster when irradiated with light than in the dark.

Although the ultraviolet portion of the solar radiation is mainly responsible for initiating weathering effects in sealants, the visible and infrared portions also contribute to degradative processes. Colored sealants are susceptible to visible radiation, since their pigments absorb visible light. Visible light absorbed by pigments can cause photolytic cleavage of bonds in the pigment or damage the polymer as a result of charge-transfer transitions or radical initiation reactions. The visible radiation

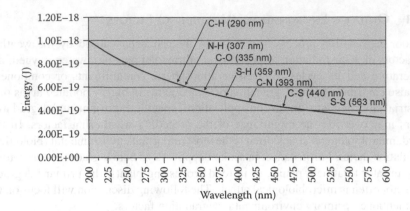

FIGURE 6.2 Energy per photon and examples of wavelengths capable of inducing cleavage of chemical bonds.

absorbed that has insufficient energy to cause bond cleavage and the infrared radiation absorbed by both colored and colorless sealants raise their temperature; the temperature is dependent on the color of the sealant. Both visible and infrared radiations can thus accelerate chemical reactions in sealants by raising their temperature. Visible light absorption may not only cause fading of the pigment but also can degrade the sealant through attack by free radicals formed in organic pigments.

Aromatic polyurethanes based on toluene diisocyanate (TDI) or methylene diphenyl diisocyanate (MDI) chain extenders absorb light in the near-ultraviolet and short-wavelength visible regions and degrade via several possible bond cleavage processes (see References 4 and 5 and literature cited therein). Polysulfides degrade via photo-oxidation of the formal group (-O-CH$_2$-O-) in a similar mechanism as poly(propylene oxide), which is the polymer backbone present in polyether polyurethanes, silicon-modified polyethers, and silicon-modified polyurethanes [6].

Silicones absorb very little ultraviolet radiation in the 300–400 nm region. When irradiated under conditions of natural photoaging, silicones are slowly oxidized. The oxidation of the hydrocarbon side groups results in the formation of carbonyl groups. Since carbonyl groups do not interact strongly, the oxidation has little effect on the mechanical properties of the sealant.

It is evident from the preceding discussion that for almost all organic sealant polymers, some form of photostabilization is essential if adequate protection against the destructive effects of solar radiation is to be achieved [7]. Organic sealants may contain up to three different stabilizers in order to achieve optimum photostabilization. Completely effective stabilization is, of course, never achieved in commercial practice. It must also be taken into consideration that organic photostabilizers do not last indefinitely but are slowly degraded by solar radiation. Thus, the damage to the sealant polymer they are supposed to prevent will ultimately occur. However, degradation effects such as yellowing, surface crazing, and cracking can be substantially delayed by addition of suitable stabilizers.

Solar radiation may cause damage to a sealant's exposed surface or its interface with the substrate. In service joints, the following changes to a sealant's surface may occur as a consequence of the exposure to sunlight and the atmospheric environment (especially to water) [8]:

- Changes in color and gloss
- Wrinkles ("elephant skin")
- Crack formation (crazing, mud cracking)
- Chalking

Figure 6.3 shows, as an example, the type of parallel cracking and mud-cracking patterns observed on a polysulfide and a polyurethane sealant, respectively, after 5000 h of accelerated artificial weathering in ATLAS Suntest accelerated weathering equipment, which utilizes an exposure to simulated sunlight and water immersion.

Changes in the color of a sealant occur when either the pigment used in the sealant formulation or the sealant itself is not sufficiently resistant to solar radiation. In the first case, the sealant's color changes without any other signs of surface degradation being evident. Organic pigments, for instance, used in some sealant products, may fade or change color to an unacceptable shade. In the second case, a degradation of the sealant's surface leads to a loss of color intensity and gloss; the appearance of an originally shiny dark brown sealant may change to a dull light brown color in the course of time.

Many pigments are very effective in absorbing solar radiation. Pigments may have either a stabilizing or destabilizing effect on the sealant with regard to its weatherability. Furthermore, photostabilizers and antioxidants may adsorb on certain pigments and, by doing so, be rendered ineffective [9]. Pigmented versions of the same sealant formulation may, therefore, show different resistance to outdoor weathering, as has been recently demonstrated for two differently colored polysulfide sealants [10].

Solar radiation may affect a sealant's adhesion to a substrate in two ways: first, for some transparent sealants, ultraviolet radiation may reach the sealant–substrate interface and cause degradation of either the sealant–substrate interface or of the substrate surface; second, for sealants applied to transparent substrates, some portions of the solar radiation transmitted through the substrate reach and may degrade the sealant–substrate interface. The first situation, for instance, may occur when a transparent sealant is applied to a painted substrate having insufficient resistance to ultraviolet radiation. The second situation occurs for any glazing sealant, since the sealant–glass interface is directly exposed to the solar radiation penetrating the glass pane. Generally, the degradation of sealant–glass adhesion occurs more rapidly if

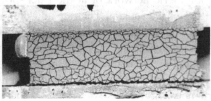

FIGURE 6.3   Crazing patterns observed on polysulfide (left) and polyurethane (right) sealants after 5000 h of exposure in accelerated xenon arc weathering machine.

ultraviolet radiation and water are allowed to act synergistically. Most standards for glazing sealants, therefore, include a test method in which the sealant–glass interface is exposed simultaneously to ultraviolet radiation and to either water immersion, water spray, or high humidity.

6.3.1.2   Water or Moisture

Water is omnipresent in our environment, whether in the form of airborne humidity, rain, or dew. Water and water vapor can exert both physical and chemical influences on sealants and sealed joints. As a result, degradation of a sealant's appearance, physical properties, or adhesion may occur. Since a sealed joint is a composite system consisting of the sealant, generally a backup filler material, and the two substrates that form the joint, water can exert influences on any of the elements forming the sealed joint. Moisture can affect a sealed joint in three distinctive ways: by degradation of

*   Properties of the bulk sealant
*   Adhesion at the sealant–substrate interface
*   Properties of the substrates

Water can affect the bulk properties of a sealant both physically and chemically. The first step in any degradation mechanism affecting the bulk properties is the sorption of water into the polymeric network structure. Immersing a cured sealant sample in water or exposing it to a high relative humidity causes water to diffuse into the sealant. The kinetics of fluid sorption in sealants typically follows the linear Fick's law, which predicts corresponding behavior between mass-gain data in initially dry specimens and weight-loss data upon drying of saturated coupons (no hysteresis). However, in some circumstances, weight-gain data for the sorption and desorption of water in sealants do not agree with the prediction of linear Fickian diffusion. Some of these anomalies can be attributed to the inherent time-dependent response of polymers (viscoelastic diffusion). Circumstances where severe deviation from Fickian diffusion is observed usually correspond to a gradual increase in weight gain without ever attaining equilibrium. This increase is generally accompanied by noticeable degradation in properties, damage growth, material breakdown, and/or mechanical failure. As will be discussed later, this behavior can be associated with irreversible damage to the polymer backbone caused by hydrolysis of chemical bonds.

Upon sorption, the water molecules may be deposited in the free volume of the polymeric network or at the filler–polymer interface, resulting in macroscopic swelling of the sealant. The rate and the degree of swelling depend on the sealant's formulation and the nature of its polymeric binder, specifically, the polarity of the sealant's polymer and plasticizer components; the hydrophilicity of its fillers, which is affected by the nature and degree of filler surface treatment; the type and level of adhesion promoters; and the crosslink density and hydrolytic stability of the elastomeric network.

Due to either the polarity or the hydrophilicity of their polymeric backbone, many organic sealants exhibit a high degree of swelling when exposed to liquid water or high humidity, while silicone sealants swell very little; the water absorption of silicone sealants generally is about 1/3 to 1/10 that of organic sealants tested under identical conditions (see Reference 11 and literature cited therein). Comyn et al. [12] studied the mass uptake and adhesion stability of polysulfide, silicone, and fluorosilicone sealants in various fluids, including water, and concluded that the best guide to the durability of these joints was not the thermodynamic work of adhesion, but the amount of fluid absorbed by the scalant.

The swelling generally is accompanied by a decrease in indentation hardness and cohesive strength, increased permeability of the sealant, and impairment of the adhesion at the sealant–substrate interface [13]. Generally, the swelling of the sealant due to water absorption is reversible; a subsequent drying period will entail the loss of water. Drying of the sealant's surface later leads to a volume contraction, which is hindered, however, since the underlying layers are still swollen. Repeated swelling and drying processes thus induce stresses in the sealant's surface that may result in subsequent cracking, especially if the surface has been embrittled by photochemical aging processes. Furthermore, repeated swelling and drying can exert stresses at the sealant–substrate interface and, subsequently, can cause bond separation. Liquid water may also leach formulation constituents such as plasticizers, pigments, or fillers from the sealant, a process that results in a loss of weight upon drying or may carry impurities into the sealant.

Water permeation into polymeric networks generally lowers the glass transition temperature of the polymer by reducing the interactions between the molecules. Sorption of water into the polymeric network lowers tensile strength and modulus while simultaneously increasing elongation at break. These property changes typically are fully reversible, unless hydrolysis of chemical bonds or leaching of formulation components has taken place.

Irreversible processes may occur when water hydrolyzes chemical bonds within the sealant's polymeric network or between the sealant and the substrate. Hydrolysis of chemical bonds within a sealant's polymeric network often occurs even at room temperature. However, for a temporary or intermittent exposure to water, hydrolysis at low temperatures is generally quite limited and does not manifest itself in substantial changes in the sealant's macroscopic properties. Beyond a certain polymer- and formulation-dependent temperature threshold, however, the hydrolytic breakdown becomes noticeable and the degradation may manifest itself by water being continuously absorbed into the sealant's polymeric network without reaching an equilibrium saturation limit.

Certain chemical linkages such as ester, urethane, amide, and urea can be hydrolyzed. The rate of attack is fastest for ester-based linkages [14]. Some polymeric materials, notably ester-based polyurethanes, will chemically change or "revert" because of hydrolysis upon continuous exposure to moisture. Reversion induced by warm, humid environments may cause ester-based polyurethane sealants to lose hardness and strength and, in the worst cases, transform the cured sealant back into a fluid.

A sealant's adhesion to a substrate results from a relatively low number of chemical bonds as compared with the total number of chemical bonds in the bulk of the material. It is, therefore, not surprising that hydrolysis of these bonds can strongly affect a sealant's adhesion. The rate at which the adhesion is lost depends on the rate of water diffusion to the sealant–substrate interface, the number of such bonds and their binding energy, the stress at the sealant–substrate interface as well as the presence or absence of a catalyst.

Water can reach the sealant–substrate interface either by "wicking" along the interface or by permeating the sealant and preferentially migrating to the interface region. Water molecules accumulating at the sealant–substrate interface may displace the sealant polymer, causing adhesion failure. There appears to exist a critical water concentration within the sealant or at the sealant–substrate interface, below which water-induced adhesion loss may not occur (see, for instance, Reference 15 and literature cited therein). This critical water concentration in the sealant may be expressed directly or as the corresponding critical relative humidity in air that is in equilibrium with the water concentration in the bulk sealant.

Leidheiser and Funke have hypothesized that the debonding process by water accumulation at the interface involves the following steps [16]:

1. A continuous or discontinuous water film is formed at the substrate–sealant interface with a thickness of a few to many molecular layers.
2. Water diffuses through the polymer matrix or through capillaries or pores in the sealant.
3. A driving force exists for directional water transport through the sealant to the interface by diffusion (driven by a concentration gradient). Osmotic force, temperature differences, and chemisorption or physisorption at the interface lead to accumulation of water at the interface.
4. Water accumulation at the interface can be attributed to the presence of nonbonded areas.
5. The local water volume grows laterally by the continued condensation of water molecules under the driving forces outlined earlier. Lateral growth is permitted by the stresses caused by water condensation.

Although the process and mechanisms of sealant failure by exposure to moisture and temperature are complex, the general hypothesis proposed by Leidheiser and Funke can adequately describe many adhesion failures.

When water reaches the interface between a sealant and an untreated high-energy substrate, the interfacial bonds attributable to secondary molecular interactions (van der Waals and hydrogen bonding) are disrupted immediately [17]. Furthermore, the work of adhesion between the polymer and a high-energy substrate, in the presence of water, is negative. In other words, the force between the two adherends become repulsive. However, it should be noted that when covalent bonding, interdiffusion, and mechanical interlocking are present, the work of adhesion cannot be used to explain adhesion failure.

Two general strategies are suggested to improve the durability of sealed (bonded) joints:

• Preventing water from reaching the interface in sufficient quantity
• Improving the durability of the interface itself

Selection of the appropriate sealant or chemical modification of the sealant can reduce the permeability and diffusivity of water into the sealed joint. Because water transport in the bulk is strongly related to free volume, the absorption of water may be reduced by introducing crystallinity or by increasing the crosslink density, both of which lower free volume. Water permeation can be reduced by decreasing the polymer's affinity for water, for example, by reduction of the number of polar or hydrophilic groups. In addition, water absorption may be reduced by adding fillers to the polymer, which may effectively act as a barrier to moisture intrusion. The durability of the interface may be improved by addition of suitable hydrophobing agents (silanes) to the sealant and by increasing the number of covalent bonds, degree of interdiffusion, and mechanical interlocking that are present at the interface [18].

Degradation of sealed metal joints by salt water generally is much more severe than that induced by pure water or moisture. Generally, the degradation mechanism involved is corrosion of the metal surface, resulting in a weak boundary layer. Since this chapter focuses on the degradation effect on the sealant material itself, this degradation mechanism will not be discussed here. However, even on noncorrodible substrates, salt water may weaken the sealant's bond strength beyond the effect of pure water, since soluble salts, deposited over time at the crack tip, may make the sealant–substrate interface more hydrophilic and, therefore, prone to more water ingress.

The acidity or basicity of water can also have a major effect on the durability of both bulk and interfacial sealant properties. Acids and bases may directly attack the sealant or the substrate. The hydrolysis of chemical bonds is substantially accelerated both in acidic and alkaline environments. Industrial applications often involve exposure of sealed joints to an acidic or alkaline aqueous medium, sometimes at the extremes of the pH range. In construction applications, acidic precipitation can be a significant source of acid. The pH of rainfall can be as low as four for some areas highly polluted with combustion by-products.

6.3.1.3 Oxygen and Ozone

Atmospheric oxygen is capable of oxidizing sealant polymers. Oxidative degradation of organic polymers generally is very slow at room temperature in the dark, but is greatly accelerated under the influence of ultraviolet radiation or elevated temperatures. Polymers containing carbon-carbon double bonds are most sensitive to oxidation, their sensitivity increasing rapidly with the degree of unsaturation. Saturated polymers are considerably less sensitive, especially when carrying bulky side groups that interfere with the oxidative attack. Tertiary carbon atoms are more susceptible to oxidation than secondary ones, unless shielded by bulky side groups, such as in polystyrene. The oxygen reactivity also increases with the polarity of the carbon-hydrogen bond. Polymers that contain activated carbon-hydrogen bonds, such as polyethers, polyesters, or polyamides, display a higher susceptibility to oxidation than regular hydrocarbon polymers.

Sealants, which are not transparent to ultraviolet radiation, for instance, because of the shading effect of fillers and pigments, experience photooxidative damage only at the outermost surface layers. As a result of the subsequent crosslinking reactions, a resinous skin forms on the sealant's surface. When exposed to mechanical strain, cracks form in the skin perpendicular to the direction of the strain. In the case of sealants exposed to outdoor weathering, rain washes off some of the soluble oxidation products, causing a loss of volume and, induced by the resulting stresses, irregular cracks in the tough surface skin. This process of leaching and cracking repeats itself, resulting in a slow erosion of the sealant (Figure 6.3 shows these two types of photooxidative surface damage).

Thermooxidation of sealant polymers by oxygen becomes noticeable at temperatures above about 70–100°C, causing embrittlement of the sealant bulk. Oxidation by ozone occurs much more rapidly and generally manifests itself already at room temperature. Silicone and aliphatic acrylate polymers show excellent resistance to photo- and thermooxidation, while polysulfide, polyisobutylene, and aromatic polyether-based polyurethane polymers are oxidized and may require the addition of antioxidants. In general, the degradative effect of autooxidation is greater the higher the polymer content of the sealant.

6.3.1.4 Temperature and Its Cycles

Temperature can trigger physical as well as chemical effects in sealants. The most important physical effect on sealants is the reversible change in their viscoelasticity with temperature. All sealants, regardless of their polymer base and formulation, gradually stiffen when the ambient temperature is lowered. The extent of this stiffening is primarily determined by the nature of the polymeric binder. If the temperature is lowered further, the sealant suddenly loses its elasticity and becomes brittle. The temperature at which this sudden embrittlement occurs is related to the glass transition temperature (T_g) or the crystallization temperature (T_c) of the polymeric binder. The second temperature is only relevant if the polymer can crystallize and is held below the crystallization temperature for a sufficiently long period of time. The higher the glass transition temperature or the crystallization temperature of the polymeric binder, the earlier the sealant will stiffen as the temperature is lowered. The influence of cold on sealants is reversible; the original flexibility returns at higher temperature. The addition of a suitable plasticizer can substantially reduce both the glass transition and crystallization temperatures of a polymer.

Below its glass transition temperature, a sealant is incapable of accommodating any movement. The glass transition temperature, therefore, is often referred to as a *key material parameter* that determines the movement capability of a sealant. The closer the glass transition temperature is to the actual service temperature range, the stronger is the temperature dependency of the viscoelastic properties in that range. Figure 6.4 shows, as an example, the temperature dependency of the secant modulus at 25% extension of four different sealants. As can be seen, some sealants stiffen substantially at lower temperatures and soften at higher temperatures. During cold weather, when the building joints expand, the stiffening increases the stress on the sealed joint and may result in either cohesive or adhesion failure of the joint.

Thermoplastic sealants soften at elevated temperatures, as can be seen for the acrylic sealant (Ac1) shown in Figure 6.4. The softening progresses with further temperature increase until the sealant starts to flow. The softening is a result of the loss of weak bonds between the polymer chains. If the heating period is brief, most of the effects on thermoplastic sealants are reversible, assuming that the temperatures in question were not so high as to initiate polymer degradation.

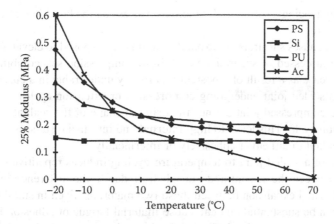

FIGURE 6.4 Temperature dependency of the secant modulus (at 25% extension) of various sealants (measured on ISO 8339 specimens with aluminum as substrate; legend: PS = polysulfide, Si = silicone, PU = polyurethane, and Ac = acrylic).

Exposure of cured elastomeric sealants at moderately elevated temperatures (up to about 70°C) generally does not lead to any degradation of the polymeric backbone. However, even moderate temperatures provide sufficient energy to the polymeric network to trigger formation of additional crosslinks, a phenomenon termed *postcure*. The postcure manifests itself in an irreversible increase in modulus and surface hardness (durometer) and a decrease in maximum elongation. The tensile strength of the sealant typically remains almost constant, since the increase in the sealants' modulus is approximately compensated by the reduction in maximum elongation. Almost all sealants exhibit a postcure phenomenon; however, the degree of postcure experienced at a certain temperature depends on the sealant's polymer type and formulation as well as on the duration of the exposure.

For polysulfide sealants, it is well established that the disulfide bonds are labile at moderate temperatures and break and re-form readily [19]. When this occurs in a stretched or compressed specimen, the newly formed bonds do not contribute substantially to the stress. However, they prevent complete recovery upon release. The amount of "permanent set" can be satisfactorily explained by the new bonds formed in the compressed or extended state, which are strained as the specimen retracts toward its initial, undeformed state. If the joint movement occurs primarily in one direction, such as a single movement caused, for instance, by the settlement of a building, this stress relaxation phenomenon can have a beneficial effect, since it reduces the stress at the sealant–substrate interface.

Even moderately elevated temperatures (ca. 60°C–80°C) can cause noticeable evaporation of plasticizers in some sealant formulations when these are exposed over prolonged periods of time, such as several weeks or months. The evaporation of plasticizer has a dual effect: First, a reduction of the plasticizer level in the formulation causes embrittlement of the sealant; second, the loss of plasticizer from the cured network results in a loss of volume (shrinkage) of the sealant. Both effects act in synergy by placing a substantially increased stress on the bondline, often causing adhesion failure.

Since heat causes substrates to expand, sealants are exposed to elevated temperatures in a compressed or sheared state. A high compression set, combined with a modulus increase as a result of a postcure, is highly undesirable in a cured-in-place sealant. In a sealed joint undergoing compression or extension cycles, the reduced width of the compressed joint defines the original shape of the sealant prior to the next extension cycle. This leads to high internal and interfacial stresses, which may cause the sealant to fail either cohesively or interfacially.

Exposure of a sealed joint to temperature cycling induces repetitive differential movement between the sealant and the substrate due to the difference in the coefficients of thermal expansion between these two materials. Even in small lap joints, this effect can be substantial and can cause material fatigue or adhesion failure. At low service temperatures, the difference in thermal expansion is especially important, since the elastic modulus of the sealant generally increases at lower temperatures. The sealant's thermal conductivity is also important in minimizing transient stresses during cooling. This is why thinner bondline and sealants with higher thermal conductivity generally provide better performance in cryogenic applications.

6.3.2 MICROBIOLOGICAL FACTORS

The susceptibility of sealants to biological attack not only depends on the nature of the polymeric binder but also on the sealant's formulation, especially the type and level of plasticizer used. Plasticizers may impart their own biological properties to an otherwise inert polymer system. The extent of their influence is governed by their susceptibility to biological attack as well as their biological availability, which in turn is influenced by their compatibility with the cured sealant's polymeric matrix, their migration rate to the sealant's surface, and their volatility. Factors that influence microbiological growth are as follows:

- Moisture
- Temperature
- Nutrients
- Antimicrobial additives

Nutrients may be provided by the sealant's bulk composition or may be deposited under use conditions on the surface of the sealant. Microbial growth occurring on sealants, in which the sealant's composition is providing the nutrient, is termed *primary infection*, while growth occurring on a nutrient that is deposited on the sealant surface, for instance, due to use conditions, is termed *secondary growth*.

Divergent conclusions are to be found in the literature on the microbiological resistance of sealants. The confusion is caused by the fact that in many publications the formulations of the sealants studied are either unknown or not disclosed, so that the effect of the polymeric binder cannot be separated from that of the other formulation ingredients.

6.3.2.1 Molds

While molds generally do not cause strong physical deterioration of modern elastomeric sealants, they detrimentally affect the service value of these sealants from aesthetic and hygienic perspectives. The discoloration of the sealant surface is only a by-product of the mold growth, caused by the dyes formed by the mold's metabolism. Once the mold has grown on the sealant surface for some time, the dyes have penetrated so deeply into the sealant's surface that conventional cleaning agents can no longer remove the stains.

6.3.2.2 Bacteria

Bacterial infection of sealants may occur simultaneously with mold growth; however, bacterial degradation appears to be of little importance for most sealant applications. No systematic study of the effect of soil bacteria on sealants used in below-grade (underground) applications has been published, although the author suspects soil bacteria to be an important degradation factor for some organic sealant types. Bacterial degradation is an important issue for most civil engineering applications, and various studies of its effect on sealants have been published (see references cited in Reference 11). Sealants installed in tunnels, sewers, or sewage treatment plants are exposed to microbiological attack by both aerobic and anaerobic bacteria. For these applications, the resistance of sealants to biodegradation is of critical decisive importance. As with molds, the biodeterioration caused by bacteria is primarily a result of their enzyme production.

6.3.3 Mechanical Stresses on Sealed Joints

A sealed joint is a system of materials that must withstand various induced or imposed stresses. Frequently, it is the mechanical stress acting on the sealed joint that is responsible for its failure. Failure of a sealed joint occurs when the stresses in the joint exceed the strength of the weakest link in the joint. Sealed joints may be exposed to movements during or after completion of the sealant cure. The various types of stresses (displacement versus load controlled, singular versus repetitive, etc.) and the roles of stress relaxation, stress concentrations, and energy dissipation in the bulk sealant have been discussed elsewhere [20]. The following discussion will focus primarily on the effect of constant versus intermittent (repetitive) mechanical stresses on a sealed joint and the synergy of these mechanical stresses with other environmental degradation factors.

6.3.3.1 Movement during Cure

During cure, sealants gradually change in their rheological characteristics from being viscous, semifluid materials to nonflowable, viscoelastic materials. Joint movements may impose mechanical strain on the sealant while it is curing, causing permanent and detrimental changes in the sealant bead geometry and performance characteristics. The speed at which a sealant cures is directly related to its ability to withstand movement during cure. A critical parameter for sealant performance appears to be the extent of cure at the onset of movement [21]. Because of their much faster and more homogeneous cure, two-part sealant systems, in general, perform substantially better than one-part systems. Sealants that cure slowly are subject to deformation of the seal geometry, void formation, surface cracking, cohesive splitting and, in some cases, adhesion failure. Movement during cure can be a substantial factor in reducing joint performance and is a real problem, which does occur in high-movement joints (see Reference 22 and literature cited therein).

6.3.3.2 Movement after Cure

Sealed joints may be damaged by movements that induce either constant (static) or repetitive (cyclic) loads. Static loads may induce cohesive or adhesion failure in sealed joints. Cohesive failure occurs when constant (fixed) loads induce creep or stress-relaxation-enhanced fatigue in sealed joints. As the load is applied, the joint undergoes an initial displacement due to the elastic response of the sealant to the load, and possibly a plastic deformation as well, depending on the viscoelastic nature of the sealant. As time passes, the sealant may continue to deflect under the constant load. This phenomenon is known as *creep*. Creep rupture occurs when the joint cannot support the load any longer and the sealant fractures. Stress relaxation involves the same deformation mechanisms as creep, but it occurs when the stress in the sealed joint decreases while the deflection remains the same. The decreased stress in the joints is a result of the steady deformation of the sealant under fixed boundary conditions. Both creep and stress relaxation may be strongly affected by temperature and moisture.

Static loads may also induce adhesion failure by deforming the sealant–substrate interphase and by placing it under stress. Once a rupture or separation of a chemical bond between the sealant and the substrate occurs, the elastic behavior of the sealant network causes a withdrawal of the polymer segment holding this bond, resulting in further separation of those molecular groups that originally made up the chemical bond. As a net result, the split bond has much less a chance to re-form than if the fragmented molecular groups had stayed in close contact. Since fragmentation of molecular bonds to the substrate surface often occurs by hydrolysis, water and temperature have dramatic effects on the adhesion failure rates under constant mechanical load.

In many applications, cyclic joint movement is considered to be one of the primary factors leading to premature sealed joint failure. Fatigue damage is caused by exposure to repeated or cyclical mechanical stress, where the stress level exceeds a limiting value called the *critical stress*. The damage is due to internal structural deformations within the material, which do not return to the original prestressed condition when the stress is removed. Fatigue damage is cumulative, resulting ultimately in failure of the component.

Durability assessments of sealants, which are used for applications involving repetitive joint movements, must include the ability of the sealant to withstand these movements as part of the service environment. It is generally recognized that in order for test methods to provide a performance assessment of such sealants, they must subject the sealant to cyclic movement and, simultaneously, to a combination of relevant environmental factors. At a minimum, such sealants should be exposed to cyclic movement and simultaneous temperature variations (see Reference 22 and literature cited therein) in order to induce changes in viscoelastic properties that occur within the expected service temperature range of the sealant.

6.4 DURABILITY OF SEALANTS: EFFECTS OF POLYMER TYPE AND FORMULATION INGREDIENTS

The performance of the various chemical types of sealants is often generalized in textbooks and articles. For example, all the sealants based on a particular polymer type may be said to be resistant to certain chemicals. Other sealant families may be listed as resistant to UV radiation, or having stability at high temperatures. While the statements may have a sound historical justification and some chemical logic, one needs to recognize that these generalizations may be misleading.

There are some simple explanations for this. The various generic classes of sealants are based on polymers of differing structures, which degrade in different ways under the influence of outdoor or artificial weathering. In the same way, within a generic family of sealants, the structure of the base polymer can differ, either at the level of polymer backbone or in the polymer fine structure or in impurities from the polymer synthesis. Different cure mechanisms used for a particular polymer type can lead to differences in the polymeric network structures of the cured sealants, and leave different curative or catalyst residues, both of which can influence aging behavior.

Also, and quite significantly, within and between the generic classes, the aging behavior is complicated by the huge number of individual formulations and types of ingredients found in commercial sealants. These have both direct and indirect effects on the way sealants perform and properties deteriorate with time. For example, two sealants of the same generic type, with different formulations, may age quite differently in real life: one softening with time, the other hardening.

The assumption that all sealants of a particular generic class are equally durable is clearly misleading. Even a specific brand product may, over many years, change in its precise formulation. Sources and grades of raw materials may change for reasons of availability or cost. In some cases, polymer content may slowly drift down with time, again for reasons of cost. In order to avoid dire consequences in terms of in-service failures, use of relevant material property information and of suitable durability test methods is essential.

6.4.1 Polysulfide Sealants

Structurally, the polysulfide polymer has some relatively weak -S-S- and -C-S- bonds in its backbone. The former are susceptible to cleavage and rearrangement at quite low temperatures (bond rearrangements occur already just above 50°C) [23]. As these reactions are base-catalyzed, basic ingredients in the formulation, such as cure accelerators, also accelerate the reaction rates of cleavage and rearrangement. The sulfur radicals formed recombine, either with each other or with sulfur radicals formed by other -S-S- cleavages (so-called *interchange*). The process manifests itself as stress relaxation (a good property in some circumstances) or compression set (an undesirable property in joints subject to fast cyclic movement). The latter results in folding and creasing in the joint, which may eventually lead to tearing. Thus, cleavage of the -S-S- bonds is a factor in the durability of polysulfide sealants in fast cycling joints. At higher temperatures (>100°C) and in the presence of oxygen, UV, and water, the formal (-O-CH$_2$-O-) groups in the backbone break down, leading to irreversible changes in the polymer network. Such temperatures are uncommon for standard building sealants but may be found in certain glazing, curtain wall, or industrial applications. Interestingly, common thermal stabilizers, antioxidants, and UV stabilizers (apart from pigments such as carbon black or titanium dioxide) are ineffective in polysulfides.

While the relatively low disulfide (-S-S-) and carbon–sulfide (-C-S-) bond energies make degradation via thermal or photolytic cleavage of these links likely, degradation in the outdoor environment occurs primarily via photooxidation of the formal group (-O-CH$_2$-O-) via a similar mechanism as in poly(propylene oxide) polymers. Mahon [6] attempted to characterize the polymers more accurately and then assign the fragmentation pathways for degradation. It was found that the polysulfide polymer was somewhat random in that the C$_2$H$_4$O and CH$_2$O units were not evenly spaced. In fact, 11 variants were found with 4 subvariants. Mahon suggested that for thiol-terminated polymers, the primary degradation commenced by cleavage at the C$_2$H$_4$O or CH$_2$O linkages. NMR studies indicated the possibility of several cleavage sites yielding fragments from formaldehyde to polymer fragments with molecular weights nearing 500.

Barton showed that the rate and extent of degradation were related to the type of curing agent used [24]. It was found that manganese dioxide, currently the most favored curing agent, also gave the most stable systems. Barton also found that the degradation pathways were similar for both ultraviolet radiation and thermal degradation. He postulated that degradation was more rapid when the system was not fully cured; that is, too little curing agent was involved to oxidize all thiol groups.

Caddy showed that use of organic peroxides as curatives impaired the thermal stability of the cured polysulfide sealant [25]. Complete liquefaction occurred after 800 h at 140°C when the polysulfide was cured with tertiary butyl hydroperoxide.

The cure system, thus, has a strong effect on the durability of polysulfide sealants. Manganese dioxide is used to impart thermal resistance. Organic curatives such as isocyanates or acrylic monomers also give good thermal stability and reduce stress relaxation [26]. The isocyanate cure produces a thioisocyanate link, which is resistant to heat, but breaks down in boiling water [27]. Manganese dioxide and sodium perborate both impart some biocidal activity to polysulfide sealants used in water [28].

High polymer content in polysulfide sealants has been shown to be important for certain applications, for example, water immersion and fuel immersion. Case histories also indicate that durability in polysulfide building construction sealants is enhanced by high polymer content [28].

Work carried out by Lowe attempted to measure the potential for degradation of polysulfide sealant adhesion to glass by water and to mathematically derive the involved risk [29,30]. The major part of this study was concerned with water attack only and not with the more severe synergistic combination of ultraviolet radiation and water. Still, the study yielded a number of important findings:

- The sealant-to-glass adhesion is weakened upon exposure to air at 95% relative humidity and 60°C.
- The amount of (visually assessed) interfacial failure for exposed test joints increases with exposure duration and is closely related to a discoloration that develops in the sealant and can be seen through the glass. This area is permanently damaged and does not recover on drying.
- The sealants absorb large amounts of water (45%–92%), but most of this is present as droplets.
- Water attacks the interface.
- The values for diffusion coefficients of water in the various polysulfide sealants are very similar. However, the rate of joint weakening is greater than can be accounted for by the rate at which water diffuses into the sealant.
- The thermodynamic work of adhesion of sealant–glass systems is positive in dry conditions (about 80 ± 20 mJm^{-2}) but is slightly negative in the presence of water (about -5 ± 25 mJm^{-2}), indicating that water may displace the sealant from glass.

Ingredients or contaminants in the curing agent that are water soluble influence the water uptake of the sealant, as has been demonstrated by Gick [31] for chromate cure and by Lowe et al. [30] for manganese dioxide cure.

6.4.2 POLYURETHANE SEALANTS

Polyurethane sealants can be formulated for good durability, as evidenced by many commercial products available in the automotive, aerospace, and to a lesser extent, construction markets. However, since polyurethane chemistry offers great formulation flexibility and polyurethane polymers readily accept high levels of filler and plasticizer loading, it is tempting to formulate these sealants for cost rather than durability. Therefore, there is a requirement, probably more than for any other class of sealants, for skillful formulation and compounding.

The durability of polyurethane sealants is influenced by the polyol type, the cure system (isocyanate type and catalyst level and type), the ratio of soft-to-hard segments, and other formulation ingredients such as water scavengers, UV stabilizers, antioxidants, plasticizers, and pigments (see Reference 32 and literature cited therein). For instance, polyurethane sealant and coating formulations based on castor oil polyols develop strong yellowing on UV exposure [33].

Within the various polyol component options, the most commonly used polyether soft segments are quite stable against moisture but are oxidized when irradiated under conditions of natural photoaging [34]. Light-colored sealants based on aromatic isocyanates have a tendency to yellow under the influence of UV radiation. This degradation has been studied in detail and involves two separate degradative pathways, that is, oxidation of the central methylene group and photo-Fries rearrangement [35]. Long-wavelength UV radiation initiates photooxidation, while short-wavelength UV radiation, absorbed by the aromatic structural components, is responsible for the photo-Fries rearrangement. The wavelengths in sunlight responsible for degradation of aromatic polyurethanes extend into the visible region to about 450 nm. The photooxidation of the polyether segments results in the formation of esters and other volatile products. There is a loss in free volume during polyurethane photodegradation, coupled with a high degree of crosslinking [36,37]. The numerous radical species formed during the photooxidation of the polyether segments may also induce the oxidation of the aromatic urethane segments on the central methylene carbon atom. For this reason, the rate of yellowing is higher for polyurethanes with a higher polyether segment content [38,39]. Although polyurethanes based on aliphatic diisocyanates do not discolor under the influence of UV radiation, they still degrade under conditions of natural photoaging [5]. Surprisingly, the process of photodegradation may be even more prevalent in polyurethanes based on aliphatic diisocyanates than in those based on aromatic diisocyanates [40]. Electron spin resonance (ESR) has been used to identify radicals formed during irradiation of polyurethanes under different light sources [41].

Residual cure catalysts can affect the stability of polyurethane sealants. Dibutyltin dilaurate is widely used as a catalyst in polyurethane sealants because of its stability, which is advantageous in that an uncured sealant will maintain its reactivity during its shelf life. However, this stability also results in an active catalyst being available in the sealant after application and cure to accelerate reverse reactions. The higher the level of Sn (IV) catalyst, the greater is the deterioration under weathering [42]. Stannous octoate (Sn (II)), which is used in some two-part polyurethane sealants, will oxidize upon exposure to air, and therefore will lose its activity in the cured sealant upon environmental exposure [42]. Catalysts retaining residual activity in the cured polyurethane sealant also contribute to the so-called *reversion problem*, a recurring problem with certain formulations, in which the sealant softens and sometimes liquefies upon aging, especially when exposed to the combined action of water, ultraviolet radiation, and heat [43]. The rate of reversion or hydrolytic instability depends on the chemical structure of the sealant polymer, the type and amount of catalyst used, and the permeability of the sealant [14].

A wide range of commercial antioxidants, UV absorbers, and hindered amine light stabilizers (HALS) are used in both one- and two-part polyurethane sealant formulations to improve the resistance against heat, UV radiation, and visible light. However, in one-part formulations, these additives often are ineffective, since the stabilizers contain active hydrogen in their molecules, which is likely to react with the isocyanate functional prepolymer. This reaction may be slow and sometimes takes place only over extended periods during the shelf life of the sealant. However, when stabilizers get attached to the main polymer matrix, they become immobilized and no longer provide the required protection. Consequently, some one-part polyurethane sealant formulations lose their stabilization properties during extended shelf life; that is, the same sealant product may display different environmental resistance at the beginning and at the end of its shelf life.

In two-component polyurethane sealants, the stoichiometry, that is, the correct balance of polymer and curative, is vital for performance and durability. Another recurring problem with certain polyurethane sealant formulations is the continuing development of hardness, probably due to continuing isocyanate-to-urethane conversion (postcure).

6.4.3 ACRYLIC SEALANTS

Acrylic sealants are based on fully preformed polymers dispersed in solvents or water, and they develop their properties through loss of solvent or water. Typically, chemical curing is not involved; their physical properties are derived from the elasticity of their polymer backbone and the physical interactions between the polymer chains.

Acrylic sealants have good resistance to environmental stresses. Their polymer backbone is based solely on the relatively thermally stable -C-C- bonds, although the side groups may be susceptible to thermooxidation and photooxidation. The aliphatic structural components do not absorb terrestrial solar radiation. It is only the oxidation products, that is, the impurities, formed during processing that absorb some solar ultraviolet radiation. Under normal operating conditions (<100°C), well-designed acrylic sealants are very durable. Temperature and water leaching are the two most important parameters affecting the durability of waterborne acrylic sealants. Temperature may improve the adhesion but may also deteriorate hardness, elasticity, and surface tack; leaching will always worsen the physical properties of the sealant. All these influences are strongly formulation dependent.

Polymer type and the specific formulation influence the outdoor stability of acrylic sealants. The pigment volume concentration (PVC) as well as the content of titanium dioxide both play important roles in the durability of acrylic sealants. Specially, surface-treated grades of the rutile modification of titanium dioxide serve as ultraviolet screening agent to reduce the damage induced by solar radiation. Less than a minimum amount of titanium dioxide (<1%) may cause in some high binder formulations greasy and tacky surfaces [44]. Certain acrylic sealant formulations that contain highly volatile plasticizers will embrittle under heat treatment as a result of the plasticizer evaporation. If the content of plasticizer in the sealant formulation exceeds the compatibility limit, the plasticizer may migrate to the sealant surface and cause surface tack and, hence, severe dirt pickup. This migration of plasticizer to the surface is also accelerated by heat.

6.4.4 SILICONE SEALANTS

Silicone sealants based on poly(dimethylsiloxane) (PDMS) polymers and plasticizers—often referred to as 100% silicone sealants—exhibit outstanding durability in a variety of environments. This behavior is a direct result of the unusual organic/inorganic hybrid nature of their siloxane polymer backbone.

Most of the siloxane polymers and plasticizers used in commercial silicone sealants are of the same chemical nature; that is, they have a PDMS backbone. One hundred percent silicone sealants, as a generic class, are therefore much better defined and behave much more consistently in terms of their performance and durability than many organic sealants, for which a variety of different polymers and plasticizers can be used. Formulation-dependent differences in the durability of 100% silicone sealants stem primarily from differences in their cure chemistry (catalyst and crosslinker), and these differences are generally only noticeable after prolonged periods of weathering [45,46].

Currently, plasticizers other than PDMS are increasingly used in "extended" silicone construction sealants available for low-performance building and construction applications, especially in China, Mediterranean Europe, Middle East, South America, and Mexico. In order to be of sufficient compatibility with the silicone polymer, these organic plasticizers have low molecular weight (generally below 350 Da) and, therefore, high vapor pressure. In a few cases, these organic plasticizers are added to improve the rheological properties of the sealant formulation for special applications; however, the majority of uses stems from the desire to reduce the overall cost of the sealant formulation. Because of the high vapor pressure and limited compatibility of the organic plasticizer, these extended silicone sealants generally exhibit higher shrinkage and often also substrate staining and dirt pickup. The volume shrinkage, resulting from the evaporation of the organic plasticizer, then places additional stresses on the sealant and the sealant–substrate interface, often resulting in adhesion or cohesive joint failure.

On the contrary, 100% silicone sealants exhibit excellent thermal stability and are considered for high-temperature applications where no other room-temperature-cured elastomeric sealant can be used. Depending on their formulation and cure chemistry, standard commercial silicone sealants can be exposed to maximum service temperatures of 120°C–180°C. Special heat-stabilized formulations are commercially available that achieve an upper service temperature limit of 250°C for prolonged exposure and of 300°C–320°C for short-term exposures. Hydrolysis and thermal decomposition, occurring at elevated temperatures, are accelerated by the presence of metal salt catalysts and aminopropylsilyl groups in the cured silicone sealant. Acids or bases also catalyze these degradation reactions. Thus, degradation may be accelerated by the presence of polymerization catalysts, fillers, additives, or cure by-products that fall into this category or form such species at elevated temperatures. Therefore, careful matching of the product performance characteristics with the specific application needs is required.

Due to the low glass transition and melting temperatures of PDMS polymer, 100% silicone sealants do not substantially stiffen at lower service temperature. The temperature range over which the elastomeric properties are maintained depends on the specific formulation of the sealant product; however, the Young's modulus of 100% silicone sealants generally is maintained within ±25% or less over a temperature range of −40°C to +80°C (see Reference 46 and literature cited therein).

Exposure of silicone sealants to water or high humidity at moderate temperatures initially results in a completion of their cure process accompanied by an overall improvement in their physical properties (increase in tensile strength) and adhesion. Prolonged exposure, especially at elevated temperatures, then results in physical and chemical processes that cause degradation in performance.

PDMS polymer absorbs only very little ultraviolet (UV) radiation in the 300–400 nm region, which is the wavelength range that causes problems with most other polymers at, or near, ground level. This is consistent with the fact that even after 20 years of outdoor weathering in sunny climates, 100% silicone sealants show only comparatively little changes in physical properties [47,48].

One hundred percent silicone sealants also show excellent resistance to the combined effect of the key weathering factors: water, heat, and UV radiation. The actual response to this environmental challenge depends on the cure chemistry and the formulation of the silicone sealant.

6.4.5 HYBRID-POLYMER-BASED SEALANTS

Some newer sealant types are based on so-called hybrid polymer technologies. Some of these hybrid polymer technologies are still fairly new, and only very limited scientific information has been published on the long-term durability of these materials.

In simplistic terms, hybrids combine the backbone of one sealant family with the reactive groups—typically positioned at the polymer terminals—of another sealant polymer type. For example, the polymercaptan polymers have either a hydrocarbon or a polyether (polyurethane) backbone coupled with mercaptan end groups. This allows cure to be carried out with conventional polysulfide curing agents, while imparting the thermal stability and other durability characteristics of the backbone to the sealant network. The sulfur-sulfur bonds introduced by oxidative cure of the mercaptan end groups are, as in regular polysulfide sealants, thermally weak. However, their concentration is much lower, and so the polymercaptans are less likely to stress relax or permanently set. The polymer backbone, however, has the thermal, thermooxidative, and hydrolytic resistance of conventional polyurethane sealants and can be protected by stabilizers, which are not effective in polysulfide sealants.

Other examples of hybrid polymer technology are silicon-modified versions of polyether (SMP), polyisobutylene (SMPIB), acrylate (SMA), or polyurethane polymers (the last polymer type is often abbreviated SPUR for silicon-reactive polyurethane, or PUH, for polyurethane hybrid). These silicon-modified polymers combine the features of organic backbones with the curability of functional silyl groups. Sealants based on these polymers are cured with conventional silicone cure systems. The siloxane linkages bring improved environmental stability to the cured sealant, but it is the hydrocarbon backbone that determines the thermal and thermooxidation resistance. The durability of the siloxane (Si-O-Si) linkage in hybrid sealants is similar to that in silicone sealants, and its hydrolytic stability is influenced by the same formulation parameters, that is, primarily catalyst and cocatalytically acting formulation ingredients.

One of the key benefits of silicon-modified hybrid polymer formulations over equivalent organic cure sealants is that "standard" stabilizers, such as UV-absorbers and antioxidants, can be used and do not interfere with the cure systems, which is especially of importance for one-part sealant systems as these additives may cause side reactions to occur during shelf life. The active hydrogen groups in stabilizer molecules, such as in hindered amine light stabilizers (HALS), do not react with the alkoxysilyl groups of the silicon-modified hybrid polymer, while they do react with certain functional organic groups such as isocyanates. These polymers, therefore, offer the formulator greater flexibility and possibilities toward effective stabilization of the end product versus their conventional organic counterparts. The addition of stabilizers in silicon-modified hybrid sealants probably accounts for most of the improvement in their stability; however, this comes at a cost, since stabilizers are expensive.

6.5 DEVELOPMENT OF PERFORMANCE-BASED DURABILITY TEST METHODS AND PRELIMINARY EXPERIMENTAL FINDINGS

As mentioned earlier, any durability test method for sealants used in applications involving joint movement should include a realistic reflection of joint movement exposure in combination with other relevant stress factors that are part of the expected operating environment. Work toward standard durability tests for sealants has progressed most in the area of building sealants, and the approach chosen for these standards is discussed here as an example.

Work on an accelerated durability test method for building sealants was started in 1989 within the International Standardization Organization Committee ISO TC59/SC8 (Work Group 6). Later, in 1994, the activity was transferred to RILEM Committee TC139-DBS on Durability of Building Sealants [49], when the ISO committee realized that the task was too complex to be completed within the 5-year time frame allowed for the development of an ISO standard. The RILEM committee work, carried out during 1994–1999, focused on a framework for assessing the effects of cyclic movement and artificial weathering on elastomeric curtain wall sealants in a laboratory-based procedure. Shortly afterward, ASTM International Committee C24 on Seals and Sealants initiated work toward a durability test standard having a similar scope. Currently, the ISO committee TC59/SC8 considers the RILEM Technical Recommendation (RTR) method for the development of a future durability test standard.

Both the prenormative and the standardization activities will be discussed here; however, since the RILEM Technical Recommendation [50] was the first performance-based test method to be developed, and is conceptually similar to the ASTM standard, its development will be reviewed in more detail.

6.5.1 RILEM DURABILITY TEST METHOD

The RILEM Technical Recommendation (RTR) "Durability test method—Determination of changes in adhesion, cohesion, and appearance of elastic weatherproofing sealants for high-movement façade joints after exposure to artificial weathering" [50] specifies a laboratory procedure for determining the effects of cyclic movement and artificial weathering on laboratory-cured, elastic weatherproofing joint sealants for use in high-movement building façade applications.

In this method, test specimens are prepared in which the sealant to be tested adheres to two parallel contact surfaces (substrates). Sealant specimens are cured either statically (no movement) or dynamically (exposed to cyclic movement). The cured sealant specimens are then exposed to repetitive cycles of artificial weathering (light, heat, and moisture) and cyclic movement under controlled environmental conditions (degradation cycles). Weathering is carried out for 8 weeks in an artificial weathering machine. This step is optionally followed by rapid mechanical fatigue cycling. The specimens are then exposed to two thermomechanical cycles as defined in ISO 9047 [51], using the full amplitude suggested as the movement range of the sealant under test.

After completion of each degradation cycle, the specimens are extended to their full-rated extension, based on the manufacturer's recommended movement

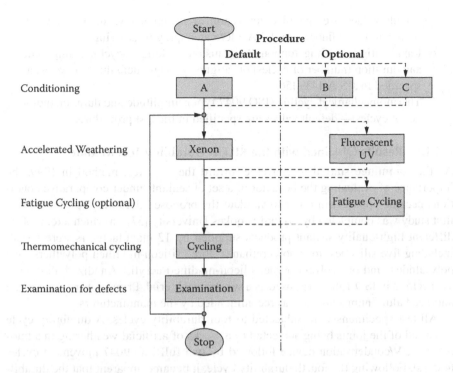

FIGURE 6.5 Schematic representation of the RILEM durability test procedure.

capability, and held there as the sealant beads are visually examined for changes in appearance, cohesion, and adhesion. The depth of any cohesive or interfacial crack is determined according to the rules provided in ISO 11600 [52], and the general condition of the sealant is reported. The weathering exposure, cyclic movement, and examination for failures constitute a degradation cycle, and it is repeated as often as desired to achieve a certain exposure. A schematic representation of the test procedure is shown in Figure 6.5.

Default test parameters and, for some procedures, alternative options are defined in the technical recommendation. In cases of dispute, the default method is the reference method. The experimenter is allowed to deviate from the default values for the following test parameters but is required to highlight any deviation from the default values in the test report:

1. Substrate—default: anodized aluminum, as specified in ISO 13640 [53]
2. Support dimensions—default: 75 × 12 × 6 mm, as specified in ISO 8339 [54]
3. Conditioning method (A, B, or C)—default: A, as specified in ISO 11600
4. Artificial light source (xenon arc, fluorescent UVA-340 lamp, etc.)—default: xenon arc, as specified in ISO 4892-2 [55]
5. Weathering procedure: duration of artificial weathering, type of moisture exposure (spraying or immersion), temperature of light exposure, temperature of moisture exposure, timing of light, and moisture/water cycle—

default values are specified for xenon arc/water spray, xenon arc/water immersion, and fluorescent UVA-340/water-spray weathering

6. Rapid fatigue cycling (optional): inclusion of fatigue cycling, amplitude and duration (number of cycles) of fatigue cycling—default: 200 cycles, as specified in JIS A 1439 [56]

7. Thermomechanical cycling (ISO 9047 type): amplitude and duration (number of cycles)—default values are specified in the test procedure.

6.5.1.1 Results Obtained with the RILEM Durability Test Method

As the committee started to converge toward the final test method in 1999, the importance of evaluating the behavior of a set of sealants under comparative conditions became apparent in order to validate the proposed durability test method. A first study was completed by Oxford Brookes University [57], in which a total of 15 different high-quality sealant products supplied by 12 manufacturers were tested, including five silicones, four polyurethanes, three silicon-modified polyethers, two polysulfides, and one solvent-borne, silicon-modified acrylic. Anodized aluminum as specified in ISO 13640 was used as a substrate material. Primers were used on the anodized aluminum substrates as recommended by the manufacturers.

All test specimens were subjected to four durability cycles. A durability cycle consisted of the joints being subjected to 8 weeks of artificial weathering in a fluorescent UV/condensation device followed by two full ISO 9047 movement cycles (4 days). Following the fourth durability cycle, it became apparent that the durability cycles did not accelerate sealant degradation as quickly as originally thought. It was, therefore, decided to incorporate 1000 fatigue cycles with an amplitude of ±25% at a rate of 5 cycles/minute into the experimental program following the fourth durability cycle. It should be noted that this procedure and the use of the UV/condensation device differ from the default method defined in the final recommendation. The results from the fatigue cycling were intended to give insight into the possibility of including fatigue into later draft versions of the durability test method. Figure 6.6 shows photos of representative test specimens at the discontinuation of the test.

Exposing the silicone sealants to the durability cycles showed no effect for the first three cycles. After the fourth durability cycle, some silicone sealants showed a minor loss of adhesion, primarily at the ends of the joints, where stresses are highest. For these sealants, further loss of adhesion during the fatigue cycling phase, but it proceeded at a different rate for each sealant. In general, no substantial differences were observed in the performance of dynamically and statically conditioned specimens, except for slight bead deformation resulting from the dynamic conditioning. The surface appearance of the silicone sealants tested was not affected by the durability cycles; no cracking, crazing, or visible discoloration occurred.

Large differences in the performance of the tested polyurethane sealants were observed. Except for taking a slight compression set, a one-part polyurethane sealant exhibited only a few signs of deterioration following the durability cycles. Another one-part polyurethane sealant lost adhesion at the ends of the joints, was discolored, and developed surface crazing. The three-component polyurethane sealant showed

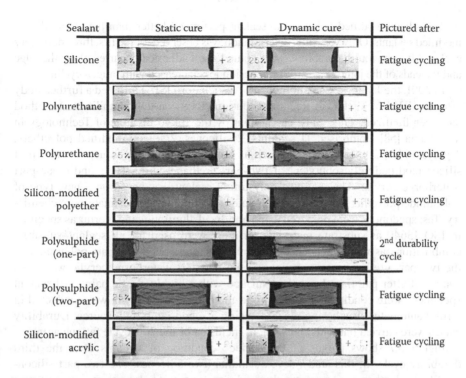

Sealant	Static cure	Dynamic cure	Pictured after
Silicone			Fatigue cycling
Polyurethane			Fatigue cycling
Polyurethane			Fatigue cycling
Silicon-modified polyether			Fatigue cycling
Polysulphide (one-part)			2nd durability cycle
Polysulphide (two-part)			Fatigue cycling
Silicon-modified acrylic			Fatigue cycling

FIGURE 6.6 Appearance of representative test specimens at the discontinuation of the Oxford Brookes University test. (*Note*: Specimens on the left have been exposed to static cure conditions [without joint movement], while specimens on the right have been subjected to movement during cure.)

severe discoloration soon after the first durability cycle, and extensive blisters and cohesive cracks developed after the second cycle. The cracks propagated into the sealant and finally led to complete cohesive failure after the fatigue cycling.

The silicon-modified polyether sealants tested were only little affected by the first two durability cycles. Surface crazing was observed for one sealant after the second cycle; the other two sealants started to craze after the third cycle. Two sealants survived four durability cycles, exhibiting no or only minor adhesion loss. All silicon-modified polyether sealants started to fail adhesively during the fatigue cycling.

The one-part polysulfide sealant cured very slowly and remained incompletely cured even after being exposed to three durability cycles. Due to the incomplete cure, the sealant developed major folds in the sealant bead, which further developed into cohesive cracks that finally resulted in complete cohesive failure after the second and third durability cycles for the dynamically and statically cured specimens, respectively. Clearly, this sealant was not able to withstand the large and rapid movements that could occur in a curtain wall joint, because of its slow cure. The two-part polysulfide sealant performed better, although cracks developed on the bead surface after the second durability cycle, which continued to propagate into the depth of the sealant during the following cycles.

The silicon-modified polyacrylate sealant performed better than the other silicon-modified sealants. Only very little deterioration was observed for the first three durability cycles. After the fourth durability cycle, some loss of adhesion occurred along the edge and the ends of the joints. The area of adhesion loss increased with fatigue cycling.

In 2001, the Japanese Sealant Industry Association (JSIA) initiated a further study after the RILEM Technical Recommendation (RTR) on the durability test method had been finalized. This study, carried out by the Tokyo Institute of Technology in Yokohama [58], comprised 11 sealants: 2 silicones, 2 silicon-modified polyethers, 2 polysulfides, 2 polyurethanes—each as one- and two-part products—1 two-part silicon-modified polyisobutylene, 1 two-part urethane-cure acrylic, and a one-part waterborne acrylic. The two-part polysulfide sealant was based on a new type of polysulfide/polyether/polysulfide copolymer and employed isocyanate-cure chemistry. Test specimens were prepared using anodized aluminum and mortar as specified in ISO 13640 as substrate materials; primers were used for all sealant–substrate combinations as recommended by the manufacturers. The two-part polyurethane, the two-part urethane-cure acrylic, and the one-part waterborne acrylic were also evaluated after painting the sealant surface with a highly elastic paint. All sealant specimens were conditioned according to method A. Weathering was conducted in a fully automated weathering machine using a xenon-lamp light source. Durability cycles were carried out with and without inclusion of the 200-cycle fatigue aging.

Figure 6.7 shows photos of the surface condition of sealants after the third durability cycle. In this study, seven sealants (two silicones, one two-part silicon-modified polyether, one one-part polysulfide, one one-part polyurethane, one two-part silicon-modified polyisobutylene, and one two-part urethane-cured acrylic) passed the first, second, and third durability cycles.

The silicone sealant specimens survived the third durability cycle without any signs of degradation—no cracks, crazing, chalking, or loss of adhesion, and the smallest compression set were observed.

Both silicon-modified polyether products showed signs of moderate chalking after the first durability cycle. An adhesion failure was observed for the one-part silicon-modified polyether sealant in the first durability cycle, which may have been caused by the strong compression set this sealant took during the ISO 9047 thermomechanical cycle (3 mm set based on 12 mm joint width after 25% compression). In spite of exhibiting moderate-to-strong chalking and crazing as well as a 2 mm compression set, the two-part silicon-modified polyether sealant passed the third durability cycle.

The one-part polysulfide sealant also passed the third durability cycle, despite a 2.7 mm compression set and signs of strong chalking and crazing. Strong chalking and crazing was observed for the two-part polysulfide sealant. Without inclusion of fatigue cycling, the sealant failed in the third durability cycle. However, with fatigue cycling included, the sealant failed in the second durability cycle.

The one-part polyurethane sealant passed the third durability cycle while exhibiting some chalking and strong crazing.

The two-part polyurethane sealant chalked and crazed badly after the first durability cycle. Due to the large depth of the surface cracks (2–3 mm), this sealant failed the ISO 11600 pass criterion both on aluminum and mortar substrates. When the same sealant was painted with a highly elastic paint prior to weathering, the sealant

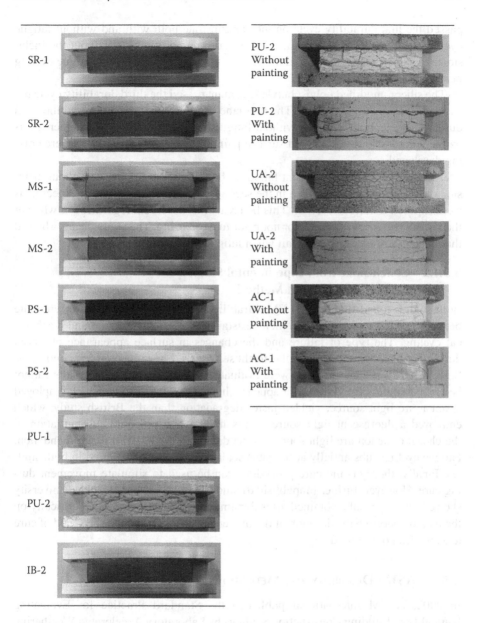

FIGURE 6.7 Appearance of test specimens exposed to additional fatigue cycling after third durability cycle (specimen PU-2 is shown after the first durability cycle). (Legend: chemical type—SR = silicone, MS = silicon-modified polyether, PS = polysulfide, PU = polyurethane, IB = silicon-modified polyisobutylene, UA = urethane-cured acrylic, AC = waterborne acrylic; –1 or –2 denotes one- or two-part product type.)

passed the first durability cycle on mortar substrate both with and without fatigue cycling. However, the sealant failed in the third durability cycle without the inclusion of fatigue cycling, and failed in the second durability cycle with fatigue cycling being included.

The silicon-modified polyisobutylene sealant passed the third durability cycle but showed signs of strong crazing. The urethane-cured acrylic sealant passed the third durability cycle, both with and without fatigue cycling, regardless of whether it was painted or not. However, without surface painting, strong chalking and severe crazing developed.

The waterborne acrylic sealant failed in the third durability cycle, without inclusion of fatigue cycling, but failed in the second durability cycle, when the specimens were exposed to fatigue cycling. This behavior was observed, regardless of whether the sealant surface was painted or not prior to weathering. The study clearly showed that sealants are more likely to fail when fatigue cycling is included.

6.5.1.2 Conclusions from Experimental Studies Using the RILEM Test Method

Both studies found that the RILEM durability test method is able to differentiate between products with regard to their resistance to accelerated aging and mechanical cycling. The type of failure and the changes in surface appearance observed during the test regime were similar to those observed in actual service conditions. Since different sealant products were evaluated in both studies, their results cannot be compared directly; however, it appears that the Japanese study, which employed a xenon arc light source, yielded faster degradation than the British study, which employed a fluorescent light source. This finding is a potential substantiation of the choice of xenon arc light source as the default in the RILEM recommendation. Fatigue cycling substantially accelerates sealant degradation, as found in both studies. Finally, the dynamic cure procedure can be used to simulate movement during cure. However, further (unpublished) studies by the Oxford Brookes University showed that the results obtained after dynamic cure conditions depend critically on the relation between the duration of the movement cycle (2.4 h) and the speed of cure (e.g., pot life) of the product.

6.5.2 ASTM Durability Test Method (with Joint Movement)

In 2002, ASTM International published the Standard Practice for Evaluating Durability of Building Construction Sealants by Laboratory Accelerated Weathering Procedures. The practice was revised in 2004 (ASTM C1519-04) [59], and currently, there is an open work item (WK5561) within ASTM Committee C24 on Seals and Sealants aimed at modifying the scope of C1519-04 by changing the standard from a practice into a test method. The long-term goal then is to replace the current accelerated weathering test, ASTM C793 [60], in the specification standard for elastomeric joint sealants, ASTM C920 [61], with the test method of C1519.

Similar to the RILEM test method, the ASTM practice describes procedures to evaluate or compare the durability of sealants when subjected to accelerated weathering and cyclic movement in a joint. As with the RILEM method, the ASTM practice acknowledges that a variety of environmental and service factors, such as sealant installation procedures, design considerations, and movement during cure, are not addressed in this test procedure. The ASTM standard also states that the results of the laboratory exposure cannot be directly extrapolated to estimate an absolute rate of deterioration caused by natural weathering, for instance, across all construction sealants or even within a chemical family of sealants, because the acceleration factor is material dependent and can be significantly different for each material and for different formulations of the same material. However, exposure of a similar material of known outdoor performance, as a control, along with the test specimens allows comparison of the durability relative to that of the control under the test conditions. Evaluation in terms of relative durability also greatly improves the agreement in test results among different laboratories.

The concept of the ASTM test is related to that of the RILEM method; therefore, the ASTM standard cites the RILEM test method as a related standard. For the ASTM procedure, specimens are prepared in which the sealant to be tested adheres to two parallel contact surfaces. The ASTM procedure uses the same type of specimens, with the same dimensions, and the same preparation and cure as described in ASTM C719. While any substrates can be specified and used, the ASTM C1519 procedure was developed with anodized aluminum substrates. Following cure, the specimens are placed in an artificial weathering chamber for 4 weeks. On removal from the weathering chamber, they are placed in a cyclic movement machine and subjected to six cyclic movements of extension and compression at room temperature according to the method of ASTM C719 (any degree of extension and compression can be used). After the movement cycles, the sealant specimen is held at the recommended extension using suitable spacers and examined for flaws. The cycle of weathering followed by movement testing and examination is repeated as often as specified. After each cycle, the number of cycles is recorded as well as the mode of failure, that is, cohesive or interfacial, extent of failure, the depth of any cracks or breaks, and other pertinent observations such as sealant deformation and bubble formation.

As with the RILEM method, ASTM C1519 allows two alternative accelerated weathering methods. Weathering is carried out either in a Fluorescent UV/Condensation Apparatus, operated in accordance with ASTM Practice C 1442 [62], Section 7.3, or in a Xenon Arc Light Apparatus, operated in accordance with Practice C 1442, Section 7.2. However, the standard states that because of the differences in test conditions, test results may differ with the type of apparatus used.

6.6 DURABILITY ASSESSMENT AND SERVICE LIFE PREDICTION

Whenever a material or component no longer performs one of its intended critical functions, it has failed and has reached the end of its service life. This failure is only a concern, if the actual service life differs from its design life. While underdesign must be avoided to prevent premature failure, overdesign is not desirable either, since it may be economically wasteful.

The service life of a product is defined as the time from putting it into service (its installation) to the time of its failure. Product failure occurs once a critical material

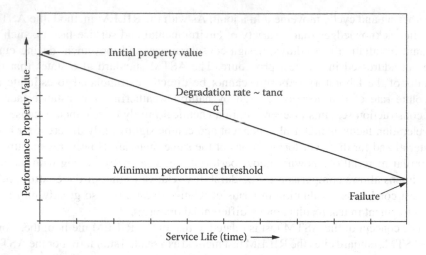

FIGURE 6.8 Simple life cycle concept based on linear degradation response.

property falls below a predefined level, which is required for the proper functioning of the product. Three important aspects must be considered in the change of the critical property value with time:

- Initial property value
- Rate of degradation
- Minimum performance threshold

Figure 6.8 shows the degradation of a property value with time for the simple case of a linear degradation response. The service life of a product is influenced by many variables related to its design, manufacture, and use. The relationship between these factors and the service life of a product is seldom well understood.

Table 6.1 shows some of the variables that influence the service life of a sealed joint. While it is generally advantageous to durability (but not necessarily to economics) to achieve high initial property values, in some situations this approach may also result in a higher rate of degradation and, ultimately, in an overall reduction in service life.

6.6.1 Service Life Prediction (SLP) Methods

Estimating the service life of a system and its components is a very complex problem as it depends on the presence, intensity, and frequency of several factors including the loading, aggressivity of the environment, maintenance, quality of materials, and workmanship (see, for instance, the discussion in Reference 63). Due to cost constraints, most service life prediction (SLP) studies do not take into account all these aspects, but focus on a rather specific application in terms of service environment and usage with a specified set of performance requirements. Because of this limitation in scope, studies are likely to neglect a number of key variables that are critical

TABLE 6.1

Variables Affecting the Service Life of a Sealant Product

Contributing Factors to Service Life	Influencing Variables
Initial performance value	• Polymer type
	• Product formulation
	• Manufacturing process
	• Shelf life of sealant prior to installation
	• Substrate quality (bulk, surface preparation)
	• Installation conditions (climate, workmanship)
	• Cure conditions (climate, movement during cure)
Degradation rate	• Polymer type
	• Product formulation
	• Manufacturing process
	• Shelf life of sealant prior to installation
	• Substrate quality (level of adhesion)
	• Service conditions (climate, movement, abrasion, other agents)
Minimum performance threshold	• Design requirements (movement capability, structural loading capacity, etc.)
	• Safety factors

to the SLP process, such as the design level, the work execution level (workmanship, climatic conditions during application, etc.), specific use conditions (e.g., abrasion, puncture, etc.), or maintenance level.

The scientific link between reference service life (RSL), which can be derived from laboratory experiments, and the estimated service life (ESL), which is to be expected under specific service conditions, is the subject of much research and debate. Various SLP methodologies have been proposed to this end, which can be classified into the following categories:

- Factorial
- Engineering (design)
- Probabilistic

6.6.1.1 Factorial Method

During the late 1980s, a factorial approach to SLP was developed in Japan [64]. It suggests that the estimated service life of a component (ESLC) can be determined by adjusting an assumed "standard life" (SLC) by multiplication with different factors that account for the use, location, and workmanship:

$$ESL = SL * A * B * C * D * E * F * G \qquad (6.1)$$

The following factors are used for adjustments:

A: Quality of components (products)
B: Design level
C: Work execution level
D: Indoor environment
E: Outdoor environment
F: In-use conditions
G: Maintenance level

The benefit—and at the same time the limitation—of the factorial method is its simplicity. While the calculation requires very little resources, the simple approach may cause users to underestimate the actual complexity of the degradation processes involved. Despite its practicality, the approach has several shortcomings [65], which include the following:

1. It is not performance based and as such provides no identification of adopted minimum performance requirements.
2. It implies a somewhat arbitrary choice of standard lives and adjusting factors.
3. It uses a deterministic approach, despite the large uncertainty and variability in the service life.

However, the Japanese publication had fostered the notion that an empirical approach to SLP was both achievable and available. The recently published ISO 15686 Standard, Building and Constructed Assets—Service Life Planning [66], recommends the factorial method in its Part 1, General Principles. The estimated service life of a component is calculated by multiplying the "reference service life" with the same set of adjustment factors. The reference service life of the component (RSLC), in turn, is determined according to ISO 15686 Part 2, Service Life Prediction Procedures [67].

6.6.1.2 Engineering Methods

Engineering methods are meant to emulate the full complexity of the SLP problem but to have their inputs linked by deterministic equations. They should be similar in format to the semiprobabilistic structural design approach [68]; that is, their basis can be probabilistic but the design approach would be deterministic. To date, no examples of pure engineering methods for SLP have been published.

Due to the current lack of engineering methods, some researchers, especially Moser [69,70], have started to work on the development of alternative methods. For example, the factorial method was given a semiprobabilistic basis by employing density functions as variables instead of plain figures. The density distributions are established using reliable and understandable engineering techniques applied in a systematic and straightforward manner.

6.6.1.3 Probabilistic Methods

Because of the stochastic nature of the variables involved in SLP, a probabilistic (or reliability-based) approach is needed if the uncertainties in predictions are to be properly assessed [71]. The approach is analogous to that for performance-based structural design, except that the load, S, and resistance, R, which are time-independent quantities in structural design, have to be considered as functions of time, $S(t)$ and $R(t)$, for a performance-based SLP. Given the time dependency and large uncertainties of the performance of building components, it is necessary to adopt a stochastic model and use stochastic process theory for the SLP. In these methods, degradation is regarded as a stochastic process and for each performance property—during each time period—a probability of deterioration is defined. Computations are based on probabilistic models such as the Markov chain [72] and require sophisticated inputs in the form of probabilities. Lounis et al. suggest that, at a later stage, it would be possible to transform the time-dependent probabilistic model into a time-independent model that has been "calibrated" through extensive performance and service life data [68]. Such a practical and reliable durability design approach should overcome the shortcomings of the factorial method while avoiding the complexities of a full time-dependent probabilistic approach. As a first step, a time-independent, semi-probabilistic approach has been developed as an intermediate solution to the design problem.

6.6.2 SELECTION AND DESIGN OF DEGRADATION STUDIES

Any performance-based approach requires the definition of minimum performance requirements that when attained would indicate the end of the "technical" service life (for a discussion of other definitions of service life such as "economical life" and "risk-based service life," see Lounis et al. [68]). Performance-based standards are, therefore, essential for SLP. These standards must be derived from quantifiable functional requirements (or "roles") of sealed joints, for example, movement accommodation, weather tightness, and aesthetic appearance. Weather tightness is a complex requirement and, depending on the type of joint, must be broken down into specific requirements such as resistance to penetration of air, moisture, and liquid water. Sealed joints can fail by a number of failure modes that are related to the loss of functional performance. The failure of a sealed joint can generally be attributed to a number of root faults [73]. The following are some of the typical root faults involved in sealed joint failures:

- Design factors
- Application factors
- Product factors
- Environmental factors
- Service factors

Root faults can be further partitioned into subfaults. For example, the root fault "design factors" may be further developed into the subfaults "joint dimensions," "accessibility of joints," and "substrate strength." The subfault "joint dimensions," in turn, may be further developed into the subfaults "irregular," "too narrow," and "too wide." This process of subdividing root faults can be continued until a basic fault is reached that cannot be further developed for the purpose of analysis. Figure 6.9 shows, as an example, a partial fault tree with some subdivisions for the failure of the sealant–substrate interface due to hydrolysis reactions (the key contributing basic faults are shown in *italic*).

Root faults are linked to a failure mode by a degradation mechanism. In the preceding example, loss of adhesion is caused by the hydrolysis of chemical bonds, which, in turn, is affected by joint preparation, sealant formulation, presence of water, elevated temperatures, joint movement, and substrate pH, which measures the acidity or alkalinity of the substrate. Fault trees thus allow tracing of specific failure modes back to the basic faults that cause these failures. It is important to note that the same failure mode, for instance, "loss of adhesion" in the previous example, may be caused by another degradation mechanism, for instance, "UV degradation" (in the case of transparent substrate or sealant). Adhesion loss caused by UV degradation would, however, be linked to a different set of basic faults, that is, sealant formulation, sunlight, presence of water, elevated temperatures, and joint movement.

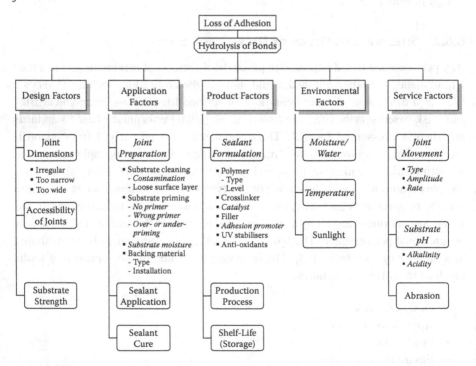

FIGURE 6.9   Example of (partial) root fault analysis.

The purpose of establishing fault trees for the various failure modes is to ensure that all basic faults have been identified. Once all basic faults have been identified and the interactions between the various root faults have been investigated, the basic faults and methods of dealing with them can be described in separate documents such as design codes, best practices for applications, and product performance and durability requirements.

Since a sealed joint may fail by various failure modes, which, in turn, are influenced by a multitude of basic faults, a rigorous durability assessment should consider all failure modes and associated basic faults. However, such an approach often is not required, since for a given set of environmental and service conditions, one failure mode usually prevails. When a sealed joint is put into service, it is common that a number of performance properties change simultaneously in response to the environmental and service conditions. The changes in performance properties therefore effectively compete with each other in causing the sealed joint to fail. The failure mode that usually prevails for a given exposure and service environment is called the *dominant failure mode*. However, it is important to note that the dominant failure mode may change with a slight change in the initial properties of the sealed joint (for instance, as a result of the installation conditions) or in the intensity of some of the weathering factors.

A key purpose of root fault analysis is the establishment of linkages between failure modes, root causes, and degradation mechanisms. Dominant failure modes resulting from specific root causes can be determined by inspecting in service joints, for which the environmental and service conditions can only be monitored but not controlled. The capability of linking specific dominant failure modes to root causes via degradation mechanisms allows the design of structured experimental protocols for accelerated laboratory tests, for which degradation of scaled joint specimens can be studied under controlled conditions. Tracing a dominant failure mode back through a root fault tree for a given degradation mechanism allows identification of the specific basic faults causing this failure. Separate experiments can be designed for a set of environmental and service conditions to address each of these linkages.

6.6.3 PRELIMINARY CONCEPTS OF AN SLP METHODOLOGY

Recently, Shephard and colleagues [74,75] took a first step in this direction by estimating the service life of simulated sealed building joints from experimental data obtained on model joints. In this study, a model joint was prepared from a one-part acetoxy-cured silicone sealant and a mill-finished aluminum substrate. A constant load peel test was used to determine the equation for the crack speed as functions of peel force, relative humidity, and temperature by exposing the peel specimen to heat and moisture during the test. From these data, a master curve was constructed, using the method of reduced variables (commonly referred to as *time-temperature super-position*). From the shift factor plot, Equation 6.2 was obtained, which describes the crack speed as a function of peel force,

$$\dot{a} = \left(\frac{G}{k}\right)^{1/n} \frac{1}{a_T a_{rh}} \tag{6.2}$$

where $\dot{a}$ is the crack speed, G is the applied fracture energy, k and n are constants, and a_T and a_{rh} are the shift factors for temperature and relative humidity, respectively.

In their experiments, Shephard and colleagues observed strong effects of relative humidity and temperature on crack growth rate and nature of failure—the failure mode changed from interfacial to cohesive when relative humidity dropped below 30%—which highlights the significance of climatic variables in establishing damage patterns.

Equation 6.2 was then used to predict the crack length as a function of time for a simulated joint exposed to known climate conditions. By using readily available climatic data, Shephard and colleagues estimated crack growth rates for identical butt joints in Miami, Florida, and Wittman, Arizona. In order to simplify the calculations, Shephard and colleagues assumed a vertical joint sheltered from direct rain and sunlight; this assumption obviated the need to correct for temperature gain or loss of the aluminum surface versus ambient conditions. The speed of crack growth greatly increased when the temperature fell below sealant application temperature (assumed to be 30°C) and the relative humidity remained above 35%. The work suggested that climates that have wide annual changes in temperature while maintaining moderate levels of humidity during cold months are likely to have the largest crack growth for the sealant–substrate system under consideration.

This analysis was carried one step further by charting the crack length for the same joint, using the (hypothetical) assumption that the sealant modulus increased with aging [76]. The analysis showed that an increase in modulus greatly accelerates the crack speed on the simulated joint by increasing the fracture energy. The authors suggested that this SLP methodology could be applied to other sealants, substrates, and shift factors in order to describe the aging effects on various sealed joint systems. For instance, cyclic crack growth data, as generated by Lacasse et al. [77], could be used to extend the method further to a fracture mechanics-based crack growth as a function of cyclic stress and environmental factors.

Lacasse, Cornick, and Shephard [78,79] have further expanded this approach of using degradation models derived from fracture mechanics tests under accelerated environmental conditions. Using the crack growth model developed by Shephard and coworkers and making the same assumptions as discussed earlier, Lacasse and Cornick [78] evaluated the importance of climatic factors on crack growth rate for a hypothetical butt joint exposed in Phoenix, Miami, Singapore, Ottawa, and Winnipeg. Applying different sealant application temperatures to their model showed that the simulated crack growth rate was highly dependent on the temperature of installation. Not surprisingly, it reached a minimum when the installation was carried out at the average annual temperature for a given location. This observation is in keeping with good installation practice. The authors also estimated the service life of the hypothetical joint exposed to the various climates by assuming failure criteria of 50 mm and 100 mm crack length. Time-to-failure periods were calculated based on average annual crack length determined from simulation, assuming identical climatic conditions (periodicity) for each year. The results of this simulation suggested that a sealant installed at the average annual temperature may last indefinitely; however, when installed at a different, especially at a higher, temperature, the service life is likely to be diminished.

The concept of evaluating the effects of aging on adhesion and mechanical properties in separate fracture mechanics experiments while combining them in an overall mathematical model together with metrological inputs is intriguing. Fracture mechanics, rather than continuum mechanics, appears to be more suitable for establishing narrow time-to-failure distributions in the laboratory, since they do not rely on an arbitrary defect in the specimen such as an air bubble or a microcrack. Dillard has previously discussed the benefits and limitations of fracture mechanics tests [80]. The transition from narrow time-to-failure distributions obtained in laboratory experiments to the broader distribution observed on actual service joints (in which failure initiation time and locus of failure are determined by these arbitrary defects) can be accounted for statistically in a more sophisticated SLP methodology. A further benefit of the currently proposed methodology is the simultaneous exposure of the test specimen to load and environmental degradation. There is a clear realization that exposing specimens sequentially to loads and degradation-accelerating environments may not yield conservative estimates of their actual service lives.

6.6.4 LIMITATIONS OF PRELIMINARY CONCEPTS

The current methodology of Shephard, Dillard, and Lacasse is based solely on accelerated weathering experiments; it does not provide comparative data for outdoor weathering exposure. This precludes the possibility of establishing degradation kinetics for these two conditions, which can be compared directly. Therefore, in their more recent paper, Lacasse et al. [79] benchmarked the simulated time-to-failure periods to a "service life of 20 years ... because this value is one that is often ascribed to high-performance products." Conducting fracture mechanics experiments under outdoor conditions while recording the microclimatic conditions at the test specimen would overcome this need for using an industry consensus benchmark.

Previous work on fracture mechanics peel tests suggests that cyclic climatic exposure may strongly accelerate crack speed when compared to exposure in static climates (see, for instance, Reference 81). Typically, these tests are carried out using rather rapid cycling techniques in order to simulate exposures that occur in aerospace or automotive applications. It appears worth studying whether this acceleration also takes place for climatic changes at rates as they occur in building applications on actual facades [82].

Exposing the test specimens to a certain climate for a given time period typically induces changes in the bulk properties. For instance, specimens exposed simultaneously to high temperature and high humidity or water immersion may undergo plasticization or chemical relaxation. To overcome these effects, the specimen may be tested under cyclic fatigue loading while exposed to the preceding conditions. Under such cyclic loads, the crack maintains its sharpness, and hence, this form of testing may be better suited than static tests for the extrapolation of service lives of sealed joints exposed to cyclic mechanical loading [83]. Applying both static and cyclic load tests over a range of humidities may allow separation of the effect of cyclic fatigue from that of stress corrosion on crack growth [84].

The question of whether peel adhesion test data can be used to model adhesion in sealed joints needs to be elucidated from both chemical and physical points of view. It is known that the cured properties and adhesion of sealants, especially of those in the one-part category, depend on the joint configuration [85]. These effects are caused by the diffusion of moisture into and diffusion of cure by-products out of the sealant joint, which results in gradients for chemical composition and network density. There may be two routes to address this question: one would be to conduct interfacial fracture mechanics experiments on sample joints, and the other to simulate cure conditions of a sealed joint with a peel specimen. This latter approach could be realized, for instance, by covering the normally freely exposed air–sealant interface of the peel specimen with a suitable material, such as poly(tetrafluoroethylene) or polypropylene, which can be easily removed after completion of cure.

The direct application of peel test fracture mechanics test data in the prediction of time to failure of a sealed joint neglects the effects of stress distribution in the joint. Local stress-concentration effects are lost, and the extrapolation may give misleading results. This limitation could be overcome by either conducting fracture mechanics experiments on sealed joint specimens that reflect the geometry of actual service joints, or by mathematically modeling the stress distribution in the service joint (for instance, by finite-element analysis) and using the stress field as an input parameter in the SLP methodology.

The viscoelastic behavior of sealants must be accounted for. Most of the studies published as part of the preliminary SLP methodology evaluation employed silicone sealants, which exhibit relatively low plastic flow behavior. This is especially true for the acetoxy-cured silicone sealant used in the previously cited studies by Shephard and colleagues. As viscous dissipation increases, linear fracture mechanics arguments are no longer valid, and the fracture mechanics equations must be based on a constitutive model that accounts for a time-dependent portion of the applied energy not being available to drive crack growth.

The effect of a "critical relative humidity" above which a pronounced loss of bond strength occurs has been previously discussed in the literature (see, for instance, Reference 86). It has been hypothesized that this characteristic may be due to a critical concentration of water being needed to cause interfacial debonding in some systems. However, it appears that for some adhesive/substrate systems, such a phenomenon does not exist, while for other systems the critical relative humidity appears to depend on the crack speed. It is, therefore, vital to investigate the behavior of a wider range of sealant–substrate systems.

Finally, the preliminary SLP methodology must be extended to account for other degradation factors such as use, location, maintenance, and workmanship, unless the factor method is to be applied for these influences. Possible extensions of the current, preliminary SLP concept for sealed joints, and proposals for overcoming its limitations are discussed in Reference 63.

6.6.5 Development of SLP Methodology by NIST

In September 2000, a Sealant Durability Workshop was held as part of the Partnership in Advancing Technologies for Housing (PATH) activities at the National Institute of Standards and Technology (NIST), Gaithersburg, Maryland [87]. The focus of the workshop was improved metrologies for accelerated determination of the service life of sealant formulations and a reliability database approach to predicting service life. As a result of this workshop, a collaboration was initiated between several government agencies and innovative sealant manufacturers. The charter of the NIST-led industrial consortium is to develop the data and associated resources required to implement a reliability-based, accurate, precise, in-service prediction for sealant materials.

The first phase of collaboration (2001–2004) focused on two main areas: creating instrumental capability and developing characterization methods. Work within this project has demonstrated the feasibility of using the mechanical motion of the sealant to both fatigue and characterize the sample. In this concept, the degradation of the sample is monitored as it is exposed to the weathering elements [88]. This greatly reduces the time required to evaluate samples. Additionally, work has been performed in developing a characterization method that accounts for movement during the cure cycle, the *Mullins effect* [89], and the strain-dependent nonlinear viscoelastic nature of the sealant samples.

During Phase I of this consortium project, a number of test devices were developed. The first device was a computer-controlled sealant-test device that uses movement to both characterize the change in the mechanical properties and fatigue standard ASTM C719 sealant samples [90]. This device records the force–response data for each discrete movement step in every deformation cycle of every sample. From this information, the maximum tensile strength, molecular weight between crosslinks, and the fatigue life are determined for each sample. Furthermore, a series of outdoor test devices was developed by the NIST and the Forest Products Laboratory. All of these devices record the stress and strain on each individual sealant sample every 5 min, thus providing a wealth of precise mechanical property data.

At one of the outdoor exposure sites at the Forest Products Laboratory at Madison, Wisconsin, a fully instrumented weather station was installed to provide metrological information suitable for service life prediction methods for sealants [91,92]. The weather station continuously measures and stores information on temperature, relative humidity, rainfall, and ultraviolet (UV) radiation at 18 wavelengths, as well as wind speed and direction. The weather data can be integrated over time to calculate the dose of the weathering factors.

NIST has also initiated studies into accelerated weathering with simultaneous mechanical cycling of the samples based on a novel light source for accelerated weathering that had been previously developed by NIST. This device, termed SPHERE (for "Simulated Photodegradation by High Energy Radiant Exposure"), provides high-flux, very uniform UV radiation in a controlled temperature and humidity environment, and allows simultaneous exposure of 300–500 sealant specimens (see Figure 6.10) [93]. Four sealant test chambers were created with independent and precise control of the mechanical deformation, UV, temperature, and humidity in each of the chambers.

FIGURE 6.10 Simulated photodegradation by high-energy radiant exposure (SPHERE) in an integrating ultraviolet light weathering device (courtesy of NIST).

The next phases of the collaboration will focus on achieving the objectives defined during the PATH workshop, that is, the development of specific degradation models that can be utilized in the service life prediction of sealed joints.

6.7 SUMMARY

Over the past two decades, the sealants industry has undergone rapid technological and structural changes. With many new products based on novel polymer types being launched, the sealants industry urgently needs a method for determining long-term performance rapidly and with assured reliability. Accelerated laboratory aging experiments are the most promising method for acquiring durability information in the shortest possible time; however, a methodology for conducting and interpreting these experiments needs to be developed that improves the predictive value of this technique.

The combination of degradation factors that act on a sealed joint under service conditions is always unique. Therefore, a careful analysis of the degradation factors operative in each individual application should be conducted. However, for applications involving direct or indirect exposure to the outdoor environment, experience has shown that four weathering factors, that is, spectral (solar) radiation, oxygen, temperature, and water or moisture, are responsible for the degradation of most sealed joints. Furthermore, cyclic mechanical stress is the most damaging service factor leading to sealant fatigue in sealed joints exposed to repetitive joint movements.

Various approaches have been proposed recently in the ASTM, RILEM, and JSIA organizations that incorporate both accelerated and outdoor exposures combined with simultaneous mechanical movement of sealed joint specimens for predicting the service life of sealants. While these test and evaluation protocols provide major improvements over current methods, they still do not address a variety of environmental and service factors.

Performance-based standards must be derived from quantifiable functional requirements of sealed joints. Sealed joints can fail by a number of failure modes, which are related to the loss of functional performance. The failure of a sealed joint can generally be attributed to a number of root faults. Conducting a thorough root fault analysis for different failure modes of sealed joints allows the development of improved accelerated testing protocols. Ultimately, these protocols will be used to develop standards, and consequently service-life prediction (SLP) models, which will enhance industry confidence in accelerated testing.

REFERENCES

1. L. Masters and E. Brandt, *Mater. Struct.*, 22, 385–392 (1989).
2. L.F.E. Jacques, L. Masters, and A.J. Lewry, in *Durability of Building Materials and Components, DBMC 7*, Ch. Sjöström (Ed.), Vol. 2, pp. 1129–1140, E & FN Spon, London (1996).
3. A.T. Wolf (Ed.), *Durability of Building Sealants (RILEM State-of-the-Art Report)*, RILEM Publications, Cachan, France (1999).
4. C. Wilhelm, A. Rivaton, and J.-L. Gardette, *Polymer*, 39, 1223–1232 (1998).
5. L. Irusta and M.J. Fernandez-Berridi, *Polym. Degrad. Stab.*, 63, 113–119 (1998).
6. A. Mahon, "Linear Polysulfides: Their Characterisation and Degradation Pathways," Ph.D. thesis, University of Warwick, U.K. (1996).
7. J.F. McKellar and N.S. Allen, *Photochemistry of Man-Made Polymers*, Applied Science Publishing Ltd., Barking, U.K. (1979).
8. A.T. Wolf, in *Durability of Building Sealants*, A.T. Wolf (Ed.), pp. 287–311, RILEM Publications, Cachan, France (1999).
9. G. Haacke, E. Longordo, J.S. Brinen, F.F. Andrawes, and B.H. Campbell, *J. Coatings Technol.*, 71 (888), 87–94 (1999).
10. P. Vandereecken, A.T. Wolf, H. Bolte, and T. Boettger, in *Durability of Building and Construction Sealants*, A.T. Wolf (Ed.), pp. 243–258, RILEM Publications, Cachan, France (1999).
11. A.T. Wolf, in *Durability of Building Sealants*, A.T. Wolf (Ed.), pp. 41–71, RILEM Publications, Cachan, France (1999).
12. J. Comyn, J. Day, and S.J. Shaw, *Intl. J. Adhesion Adhesives*, 17, 213–221 (1997).
13. P.J. Hanhela, R.H.E. Huang, and D.B. Paul, *Ind. Eng. Chem. Prod. Res. Dev.*, 25, 328–336 (1986).
14. E.M. Petrie, *Handbook of Adhesives and Sealants*, McGraw-Hill, New York (2000).
15. E.P. O'Brien, C.C. White, and B.D. Vogt, *Adv. Eng. Mater.*, 8 (1–2), 114–118 (2006).
16. H. Leidheiser and W. Funke, *J. Oil Colour Chem. Assoc.*, 70 (5), 121–132 (1987).
17. T. Nguyen, E. Byrd, and D. Bentz, *J. Adhesion*, 48, 169–194 (1995).
18. K.L. Mittal (Ed.), *Silanes and Other Coupling Agents*, Vol. 3, VSP/Brill, Leiden (2004); K.L. Mittal (Ed.), *Silanes and Other Coupling Agents*, Vol. 4, VSP/Brill, Leiden (2007).

19. M.D. Stern and A.V. Tobolsky, *J. Chem. Phys.*, *14*, 93–100 (1946).
20. N.E. Shephard, J.M. Klosowski, and A.T. Wolf, in *Durability of Building Sealants*, A.T. Wolf (Ed.), pp. 107–135, RILEM Publications, Cachan, France (1999).
21. C.C. White, D.L. Hunston, and R.S. Williams, in *Proceedings of the 27th Annual Meeting of the Adhesion Society*, K.M. Chaudhury (Ed.), pp. 112–114, Adhesion Society, Blacksburg, VA (2004); C.C. White and D.L. Hunston, in *Proceedings of Annual Technical Conference—Society of Plastics Engineers*, Vol. 62–I, pp. 986–990 (2004).
22. T.G.B. Jones and M.A. Lacasse, in *Durability of Building Sealants*, A.T. Wolf (Ed.), pp. 73–105, RILEM Publications, Cachan, France (1999).
23. M. Mochulsky and A.V. Tobolsky, *Ind. Eng. Chem.*, *40*, 2155–2163 (1948).
24. Z. Barton, "The Photo-degradation and Curing Mechanism of Sulfur Based Polymers," M.Sc. thesis, University of Warwick, Warwick, United Kingdom (1992).
25. M. Caddy, "Liquid Polysulfides—Curing and Degradation Studies," Degree Course Work, University of Warwick, U.K. (1997).
26. T.C.P. Lee, *Properties and Applications of Elastomeric Polysulfides*, Report 106, RAPRA Review Reports, *9* (10), RAPRA Technology Ltd., Shawbury, U.K. (1999).
27. E.C. Wollard, in *Polymers in Extreme Environments*, Institute of Materials, London (1994).
28. T.C.P. Lee, T. Rees, and A. Wilford, in *Science and Technology of Building Seals, Sealants, Glazing and Waterproofing*, C.J. Parise (Ed.), ASTM STP 1168, pp. 47–56, ASTM International, West Conshohocken, PA (1992).
29. G.B. Lowe, "The Durability of Adhesion of Polysulfide Sealants to Glass," Ph.D. thesis, De Montfort University, Leicester, U.K. (1992).
30. G.B. Lowe, T.C.P. Lee, J. Comyn, and K. Huddersman, *Intl. J. Adhesion Adhesive, 14*, 85–92 (1994).
31. M. Gick, "The Diffusion of Aviation Fuel and Water in Polysulfide Sealants," Ph.D. thesis, Polytechnic of North London, London (1998).
32. T.C.P. Lee and A.T. Wolf, in *Durability of Building Sealants*, A.T. Wolf (Ed.), pp. 203–223, RILEM Publications, Cachan, France (1999).
33. R.P. Singh, N.S. Tomer, and S.V. Bhadraiah, *Polym. Degrad. Stab.*, *73*, 443–446 (2001).
34. C. Wilhelm, A. Rivaton, and J.-L. Gardette, *Polymer*, *39*, 1223–1232 (1998).
35. M.P. Luda, R. Tauriello, and G. Camino, *Eur. Coat. J.*, (1), 74, 76–79 (2000).
36. R. Zhang, P.E. Mallon, H. Chen, C.M. Huang, J. Zhang, Y. Li, Y. Wu, T.C. Sandreczki, and Y.C. Jean, *Prog. Org. Coat.*, *42* (21), 244–252 (2001).
37. R. Zhang, P.E. Mallon, H. Chen, C.M. Huang, J. Zhang, Y. Li, Y. Wu, T.C. Sandreczki, and Y.C. Jean, *Polym. Mater. Sci. Eng.*, *85*, 109–110 (2001).
38. C.E. Hoyle, H. Shah, and K. Moussa, *Adv. Chem. Ser.*, *249*, 91–111 (1996).
39. E. Govorcin Bajsic, V. Rek., A. Sendijarevic, V. Sendijarevic, and K.C. Frisch, *Polym. Degrad. Stab.*, *52*, 223–233 (1996).
40. V. Rek, E Govorcin, A. Sendijarevic, V. Sendijarevic, and K.C. Frisch, in *Proceedings of 34th Annual Polyurethane Technical/Marketing Conference, October 21–24, 1992*, Society of Plastics Industry, Polyurethane Division, New York (1994).
41. Y. He, J.P. Yuan, H. Cao, R. Zhang, Y.C. Jean, and T.C. Sandreczki, *Prog. Org. Coat.*, *42*, 75–81 (2001).
42. M.D. Harper and P.E. Cranley, in *Proceedings of 35th Annual Polyurethane Technical/Marketing Conference, October 9–12, 1994*, Society of Plastics Industry, Polyurethane Division, New York (1994).
43. T.J. Bridgewater and L.D. Carbary, in *Science and Technology of Building Seals, Sealants, Glazing and Waterproofing*, J.M. Klosowski (Ed.), Vol. 2, ASTM STP 1200, pp. 45–63, ASTM International, West Conshohocken, PA (1992).
44. W. Haller and H. Loth, in *Durability of Building Sealants*, A.T. Wolf (Ed.), pp. 275–286, RILEM Publications, Cachan, France (1999).

45. M.J. Owen and Klosowski, J.M., in *Adhesives, Sealants and Coatings for Space and Harsh Environments*, L.-H. Lee (Ed.), pp. 281–291, Plenum Publishing Corp., New York (1988).

46. A.T. Wolf, in *Durability of Building Sealants*, A.T. Wolf (Ed.), pp. 253–273, RILEM Publications, Cachan, France (1999).

47. P.D. Gorman, in *Science and Technology of Building Seals, Sealants, Glazing and Waterproofing*, D.H. Nicastro (Ed.), Vol. 4, ASTM STP 1243, pp. 3–28, ASTM International, West Conshohocken, PA (1995).

48. D. Oldfield and T. Symes, *Polym. Test.*, *15*, 115–128 (1996).

49. A.T. Wolf, *Mater. Struct.*, *31*, 149–152 (1998).

50. Anonymous, *Mater. Struct.*, *34*, 579–588 (2001).

51. Anonymous, "ISO 9047—Building Construction—Jointing Products—Determination of Adhesion/Cohesion Properties of Sealants at Variable Temperatures," International Standardization Organization, Geneva (2001).

52. Anonymous, "ISO 11600—Building Construction—Jointing Products—Classification and Requirements for Sealants," International Standardization Organization, Geneva (2002).

53. Anonymous, "ISO 13640—Building Construction—Jointing Products—Specification for Test Substrates," International Standardization Organization, Geneva (1999).

54. Anonymous, "ISO 8339—Building Construction—Sealants—Determination of Tensile Properties (Extension at Break)," International Standardization Organization, Geneva (2005).

55. Anonymous, "ISO 4892-1-3 Plastics—Methods of Exposure to Laboratory Light Sources—Part 1: General Guidance" (1999), "Part 2: Xenon Lamps" (2006), "Part 3: Fluorescent UV Lamps" (2006), International Standardization Organization, Geneva.

56. Anonymous, "JIS A 1439 Test Methods of Sealants for Sealing and Glazing in Buildings," Japanese Industrial Standards Committee, Technical Regulation, Standards and Conformity Assessment Policy Unit, Ministry of Economy, Trade and Industry, Tokyo (1997).

57. T.G.B. Jones, A.R. Hutchinson, and A.T. Wolf, *Mater. Struct.*, *34* (5), 332–341 (2001).

58. H. Miyauchi, N. Enomoto, S. Sugiyama, and K. Tanaka, in *Durability of Building and Construction Sealants and Adhesives*, A.T. Wolf (Ed.), STP 1453, pp. 206–212, ASTM International, West Conshohocken, PA (2004).

59. Anonymous, "ASTM C1519–04—Standard Practice for Evaluating Durability of Building Construction Sealants by Laboratory Accelerated Weathering Procedures," ASTM International, West Conshohocken, PA (2004).

60. Anonymous, "ASTM C793–05—Standard Test Method for Effects of Laboratory Accelerated Weathering on Elastomeric Joint Sealants," ASTM International, West Conshohocken, PA (2005).

61. Anonymous, "ASTM C920–05—Standard Specification for Elastomeric Joint Sealants," ASTM International, West Conshohocken, PA (2005).

62. Anonymous, "ASTM C1442–03—Standard Practice for Conducting Tests on Sealants Using Artificial Weathering Apparatus," ASTM International, West Conshohocken, PA (2003).

63. A.T. Wolf, in *Durability of Building and Construction Sealants and Adhesives*, A. T. Wolf (Ed.), ASTM STP 1453, pp. 385–400, ASTM International, West Conshohocken, PA (2004); also published in *J. ASTM Int.*, *1* (3), Paper ID JAI 10983 (2004) (paper available online at www.astm.org/).

64. Anonymous, *Principle Guide for Service Life Planning of Buildings*, Architectural Institute of Japan, Tokyo, Japan (1993).

65. Z. Lounis, M.A. Lacasse, D.J. Vanier, and B.R. Kyle, in *Roofing Research and Standards Development*, T.J. Wallace and W.J. Rossiter, Jr. (Eds.), Vol. 4, ASTM STP 1349, pp. 3–18, ASTM International, West Conshohocken, PA (1999).

66. Anonymous, "ISO 15686—Building and Constructed Assets—Service Life Planning, Part 1: General Principles and Part 2: Service Life Prediction Procedures," International Standardization Organization, Geneva (2000 and 2001).

67. P.J. Hovde, in *Performance Based Methods for Service Life Prediction—State of the Art Reports*, compiled by CIB W080/RILEM 175-SLM Service Life Methodologies Prediction of Service, available at: http://cibworld.xs4all.nl/pebbu_dl/resources/literature/downloads/05AdditionalThemes/01Pub294.pdf.

68. Z. Lounis, M.A. Lacasse, A.J.M. Siemes, and K. Moser, in *Proceedings of Construction and the Environment—14th CIB World Building Congress: Materials and Technologies for Sustainable Construction, Service Life and Durability, Gavle, Sweden 7–12 June 1998*, Ch. Sjöström (Ed.), Symposium A, Vol. 1, pp. 315–324, International Council for Building Research Studies and Documentation (CIB), Rotterdam, The Netherlands (1998).

69. K. Moser, in *Durability of Building Materials and Components—8DBMC*, M.A. Lacasse and D.J. Vanier (Eds.), pp. 1319–1329, CISTI—NRC's Canada Institute for Scientific and Technical Information, Ottawa, Canada (1999).

70. K. Moser, in *Performance Based Methods for Service Life Prediction—State of the Art Reports*, compiled by CIB W080/RILEM 175-SLM Service Life Methodologies Prediction of Service, available at http://cibworld.xs4all.nl/pebbu_dl/resources/literature/downloads/05AdditionalThemes/01Pub294.pdf.

71. P.D.T. O'Connor, *Practical Reliability Engineering*, Heyden, London (1981).

72. D.L. Isaacson and R.W. Madsen, *Markov Chains Theory and Applications*, John Wiley & Sons, New York (1976).

73. A.T. Wolf, in *Durability of Building Sealants*, A.T. Wolf (Ed.), pp. 343–364, RILEM Publications, Cachan, France (1999).

74. N.E. Shephard, "Measuring and Predicting Sealant Adhesion," Ph.D. thesis, Virginia Polytechnic Institute and State University, Materials Engineering Science, Blacksburg, VA (1995).

75. N.E. Shephard and J.P. Wightman, in *Proceedings of the 19th Annual Meeting of the Adhesion Society*, T.C. Ward (Ed.), pp. 488–490, Adhesion Society, Blacksburg, VA (1996).

76. N.E. Shephard, J.M. Klosowski, and A.T. Wolf, in *Durability of Building Sealants*, A.T. Wolf (Ed.), pp. 107–135, RILEM Publications, Cachan, France (1999).

77. M.A. Lacasse, J.C. Margeson, and B.A. Dick, in *Durability of Building Sealants,* J.C. Beech and A.T. Wolf (Eds.), pp. 1–16, E&FN Spon, London (1996).

78. M.A. Lacasse and S.M. Cornick, in *Proceedings of International Conference on Building Envelope Systems and Technologies (ICBEST-2001)*, B.A. Baskaran (Ed.), pp. 433–438, National Research Council Canada, Institute for Research in Construction, Ottawa, Canada (2001).

79. M.A. Lacasse, S.M. Cornick, and N.E. Shephard, in *Proceedings of 9th International Conference on Durability of Building Materials and Components*, S. Burn (Ed.), Paper 247, pp. 1–11, CSIRO—Commonwealth Scientific & Industrial Research Organization, Brisbane, Australia (2002).

80. D.A. Dillard, in *Proceedings of the 20th Annual Meeting of the Adhesion Society*, L.T. Drzal (Ed.), pp. 243–245, Adhesion Society, Blacksburg, VA (1997).

81. C.A. Creegan, D.A. Dillard, N.E. Shephard, B.L. Holmes, H.A. Pavatareddy, and J.G. Dillard, in *Proceedings of the 19th Annual Meeting of the Adhesion Society*, T.C. Ward (Ed.), pp. 137–139, Adhesion Society, Blacksburg, VA (1996).

82. A.R. Hutchinson, T.G.B. Jones, M.A. and Lacasse, in *Durability of Building Sealants*, A.T. Wolf (Ed.), pp. 1–35, RILEM Publications, Cachan, France (1999).

83. A.J. Kinloch, in *Proceedings of the 20th Annual Meeting of the Adhesion Society*, L.T. Drzal (Ed.), pp. 327–329, Adhesion Society, Blacksburg, VA (1997).

84. J.E. Ritter, A.D. Huseinovic, G.S. Jacome, T.J. Lardner, and T.P. Russell, in *Proceedings of the 23rd Annual Meeting of the Adhesion Society*, G.L. Anderson (Ed.), pp. 100–102, Adhesion Society, Blacksburg, VA (2000).

85. F. De Buyl, J. Comyn, N.E. Shephard, and N.P. Subramanian, *Intl. J. Adhesion Adhesives*, *22*, 385–393 (2002).

86. R.S. Jackson, A.J. Kinloch, L.M. Gardhan, and M.R. Bowditch, in *Proceedings of the 19th Annual Meeting of the Adhesion Society*, T.C. Ward (Ed.), pp. 147–151, Adhesion Society, Blacksburg, VA (1996).

87. Anonymous, *J. Res. Nat. Inst. Stand. Techn.*, *105*, 937 (2000).

88. K.T. Tan, D.J. Benatti, C.C. White, and D.L. Hunston, in *Proceedings of the 29th Annual Meeting of the Adhesion Society,* G. Anderson (Ed.), pp. 344–346, Adhesion Society, Blacksburg, VA (2006).

89. C.C. White and D.L. Hunston, in *Durability of Building and Construction Sealants and Adhesives,* A.T. Wolf (Ed.), STP 1453, pp. 325–334, ASTM International, West Conshohocken, PA (2004).

90. C.C. White, N. Embree, C. Buch, and R.S. Williams, *Rev. Sci. Instrum.*, *76*, paper 045111 (2005).

91. R.S. Williams, A. Sanadi, C. Halpin, C.C. White, in *Woodcoatings—Foundations for the Future 28–30 October 2002 The Hague, The Netherlands—Conference Papers*, Paper 17, The Paint Research Association, Teddington, Middlesex, U.K. (2003), available at: http://www.treesearch.fs.fed.us/pubs/23847.

92. R.S. Williams, S. Lacher, C. Halpin, and C.C. White, in *Proceedings of the Third International Symposium on Service Life Prediction*, Federation of Societies for Coatings Technologies, Blue Bell, PA (2005).

93. J. Martin and J. Chin, U.S. Patent 6,626,052 (2003).

7 Adhesion Testing of Sealants

W (Voytek). S. Gutowski and A.P. Cerra

CONTENTS

7.1   INTRODUCTION

Elastomeric sealants are an important class of specialty adhesives used by the building, automotive, aerospace, electronics, and other industries for joining, gap filling and, frequently, structural bonding of a variety of engineering materials. To fulfill its designated functions, the sealant needs to retain cohesive integrity as well as adhere to the joined substrates throughout the lifetime of the sealed or bonded structure.

During their service, sealants are typically exposed to a range of static and dynamic stresses imposed by external loads; cyclic compressive–tensile forces arising from the structure's movements and deformations; and also caused by environmental factors such as solar radiation, temperature fluctuations, moisture ingress, and others.

A combination of these factors frequently leads to changes in the material's internal structure and properties such as crosslink density and chain mobility, and leads to changes in its rheological and mechanical properties. The ensuing increase in the elasticity modulus and the reduction of the material's elongation capability may lead to the increase of stresses within the sealant bead and, more importantly, at the sealant–substrate interface. Once stresses exceed the strength of interfacial or internal bonds, cracks are initiated. The onset of crack formation in the continuous presence of cyclic joint movement and mechanical and environmental stresses is typically followed by gradual crack propagation along the path of the weakest element of the joint. In most cases, this is the sealant–substrate interface.

To withstand the detrimental influence of stresses and degradation-inducing factors, sealants are formulated to exhibit the following long-term properties:

- Reliable and stable adhesion to a variety of engineering substrates
- Stable bulk properties, for example, cohesive strength, modulus of elasticity, and elongation capability, which are not adversely affected by ultraviolet or infrared radiation from solar exposure or by other environmental factors, and which hence allow accommodation of all the thermal and structural movements of the bonded elements

This chapter addresses the following fundamental and engineering issues concerning in-service performance of elastomeric sealants with emphasis placed on the sealant–substrate adhesion:

1. Stresses and degradation factors during the service life of sealants
2. Fundamental factors influencing stresses at the sealant–substrate interface, and adhesion of sealants
3. Methods for determining the strength of adhesion between the sealant and adjacent substrates
4. Industry standards and protocols for determining sealant performance

The experimental data discussed, and the ensuing performance–property relationships developed in this chapter are the result of an extensive research program carried out by the CSIRO research team in collaboration with key suppliers of structural and nonstructural elastomeric sealants such as silicones, polyurethanes, polysulfides, epoxy–polyurethane hybrids, and others.

7.2 TYPICAL FUNCTION AND GEOMETRY OF THE SEALANT BEAD IN A JOINT

In typical construction applications, sealants are used as weather seals to prevent the ingress of water, air, and other undesirable elements. In another key role as structural adhesives, they are designed to permanently attach panes of glass or cladding panels to the structural members of a façade. Figure 7.1 illustrates examples of joint designs for these two typical applications of sealants.

7.3 DEGRADATION FACTORS AND STRESSES IMPOSED ON THE SEALANTS DURING SERVICE LIFE

Throughout the service life of a sealed or bonded structure, the sealant and its interfaces with adjacent substrates are continually subjected to a set of complex and concurrently imposed environmental and mechanical stresses and degradation factors, such as the following:

- *Load-controlled stress,* imposed by external loads and forces, for example, deadloads, wind pressure, and structural load deformation.
- *Displacement-controlled stress,* caused by cyclic or permanent variations in the joint width. These are typically imposed by thermal movements of adjacent cladding panels, for example, in building facades or the external body panels of vehicles such as buses, trams, and boats.
- *Hydrothermal stresses,* caused by the ingress of water molecules (liquid water or vapors) to the sealant–substrate interface and by increase of reaction rate (e.g., hydrolytic chain scission) in response to increased ambient or service temperature.
- *Photochemical factors,* for example, photoinduced chain scission or crosslinking, and catalytic reduction of activation energy of reactions at the sealant–substrate interface.

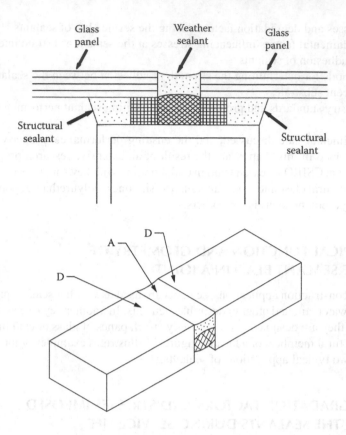

FIGURE 7.1 Typical designs of joints with elastomeric sealants: (a) structural sealant in a joint, also performing the role of a weather seal [1]; (b) weather seal *A* between façade panels *D* [2]. (Adapted from American Society for Testing and Materials (ASTM) Standard No C1193-05: "Standard guide for use of joint sealants," 2005.)

The loss of adhesion through delamination between the sealant bead and the substrate occurs when the fracture energy of the interphase, comprising an array of molecular bridges across the sealant–substrate interface, is lower than the fracture energy of the bulk sealant material. Hence, the configuration of the test specimen must be selected and optimized to maximize the stress at the interface, while other test conditions such as strain or deformation rate and environmental stress factors mimic realistic conditions imposed on the actual joint during its service life.

7.4 PRINCIPLES FOR ASSESSING THE ADHESION OF SEALANTS

As outlined in Section 7.2, in most typical applications, sealants perform the role of structural adhesives or weather seals preventing the ingress of liquid water or vapors, air, and other undesirable elements to the sealed structure. The following are the key factors essential for the satisfactory performance of sealants in any designated role:

1. Retention of cohesive integrity of bulk sealant material
2. Retention of adhesion between the sealant and substrate material

Most of the current standardized approaches for determining sealant performance require demonstration of cohesive integrity of the sealant after exposing it sequentially or concurrently to a number of static or cyclic degradation factors and stresses. While the foregoing criterion (retention of cohesion) is frequently satisfied, it is the strength of the array of chemical bonds across the sealant–substrate interface that plays an equal or more important role for ascertaining satisfactory joint performance.

To demonstrate satisfactory adhesion, one needs to subject the interface to a set of conditions likely to promote disruption of interfacial bonds before the cohesive failure of bulk sealant may occur. To prevent the latter from occurring first, the task needs to be accomplished by the concurrent use of degrading conditions together with mechanical stress of the magnitude below the cohesive strength of the sealant material.

The following are the most significant difficulties in accomplishing this demanding task:

1. How to reliably detect substandard adhesion at the sealant–substrate interface when strength and environmental stability can be masked by the elastomeric nature and premature cohesive failure of the sealant material.
2. How the joint dimensions, loading conditions, and other factors influence the physical properties of the sealant and the magnitude of interfacial stress.
3. How the results relate to the long-term durability of the sealant in service.
4. How to simulate a realistic set of degradation factors most likely to: (i) induce sealant aging, and (ii) promote adhesion failure, in order to reliably predict long-term adhesion and cohesion durability.

7.5   STRESS AT THE SEALANT–SUBSTRATE INTERFACE

In accordance with the principles of continuum mechanics, for sealant deformations to remain within the region of the material's elastic deformation, the magnitude of stress, including that at the interface, is directly related to the material's properties such as modulus of elasticity ([E]: Young's modulus) and strain [ε] through the following relationship:

$$E = \frac{stress\,(\sigma)}{strain\,(\varepsilon)} \tag{7.1}$$

Equation 7.1 uses an average stress value to determine the criterion for joint failure and completely ignores stress concentration near the edges of the stretched sealant or other loci resulting from intrinsic material defects such as air bubbles or cracks.

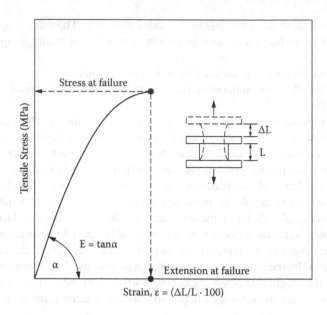

FIGURE 7.2 Typical stress–strain curve and deformation of a viscoelastic sealant material in a joint subjected to tensile force (L is the initial width of the sealant bead, ΔL is the sealant elongation, and E is the Young's modulus of the sealant).

In practice, due to the sealant's rheological properties leading to bead deformation with increasing strain, as schematically illustrated in Figure 7.2, edge effects occur with stress concentrating at the edge of the joint. The latter typically becomes the point (or line) where the crack initiates before propagating along the interface (in the case of poor adhesion) or veering into the bulk sealant when adhesion is satisfactory and cohesive strength controls the overall joint performance.

The stress peak value at the edge of the deformed joint, which can be computed using the finite elements method (FEM) or finite differences analysis (FDA) exceeds the average stress estimated through Equation 7.1. This difference increases nonlinearly with increasing deformation.

It is noticeable from Equation 7.1 that, for a constant deformation, that is, strain (ε) exerted on the sealant, the resultant average stress (σ) at the sealant–substrate interface increases with increase in the modulus of elasticity (E). This fact (the increase of interfacial stress with the modulus increase) is practically utilized in our test methods for determining the strength and quality of sealant adhesion. It is comprehensively demonstrated in Sections 7.5.1 and 7.7.1.

7.5.1 Maximizing Stress along the Interface of Peel Test Specimens

7.5.1.1 Mechanics of Peel Test

One of the most common techniques for testing sealant adhesion is the peel test (see Figure 7.3), the principles of which are discussed in detail in References 3–6.

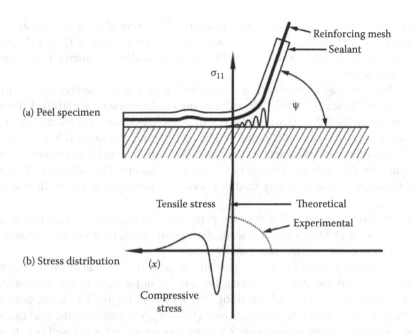

FIGURE 7.3 Distribution of tensile stresses σ_{11} at the sealant–substrate interface ahead of the advancing peel front (x). ψ is the peel angle. (From D.H. Kaelble, *Trans. Soc. Rheology*, 4, 45, 1960.)

According to Kaelble [3], the distribution of tensile stress σ_{11} at a distance x from the peel front is expressed by the following relationship:

$$\sigma_{11} = \sigma_0 \left(\cos\beta_p x + \kappa \sin \beta_p x\right) \exp(\beta_p x), \tag{7.2}$$

where

$$\beta_p = \left(\frac{E_a b}{4EIh_a}\right)^{1/4} \tag{7.3}$$

$$\kappa = (\beta_p m/\beta_p M + \sin \psi), \tag{7.4}$$

where σ_0 is tensile stress at $x = 0$; E_a is tensile modulus of the sealant material; E is tensile modulus of the flexible backing mesh material; h_a is sealant layer thickness between the substrate and flexible backing mesh material; I is moment of inertia of the peeled sealant strip; M is moment of the peel force; ψ is the peel angle.

According to Kaelble's theory [3–5] the peel force [F] per unit width [b] of the specimen is given by

$$\frac{F}{b} = \frac{h_a \ \kappa\sigma^2}{2 E \left(1 - \cos \psi\right)} \tag{7.5}$$

A drawback of Kaelble's equation (Equation 7.5) is that it does not consider the influence on the peel force of the following factors: (1) strain rate (peel-off speed), and (2) variability of elasticity modulus with the sealant geometry (e.g., sealant layer thickness).

To close this gap, experiments were carried out in CSIRO laboratories to investigate the influence of strain rate and sealant bead geometry (controlled through varying the sealant thickness) on the magnitude of resultant peel force. The strain rate was controlled within the range of 0.05 to 100 mm/min (0.05; 0.5; 1; 2.5; 5; 10; 25; 50; 100 mm/min) while sealant thickness was varied from 0.1 to 0.6 mm. Graphs in Figure 7.4 illustrate the outcome of these experiments. The influence of sealant bead thickness on the elasticity modulus was also investigated and is illustrated in Figure 7.5.

An analysis of Equation 7.5 from the perspective of experimental data illustrated in Figures 7.4 and 7.6 leads to a new insight into the mechanism of stress control at the peel front of the sealant–substrate interface.

It is noticeable from Figure 7.4 that an increase in the sealant layer thickness between the rigid and flexible substrates results in an increase in the peel force, as earlier reported by Gent and Hamed [6]. The graphs in Figure 7.4 also demonstrate that the peel force (and hence fracture energy) strongly depends on the peel rate (ε'). Hence, Kaelble's equation (Equation 7.5) requires correction for the peel rate, that is,

$$\frac{F}{b} = \frac{h_a \, \kappa\sigma^2}{2\,E\left(1 - \cos\psi\right)} \cdot f(\varepsilon') \tag{7.6}$$

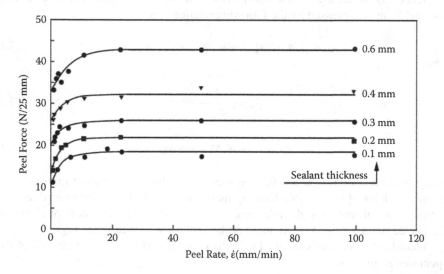

FIGURE 7.4 The influence of peel rate and sealant bead thickness on the magnitude of peel force (sealant type: Dow Corning 795; temperature: 20°C; failure mode in all cases: 100% cohesive failure within the sealant).

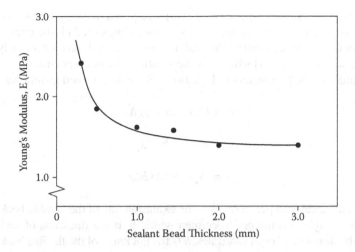

FIGURE 7.5 The relationship between sealant (Dow Corning 983) bead thickness and modulus of elasticity (Young's modulus) at 20°C. (From W.S. Gutowski, L. Russell, and A. Cerra, in *Science and Technology of Building Seals, Sealants, Glazing and Waterproofing, Volume 2*, ASTM STP 1200, J.M. Klosowski (Ed.), p. 87, American Society for Testing and Materials, Philadelphia, PA, 1992.)

It follows from Equation 7.4 that $\kappa = 1$ for the peel angle $\psi = 180°$ used in our experiments. Considering this and using Equation 7.6, the magnitude of tensile stress σ at the peel front can now be determined as

$$\sigma = \left[\frac{FE}{bh_a} \cdot \frac{1}{f(\dot{\varepsilon})} \right]^{1/2} \tag{7.7}$$

Equation 7.7 demonstrates that, for a constant specimen width b, the tensile stress at the peel front is proportional to the experimentally determined peel stress F/b and the sealant's elasticity modulus E, and inversely proportional to both the thickness of the sealant layer h_a and the function $f(\dot{\varepsilon})$, which describes the relationship between the peel force F and peel rate $\dot{\varepsilon}$. It can also be seen from Equation 7.7 that the stress at the peel front increases in direct proportion to the reduction in the thickness of the sealant layer between the rigid and flexible substrates. It means, for example, that a fivefold reduction in the sealant thickness increases the stress level at the peel front five times in comparison to that achievable at the original thickness.

As discussed in detail References in 7–10, the peeling of a flexible layer of sealant (backed by an elastic backing tape) from a rigid substrate at a constant peel force F and a peel angle ψ involves energy changes as the crack propagates by an increment Δx through the sealant or along the interface. These energy changes must be balanced in accordance with the energy conservation principle, that is,

$$W + \Gamma + V = 0 \tag{7.8}$$

where W is the work of peeling, Γ is the energy required to create a new fracture surface, and V is the peel deformation energy (comprising stored elastic energy and the work associated with deforming the sealant along the path Δx) for a newly created interfacial area $A = [(b) . \Delta x]$ where b is the width of the peel specimen.

The values of the parameters in Equation 7.8 are determined as follows:

$$W = F\,(\lambda - \cos\psi)\,\Delta x \tag{7.9a}$$

$$\Gamma = -\,G\,.\,b\,.\,\Delta x \tag{7.9b}$$

$$V = (S_a h_a + Sh)\,b\Delta x \tag{7.9c}$$

where F is the measured peel force, λ is the extension rate of the flexible backing tape, ψ is the peel angle, b is the peel specimen width, h_a is the thickness of sealant layer between the flexible and rigid substrates, h is the thickness of the flexible backing tape, S_a is the deformation energy per unit volume of sealant, S is the deformation energy per unit volume of backing tape, and G is fracture energy per unit interfacial area.

The total fracture energy G per unit interfacial area in Equation 7.9b comprises the following two components [9]:

$$G = G_o + Uh_a \tag{7.10}$$

where G_o is the intrinsic interfacial fracture energy equal to the actual magnitude of intermolecular forces at the sealant–substrate interface, and Uh_a is the plastic dissipation energy irreversibly lost by yielding in the zone around the deformed tip of the propagating crack (U = dissipation energy per unit volume of sealant, h_a = thickness of the sealant layer between the flexible and rigid substrates).

Substitution of Equations 7.9a and 7.9c into Equation 7.8 gives the general formula [10] for fracture energy (determined from peel test), which is valid for any viscoelastic adhesive such as an elastomeric sealant:

$$G = \frac{F}{b}\,(\lambda - \cos\psi) + S_a h_a + Sh \tag{7.11}$$

If the flexible backing tape can be considered inextensible at the prevailing peel forces, then $\lambda = 1$ and $S = 0$. For this case, Equation 7.11 simplifies as follows:

$$G = \frac{F}{b}(1 - \cos\psi) + S_a h_a \tag{7.12}$$

For a slightly extensible backing tape, Equation 7.12 becomes

$$G = \frac{F}{b}\left(\frac{1+\lambda}{2} - \cos\psi\right) + S_a h_a \tag{7.13}$$

Equations 7.12 and 7.13 can be used for determining the fracture energy from peel tests carried out at a constant peel rate [ε˙] and any peel angle.

It is clear from Equations 7.12 and 7.13 that the adhesion forces exerted across the sealant–substrate interface (and related to fracture energy G of the interface through these equations) can be most accurately determined when the plastic dissipation energy is minimized through the reduction of S_a and h_a terms. This is effectively achieved by minimizing the thickness of the sealant layer, which, in turn, provides the desired reduction in the deformation energy per unit volume of sealant due to the increase in the material's modulus of elasticity with decreasing thickness, as shown previously. The reduction of peel angle ψ leads to a further reduction in the plastic deformation losses.

7.5.1.2   Relationship between the Elasticity Modulus and Sealant Thickness

It has been demonstrated earlier [11] that the apparent modulus of elasticity of elastomeric sealants strongly depends on the thickness of the material bonded between rigid substrates. Figure 7.5 illustrates an example of this relationship for Dow Corning two-component DC 983 sealant.

It shows that the modulus of elasticity of the sealant remains constant when the thickness of the sealant layer is reduced from 3 mm to 1 mm, but subsequently starts to increase exponentially when the thickness becomes less than approximately 0.5 mm. At 0.25 mm, the modulus is double that at 2 mm.

7.5.1.3   Effective Stress Control at the Sealant–
Substrate Interface in Peel Specimens

Considering the experimental data on peel force and elasticity modulus presented in Figures 7.4 and 7.5, respectively, the magnitude of the stress at the peel front in relation to the thickness of the sealant bead can be determined from Equation 7.7, as illustrated in Figure 7.6.

It can be concluded from the foregoing analysis of the theory and experimental data that the magnitude of the tensile stress at the sealant–substrate interface can be effectively maximized by reducing the thickness of the sealant layer between the bonded substrates. This result has been successfully applied in the experimental design of the specimen geometry and test protocol for preferentially assessing the adhesion between the sealant and substrate rather than the cohesive strength. It is discussed in detail in Section 7.7.1.

7.5.2   Joint Geometry Parameters Controlling Stress at the
Sealant–Substrate Interface under Tensile Strain

7.5.2.1   Alternative Utilization of the Joint Geometry Design

The average tensile stress σ at the sealant–substrate interface depends, according to Equation 7.1, on the strain ε, and increases with the sealant's modulus of elasticity E.

A detailed analysis of the influence of joint geometry on the behavior of a sealant within the joint can be effectively utilized as follows:

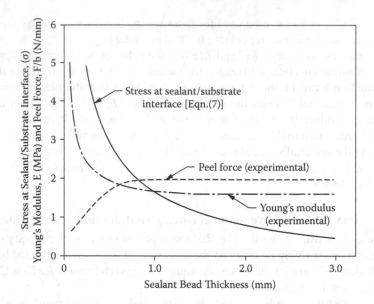

FIGURE 7.6 Tensile stress (σ) at the peel front at the sealant–substrate interface as a function of sealant bead thickness in accordance with Equation 7.7. (From W.S. Gutowski, L. Russell, and A. Cerra, in *Science and Technology of Building Seals, Sealants, Glazing and Waterproofing, Volume 2*, ASTM STP 1200, J.M. Klosowski (Ed.), p. 87, American Society for Testing and Materials, Philadelphia, PA, 1992.)

- In the engineering design of sealed joints to minimize the stress in structurally bonded systems and to ensure adequate movement capability when sealants are applied as weather seals, or
- In the design of test joints to maximize the stress at the sealant–substrate interface, thereby preferentially generating interfacial failure

7.5.2.2 Dependence of Elasticity Modulus on the Sealant Dimensions in Joints Subjected to Static or Dynamic Tensile Stresses

It has been demonstrated in Section 7.5.1.2 (see Figure 7.5) that the apparent elasticity modulus of elastomeric sealants is inversely proportional to the sealant bead thickness when subjected to dynamic tensile forces. This may be indicative of a similar profile of the relationship between the apparent tensile modulus of a sealant and the joint "aspect ratio" (ratio of depth to width).

To answer this question and to address its implications for the design of test protocols for assessing adhesion strength when subjecting the sealant–substrate interface to static or cyclic tensile stress, a set of experiments were carried out employing the tensile joints schematically illustrated in Figure 7.7.

The joint geometry was controlled by varying the sealant bead dimensions within the following range of parameters:

- Sealant bead thickness (*G*: glueline): 6–12 mm
- Sealant bead width (*B*: bite): 6–20 mm.

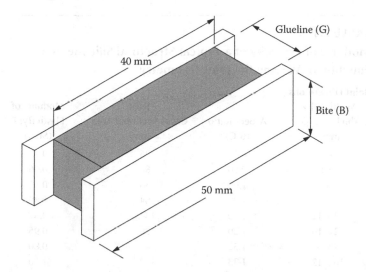

FIGURE 7.7 Geometry of tensile specimens for determining the modulus of elasticity, tensile strength, and elongation at failure of structural and weather sealants. (*Note:* The terms *glueline* and *bite* are commonly used by the building and construction industry instead of those typically used in adhesion science, that is, G = bondline thickness or sealant bead thickness or glueline, and B = joint width or sealant bead width or bite.)

The resultant joint aspect ratio [B/G], defined here as

$$Aspect\ ratio\ [AR] = \frac{[B]\ bead\ width}{[G]\ glueline\ thickness}$$

was controlled within the range of [AR] = 0.67 to 3.33. This represents changing the joint configuration from "oblong" to "flat," the latter exhibiting a bead width 3.33 times greater that the bead thickness.

The results for modulus of elasticity E for a range of joint geometries as investigated in this experiment for a typical structural silicone are provided in Table 7.1.

Figure 7.8 depicts the relationship between the elasticity modulus E and relative thickness of the sealant bead represented by the joint's aspect ratio B/G. It can be seen from a comparison of Figures 7.5 and 7.8 that the trend observed for the sealant in tensile joints follows that earlier observed for the peel joint configuration, that is, that the material's modulus of elasticity (in tensile mode) increases with the reduction of the relative thickness of the sealant layer or increasing aspect ratio.

This result can be used to design tensile test joints that preferentially fail interfacially rather than cohesively.

TABLE 7.1

Modulus of Elasticity of a Typical Structural Silicone as a Function of Varying the Joint Geometry

Joint Dimensions Width [B] × Thickness [G] (mm)	Aspect Ratio (B:G)	Joint Cross-Sectional Area (mm²)	Modulus of Elasticity: E (MPa)
8 × 12	0.67	96	0.75
8 × 10	0.80	80	0.66
6 × 6	1.00	36	0.82
8 × 8	1.00	64	0.89
12 × 12	1.00	144	0.87
12 × 10	1.20	120	0.95
8 × 6	1.32	48	0.80
16 × 12	1.33	192	1.10
12 × 8	1.50	96	1.25
16 × 10	1.60	160	1.15
20 × 12	1.67	260	1.06
12 × 6	2.00	72	1.45
16 × 8	2.00	128	1,26
20 × 10	2.00	200	1.34
16 × 6	2.66	96	1.27
20 × 6	3.33	120	1.41

Note: Test conditions: T = 20°C; strain rate = 50 mm/min.

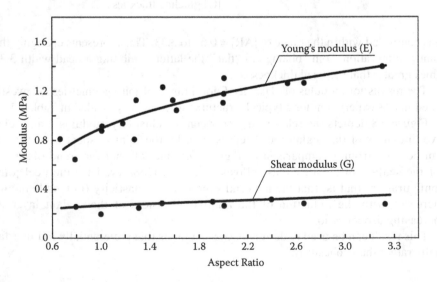

FIGURE 7.8 The relationship between elasticity modulus E and relative thickness of sealant bead represented by sealant bead aspect ratio (width:thickness).

7.6 OTHER FACTORS INFLUENCING THE MECHANICAL AND RHEOLOGICAL PROPERTIES OF SEALANTS

7.6.1 STRAIN RATE AND TEMPERATURE

7.6.1.1 Strain Rate in Conventional Sealant Joints

Sealants in applications such as building facades and other structures undergo cyclic strain imposed by daily and seasonal fluctuations of ambient temperature, which causes thermal movement in cladding panels. The observed sealant extension–compression cycling depends on the rate and extent of the temperature variation, substrate dimensions, and coefficient of thermal expansion. The typical range of strain rates is as follows [12]:

- Daily cycling rate: 10^{-1}–10^{-2} mm/min
- Seasonal cycling rate: 10^{-5}–10^{-6} mm/min

Consequently, test protocols for determining a sealants' engineering properties, including adhesion, must be determined at these strain rates and carried out within the range of actual service temperatures.

7.6.1.2 The Influence of Strain Rate and Temperature on the Tensile Strength of Sealants

It has been shown by Ferry [13] that the effects of strain rate and temperature on the rheological and mechanical properties of a polymer, for example, sealant tensile strength, are equivalent and thus can be superimposed by shifting each individual temperature-related property curve to a collective one to give a single master curve. The resultant master curve is described [13] by Equation. 7.14:

$$\frac{ST_0}{\varepsilon Ta_T} = \int_{-\infty}^{\infty} M(\tau) \cdot \tau \left[1 - e^{-\varepsilon / \dot{\varepsilon} a_T} \right] \cdot d\ln\tau \qquad (7.14)$$

where S is the stress; $\dot{\varepsilon}$ is the strain rate; T_0 is arbitrary reference temperature, for example, the glass transition temperature of the polymer; T is the actual test temperature at which the stress–strain curve (S versus ε) is determined; τ is time; $M(\tau)$ is relaxation distribution function; ε is strain; and a_T is the temperature shift factor.

The value of the temperature shift factor a_T is estimated through an Arrhenius approach using the Williams–Landel–Ferry (WLF) equation (Equation 7.15a):

$$\log a_T = \frac{C_1(T - T_0)}{C_2 + T - T_s} \qquad (7.15a)$$

where T is the test temperature; T_0 is the reference temperature, for example, the glass transition temperature of the polymer, and C_1, C_2 are experimental constants.

According to Ferry [13], in his first application of the WLF equation, the average values of C_1 and C_2 were obtained by fitting experimental data to a range of polymers and were estimated to be 17.44 and 51.6, respectively [14]. It was, however, pointed out by Ferry that the actual variations of C_1 and C_2 from one polymer to another were too great to permit the use of these constants as "universal" values, as is often done by many researchers. Ferry pointed out that, in a somewhat better approximation, fixed values of $C_1 = 8.86$ and $C_2 = 101.6$ should be used in conjunction with a reference temperature T_0, which was allowed to be an adjustable parameter but generally fell about 50°C above the T_g, consequently giving rise to Equation 7.15b:

$$\log a_\mathrm{T} = \frac{8.86\,(T - T_g - 50)}{101.6\,(T - T_g - 50)} \tag{7.15b}$$

The WLF formula (Equation 7.15b) gives satisfactory results over the temperature range: $T_g < T < T_g + 100°C$.

An example of the master curve construction using the WLF approach is illustrated in Figure 7.9 for the tensile strength of joints made using polybutadiene-styrene elastomeric adhesive, PB-SR ($T_g = -40°C$).

Once converted into a master curve, the rheological or mechanical properties of the sealant can be easily estimated over the range of more than ten decades of the reduced test rate ($\varepsilon\,a_\mathrm{T}$), which is easily converted into real time and strain rate.

Figure 7.10 illustrates the master curve, as developed by us earlier [12], representing the tensile strength of the Dow Corning 795 sealant, which in addition to temperature T and strain rate ε, also considers cross-sectional area A of the sealant bead in the joint.

7.7   TEST METHODS FOR DETERMINING ADHESION OF SEALANTS

7.7.1   PEEL TEST METHODS

Historically, one of the most common tests for studying the adhesion of sealants has been the peel test, commonly conducted at a peel angle of 180° and strain rate (peel rate) of 50 mm/min (e.g., ASTM C-794 [16]). After appropriate artificial aging of the specimen (water immersion, elevated temperature, freezing, etc.), the peel force and failure mode are recorded. Although the specimen geometry does not realistically resemble that of a typical joint, peel tests can reveal underlying adhesion problems *if* the factors controlling the failure mode are well understood.

A thorough understanding of the mechanics of peel testing (see Section 7.5.1.1), especially the use of appropriate specimen geometry (i.e., sealant bead thickness) and test conditions is essential for promoting the fracture of interfacial bonds rather than cohesive failure of the bulk sealant. This, of course, is essential for studying adhesion behavior.

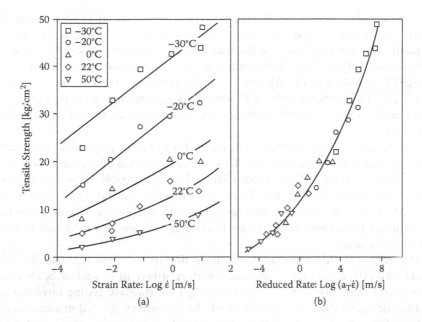

FIGURE 7.9 (a) Tensile strength versus test rate curves for joints comprising PET (Mylar) substrate and PB-SR elastomeric adhesive tested at various temperatures: −30°C to +50°C, (b) Master curve for the tensile strength of joints tested in (a) versus reduced test rate ($\varepsilon \cdot a_T$) [15]. (From A.N. Gent, *J. Polym. Sci. A-2, 9,* 283, 1971.)

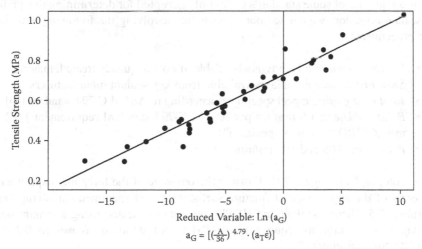

FIGURE 7.10 Master curve for the tensile strength of Dow Corning 795 silicone sealant. Range of applicability: Temperature T = −20°C to +80°C; strain rate: $\varepsilon' = 0.05$–250 mm/min; joint cross-sectional area: 36–420 mm². Model statistics: coefficient of correlation R = 0.968, standard deviation S = 0.046 MPa, average error = ±6.2%. (From W.S. Gutowski, P. Lalas, and A.P Cerra, in *Science and Technology of Building Seals, Sealants, Glazing and Waterproofing, Volume 5,* ASTM STP 1271, M. Lacasse (Ed.), p. 97, American Society for Testing and Materials, Philadelphia, PA, 1995.)

A disregard of these principles may lead to erroneous results and, consequently, to wrong conclusions regarding the quality of adhesion. This has been demonstrated through a comparative study on the adhesion of a structural silicone sealant to a high-density polyethylene (HDPE) substrate using the following two test protocols: (1) ASTM C-794 standard [16], and (2) CSIRO protocol involving "thin peel specimens" [11]. The former specifies a 1.6-mm-thick sealant layer in the peel specimens tested at strain rate [ε˙] of 50 mm/min, while the latter uses a sealant layer thickness of 0.2 mm.

HDPE, and polyolefins in general, are well known for poor adhesion to adhesives, sealants, or surface coatings. This is a consequence of their chemical inertness arising from the lack of surface chemical groups capable of creating chemical bonds with adhesives or sealants. Considering these any assessment of the adhesion strength of untreated HDPE to any sealant should reveal poor adhesion, which, in turn, should be demonstrated by interfacial delamination of the sealant from the substrate surface.

Regardless of this expectation, we have discovered, however, that the joints made with HDPE substrates and sealed with relatively thick sealant beads (thickness greater than 3 or 6 mm) frequently passed the standard testing involving peel and tensile specimens. This, in turn, might have inadvertently led to a catastrophic failure of structural joints involving this particular combination of materials under real-life service conditions regardless of the fact that the requirements of typical standards (e.g., ASTM or ISO) were satisfied.

To demonstrate the seriousness of the foregoing problem pointing out the inherent inadequacies of some standards commonly accepted for determining the quality of sealant adhesion, we carried out experiments involving the following materials and procedures:

- *Substrate material*: HDPE (nonbondable, if not adequately treated, material)
- *Sealants*: structural sealants available from key sealant manufacturers
- *Joint configuration*: peel specimens according to ASTM C-794 standard [16]
- *Bead thickness*: 1.6 mm (as per ASTM C-794 standard requirements) 0.2 mm (CSIRO "thin peel specimen")
- *Peel rates*: 0.05 and 50 mm/min.

The photographs in Figure 7.11 illustrate the outcome of the foregoing experiments in terms of the appearance of fracture surfaces of peel specimens involving Dow Corning 795 silicone sealant and HDPE substrate fabricated using a sealant bead thickness of 1.6 mm, as required by ASTM C-794 standard, as well as 0.2 mm (CSIRO "thin specimens").

It is evident from the photo in Figure 7.11a that the ASTM C-794 "thick peel specimen" exhibits 100% cohesive fracture within the sealant regardless of the fact that its interfacial delamination from the HDPE substrate has been expected. It can also be seen from this figure that lowering the peel rate by up to 2 orders of magnitude (from 5 to 0.05 cm/min) was not sufficient to reveal any potential deficiency of adhesion between the sealant and HDPE.

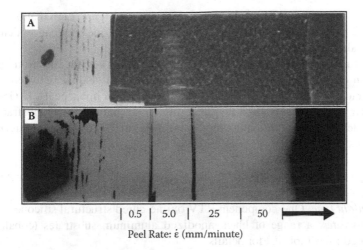

| 0.5 | 5.0 | 25 | 50 ➡

Peel Rate: ε̇ (mm/minute)

FIGURE 7.11 Appearance of fracture surfaces of 180° peel specimens prepared using a structural silicone sealant and HDPE substrate at the following two sealant thicknesses: (a) 1.6 mm (as per ASTM C-794 standard), and (b) 0.2 mm. Specimens were tested using peel rates of 0.5, 5, 25, and 50 mm/min.

The analysis of the fracture surface of the "thin peel specimen" illustrated in Figure 7.11b demonstrates, on the other hand, that 100% interfacial delamination occurs independently of the peel rate, thus confirming inadequate adhesion of the silicone sealant used in these experiments to HDPE.

The following becomes evident from an analysis of the fracture surfaces in Figure 7.11:

a. *Specimen (a)—thick sealant layer (1.6 mm)*: The use of a relatively thick sealant layer between the HDPE substrate and flexible backing tape leads to a "masking" of the expected poor adhesion by "producing" 100% cohesive failure within the sealant. This implies that the quality of adhesion satisfies the requirements of the ASTM C-794 standard, and hence, the sealant could be inappropriately used in structural applications.

   It is apparent that 100% cohesive failure has been observed not only for the standard-recommended strain rate of 50 mm/min but also for the rate as low as 0.5 mm/min, that is, 100 times slower. Typically, reducing the strain rate by two orders of magnitude should allow for the detection of poor interfacial adhesion through a change in failure mode from cohesive to 100% delamination along the substrate-sealant interface.

b. *Specimen (b)—thin sealant layer (0.2 mm)*: The use of a thin sealant layer between the HDPE substrate and backing tape creates conditions that promote the disruption of the interfacial bonds before cohesive failure can occur (see Section 7.5.1 for a detailed discussion of underlying phenomena).

Thus, this test configuration (thin sealant bead layer) yields results that are expected for a silicone sealant on an untreated HDPE, that is, poor adhesion. Also it is noticeable from the analysis of the fracture surface in Figure 7.11b that even a 100-fold increase in the strain rate from 0.5 to 50 mm/min does not cause a change in failure mode from interfacial to cohesive.

To further demonstrate how the control of the two key factors promoting the disruption of interfacial bonds, that is, a thin layer of sealant between the substrates, and low strain (peel) rate applied during specimen peel-off (180° peel angle), can be effectively used for detecting inadequate sealant–substrate adhesion, we designed an additional experiment utilizing the following materials and procedures:

- *Sealant bead thickness:* (a) "thick" peel specimen": 1.6 mm (as per ASTM C-794), (b) "thin peel specimen": 0.2 mm (CSIRO specification)
- *Sealant type:* One-component RTV oxime-cured structural silicone
- *Substrates:* a range of black anodized aluminum substrates (cobalt salt sealed): see Table 7.2 for details
- *Peel rate*: 0.05 and 50 mm/min.
- The complete set of data on the failure mode (% interfacial delamination) and peel force recorded for both "thick" and "thin" peel specimens are provided in Table 7.2 (data originally were reported in Reference 11). Similar trends have been observed for clear and bronze-colored anodized aluminum [11].

The data in Table 7.2 clearly indicate that the thick (1.6 mm) peel specimens are not suitable for testing the quality of adhesion between the one-component RTV oxime-cured structural silicone and a range of anodized aluminum substrates, as this configuration is insensitive to inherent adhesion problems. However, specimens with a thin sealant layer provide the desired stress concentration at the sealant–substrate interface to create adhesion failure (interfacial delamination) relatively independent of the applied peel rate (0.05 to 50 mm/min).

An analysis of Equations 7.12 and 7.13 in Section 7.5.1.1 provides further insight into the outcome of the foregoing experiments. It is seen from these equations that the strength of adhesion at the sealant–substrate interface (or fracture energy G of the interface) is most accurately determined when the term quantifying plastic dissipation energy is minimized through the reduction of factors S_a and h_a in these equations. This is effectively achieved by minimizing the thickness of the sealant layer. In turn, this provides the desired reduction in deformation energy per unit volume of sealant due to an increase in the material's modulus of elasticity with decreasing thickness (see Section 7.5.1.2). The reduction of peel angle ψ leads to further minimization of the plastic deformation losses.

The foregoing observations, originally reported in our earlier publications [7,11], were subsequently confirmed by Shephard [17]. He also provided comprehensive experimental evidence [18] demonstrating a significant increase in the sensitivity of peel tests carried out at 45° peel angle. This test configuration led to complete interfacial delamination of the sealant that originally exhibited 100% cohesive failure when tested at identical strain rates but peeled at a standard 180° angle.

TABLE 7.2
Comparison of the Failure Mode (Percentage of Interfacial Delamination) and Peel Force for 180° Peel Specimens Comprising a Range of Black-Snodized Aluminum Substrates and a One-Component RTV Oxime-Vaured Structural Silicone

| | Thick Specimen: 1.6 mm Peel Rate: 50 mm/min | | Thin Specimen: 0.2 mm | | | |
| | | | Peel Rate: 50 mm/min | | Peel Rate: 0.05 mm/min | |
Substrate[a]	Peel Force (N/mm)	Percentage of Interfacial Delamination	Peel Force (N/mm)	Percentage of Interfacial Delamination	Peel Force (N/mm)	Percentage of Interfacial Delamination
25 μm/HS	3.04	0	1.00	95	0.24	100
25 μm/CS	1.60	60	0.52	100	0.16	100
25 μm/AS	2.48	0	0.92	60	0.36	100
15 μm/HS	3.24	0	0.80	95	0.18	100
15 μm/CS	3.00	5	0.88	95	0.18	100
15 μm/AS	3.20	0	1.04	95	0.20	100

[a] Anodized aluminum (6063 alloy): 15/25 μm anodized layer thickness; HS—hot sealed; CS—cold sealed; AS—accelerated sealing.

7.7.2 TEST METHODS BASED ON TENSILE SPECIMENS

7.7.2.1 Deficiencies in Existing Test Methods Involving Tensile Specimens

Test methods based on the use of a tensile joint configuration typically employ specimens such as those depicted in Figure 7.7. In most cases the width of the joint, B, is identical to the glueline thickness G, and typically set at 10 or 12 mm.

As discussed in depth [19–21], most test protocols and standards employing a tensile test configuration provide poor discrimination in adhesion levels between sealants and substrates. Typical approaches involve preconditioning the specimens by controlled exposure to static environmental conditions such as (1) heat–freeze cycling at the extremes of service temperatures (e.g., −29°C and +88°C) (2) immersion in water, or application of water spray, and (3) UV radiation (xenon arc, fluorescent UV-A lamps, or carbon arc). Subsequent mechanical testing provides data on the joint strength, elongation at failure, and percent delamination at the interface.

Mechanical testing, when carried out *after* artificial exposure, rarely yields significant information on the quality of adhesion except in cases when 100% delamination occurs. Frequently, partial delamination (e.g., 10%–15%) of sealant is initiated at the location of the highest stress concentration, that is, along the edge of the extended sealant bead, followed by the crack tip veering off from the interface into the bulk sealant and propagating there with increasing strain. Eventually, the interfacial stress is relieved, and the joint appears to have failed in a predominantly cohesive failure mode.

As first noted by Cerra [19], standard tensile-based tests can often lead to ambiguous results as follows:

a. A mixed failure mode (partial interfacial delamination and partial cohesive failure of the sealant) results when the crack path deviates from the interface to the bulk material, masking any poor adhesion due to the resultant stress relief. This leads to the release of interfacial stresses as the subsequent strain energy dissipation then continues through the deformation of the bulk sealant material instead of concentrating at the sealant–substrate interface.

b. Most test protocols for determining sealant adhesion fail to recognize that, in actual service, joints fail due to the concurrent action of three key degradation factors that adversely affect adhesion, namely,
   • Mechanical stress
   • Sealant strain
   • Moisture present at the interface in the form of a condensed film and/or water vapor.

The combined action of these factors results in the irreversible disruption of the interfacial bonds even with silicone sealants, which can usually recover when individual stresses are removed. The presence of moisture together with tensile stress imposed by the joint strain leads to the rupture of individual bonds, while the resultant crack opening prevents them from being reconstituted, as would have been the case in absence of strain.

7.7.2.2 CSIRO Test Method for Determining Adhesion of Sealants

7.7.2.2.1 Fundamental Principles of the Method

Molecular bridges, which may initially provide good adhesion across the sealant–substrate interface, can be effectively disrupted under the combined action of mechanical stress, strain, and water molecules. If the strain energy results in the interfacial stress concentration at the joint's edge exceeding the bond's critical fracture energy, fracture of the bonds bridging the interface occurs. Under test conditions that apply continuous and high-enough stress levels and sustained strain, stress relaxation is prevented, and a sustained crack propagation along the interface takes place.

The principle of applying simultaneous, rather than sequential, mechanical, and hydrothermal stresses to the interface gave rise to the novel test procedure developed by Cerra [19] for determining the strength and durability of adhesion between elastomeric sealants and rigid substrates. The magnitude of tensile stress applied to the interface (and combined with immersion in water or exposure to close to 100% humidity) is well below the cohesive strength of the sealant material. Typically, it does not exceed 0.138 MPa (20 psi), which is the design stress for structural bonding applications of sealants in façade engineering.

Another important phenomenon concerning the exposure of elastomeric sealants to constant stress (as applied in the CSIRO test procedure) has been noted by the

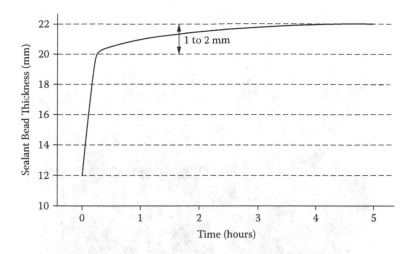

FIGURE 7.12 Creep deformation of Dow Corning 795 structural sealant subjected to constant stress under water immersion at 55°C.

authors of this paper and is also relevant to this protocol, that is, to their viscoelastic nature, sealants subjected to appropriate levels of tensile stress will deform through progressive creep, as illustrated in Figure 7.12 for Dow Corning 795 sealant.

In this example, the material starts deforming through creep after approximately 20 min. The strain rate during this stage of sealant deformation was found to be in the range of 0.002 to 0.003 mm/min. This is similar to the strain rates that building façade components experience due to daily and annual temperature fluctuations that lead to extension–compression cycling of sealants in the façade's joints.

Based on the foregoing results, it has been demonstrated [12] that the adhesion and cohesive strength of the sealants, as determined through the CSIRO test protocol, reflect the material's properties under similar service conditions that might occur in curtain walls comprising 2500-mm-long aluminum frame and glass panels bonded with a structural silicone sealant.

7.7.2.2.2 Technical Details of the Method

The full details of the CSIRO method are described in References 19 and 20. The quality of sealant–substrate adhesion is determined using tensile specimens (see Figure 7.7) with sealant bead dimensions of 12 × 12 × 40 mm. These are subsequently subjected to creep-load conditions under a combination of mechanical and hydrothermal stresses. This is achieved by exposing the specimens to water immersion at 50°C and simultaneously applying a predetermined level of stress to the bondline. Figure 7.13 illustrates the general outline of the CSIRO creep exposure apparatus.

The specimens are inserted into a row of stainless steel clamps at the bottom of the water baths below the water level.

The broad range of fracture stress is initially determined by increasing the stress in 69 kPa (10 psi) increments until failure occurs. The permanent load applied to the sealant–substrate interface is retained for a predetermined period,

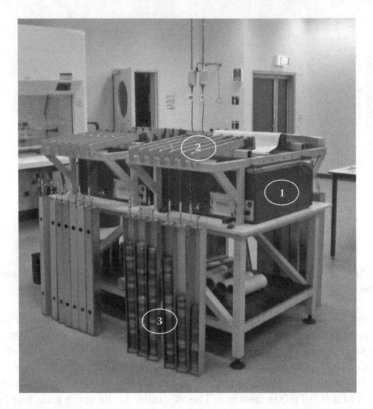

FIGURE 7.13 A view of the CSIRO creep apparatus comprising the following key components: (1) water bath preset to desired temperature, for example, 50°C; (2) system of levers that transpose the load (3) to tensile specimens, hence applying a predetermined tensile stress level to the sealant–substrate interface.

for example, 24 h. If the bonded specimen does not fail during this time a new specimen is inserted and the load increased by the aforementioned incremental value. The cycle, always employing a new specimen, is repeated up to the stage when the failure stress resulting in specimen delamination or cohesive failure is achieved. Subsequently, multiple replicates are tested at gradually increasing stress levels (13.8 kPa [2 psi]). Ultimately, the accurate value of the true strength of adhesion is determined through the use of a statistical procedure known as the "Dixon up-and-down" protocol [22]. All failed specimens are also visually inspected to determine percent delamination and cohesive failure, and the procedure requires about 15 specimens to achieve a statistically significant result.

7.7.2.2.3 Comparison of Results from the CSIRO Creep Test Method and Standard Tensile Tests

A comprehensive analysis of multiple experimental data sets demonstrating the advantages of the CSIRO test method over currently known test methods is provided in References 19 and 20. A comparison of typical results is discussed in this section to demonstrate that the standard tensile test carried out after accelerated exposure may not discriminate the adhesion properties of sealant–substrate combinations.

Table 7.3 below provides details of substrates bonded in one of our experiments using a 1-component oxime-cured structural silicone sealant commonly used in façade engineering for structural bonding of large glass panes to various substrate materials.

Table 7.4, in turn, provides results of adhesion tests carried out after the following conditioning [denoted (1) to (3)] of the tensile specimens (12 × 12 × 40 mm sealant bead) fabricated using the one-component oxime-cured structural silicone sealant and substrates listed in Table 7.3.

1. "Dry" tensile specimens (1 month sealant cure) without any conditioning
2. Specimens immersed in 20°C water for 7 days without stress and strain
3. CSIRO test procedure [see Section 7.7.2.2]: simultaneous application of tensile stress and water immersion (at 50°C).

It is seen from the data presented in Table 7.4 that tensile testing of dry specimens, or even those immersed (without prestressing) for 7 days in room temperature water prior to tensile testing, always results in 100% cohesive failure of the sealant without indication of any inherent adhesion problems.

However, as seen from the last column of Table 7.4, the CSIRO test protocol involving the simultaneous application of hot water immersion, stress, and strain provides clear discrimination in the quality of adhesion for the sealant–substrate combinations studied. The results show that only the coated glass displays no adhesion failure.

Another set of experiments involved a range of generic types of structural and weather sealants and a diversified range of substrates including anodized aluminum, concrete, granite, and compressed fiber-cement sheet as used in building and construction applications. A description of the individual materials is provided in Table 7.5, while Table 7.6 provides the comparative test data for tensile specimens

---

TABLE 7.3
Details of Anodized Aluminum and Glass Substrates Bonded with a One-Component RTV Oxime-Cured Structural Silicone Sealant

Substrate Code	Substrate Description
G 10	Float glass with stainless steel coating
B 9	Black anodized *"accelerated-sealed"* Al
EB 3	Black anodized Al ("Chemel" process[a])
EA 1	Clear anodized Al ("Chemel" process [a])
B 9	Black anodized "accelerated-sealed" Al

[a] Anodized (sulfuric acid) and electrolytically colored (using a tin salt) aluminum.

TABLE 7.4

Results of Sealant Adhesion Tests in Terms of Tensile Strength Data after the Following Sample Conditioning Protocols: (1) No Conditioning; (2) 7-Day Immersion in 20°C Water; (3) CSIRO Procedure: Concurrent Stress and 50°C Water Immersion

Substrate Code	Conditioning (1) "Dry Specimen"		Conditioning (2) 7-Day Immersion in 20°C Water		Conditioning (3) "CSIRO Protocol" (Stress + 50°C Water)	
	Strength (MPa)	Interfacial Delamination (%)	Strength (MPa)	Interfacial Delamination (%)	Strength (MPa)	Interfacial Delamination (%)
G 10	0.85	0	0.85	0	0.65	0
B 9	0.83	0	0.80	0	0.52	20
EB 3	0.94	0	0.85	0	0.12	93
EA 1	0.91	0	0.74	0	0.12	92
B 9	0.92	0	0.72	0	0.16	78

TABLE 7.5

Details of Sealants and Substrates Used in THE Comparison of Various Exposure Conditions Prior to Testing Sealant Adhesion Using a Tensile Joint Configuration

Sealant		Substrate	
Sealant Code	Generic Type	Substrate Code	Substrate Details
A	One-part silicone (alkoxy-cured)	1	Anodized Al (dark bronze)
G	Two-part polyurethane	7	Granite (polished)
H	One-part polyurethane	8	Compressed fiber-cement
G	One-part polyurethane (silicone-modified)	10	Concrete (prepared as per ASTM C719)

tested as mentioned, that is, dry, after 7-day water immersion at 20°C and using the CSIRO test procedure.

The results presented in Table 7.6 provide further evidence that the CSIRO test procedure is better able to discriminate adhesion quality than the more conventional methods.

For more comprehensive information, readers are referred to complete sets of experimental data provided in References 19 and 20.

TABLE 7.6
Results of Sealant Adhesion Tests in Terms of Tensile Strength Data for a Range of Sealant–Substrate Combinations (Structural and Weather Sealants) Listed in Table 7.5

Substrate Code	Conditioning (1) "Dry Specimen"		Conditioning (2) 7-Day Immersion in 20°C Water		Conditioning (3) "CSIRO Protocol" (Stress + 50°C Water)	
	Strength (MPa)	Interfacial Delamination (%)	Strength (MPa)	Interfacial Delamination (%)	Strength (MPa)	Interfacial Delamination (%)
Structural Sealant: Silicone						
A1	0.62	0	0.82	12	0.19	25
A7	0.74	0	0.52	0	0.35	5
Weather Sealant: Two-Part Polyurethane						
G1	0.95	47	0.31	0	0.05	100
G7	1.00	0	0.47	30	0.05	100
G8	1.03	0	0.30	3	0.13	2
G10	1.03	0	0.49	0	0.05	100
Weather Sealant: One-Part Polyurethane						
H1	0.41	0	0.36	17	0.137	20
H7	0.37	0	0.34	0	0.11	40
H8	0.35	0	0.33	100	0.14	40
H10	0.42	0	0.33	10	0.95	100
Weather Sealant: One-Part Polyurethane (Silicone-Modified)						
J1	0.74	0	0.66	0	0.08	1
J7	0.76	0	0.53	0	0.05	100
J8	0.73	0	0.24	77	0.05	100
J10	0.77	100	0.35	100	0.05	100

7.8 CONTEMPORARY STANDARDS FOR DETERMINING SEALANT PERFORMANCE

The first standards [23–25] for evaluating the performance of elastomeric sealants were developed for the assessment of weather sealants and focused on the retention of bond and cohesive integrity rather than load-bearing capacity such as in structural adhesives.

In the mid-1990s, the International Union of Laboratories and Experts in Construction Materials, Systems and Structures, RILEM (Reunion Internationale des Laboratoires et Experts des Matériaux, Systèmes de Constructions et Ouvrages) recognized the deficiencies of the existing protocols for the accelerated evaluation of the performance of elastomeric sealants, particularly with regard to adhesion and long-term durability. The task of designated Technical Committee TC139-DBS

(Durability of Building Sealants) was to prepare an internationally recognized protocol for the accelerated determination of the performance of elastomeric joint sealants for use in high-movement building façades, which are subjected to continuous cyclic movements and severe weathering throughout their service life.

7.8.1 RILEM ACCELERATED DURABILITY TESTS

The protocol developed by TC 139-DBS [26] introduces "static" or "dynamic" conditioning of sealant specimens ($50 \times 12 \times 12$ mm) during the 28-day curing at $23 \pm 2°C$ and $50\% \pm 5\%$ RH. The dynamic cure simulates the variation of the working glueline thickness of the sealant bead in the building façade during cure.

After the completion of 28-day (static or dynamic) cure, the sealant specimens are exposed to three cycles of accelerated degradation as follows:

a. Artificial weathering exposure
b. Thermomechanical cycling comprising two consecutive cycles of low-temperature extension ($-20 \pm 2°C$) and high-temperature compression ($70°C \pm 2°C$), both carried out at the extremes of the sealant's rated movement capability (e.g., $\pm12.5\%$, $\pm20\%$, or $\pm25\%$)

After completion of each degradation cycle (a and b) the specimens are extended to their rated extension capability and then examined for the following signs of failure:

1. Percent adhesion loss (delamination between sealant and substrate)
2. Loss of cohesion
3. Whether the locus of failure was at the sealant–substrate interface or in the bulk of the sealant

7.8.1.1 Static Conditioning Protocol

RILEM TC-139 "static conditioning" [26] comprises three cycles of the following consecutive steps:

1. 3-Day oven exposure at $70°C \pm 2°C$
2. 1-Day immersion in distilled water at $23°C \pm 2°C$
3. 2-Day oven exposure at $70°C \pm 2°C$
4. 1-Day immersion in distilled water at $23°C \pm 2°C$

7.8.1.2 Dynamic Conditioning Protocol

RILEM TC-139 "dynamic conditioning" [26] recognizes the fact that, in real life, particularly in the building (e.g., in building facades) or shipbuilding industry, the joint width varies cyclically after installation due to thermal movement in the substrates. Consequently, the freshly fabricated specimens are inserted into a cyclic movement device, allowing joint extension or compression at a controlled rate of 70 ± 20 mm/min.

The "dynamic conditioning" involves an overall 28-day sealant cure as follows:

a. A total of 14 daily cycles involving the following steps carried out at 23°C ± 2°C and 50°C ± 5 % RH:
   1. Extension of the freshly fabricated specimen by 7.5% within 5 min of fabrication
   2. Holding for 2.4 h at 7.5% extension
   3. Compression of the specimen by 7.5%
   4. Holding for 2.4 h at 7.5% compression
   5. Returning the joint to its initial width of 12 mm for the remainder of the 24-h cycle.
b. 14-day cure in a static state (joint returned to its initial width of 12 mm) at 23°C ± 2°C and 50% ± 5% RH

7.8.2 RILEM PROTOCOLS INVOLVING ARTIFICIAL WEATHERING

After the completion of the cure under "static" or "dynamic" conditions, the specimens are subjected to artificial weathering [26–30] involving one of the following:

- *Xenon arc (340 nm lamps at 0.5 W/m² nm) automatic weathering* [27] involving 8-week cycling as follows:
  - 102 min of "dry heat" exposure at 65°C ± 2°C and 65% RH
  - 18 min of water spray or immersion in water of temperature less than 40°C, or
- *Fluorescent UVA-340/water spray cyclic exposure* [28] involving 8-week cycling as follows:
  - 8 h of "dry heat" exposure at 65°C ± 2°C
  - 4 h of UV radiation and water spray (water temperature less than 40°C)

7.8.3 THERMOMECHANICAL CYCLING AND DETERMINATION OF SEALANT DURABILITY

After the artificial weathering outlined in Section 7.8.2, the specimens are subjected to two cycles of thermomechanical stress involving low-temperature extension and high-temperature compression at the sealant's rated movement capability as follows:

- *Day 1*: The specimen is conditioned for 3 h at –20°C ± 2°C, then extended to its rated movement capability (e.g., +25%) for a period of 21 h.
- *Day 2*: After release of extension, the specimen is conditioned for 3 h at 70°C ± 2°C, then compressed to its rated movement capability (e.g., –25%) for a period of 21 h.
- *Day 3*: Repeat of procedure for Day 1.
- *Day 4*: Repeat of procedure for Day 2.

- *Day 5–7*: Release of compression and 3-day storage at 23°C ± 2°C and 50°C ± 5% RH.

The specimens are then assessed for

1. Percent adhesion loss (delamination between sealant and substrate)
2. Loss of cohesion
3. Whether the failure occurred at the sealant–substrate interface or in the bulk of the sealant

The standard RILEM TC-139 protocol [26] requires exposure of the specimens to at least three cycles of artificial weathering exposure and thermomechanical cycling, as described earlier. It is recommended, however, that the number of degradation cycles should be such as to induce substantial (visible) degradation in the least stable material in the group of candidate sealants.

7.8.4 MOST RECENT RILEM RECOMMENDATIONS FOR SEALANT DURABILITY ASSESSMENT

The outcome of the most recent work carried out by the RILEM working group TC 190-SBJ ("Service Life of Building & Construction Joints") provides further improvements over the earlier protocols described in Sections 7.8.1 to 7.8.3, by adopting test and cycling conditions to better reproduce local climatic and service conditions. The Group, under the direction of Wolf and Enomoto, has recently prepared a draft document [31] outlining the details of a proposed "RILEM Technical Recommendation."

The broad details of the new protocol [31] are as follows:

1. Sealant specimens, 100 × 20 × 15 mm (length × width × thickness) are first cured and conditioned "statically," that is, without any joint movement for 28 days at 23°C ± 2°C and 50% ± 5% RH.
2. The fully cured specimens are then exposed to repetitive exposure cycles (6- week cycles, as described in Section 7.8.2) comprising
   - UV light (xenon arc, or fluorescent UVA-340 nm, or carbon arc), heat and moisture (water spray or complete water immersion)
   - Repetitive cyclic movement—manual or automated (6 weeks) carried out during or after the accelerated weathering period (see earlier) comprising mechanical cycling up to the sealant's rated movement capability (i.e., ±12.5%, ±25%, or ±50%) with the following cycles:
     - One extension and compression cycle (at a rate of 5.5 ± 0.7 mm/min) per 24 h for 4 consecutive days
     - Specimen at rest (no joint movement) for 3 days
3. After each exposure cycle (item 2 in this list) the specimens are extended to their full rated movement capability (i.e., +12.5%, +25%, or +50%), and the joint is examined for signs of failure.

4. This procedure is repeated as often as required to achieve a visible degree of degradation.

It is noteworthy that the proposed RILEM TC 190-SBJ protocol broadly resembles the procedure adopted by the ASTM C 1519 standard [25], which recommends the following test procedure:

- Sealant specimens (12.5 × 12.5 × 50 mm) are cured for 21 days under the following conditions: (a) 7 days at 23°C ± 2°C and 50% ± 5% RH; (b) 7 days at 38°C/95% RH; and (c) 7 days at 23°C ± 2°C and 50% ± 5% RH.
- 4-week artificial weathering cycle (xenon arc or Fluorescent UVA-340 nm/ water spray cyclic exposure: 2 h UV light/2 h water spray), followed by 6 cycles of extension and compression at room temperature.
- Joint examination for signs of failure (including quantification).
- Repeat cycles until some degradation is visible.

7.9   SUMMARY

This chapter provides a comprehensive analysis of the theoretical and practical aspects of the principal factors and mechanisms controlling interfacial failure between sealants and substrates. The factors influencing the properties and in-service performance of elastomeric sealants are analyzed with emphasis on their short- and long-term adhesion.

An in-depth review and analysis of contemporary test methods for determining the adhesion behavior of sealants indicated a number of deficiencies in the existing test protocols and standards. It is shown through numerous practical examples that in order to demonstrate satisfactory strength and durability of adhesion, one needs to subject the interface to a set of conditions likely to promote disruption of interfacial bonds before the cohesive failure of bulk sealant occurs. It is shown that, to achieve this, it is necessary to simultaneously apply both mechanical and accelerated environmental stress, the former at the levels below the cohesive strength of sealant material.

A practical guide on reliable test protocols designed to ascertain satisfactory long-term performance of sealants in service is also provided.

REFERENCES

1. American Society for Testing and Materials (ASTM) Standard No. C1401-02: "Standard guide for structural glazing" (2002).
2. American Society for Testing and Materials (ASTM) Standard No C1193-05: "Standard guide for use of joint sealants" (2005).
3. D.H. Kaelble, *Trans. Soc. Rheology*, *4*, 45 (1960).
4. D.H. Kaelble, *Trans. Soc. Rheology*, *3*, 161 (1959).
5. D.H. Kaelble, *Trans. Soc. Rheology*, *9*, 135 (1965).
6. A.N. Gent and G.R. Hamed, *J. Appl. Polym. Sci*, *21*, 2817 (1977).
7. W.S. Gutowski and E.R. Pankevicius, *Fatigue Fract. Eng. Mater Struct.*, *17*, 351 (1994).
8. K. Kendall, *J. Phys. D.: Appl. Phys.*, *8*, 1449 (1975).

9. S. Wu, *Polymer Interface and Adhesion*, p. 535, Marcel Dekker, New York (1982).
10. E.J. Ripling, S. Mostovoy, and H.T. Corten, *J. Adhesion*, *3*, 107 (1971).
11. W.S. Gutowski, L. Russell, and A. Cerra, in *Science and Technology of Building Seals, Sealants, Glazing and Waterproofing, Volume 2*, ASTM STP 1200, J.M. Klosowski (Ed.), p. 87, American Society for Testing and Materials, Philadelphia, PA (1992).
12. W.S. Gutowski, P. Lalas, and A.P Cerra, in *Science and Technology of Building Seals, Sealants, Glazing and Waterproofing, Volume 5*, ASTM STP 1271, M. Lacasse (Ed.), p. 97, American Society for Testing and Materials, Philadelphia, PA (1995).
13. J.D. Ferry, *J. Am. Chem. Soc.*, *72*, 3746 (1950).
14. M.L. Williams, R.F. Landel, and J.D. Ferry, *J. Am. Chem. Soc.*, *77*, 3701 (1955).
15. A.N. Gent, *J. Polym. Sci. A-2*, *9*, 283 (1971).
16. American Society for Testing and Materials (ASTM) Standard No C794-06: "Standard guide for structural glazing" (2006).
17. N.E. Shephard, in *Durability of Building Sealants*, RILEM Proc., Volume 37, A.T. Wolf (Ed.), p. 161, E&FN Spon, London (1999).
18. N.E. Shephard and J.P. Wightman, in *Science and Technology of Building Seals, Sealants, Glazing and Waterproofing, Volume 5*, ASTM STP 1271, M. Lacasse (Ed.), p. 226, American Society for Testing and Materials, Philadelphia, PA (1995).
19. A.P. Cerra, *J. Testing Eval.*, *23*, 370 (1995).
20. A.P. Cerra and W.S. Gutowski, in *Science and Technology of Building Seals, Sealants, Glazing and Waterproofing, Volume 5*, ASTM STP 1271, M. Lacasse (Ed.), p. 209, American Society for Testing and Materials, Philadelphia, PA (1995).
21. J. Iker, and A.T. Wolf, in *Proceedings of the Symposium on Building Sealants: Materials, Properties and Performance*, ASTM STP 1062, T.F. O'Connor (Ed), p. 67, American Society for Testing and Materials, Philadelphia, PA (1990).
22. W.J. Dixon, *J. Am. Stat. Assoc.*, *69*, 967 (1965).
23. American Society for Testing and Materials (ASTM) Standard No. C794-80, "Standard test method for adhesion-in-peel of elastomeric joint sealants" (1980).
24. American Society for Testing and Materials (ASTM) Standard No. C719-05, "Adhesion and cohesion of elastomeric joint sealants under cyclic movement (Hockman cycle)" (2005).
25. American Society for Testing and Materials (ASTM) Standard No. C1519-04, "Evaluating durability of building construction sealants by laboratory accelerated weathering procedures" (2004).
26. RILEM Standard: RILEM TC 139-DBS: "Durability Test Method—Determination of changes in adhesion, cohesion and appearance of elastic weatherproofing sealants for high movement façade joints after exposure to artificial weathering," *Mater. Struct.*, *34*, 579 (2001).
27. ISO 4892-Part 2: 2006 Plastics, "Method of exposure to laboratory light sources—Part 2: Xenon lamps" (2006).
28. ISO 4892-Part 3: 2006 Plastics, "Method of exposure to laboratory light sources—Part 3: Fluorescent UV lamps" (2006).
29. (a) ISO 9047: 1989, and (b) ISO 9047: 2001. "Building Construction—Jointing Products—Determination of adhesion/cohesion properties of sealants at variable temperature" (1989 and 2001).
30. ISO/DIS 11600: 2002, "Building Construction—Jointing Products—Classification and Requirements for Sealants" (2002).
31. A.T. Wolf, and N. Enomoto, "Durability Test Method—Determination of changes in adhesion, cohesion and appearance of elastic weatherproofing sealants after exposure of statically cured specimens to artificial weathering and mechanical cycling," in *RILEM Technical Committee 190-SBJ: Document No. NO15*, Proposed RILEM Technical Recommendation (RTR) Accelerated Weathering Test Method—Work Group Draft Version 2006-08031 (2006).

8 Answering the LEED™ Challenge to Sealant and Weatherproofing Products

Michael Schmeida

CONTENTS

8.1 INTRODUCTION

Green building is arguably the newest trend in the construction industry. As our society has paid the price for decades of unrestrained use of natural resources such as crude oil, water, and clean air, we are faced with increasing needs to conserve such resources and to reduce the impact on the environment while maintaining and pushing forward with the conveniences of our modern lifestyle.

Before we can discuss the trends, practices, and direction of green building, we need to define this term. Green building is the practice of designing and constructing occupied space that uses less energy, water, and nonrenewable natural resources and allows for the convenient and continued use of methods by the building occupants [1]. While this is a very generic definition, as we proceed through this discussion, each of these points will be further elaborated.

Sealants and waterproofing have, even before green building became a trend, functioned in a "green" manner. Sealants have always been and by definition are used to seal gaps in the building envelope to eliminate the infiltration of air and water, and thus, one can deduce that these materials reduce the energy consumed to heat and/or cool buildings [2]. Today, however, "green" takes on a whole new meaning not only as it relates to traditional sealant and waterproofing applications and uses but also how these materials are made, how they are used, and even which products are used.

The most notable guide for green building is the United States Green Building Council's (USGBC) Leadership in Energy and Environmental Design (LEED) program. LEED is a systematic approach to building construction, design, and use that makes the practice of green building easy to understand and achieve. LEED has evolved from one single program—LEED for New Construction and Major Renovations (known as LEED-NC)—into many programs and even subprograms encompassing renovation of an existing structure, programs for leased and tenant-occupied spaces, homes, and even specific subprograms for specific use structures such as schools and hospitals. Still, the most common LEED program in use today is LEED-NC. Most discussions, including this one, utilize LEED-NC as the basis for discussing green building and defining possible practices for constructing a truly eco-friendly building [3]. It should be noted that, regardless of whether you are under the guise of the USGBC or another GBC (there are over two dozen GBCs throughout the world), all LEED programs are currently based on those established by the USGBC. For reference, here are the other LEED programs and what construction segments they are designed to fit into [4]:

- LEED for Existing Buildings—designed for major remodeling of buildings over 5 years old
- LEED for Core and Shell—designed for landlords and developers to utilize LEED for unfinished, leased commercial space
- LEED for Commercial Interiors—designed for tenants, being a complement to the LEED for Core and Shell
- LEED for Homes—designed for single-family home construction
- LEED for Housing Developments—designed for allotments and condominium development

Regardless of the LEED program that is applicable to your particular construction type, there are basic goals common to all of these programs. First, both construction-phase and long-term occupancy impacts on the surrounding ecosystem must be minimized. Second, water conservation and impacts on the aquifer are minimized. Third, energy efficiency is addressed. Fourth, the classic "reduce, reuse, and recycle" principles come into play through the reuse and recycling of materials as well as the emphasis on renewable and locally harvested materials. Fifth, the natural light and ventilation for both energy reduction as well as occupant health are emphasized. Finally, LEED programs reward those going the extra step and those who utilize the knowledge of LEED Accredited Professionals (LEED-APs). In just this high-level overview, we can see how building sealing can be primarily impacted by and also have an impact on green design [5].

8.2 WHY CARE ABOUT LEED/GREEN BUILDING?

The classic question anyone asks regarding any topic is, "Why should I care at all?" This is a fair question. There really is only one answer, "Because you have to care or be at a disadvantage." Green building is no exception. It has begun to play a significant role in the sealant and weatherproofing industry.

The answer is a little more complicated than "Because." There are a number of reasons detailed on the USGBC web site [6]:

- LEED is being adopted globally. Over two dozen countries have GBCs, which are in essence just "branches" of the USGBC. These include Canada, Mexico, most countries in Western Europe, Australia, New Zealand, India, China, and Chile.
- LEED is being integrated into law, be it in building codes or traditional legislation. California has mandated that all government buildings meet LEED requirements for certification; Boston has done the same and, most recently, the USGBC and ICC have signed a letter of understanding to integrate basic green building design requirements into future versions of the code. Further, the U.S. Federal government requires sustainable design requirements on new government buildings.
- Because of these requirements, many green building structures tend to be very high profile projects such as convention centers, arenas, and municipal buildings.
- There is currently nearly 1 billion square feet of commercial construction either registered or certified with the USGBC. This does not account for green building projects outside of the United States. It is unknown how many square feet are unregistered or uncertified that are being or were built with green building requirements in mind.
- Of the nearly 1 billion square feet that has been registered, only 12% of the buildings are certified, meaning that the number of in-progress buildings is dramatically higher than in previous years.
- Financial reasons such as tax breaks for high-efficiency buildings that minimize stress on the infrastructure, and lower utility bills including water, gas and electric, are becoming more prevalent.
- Approximately 80% of a company's operating costs are in its employees. In high-efficiency green buildings, incidents of absence due to allergies, asthma, and colds are reduced. Further, since these buildings have generally pleasant lighting and emphasize occupant comfort, related employee complaints are reduced on average by 16%, and work output is increased similarly.
- Architects, engineers, and contractors are being required to know about green building practices. There are over 35,000 LEED APs and an estimated 90,000+ design professionals currently involved in LEED-registered products in the United States.

Now that we know the reasons to consider green building, let us examine LEED-NC to see more specifically how the sealant industry is impacted. However, it should be noted that this chapter is not meant to be a guide to the LEED-NC system. This discussion will not cover every credit but only those where there is or potentially could be some impact on the building-sealing industry, be it either in the use or manufacturing of such products. That includes sealants, waterproofing, coatings, and associated products sold and marketed by companies in the sealant and weatherproofing industry.

8.3 SEALANTS/WEATHERPROOFING IN LEED

8.3.1 SUSTAINABLE SITES CREDITS

The first section of LEED-NC is Sustainable Sites Credits. The overall spirit of this series of credits is to reduce the impact of the construction as well as the long-term existence of the building on the local ecosystem. This plays right into the old cliché, "Think globally, act locally." By changing actions on the local level, the impact will be global.

Arguably, the most noticeable effect on the local ecology, aside from the change to the landscape itself, is the resulting increase in what is called the heat island effect. The removal of natural landscape in favor of concrete, gravel or, even worse, blacktop has resulted in noticeable changes in local temperature referred to by this name. As an example, I can drive from my suburban home just seven highway miles from downtown Akron to the center of the city, a city of less than 250,000 people, and the thermometer in my car will consistently read 3°F to 4°F higher downtown regardless of time of year. This is because concrete and blacktop absorb light as heat energy, which subsequently is given back as heat. In fact, according to studies done at the Lawrence Livermore National Laboratory cited by both the U.S. Department of Energy and USGBC, a pure black surface can reach a temperature 90°F higher than ambient on a sunny summer afternoon. On a hot July day in a city such as Phoenix, Arizona, or Las Vegas, Nevada, this means the surface can be over 200°F!

Most simplistically, the easiest way to alleviate the heat island effect is to get rid of the concrete and asphalt. Of course, concrete is relatively inexpensive and very versatile as a building material. As such, just getting rid of it is not feasible. This is where waterproofing and sealants organizations come into play in the LEED program. There are two remedies such organizations can provide to help alleviate this effect:

- White coatings on rooftops, parking areas, and other exposed concrete and asphaltic surfaces
- Green roofs

White is the most reflective of all colors because it is comprised of all colors. Since what we perceive as color is the reflected wavelength of light, and light itself is "white," a white coating in theory reflects all light. Because it reflects all light, none is absorbed and converted to heat energy. Conversely, "black" is an absorption of all light. Thus, darker colors lead to the largest amount of absorbed light to be converted to heat [7].

LEED requires a white coating meeting the Energy Star (a government system for rating the energy-saving properties of a product) requirements for reflectivity and emissivity. It also requires that such a coating be used over minimal percentages of the roofing or parking surface, which vary depending on whether other antiheat island effect techniques are utilized in design—most notably the installation of green roofs. It should also be noted that, in the case of all metal structures or those that are being rehabilitated with very low load-bearing capacity where green roofs are not an option, a white coating is really the only option for heat island effect reduction.

For those in more traditional sealant organizations who may not be 100% familiar with what a green roof is, such a system is that which encompasses waterproofing, insulation, drainage, water retention, and growth media for plants on the top deck of a building (Figure 8.1). More simply, it is a plaza deck or rooftop that has growth media and vegetation in place of a topping slab of concrete or asphalt, or in place of a traditional roofing membrane. While such systems are available in a variety of designs and configurations from numerous manufacturers, they all have one thing in common: they all waterproof and seal the building off from water infiltration while providing a viable environment for plant growth.

LEED is very generous to the sealant and waterproofing industry when it comes to sustainable sites credits. Green roofs and white coatings under the sustainable sites section are among a very few instances where, simply by installing a product/system, building owners qualify for LEED points.

Of course, this last statement implies that there are less obvious yet certainly viable opportunities within LEED for the sealant and weatherproofing industry. One example is that, under sustainable sites, a point is possible for redeveloping a brownfield. Brownfields are areas that may be contaminated/polluted due to previous development. Many manufacturers offer waterproofing products and sealants approved and tested to prevent the infiltration of the building envelope by possible pollutants at such sites such as methane, petroleum distillates/solvents, and even

FIGURE 8.1 A typical green roof. Green roofs have many ecological benefits including heat island reduction and storm water management, and the plants convert carbon dioxide into oxygen.

fuels. Another "soft" use for waterproofing products under the sustainable sites credits involves solutions that manufacturers offer for sealing the below-grade area in lagging applications. By utilizing lagging, builders require a smaller excavation to build the building and subsequently leave a larger amount of a property undisturbed, which can garner a point in LEED.

Other areas within Sustainable Sites include:

- Site Selection—choosing a site that is not farmland, a flood plain, or in a wildlife area, etc.
- Development Density
- Alternative Transportation—being located near mass transit and having facilities for alternative transport methods such as locker rooms for bicycle riders and special parking and facilities for alternative fuel vehicle and carpool drivers
- Light Pollution Reduction—Minimizing the amount of artificial light that reaches beyond either the building envelope or the immediate used property (i.e., beyond the parking lot, beyond walkways, etc.).

8.3.2 WATER EFFICIENCY CREDITS

There are many major environmental concerns in our world today. Global warming is on everyone's list and, consequently, so are ways to reduce greenhouse emissions. Certainly, the reduction in the use of fossil fuels, particularly those that are petroleum-based, not only fits into this but also has an economic and political tinge to it as well. Another issue is water—clean, untainted fresh water for potable use, as well as for aqueous plant and animal life to survive in, is a scarce commodity. The Colorado River does not flow to the Gulf of California in dry months anymore and, in the Midwest, the Great Lakes are being explored as a source for the booming populations of desert cities such as Phoenix and Las Vegas [8,9].

LEED addresses the need for preserving water resources. However, much as with brownfields, and as we will see, with many more examples in LEED, there are not always direct opportunities for the sealant and weatherproofing industry, but indirect ones do exist. The same green roofs we mentioned earlier, when designed and planted for minimal watering, count towards Water-Efficient Landscape calculations. As an example, if a building's footprint were to cover 50% of the property, and the entire roof was a low or even zero water-consuming green roof, then 50% of the landscape would contribute towards this credit.

Another more abstract impact on the industry is in the collection of water for use in nonpotable situations such as flushing of toilets and landscape watering. Manufacturers of prefabricated drainage (Figure 8.2) have an ideal tool for gathering runoff to be subsequently transported and stored in cisterns, retention ponds, etc., for such uses, thus helping to meet the other two credits in this section: Water Use Reduction and Innovative Wastewater Technologies. Additionally, roof coatings that are approved for the catchment of potable water by a government agency are ideal for collecting rainwater for human consumption. These coatings in white (common in the Caribbean) also have the added heat island reduction function previously discussed.

FIGURE 8.2   A prefabricated drainage medium. Such a medium can be used to collect rainwater, which can subsequently be used in place of potable water for watering plants, flushing toilets, and other nonpotable applications.

8.3.3 ENERGY AND ATMOSPHERE CREDITS

If water consumption and preservation is among one of the top three ecological concerns we face, then energy consumption is a second member of that list and quite possibly the top entry. We are all aware that, in the year 2008, the United States and Canada, in particular, are in the midst of an energy crisis to rival that of the 1970s. We only now truly comprehend that petroleum-based energy sources such as heating oil, propane, and natural gas are finite and will only become increasingly scarce in the coming decades and increasingly more costly. In addition to limited quantities of these energy sources, the long-term burning of all fossil fuels, including those listed as well as coal, has released significant quantities of gases now linked to global warming [10].

While the extent of the impact of global warming is in itself a topic capable of filling volumes of books, it is nonetheless one that the construction industry is now facing as 40% of all energy resources consumed in the United States are for lighting, heating/cooling, and maintaining buildings. In addition, to relate this back to LEED, approximately 25% of the points needed to qualify for certification can be obtained under one credit of Energy and Atmosphere, the Optimized Energy Performance credit.

Fortunately, sealant and weatherproofing product manufacturers have always been working to reduce energy consumption related to heating, ventilating, and air-conditioning (HVAC) in particular since that is by definition what a sealant or an air barrier is supposed to do.

Products in the sealant and weatherproofing industry can assist in green building by helping to minimize energy consumption in many ways:

- Using sealants to seal all gaps in duct work and insulating products in all fenestration openings and around HVAC units, conduits, etc., eliminates the leakage and energy loss of HVAC systems and reduces energy consumption of these units as a result. Sealants are ideal for this as they are flexible, durable and, in the case of the butyl technologies commonly seen in duct work, can be easily removed and reapplied should work need to be done to the HVAC system.
- Using air barrier systems (Figure 8.3), including the air barrier itself and the detailing materials, on the above-grade building envelope, eliminates the transfer of air between the inside and the outside of the envelope. Some studies by the U.S. Department of Energy have shown the reduction of HVAC can be as much as 40% depending on climate, design, and other factors. Regardless of energy savings, this is a direct financial gain that cannot be ignored.
- Alternative roofing solutions help reduce temperatures. As discussed earlier, traditional roofing can become extremely hot in the sun. Through the use of high-reflectivity/low-emissivity white roofing, this heat buildup is avoided, and subsequent demand on HVAC is reduced. On this same note, green roofs are also less heat absorptive than conventional roofing and have the additional benefit of being insulating, again reducing heating and cooling requirements.
- Alternatives to conventional drainage have several advantages. Many manufacturers now offer prefabricated drainage for below-grade building envelopes that also acts as an insulating layer in addition to fulfilling the role of

FIGURE 8.3 Spray application of an air barrier membrane. In combination with sealants and flashings, a complete air barrier system can eliminate air loss in the building façade, reducing HVAC use by up to 40%.

drainage. As a result, the requirement for heating subgrade occupied space is reduced due to both the presence of insulation and removal of thermally conductive moisture from contact with the below-grade wall.

Reduced energy consumption is not the only credit that those in the sealants industry should find important. Manufacturers of aerosolized products, particularly aerosolized foams, not only are providing a product that is insulating and gap filling but also are utilizing next-generation propellants that are free of ozone-depleting chemicals, which helps their customers attain points under the Ozone Depletion Credit. Finally, the sum of our efforts in reducing energy consumption, combined with other methods such as on-site solar panels, help in attaining Green Power Credits by allowing building owners to not only rely less on power from the grid but, in the case of those with production capabilities exceeding their consumption, they can place energy onto the grid for others.

Other credits within this section include the following:

- Renewable Energy—utilizing on-site alternative energy sources such as windmills or solar panels
- Additional Building Commissioning—training of building occupants on how the building systems work to ensure they maintain their ecological benefits on a long-term basis
- Measurement and Verification
- Green Power

8.3.4 MATERIALS AND RESOURCES

So far we have examined two major ecological issues today and how building seals and sealants can have a significant impact in reducing the problem through green building practices. A third major issue is materials. Everything that goes into building a structure falls into materials. When it comes to materials in the sealants and weatherproofing industry, one is most concerned about their source, disposal, and transport. As it relates to source, are the materials recycled or are they virgin? If they are virgin, are they renewable or nonrenewable? Regarding disposal, can materials or their containers be reused or recycled in some fashion as opposed to being sent to the garbage dump? Finally, how can we minimize the environmental impact of transporting goods by greenhouse-gas-emitting and fossil-fuel-consuming means such as truck or train?

Construction waste management, or disposal, is by far the most visible of the three areas regarding materials and resources within LEED. It is easy to visualize dumpster after dumpster full of packaging material, applicator tools, and general waste being removed from a job site and consequently being placed in a landfill. Diverting as much waste as possible from the landfill to some other reuse is what is being emphasized in LEED. Fortunately, sealants and weatherproofing products tend to arrive in reusable or recyclable packaging. Packages such as plastic tubes, cans, aluminum foils, and buckets/pails are generally recyclable so long as they are free of toxic materials. Drums, totes, and even tankers are considered, in most cases, to be recyclable/reusable. Another benefit in most of these cases is that these containers can be sold for scrap or returned to obtain container deposits from some manufacturers. Also, even if the packaging cannot be diverted from the landfill, new packaging options, such as foil wrapping, which are completely crushed, reduce the volume of material being disposed of in a landfill. In the LEED process, the amount of diverted waste is calculated by weight or volume. By encouraging not only packaging reuse but endorsing the adoption of waste diversion, sealant and weatherproofing product manufacturers are certainly helping customers achieve the points for this credit.

Similarly, LEED places emphasis on the recycled content of materials used in the construction process. After all, it is not just the prevention of materials from the job making their way into a landfill, but reusing materials that were themselves diverted from disposal that truly reflects the spirit of ecologically friendly building. Historically, many of the products used in sealing the building envelope have utilized virgin materials. Being sophisticated, well-engineered, chemically reactive products in a host of cases (e.g., caulks, adhesives, coatings, waterproofing membranes, etc.), these products are too sensitive to the unknown contents often found in recycled raw materials. Recent advances have made it possible to use some reclaimed materials such as fly ash and ground rubber, but the uses are still being discovered for these materials. Though sealant and weatherproofing products may not have any recycled content, LEED acknowledges the recycled content in the packaging itself as well as cases/boxes and pallets. While packaging generally only accounts for a few percent of the product by weight, the points are awarded as follows:

- The total amount of recycled content by weight of the product is multiplied by the value of the product.
- These values are then added up, and the points are then awarded based upon the total dollar value of recycled content for the job.

As sealants and weatherproofing products generally account for less than 2% of the total construction costs of the job, even products with minimal recycled content, which constitute the vast number of the products in the sealants field, have an overall minimal negative impact on average for the entire job. Though some sheet-applied membranes and most prefabricated drainage media have a much higher recycled content, the impact is still quite small [11].

The third point category within Materials and Resources that affects the sealants and waterproofing industry is Regional Materials. These are defined as:

- Locally manufactured materials have a final assembly and/or filling within 500 straightline miles of the job site.
- Locally harvested materials are those that are harvested (harvested meaning grown, mined, or similarly "picked") within 500 straightline miles of the job site.

If a job is within 500 miles of your plant, that means you are making a positive contribution towards obtaining the locally manufactured point. If you are not so lucky, though, it is not, nor should it be, a make-or-break issue as, again, the point is calculated based on dollar value of the material and this is a small part of the finished building. The same can be said for the locally harvested portion of this credit. This point has a bigger twist, though, as most, if not all, manufacturers have several suppliers for each raw material. In turn, each of these suppliers likely has more than one source for its raw materials, and so on. Consequently, sealant and weatherproofing product manufacturers have very little chance of knowing which point source the raw materials originated from for any given production run.

The final credit area of concern in this part of LEED is the use of Rapidly Renewable Materials. Many of the products in the sealant and weatherproofing industry have roots in petroleum-based chemicals, which are becoming scarce. This emphasizes the need to explore chemicals from renewable sources such as natural-based chemicals derived from sources such as corn, soy, or peanuts. It also leads to the need to explore additional sources of raw materials such as plant-based fillers. Again, while a small part of the building, good environmental stewardship, and the eventual rise in the cost of limited resources should be forcing sealant and weatherproofing manufacturers into exploring these options, the bigger challenge is maintaining product quality while incorporating eco-friendly technology and maintaining price.

Additional credits within this section of LEED include:

- Building Reuse
- Resource Reuse
- Certified Wood—wood used in the construction that is certified as being harvested and grown under the guidelines set forth by the Forestry Stewardship Council

8.3.5 INDOOR ENVIRONMENTAL QUALITY

Another aspect of LEED, besides the concerns over environmental impact, is the emphasis placed on providing a pleasant and healthy environment for building occupants and construction workers. Employees are the single largest dollar investment for any organization, and providing an environment in which they can be more productive ensures a good return on investment.

First, volatile organic compound (VOC) content is directly addressed in LEED via the Low-Emitting Materials Credit. For over a decade since the EPA passed the first federal laws in 1997, the industry has been working to reformulate products to lower VOC levels while maintaining or improving performance. Since 1997, stricter rules have come into existence, most notably those in California. In particular, the South Coast Air Quality Management District (SCAQMD), encompassing the Greater Los Angeles Area, and the Bay Area Air Quality Management District (BAAQMD), encompassing the San Francisco and Oakland areas, have traditionally set the strongest VOC regulations not only in the United States but in the world.

There are two basic requirements for sealants and adhesives outlined in LEED:

- Such products must adhere to the requirements set forth in SCAQMD
  Rule 1168.
- Such products must adhere to BAAQMD Regulation 8, Rule 51.

To do business in SCAQMD- and BAAQMD-regulated areas, most manufacturers have already made significant advancements in sealant technologies to produce compliant products. Similarly, coating systems used in interior work areas, utility closets, interior parking, and similar areas must also meet VOC rules established by the SCAQMD in LEED situations. Yet again, this has already been accomplished by most manufacturers so that they can continue to do business in the Los Angeles area.

As we have seen in other sections of LEED, the impact of/on the sealant and weatherproofing industry is not always so obvious in the Indoor Environmental Quality section. However, by looking at the basic function of the products manufactured, we can see the link.

The emphasis on Daylight and Outdoor Views is another area in Indoor Environmental Quality that is specifically designed for occupant comfort. The premise is that by eliminating the "cube world" of fluorescent lights and no windows, occupants will feel better emotionally. In addition, by allowing natural light in to illuminate space, electricity is saved, as additional lighting is not required. The use of more windows means that better glazing and curtainwall sealing must be provided to minimize HVAC consumption. Regarding HVAC, Ventilation Effectiveness (air exchange effectiveness) is also taken into account in LEED. Again, sealants have always been used to maximize the efficiency of HVAC systems.

While we are on the topic of sealing in/out, LEED places emphasis on Indoor Chemical and Pollutant Source Control. The intent of this credit is to keep the necessary evils of occupying and maintaining a structure (janitorial closets, restrooms, etc.) from contaminating office and/or living space. Again, sealants by definition have always been used to control the movement of air and water, and this is no exception. Even products not considered as typical sealants, such as fire-stopping sealants (Figure 8.4), work to ensure indoor environmental quality even while functioning to ensure occupant health in a much different circumstance from the norm.

Additional credit areas in this section include the following:

- Carbon Dioxide Monitoring
- Controlability of Systems
- Thermal Comfort

8.3.6 INNOVATION AND DESIGN PROCESS

Finally, LEED has what can best be described as a "free answer" section. There are two credits in this section:

- Innovation in Design—This is awarded if the USGBC determines that a
  project has in some way gone above and beyond what was required for some
  aspect of LEED.

FIGURE 8.4   A fire-stopping sealant at a through-wall penetration. While the primary focus of the sealant is to prevent fire and smoke migration to adjacent areas in the case of a fire, it has the secondary function of preventing everyday air flow, isolating possible pollutants and irritants in one room from another.

- LEED AP—This is awarded if a member of the project team holds the LEED AP title.

The first credit is possibly the most subjective of the credits in the LEED system. By providing consulting services to customers, any manufacturer could arguably help meet criteria for the former credit, particularly if the manufacturer has a LEED AP on staff. As LEED becomes more and more widespread, criteria for answering these questions may be more clearly defined.

8.4   WHAT CAN WE DO?

Undoubtedly, every manufacturer and, probably, everyone within each organization has a different view on what he or she could or should be doing to address the concerns of green building. Opinions range from the very enthusiastic and outright sold on marketing the idea to the hilt to those who think it is a fad and deserves minimal attention, if any at all.

Ultimately, there is only one answer to this question that we can probably all agree on—whatever it takes to help the customer. We should all be willing to help answer any questions and provide any information required. There are definitely a few questions we will all get:

- What is LEED?
- What is a VOC?
- What is the VOC content of this product?

- What is the recycled content of said product (both postconsumer and postindustrial)?
- Where is this product made?
- Where did the raw materials used in making this product come from?
- Is this product made with any renewable materials? If so, what percentage is the content? Where were these materials harvested?

How a manufacturer decides to not only answer these questions but proceed further with pursuing green building is going to determine the questions he or she is asked and how they will be seen in the marketplace. Having just said that, there are a few things we should all be doing:

- Provide basic product information related to LEED in a simple format to allow customers to fill out their registration documents.
- Understand the basics of the LEED system, including what the most pertinent systems are to the sealant and weatherproofing customer, how sealant and weatherproofing products are affected and, more importantly, how customers are affected.
- Have a plan to address not only LEED but other sustainable design issues related to the industry.

8.5 SUMMARY

Ultimately, our society must deal with the environment of which we are a part. Given that building construction and occupancy have a significant impact on the environment, it is only logical that ecologically friendly building design and construction practices can help to improve our environment. LEED in particular has made accomplishing such construction practices easy and has thus led to its greater acceptance. As sealant and weatherproofing product manufacturers, embracing this change and adapting ourselves to it not only makes sound business sense but common sense as well.

REFERENCES

1. http://en.wikipedia.org/wiki/Green_building; accessed July 10, 2007.
2. C.M. Harris (Ed.), *Dictionary of Architecture and Construction*, p. 430, McGraw-Hill, New York (1975).
3. *LEED for New Construction*, Version 2.2; United States Green Building Council, Washington, D.C. (2005).
4. *LEED for New Construction*, Version 2.2 pp. 3 and 4; United States Green Building Council, Washington, D.C. (2005).
5. http://www.usgbc.org/DisplayPage.aspx?CategoryID=19; accessed July 27, 2007.
6. http://www.usgbc.org/DisplayPage.aspx?CMSPageID=1442; accessed July 26, 2007.
7. R.C. Weast (Ed-in-Chief), *CRC Handbook of Chemistry and Physics*, 70th ed., p. E-405; CRC Press, Boca Raton, FL (1989).
8. http://www.savemiwater.org/images/Water-Diversions-or-Excursions-09-27-01.pdf; accessed July 27, 2007.

9. http://www.msnbc.msn.com/id/17276693/; accessed July 27, 2007.
10. http://zebu.uoregon.edu/1999/ph161/l10.html; accessed July 19, 2007.
11. http://www.librisdesign.org/docs/CostEstimatSimp.pdf; accessed July 27, 2007.

Section III

Various Types of Sealants
and Their Applications

9 Hot Melt Sealants

Ju-Ming Hung

CONTENTS

9.1 INTRODUCTION

Sealants should have good adhesion properties to different substrates and also provide the necessary physical properties to fill the gap between substrates. They serve as a barrier to exclude dust, dirt, moisture, chemicals, and air in end-use environments. Sealants can also serve to maintain a pressure differential between internal and external environments to attenuate mechanical shock and vibration. Since the applications of adhesives and sealants are different, the requirements for the mechanical strength of sealants are generally lower than those for adhesives. Therefore, sealant formulations usually contain large amounts of filler for cost reduction and gap-filling purposes.

Several very important industrial areas that use significant amounts of sealants are transportation, commercial, home construction, and repair industries. When materials having highly different thermal expansion coefficients are joined together for construction purposes, exterior commercial construction sealants need to accommodate large changes in joint width resulting from temperature cycling, wind load, rain, heat, ultraviolet light, oxygen, and ozone. Substrates such as stone, marble, aluminum, steel, glass, plastics, and wood are frequently utilized for construction applications. Only high-performance sealants, such as silicones, polyurethanes, and polysulfides, can be used in commercial buildings and construction for exterior sealing. Hardness is one of the key characteristics of a sealant. For most sealants, hardness values required are between Shore 10A and 85A. However, several rigid sealants, such as epoxy, acrylic, and polysulfide, have a higher hardness than Shore 85A. These types of sealants are commonly used when both sealing and joining functions are necessary [1].

One-component hot melt sealants have been available for decades but were not ubiquitous until the past few years. Their use has been rather limited because of well-known deficiencies such as sag at elevated temperatures, equipment costs, and low bond strength as compared to thermoplastic hot melt adhesives.

In order to assemble newer, more complex design components and to overcome weaknesses in meeting current requirements of the construction and transportation industries, sealant manufacturers have to provide better-quality products. These new sealants should possess higher tensile and shear strengths. One method of reducing industry cost is to increase productivity on assembly lines, resulting in cost-effective hot melt sealants.

Developments over the last several years indicate that these materials will grow in market share [2]. The following are the primary advantages of hot melt sealants:

1. They are well suited to high-speed production.
2. They require less energy per unit production than solvent- or water-based sealant systems.

With the cost of energy going up, more and more emphasis is being placed on productivity by sealant users. The demand for hot melt sealants is likely to increase significantly because of the geopolitical concerns relative to petrochemicals supply.

Traditional markets that have made maximum use of hot melt sealants to date are the following:

- The insulating glass window industry (its sealants are sometimes known as *hot flow sealants*)
- Sealing and insulating electrical appliances
- Sealing automobile, mobile home, and truck components
- Sealing prefabricated windows and panels in field installations

The largest application of hot melt sealants (i.e., the manufacture of insulating glass units) has an estimated annual consumption of hot melt sealants of over 30 million pounds [3].

9.1.1 Hot Melt Technology

There are several technologies that have been used by the industry to improve the green strength of both adhesives and sealants, such as two-part systems, one-component systems with either UV cure or e-beam cure, and one-component hot melt systems. Among these different technologies, hot melt adhesives and sealants can be used without requiring additional mixing pumps or additional curing devices as are required for either UV or e-beam curing systems. Therefore, hot melt systems are the most convenient for industrial assembly applications.

Hot melt sealants are solvent-free, one-part solids at room temperature. These materials are heated to their molten state prior to use, then pumped and dispensed onto construction pieces in a hot and tacky state. When the assembled substrates with sealants cool to ambient temperature, the bond strength builds up rapidly due to the solidification of the sealants. Therefore, the speed of the commercial assembly process can be increased, and manufacturing cost reduction through improved production rate can be achieved.

The general approaches to improve the initial strength of hot melt materials are to adjust the molecular weight, the T_g (glass transition temperature) or T_m (melting temperature) of one or more of the ingredients in the sealant formulation. In the final formulation, a careful balance needs to be maintained between the hard portion that delivers high initial strength and the soft portion that provides good sealant properties.

9.1.2 THE CHALLENGE OF FORMULATING HOT MELT SEALANTS

Hot melt sealants are unique materials and can provide special physical characteristics to fill needs in two different areas: processing and final adhesion properties. The process requirements, which differ depending on the application, may include process temperature, melt viscosity, nip pressure, open time, initial strength, melt stability, and sag resistance. The final adhesion properties of a hot melt sealant should provide good adhesion to different substrates in addition to having a wide variety of physical properties such as excellent tensile qualities with high mechanical strength and acceptable UV resistance when the sealant is used for exterior application.

Softness is also a key characteristic of many kinds of sealants. This property is highly desired for materials to absorb vibrations and impacts. Gap filling between the substrates is also necessary to overcome significant differences in thermal expansion coefficients. In order to increase the initial strength while maintaining adequate softness of the sealant, the formulation window needs to be reduced significantly. The invention of new technology or new raw materials in this area is necessary to broaden the formulation window.

9.2 HOT MELT SEALANT CLASSIFICATION

Hot melt sealants can be easily classified into two groups: nonreactive and reactive types. The nonreactive hot melt sealants (thermoplastic type) can provide good initial strength. However, these materials can remelt and lose holding strength when the assembled substrates are exposed to high-temperature environments. As the heat resistance of a nonreactive hot melt sealant is very poor, this type of sealant should not be used for exterior applications.

However, a reactive hot melt sealant (thermoset type) can react after the sealant has been dispensed. The sealant changes from a thermoplastic to a thermoset after the sealant is cured. The heat resistance of this type of sealant is significantly increased as the molecular weight of the sealant builds through the crosslinking reaction. The melt and remelt cycle by heat exposure is no longer a problem. Therefore, reactive

TABLE 9.1

Advantages and Disadvantages of Different Sealant Systems

Type of Sealant	Advantages	Disadvantages
Two-part system	1. Good initial strength 2. Room temperature processing	1. Two pumps and a mixing system are needed. 2. Mixing or metering could be complicated and difficult.
One-part, nonreactive hot melt sealant	1. Excellent initial strength 2. Only one pump, and no metering or mixing system, is required. 3. Special reactor and dispenser equipments are not needed.	1. High-temperature processing. 2. Poor heat resistance. 3. Good for interior application only.
One-part, reactive hot melt sealant	1. Excellent initial strength. 2. Only one pump, and no metering or mixing system, is required. 3. Good for both interior and exterior applications. 4. Excellent heat resistance.	1. High-temperature processing. 2. Special reactor and dispenser equipment are necessary.

hot melt sealants are suitable for both indoor and exterior applications. Among them, urethane and silicone hot melt sealants are the only two that can be cured directly through moisture reaction and that provide excellent mechanical properties.

Other types of sealants that can produce fast initial strength are two-component systems. The advantages and disadvantages of the different sealant systems are listed in Table 9.1.

9.2.1 BUTYLS

With the lowest moisture vapor transmission rate (MVTR), butyl sealants have been widely used to bond the double panels of insulated glasses as an insulation sealant. Such sealants are available as a one-component product in solvents as well as a thermoplastic hot melt. Low- and medium-performance solvent systems tend to have low cohesive strength; therefore, they easily string or drip during application and cause difficulties in final process assembly.

Thermoplastic butyl hot melt systems are used more and more as insulation sealants for double-panel windows and doors. Table 9.2 shows the formulation of a simple butyl hot melt sealant. All the process properties for dispensing and the final adhesion properties of the butyl hot melt sealant are determined by these three key ingredients.

Butyl rubber, which is a copolymer of isobutylene and isoprene, is readily available for the formulation of butyl hot melt sealants. The copolymer can be crosslinked by p-quinone dioxime (see the following structure) and an oxidizer to increase heat resistance. Crosslinking butyl elastomer is one of the methods used to improve heat resistance of butyl hot melt sealants. Bayan [4] developed a thermoplastic halo-butyl

TABLE 9.2
Formulation of Butyl Hot Melt Sealants

Ingredient	Wt%
Thermoplastics (butyl copolymers with block copolymers)	25–40
Tackifier resins and plasticizers	35–50
Fillers, pigments, and antioxidants	30–50

rubber elastomer system that could be cured by dynamic vulcanization with a multi-functional amine-curing agent. Zaharescu et al. [5] studied how an ionizing radiation induced various changes in the molecular structures of butyl, and chlorinated and brominated butyl, elastomers. Newly formed products and alteration of the initial chemical structure are the results of free-radical reactions, which cause crosslinking and/or depolymerization of tested materials.

p–quinone dioxime

It was found that the initial strength of butyl hot melt sealant could be improved when 0.5 to 10 wt% of a styrene–butadiene–styrene block copolymer or styrene–iso-prene–styrene block copolymer were added into the formulation [6]. Based on a similar type of approach, another new technology has been identified that can further improve high-temperature heat resistance of butyl hot melt sealants. Recently, Kraton Polymers LLC (Houston, Texas), announced the availability of a new technology created for hot melt butyl sealant manufacturers that allows for an increase in the range of product service temperatures as well as the potential for expanded use into new markets, including insulated glass windows for warmer climates. The addition of Kraton® G (hydrogenated styrene block copolymer) (HSBC) to hot melt butyl sealants improves both high-temperature slump and creep resistance while maintaining excellent process ability.

The physical properties of butyl hot melt sealants can also be improved by the addition of other raw materials, such as amorphous polyolefins, that are the side products of either polyethylene or polypropylene production. These amorphous materials include propylene homopolymers and copolymers of propylene with ethylene or hexane, and were found to be a suitable substitute for block copolymers of styrene–ethylene-butadiene–styrene (Kraton products) in butyl hot melt sealant formulations to reduce both raw material cost and process temperature [7].

9.2.2 POLYSULFIDES

Polysulfides, which are considered one of the high-performance sealants, provide better fuel resistance than any other elastomeric sealants. They are second only to polyurethanes in alkaline chemical resistance and only slightly worse than butyl sealants in permeation by water vapor. Polysulfides are the most widely used sealants for airplane fuel tanks, curtain-wall construction, glazing, insulating glass windows, marine deck, and joints in airport runways, highways, and canals.

The liquid polysulfide is prepared by the reaction of an excess of sodium polysulfide with bis(2-chloroethyl) formal [8]. The crosslinking reaction can be introduced by adding 1,2,3-trichloropropane. The reaction occurs at 100°C and produces a mixture of chain lengths in which sulfur is present as -C-S-S-C-, -C-S-S-S-C-, and even as groups with higher sulfur content. The mixture can be reacted with sodium sulfite to reduce most of the polysulfide groups to disulfides and end groups with thiols (-S-H).

Morton Thiokol has produced a series of LP (LP-2, LP-3, LP-12, LP-32, and LP-33) products. These are mercaptan-terminated liquid polymers (HS-R-SH) with a molecular weight of 1,000–7,000 and crosslink density from 0.5 to 2.0 mol percent, and can be polymerized to elastomers by treatment with an oxidizing agent. The model structures of LP series are as follows:

$$H\text{-}S\text{-}[R\text{-}S]_{n/2}\text{-}S\text{-}(R')\text{-}S\text{-}[S\text{-}R]_{n/2}\text{-}S\text{-}H$$

$$\text{where } R = \text{-}C_2H_4\text{-}O\text{-}CH_2\text{-}O\text{-}C_2H_4\text{-}$$

$$R' = \text{-}CH_2\text{-}CH\text{-}CH_2\text{-}$$
$$|$$

$$n = \text{about 6 for LP-3; 24 for LP-32}$$

Several approaches have been shown to improve the physical properties of polysulfide sealants, such as the following:

1. A polyoxypropylene urethane backbone with mercaptan termination has been synthesized. With this material, the molecular weight of the resulting polysulfide is controlled more easily, and impurities from side reactions are minimized [9].
2. Polysulfides with polythioether backbone have also been prepared. These polymers replace the weakest -S-S- linkage with a thioether group and show excellent fuel and chemical resistance, and better heat resistance, than that of conventional polysulfides. Mercaptan, hydroxyl, silyl, and nonreactive end groups can be selected as end groups for this type of new polysulfides [10].
3. Chemical modification with dithiol liquid polysulfides shows better heat resistance, and lower liquid and solvent permeations. These modified polysulfides have a wider compatibility window to common formulation ingredients, such as plasticizers, pigments, and fillers, than regular polysulfides [11].

High-molecular-weight polysulfide elastomers should be used for hot melt applications. The molecular weight of polysulfides can be increased by using higher percentages of the crosslinker 1,2,3-trichloropropane, or with higher-functionality polyoxypropylene urethane backbone liquid polysulfides. Fettes and Gannon reported the reaction of liquid polysulfides with epoxy also [12].

Liquid polysulfide can be formulated to make a stable single-component polysulfide sealant with dried metal peroxide or sodium perborate. An oxidizing agent is reacted with moisture to generate initiator for polymer chain extension and crosslinking to solidify the whole system. However, the curing rate of one-part polysulfide systems is much slower than both silicone and polyurethane sealants.

Mercaptan-terminated liquid polymers can be polymerized to rubbery solids with oxidizing agents. This type of curable liquid polymer has been applied by hot extrusion [13–14]. Vietti et al. [15] prepared a thermoplastic elastomer by the reaction of an epoxy resin with a liquid mercapto-terminated polysulfide to increase the mechanical strength of polysulfide hot melt sealant. While this reactive elastomer has residual -SH groups, which with additional dried metal peroxide or sodium perborate initiators can be air-curable.

The families of block copolymers of styrene sold by Shell under the Kraton trademark are suitable for use as coresins for polysulfide hot melt sealants also to adjust tackiness, adhesion, hardness, and other properties. Representatives of the Kraton family include styrene–butadiene–styrene (SBS), styrene–isoprene–styrene (SIS), and styrene–ethylene-butylene–styrene (S-EB-S), copolymers.

Since polysulfides have very MVTR and high initial strength as hot melt elastomers, they can be appropriately used as hot melt sealants for insulated window applications.

9.2.3 ACRYLICS

Most acrylic sealants are prepared by latex polymerization and used in residential construction and home repair. This type of sealant exhibits medium or low performance because it shrinks on curing and has limited flexibility. However, acrylic polymers can be prepared by different methods using other raw materials to improve their properties as a hot melt sealant and to increase the mechanical strength of the resulting sealant.

High-molecular-weight acrylic elastomers have been prepared and used as the key ingredient in acrylic hot melt pressure-sensitive adhesive (PSA) formulations. By simple modification of the glass transition temperature and molecular weight of the acrylic elastomer, an acrylic hot melt PSA can be altered to behave as a hot melt sealant. Depending on end-use applications, elastomers can be mixed with appropriate tackifiers and fillers to obtain the desired properties. In addition, UV or radiation cure can be utilized to increase the molecular weight of acrylic elastomers and improve the heat resistance of the final product.

Copolymers of ethylene/acrylic have been formulated as hot melt sealants to increase the adhesion for glass-to-metal applications. If acrylic acid (AA) or methacrylic acid (MAA) is used instead of an ester as a comonomer, the materials have high bond strength to metals but tend to be corrosive. Tapes prepared from acrylic hot melt sealants have been widely used in automotive applications.

In the use of hot melt acrylic sealants, the vapors of residual acrylic monomers are a major health and safety concern. Adequate ventilation is always necessary during application.

9.2.4 STYRENE BLOCK COPOLYMERS

Styrene tri-block copolymers (SBCs) are composed of an elastomeric block in the center and rigid thermoplastic blocks on each end (e.g., styrene–ethylene-butylene–styrene). Hot melt sealants have been developed, based on these block copolymers, with large proportions of oils, waxes, plasticizers, resins, and fillers since the base polymers are quite tough. Such materials can be combined with a butyl hot melt sealant as one of the key components for insulated glass applications to improve both set strength and heat resistance. Poly(α-methylstyrene) can be used as a reinforcing resin to further improve mechanical strength as well as heat resistance.

The rigid phase and rubber phase exist as two incompatible phases but form a reinforcing network. These block copolymers can be formulated as hot melt sealants for insulated glass sealing with not only low nitrogen and water vapor permeabilities but also good ozone, UV, and chemical resistances. Typical formulation ranges are listed in Table 9.3.

Kraton Polymers LLC (Houston, Texas), a manufacturer of SBCs, has developed a technique that can enhance the high-temperature performance of certain adhesives and sealants as well as provide increased resistance to solvents and prevent leaching out of plasticizers. The technique works by crosslinking acid-functional SBCs with aluminum acetylacetonate (AlAcAc). Three crosslinking options (i.e., ultraviolet light, electron beam, and chemical curing) have been offered.

The high cohesive strength of acid-functional SBCs at ambient temperatures is due to their interacting network structure. However, the interactions are physical, not chemical, and therefore are reversible; when the polymer is heated above nearly 90°C, it begins to soften, weaken, and slump. The physically interacting network can be reinforced further with ultraviolet light, electron beam, or chemical crosslinking between the acid groups on the SBC and the AlAcAc. This crosslinking minimizes softening at high temperatures. However, use of modified waxes, modified asphalts, or oil gels can increase the softening points of the blends significantly.

9.2.5 POLYOLEFINS

Amorphous polyolefins (APOs) are used to enhance tack and adhesion and to improve the flexibility of formulations used for paper, film, and foil laminations as

TABLE 9.3
Formulation of Styrene Block Copolymers as Hot Melt Sealants

Ingredient	Wt%
Block copolymer thermoplastic	15–35
Tackifier resins and plasticizers	30–40
Reinforcing resin poly(α-methylstyrene)	10–20
Fillers, pigments, UV stabilizers, and antioxidants	30–50

well as ingredients in hot melt adhesives and sealants. They are also used in rubber compounding, in waterproofing compounds for wire and cable applications, and as a modifier for asphalt hot melt sealants in single-ply roofing.

For their strength properties, hot melt sealants commonly employ base polymers that include butyl elastomers and SBCs. The types of APOs used include propylene homopolymers and propylene copolymers of ethylene and butene. These polymers are blended with various additives, which may include tackifiers, oils, inorganic fillers, pigments, and antioxidants. End-use product applications for these formulated polymers include extruded preformed tapes, and hot melt sealants for insulated glass [16].

9.2.6 Ethylene Vinyl Acetate Copolymers

Ethylene vinyl acetate (EVA) is one of the most common polymers used in hot melt sealants. Vinyl acetate levels used in hot melt sealants can range from 5–80 wt%. With different molecular weights and VA contents, the melt flow index of this type of polymers can be varied from 1 to 3000 (g/10 min). When an EVA polymer contains a high amount of vinyl acetate, the material has increased polarity as well as softness. At about 45 wt% vinyl acetate, all crystallinity of the EVA is lost. Due to its good compatibility with a wide range of tackifiers and waxes, EVA can be formulated to fulfill very broad performance requirements. A low-cost and wide-formulation window provides excellent advantages for EVA to compete with some other types of hot melt sealants. A typical formulation range for EVA-based hot melt sealant is shown in Table 9.4.

In most EVA-based hot melt sealants, tackifiers are a very important ingredient and are present in very high amounts. They function by reducing viscosity and adjusting the T_g of the hot melt sealants. With the right tackifier choice, desired physical properties of the hot melt sealant, such as wet-out, hot tack, open time, set speed, and heat resistance, can be achieved. Tackifiers are high-T_g, low-molecular-weight amorphous materials. They act as solid solvents for the other ingredients in hot melt sealants.

Tackifier resins can be naturally derived or petroleum based. Naturally derived tackifiers include rosins and polyterpenes. Rosins are derived from pine trees and are a blend of various abietic acid analogs. Rosin esters are more commonly used tackifiers than rosin itself, and are produced by reacting rosin with polyols such as glycerol and pentaerythritol. Terpene monomers are predominantly a by-product of the extraction of oils from citrus fruits. The petroleum-based tackifiers are a by-product of oil refining processes. These monomers are polymerized into

TABLE 9.4
Formulation of an EVA Hot Melt Sealant

Ingredient	Wt%
EVA polymer	10–50
Tackifier resins and waxes	30–80
Additives, fillers, and antioxidants	10–50

low-molecular-weight polymers with molecular weight ranging from several hundred to around a thousand grams per mole. A variety of petroleum-based tackifiers have been developed in the past 30 years.

Wax is another critical ingredient in EVA-based hot melt sealants. Waxes can reduce viscosity, improve the set speed, and increase the heat resistance of hot melt sealants. This type of raw material falls into two categories: petroleum fractions and synthetic waxes. Paraffin waxes and microcrystalline waxes are obtained from a dewaxing process in the refining of crude oils, and generally, low-melting-point materials are produced. Synthetic waxes include Fischer–Tropsch, polyethylene, and specialty waxes. Synthetic waxes are used instead of paraffin where higher heat resistance is required.

Poly(ethylene–vinyl acetate) copolymers can also be blended with some other elastomers or can be used alone as hot melt sealants. EVA copolymers have been blended with polyurethane hot melt sealants also to control the balance of initial strength and open time. Sometimes, these copolymers can be considered as a means for raw material cost adjustment in the formulation of reactive polyurethane hot melt sealants.

9.2.7 BITUMENS

Bitumens or asphaltic materials have been used as hot melt sealants for many centuries. Bitumen is a category of organic liquids that are highly viscous, black, sticky colloids, completely soluble in carbon disulfide. These dark-colored sealants are used for joints in highways and buildings. Their adhesion and waterproof properties, combined with low cost, make them useful in roofing, siding, and highway construction. They have also been widely used as adhesives. Combined with other materials, bitumens may be used as vapor barriers and as agents to waterproof and damp-proof structures.

Asphalt and tar are the most common forms of bitumen. Bitumen is altered physically if this colloid system is disturbed, as demonstrated when bitumen is overheated. Even after bitumen has been successfully applied, the bond strength can be decreased or even destroyed by the ingress of water at the bitumen–solid interface. The presence of water soluble salts in any quantity will result in a large capacity for water absorption by osmosis. For this reason, oil refineries desalt the crude oil before refining it. Fillers can also absorb certain quantities of water, the amount varying with the composition and granular size of the filler material [17].

The viscous or flow properties of bitumens are important, both at the high temperatures encountered in processing and application as well as at the low temperatures that bitumens are subjected to in service. Flow properties are complex due to changes in the colloidal nature of the bitumens that occur with heating. When chilled sufficiently, all bitumens lose their viscous properties and become brittle solids. The interval between the softening point temperature and the temperature at which the brittle condition is reached gives a measure of the temperature susceptibility of the material. This can vary a great deal, depending on the crude oil source and bitumen processing. Successful use of bitumen usually results if the material has been chosen that will be subjected in service to temperatures well within the limits defined by its brittle condition and softening point. Bituminous hot melt compositions usually contain scrap rubber or neoprene, which act as a flexibilizing agent.

TABLE 9.5
Formulation of a Bitumen Hot Melt Sealant

Ingredient	Wt%
Asphalt	50–70
Ground rubbers	20–30
Cyclic hydrocarbon compounds	10–30
Reinforcing synthetic polymers	10–30

An obvious advantage of emulsified bituminous products is that they are easy to handle, addition of water being all that is necessary to decrease their viscosity. Curing involves, primarily, a loss of water by evaporation; its stability, however, depends on many factors such as asphalt concentration, size and distribution of asphalt droplets, freezing of the water, and the nature of the stabilizing agent. These emulsion-type bituminous sealants are widely used for protecting the surface of residential driveways.

Modified bituminous materials are beneficial for highway maintenance and construction by providing longer-lasting roads and by savings in total road maintenance costs. The polymer additives do not chemically combine with or change the chemical nature of the bitumen being modified. They alter the physical nature of bitumens by modifying such physical properties as the softening point and the brittleness of the bitumen. Elastic recovery and ductility can also be improved [18].

The basic properties of bitumens can be modified by the addition of highly viscous oils or volatile oils to produce bitumens of various grades. These grades are specified by their viscosity and softening point. Styrenic block copolymers (SBCs) are useful for bitumen modification. These synthetic polymers were originally developed for use in the production of tires and shoe soles but are also suitable for the modification of bitumen. However, this type of system is thermodynamically unstable, and phase separation may occur under the influence of the gravitational field. Phase separation can be accelerated by hot storage [19].

The most common grade of EVA for bitumen modification for road pavement materials is the classification "150/19." This classification means that it has a melt flow index of 150 and a vinyl acetate content of 19 wt%. In addition, epoxy resins are frequently used to reduce the cold flow of bitumen hot melt sealants.

A typical formulation of a highway sealant is given in Table 9.5.

9.2.8 SILICONES

Silicone sealants, which are high-performance sealants, have excellent resistance to environmental extremes. They are often used in exterior applications that may experience vibration and flexing. The sealants are based on mixtures of fillers, silica, silicone polymers, crosslinking agents, and catalysts. Most commercial silicones are based on poly(dimethylsiloxane) (PDMS) polymers. Since these types of polymers are very low glass-transition temperature (T_g) materials, it is necessary to prepare very high molecular weight materials to build up enough mechanical strength for end-use applications.

PDMSs are obtained by the hydrolysis of dimethyldichlorosilane in the presence of an excess of water according to the following structure:

$$x\ Me_2SiCl_2 \xrightarrow[\ -HCl\ ]{\ +H_2O\ } y\ HO(Me_2SiO)_nH + z(Me_2SiO)_m$$

$$\text{linears} \qquad \text{cyclics}$$

with n = 20 – 50 and m = 3, 4, 5 (mainly 4)

However, the linear and cyclic oligomers obtained by hydrolysis of dimethyldichlorosilane have too short a chain that cannot provide sufficient mechanical strength for most applications. They need to be condensed (as linear) or polymerized (as cyclic) to provide macromolecules of sufficient length. Unfortunately, in most cases, the low-molecular-weight oligomers cannot completely be converted into high-molecular-weight polymers, and therefore diffuse out to contaminate the surfaces of the adherends. This is one of the disadvantages of the silicone sealants.

The PDMS polymer, containing a siloxane (-Si-O-Si-) backbone with alkoxy, acetoxy, amine, or oxime pendant groups can be cured by moisture. These groups are readily hydrolyzed to silanol groups (Si-OH) via the loss of an alcohol, acetic acid, amine, or oxime, as shown in the following structure:

$$X = \ \underset{\underset{H}{|}}{-N-R} \quad \text{Amine} \qquad X = \underset{\underset{O}{\parallel}}{-O-C-CH_3} \quad \text{Acetate} \qquad X = -O-N=C\overset{\displaystyle R}{\underset{\displaystyle R}{\diagdown}} \quad \text{Oxime}$$

$$X = \ -O-CH_2-CH_3 \quad \text{Alcohol}$$

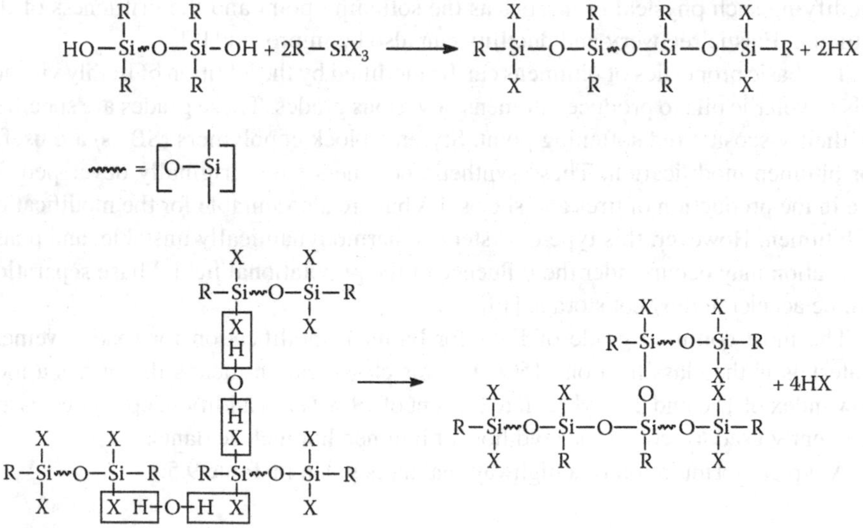

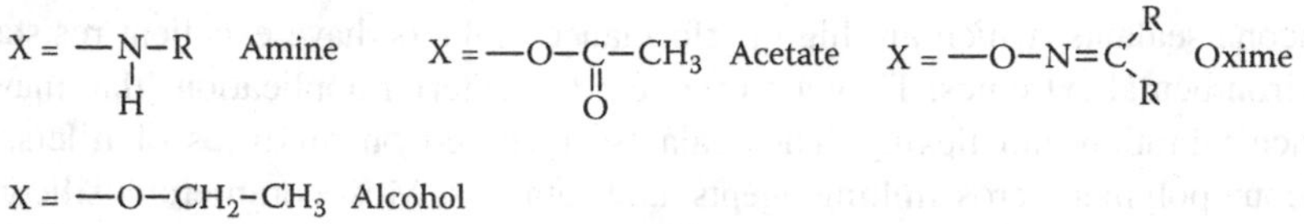

Trifunctional silanes, such as trimethoxymethylsilane or triacetoxymethylsilane, can be used as crosslinking agents during cure to increase the molecular weight and crosslink density of the final cured product.

One-part moisture-cure silicone elastomeric sealants can be prepared by mixing together a trifunctional or tetrafunctional silane crosslinker and a difunctional silane chain extender with alkoxy functional polydiorganosiloxanes, which have alkoxysilethylene ends. The end structure of this particular moisture-cure difunctional silicone sealant is as follows:

$$(R"O)_{3-a} - \underset{\underset{(R)_a}{|}}{\overset{(R)_a}{Si}} - [\, \underset{\underset{H}{|}}{\overset{\overset{H}{|}}{C}} - \underset{\underset{H}{|}}{\overset{\overset{H}{|}}{C}} -(\, \underset{\underset{R}{|}}{\overset{\overset{R}{|}}{Si}}\text{-}O)_c - \underset{\underset{R}{|}}{\overset{\overset{R}{|}}{Si}}\,]_b -$$

where R is methyl, ethyl, propyl, phenyl, or trifluoropropyl, R" is methyl or ethyl, a is 0 or 1, b is 0 or 1, and c is from 1 to 6.

The sealant composition is stable upon storage in the absence of moisture [20]. Fillers can be added for reinforcement in order to create high-strength one-part RTV (room-temperature-vulcanized) sealants. The reinforced polymer is modified to change the terminal functionality to a more stable form in order to have longer storage life [21]. The final physical properties of one-part sealants can be affected by the method of preparation [22].

Since high-green-strength silicone hot melt sealants cannot be obtained by either the crystallization of high Tm polyester or the solidification of high-Tg polymer when the molten sealant cools down, it is necessary to prepare extremely high melt viscosity silicone hot melt sealants to build up the initial strength needed for assembly process requirements.

In order to improve the melt viscosity of silicone hot melt sealants, trifunctional groups can be grafted onto the backbone of PDMS. This increases both the molecular weight and melt viscosity, and results in a higher initial strength. The degree of crosslinking is controlled by the amount of trifunctional silane, such as trimethoxymethylsilane or triacetoxymethylsilane, used to produce the prepolymers. Desired process and final properties of cured silicone hot melt sealants can also be obtained by adjusting the ratio of monofunctional, difunctional, trifunctional, or tetrafunctional silanes as well as by mixing with reinforcing fillers.

Since the Tg of PDMS is extremely low, silicone hot melt sealants provide superior softness and elasticity, and they can be used over a wide temperature range (−50°C–80°C) with excellent adhesion to glass. Most PDMS polymers do not contain aromatic groups, and therefore, silicone hot melt sealants provide very good UV and ozone resistance. Their disadvantages include low tear strength, poor abrasion resistance, high water-vapor transmission rate, poor resistance to organic fluids, cold flow problems even when cured, and poor paintability. Neither cured nor uncured silicone sealants are compatible with most types of sealants or adhesives. Repair operations may cause some problems due to poor bonding of silicone to nonsilicone sealants and vice versa if different types of sealants are used.

9.2.9 POLYURETHANES

With excellent mechanical strength, superior abrasion resistance, and low percent shrinkage during cure, polyurethane sealants are considered to be one of the high-performance sealants that can be used for exterior applications. The backbones of polyurethane sealants are formed by the reaction of excess isocyanates with organic polyols (see the following structure). Based on the variety of isocyanates and polyols available to the formulator, polyurethanes can be tailor-made to nearly any desired combination of properties.

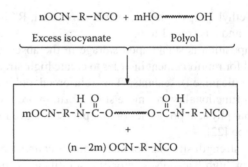

For polyurethane hot melt sealants, polyols are mixed with polyether and polyester diols to improve initial strength while maintaining flexibility and cold-temperature resistance. Sometimes, thermoplastics, such as PVC, EVA, SBCs, acrylics, and styrene copolymers are incorporated into the formulations to balance the open time and green strength (initial strength). Depending on the end-use application, fillers and thixotropic agents may be added to control process characteristics and physical properties of cured materials.

Isocyanates can react with any material containing active hydrogen, such as alcohols, amines, or water, to form either polyurethanes or polyureas (see the reactions in the following structure). Polyurethane hot melt sealants with excess isocyanate will react with moisture at either ambient temperature or process temperature. Therefore, any good commercial reactive hot melt sealant should be designed to have stable melt viscosity and long shelf life. By careful selection of catalyst, polyurethane hot melt sealants or adhesives can exhibit significantly increased curing rate while maintaining very stable melt viscosity [23].

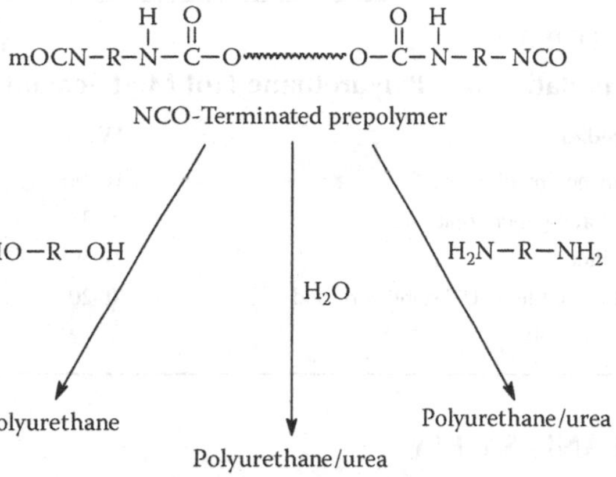

$$\text{mOCN}-R-\overset{\overset{\displaystyle H}{|}}{N}-\overset{\overset{\displaystyle O}{\|}}{C}-O\text{\small\textasciitilde\textasciitilde\textasciitilde}O-\overset{\overset{\displaystyle O}{\|}}{C}-\overset{\overset{\displaystyle H}{|}}{N}-R-\text{NCO}$$

NCO-Terminated prepolymer

HO−R−OH

H_2O

$H_2N-R-NH_2$

Polyurethane

Polyurethane/urea

Polyurethane/urea

The isocyanate and moisture reaction forms urea linkages and generates carbon dioxide (see the reactions in the following structure). As a result, cured bondlines can have foaming problem if the shape and fit of assembled parts and the sealant thickness are not well designed. This problem can be solved by appropriate formulation.

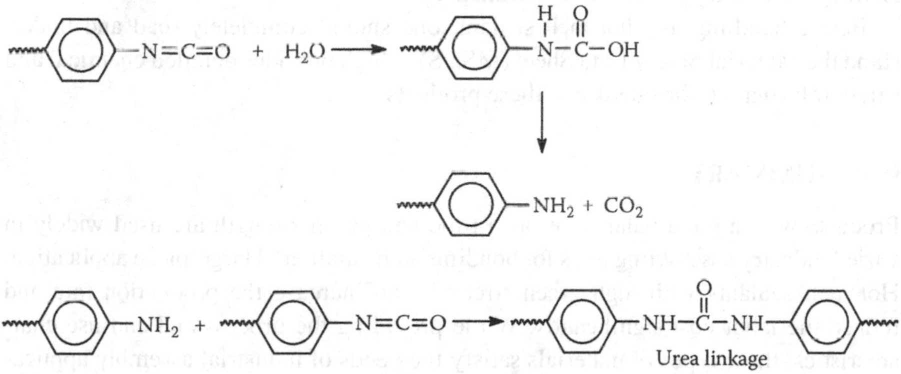

$$\text{\small\textasciitilde\textasciitilde\textasciitilde}\langle\bigcirc\rangle-N{=}C{=}O \;+\; H_2O \longrightarrow \text{\small\textasciitilde\textasciitilde\textasciitilde}\langle\bigcirc\rangle-\overset{\overset{\displaystyle H}{|}}{N}-\overset{\overset{\displaystyle O}{\|}}{C}-OH$$

$$\text{\small\textasciitilde\textasciitilde\textasciitilde}\langle\bigcirc\rangle-NH_2 \;+\; CO_2$$

$$\text{\small\textasciitilde\textasciitilde\textasciitilde}\langle\bigcirc\rangle-NH_2 \;+\; \text{\small\textasciitilde\textasciitilde\textasciitilde}\langle\bigcirc\rangle-N{=}C{=}O \longrightarrow \text{\small\textasciitilde\textasciitilde\textasciitilde}\langle\bigcirc\rangle-NH-\overset{\overset{\displaystyle O}{\|}}{C}-NH-\langle\bigcirc\rangle\text{\small\textasciitilde\textasciitilde\textasciitilde}$$

Urea linkage

Aromatic isocyanates provide better mechanical strength, faster cure rate, and higher melt viscosity with lower cost than aliphatic isocyanates. Therefore, industries use a greater amount of aromatic isocyanates than aliphatic. However, aromatic rings can generate chromophors very easily under UV light exposure. UV stabilization packages can be added into any aromatic isocyanate sealant system, thereby reducing or delaying the degree of color change and mechanical strength loss.

Polyester diols are very common raw materials that can be used in urethane hot melt sealants to improve initial strength. However, this type of raw materials can be hydrolyzed easily. Recently, several new approaches have been developed to solve this problem [24–29].

A model formulation of a polyurethane hot melt sealant is presented in Table 9.6:

TABLE 9.6

Formulation of a Polyurethane Hot Melt Sealant

Ingredient	Wt%
Urethane prepolymer	10–90
Reinforcing thermoplastics	5–35
Tackifiers	0–30
Additives, fillers, UV stabilizers, and antioxidants	0–20

9.3 HEALTH AND SAFETY

A great variety of polymers, plasticizers, and fillers are used in a sealant composition. Since hot melt sealants are processed and dispensed at high temperature, the use of appropriate personal protective equipment (PPE), such as chemical-resistance and heat-protective gloves, lab coat, safety glasses, and protective boots, is necessary during the manufacturing and cleaning operations. In addition, adequate ventilation is always necessary during such operations.

Before handling any hot melt sealant, one should completely read and understand the Material Safety Data Sheet (MSDS), which provides detailed chemical and safety information for safe use of these products.

9.4 SUMMARY

Products with a good balance of open time and green strength are used widely in varied industry assembling lines for bonding both small- and large-piece application. Hot melt sealants with high green strength can increase the production rate and reduce the labor cost significantly. While providing the process and end-use characteristics, these types of materials satisfy the needs of industrial assembly applications. Because they can significantly reduce manufacturing cost for any assembly, the amount of hot melt sealants used would continue to grow in the future.

For outdoor application, use of a reactive-type hot melt sealant is a good choice. The heat resistance of this type of sealant can be improved after the backbones of prepolymers have reacted and crosslinked. Through certain chemical reactions, reactive hot melt sealants change from thermoplastics to thermosets; therefore, they will not be able to flow any further. Only reactive hot melt sealants, which can react with either moisture or some other components, are suitable for both indoor and exterior applications.

REFERENCES

1. E. M. Petrie, *Handbook of Adhesives and Sealants*, Chapter 12, McGraw-Hill, New York (2000).
2. E. M. Petrie, Hot Melt Sealant Market (April 7, 2003).

3. J. S. Armstock, *Handbook of Adhesives and Sealants in Construction*, Chapter 5, McGraw-Hill, New York (2001).
4. G. Bayan, U.S. Patent 4,810,752 (1989).
5. T. Zaharescu, C. Postolache, and M. Giurginca, *J. Appl. Polym. Sci. 59*, 969 (1996).
6. H. Haverstreng, U.S. Patent 4,032,489 (1977).
7. The technical information on Eastoflex, Product of Eastman Chemical Company, Publication WA-12B (June 1998).
8. J. Panke, in *High Polymers,* N. Gaylord (Ed.), pp. 13, 145, John Wiley & Sons, New York (1962).
9. J. Hutt and H. Singh, U.S. Patent 3,923,748 (1975).
10. J. Hutt , H. Singh, and M. Williams, U.S. Patent 4,366,307 (1982).
11. L. Morris and H. Singh, U.S. Patent 4,623,711 (1986).
12. E. M. Fettes and J. A. Gannon, U.S. Patent 2,789,958 (1957).
13. E. R. Bertozzi, U.S. Patent 4,165,452 (1979).
14. E. G. Millen, U.S. Patent 4,314,920 (1982).
15. D. E. Vietti, K. B. Potts, and K. A. Leone, U.S. Patent 5,610,243 (1997).
16. R. A. Millers, TAPPI Hot Melt Symposium Proc., pp. 121–130 (2001).
17. P. M. Jones, Canadian Building Digests, CBD-38. "Bituminous" (February 1963).
18. C. J. Summers, The Idiots' Guide to Highways Maintenance, "Modified Bitumen and Bituminous Materials" (May 2000).
19. X. Lu, U. Isacsson, and J. Ekblad, *J. Mater. Civil. Eng. 11*, 51–57 (1999).
20. R. G. Chaffee and L. F. Stebleton, U.S. Patent 4,687,829 (1987).
21. C. Lee and M. T. Maxson, U.S. Patent 4,711,928 (1987).
22. R. P. Kamis, J. M. Klosowski, and L. D. Lower, U.S. Patent 4,888,380 (1989).
23. J. M. Hung and M. Graham, U.S. Patent, 5,550,191 (1996).
24. D. T. Dumack, U.S. Patent, 6,613,836 (2003).
25. J. M. Hung, Y. S. Zhang, and W. K. Chu, U.S. Patent, 7,037,402 (2006).
26. J. M. Hung, W. K. Chu, and Y. S. Zhang, U.S. Patent Application 20040072953A (2004).
27. J. M. Hung, W. K. Chu, Y. S. Zhang, and I. Cole U.S. Patent Application 20040198899A (2004).
28. J. M. Hung, Y. S. Zhang, Y. Zhang, and Z. Li, U.S. Patent Application 7,138,466 (2006).
29. Y. S. Zhang, J. M. Hung, and W. K. Chu, U.S. Patent, 7,112,631 (2006).

10 Intumescent Sealants

Pamela K. Hernandez

CONTENTS

10.1 INTRODUCTION

In any nonresidential building, results of fire are catastrophic. The potential for loss of life, property damage, and loss of revenue is tremendous. A classic example is the MGM Grand Hotel fire in 1980. At the time, it was the second-largest loss of life in a hotel fire in U.S. history, with 84 deaths and 785 injuries from smoke and toxic fume inhalation [1]. A small fire spread rapidly through elevator shafts and stairwells. At the time, sprinkler systems were the only line of defense, but they were inadequate to stop the spread of fire. Since that time, much effort in fire prevention, both public and private, has gone into protecting nonresidential structures, and the results have been highly effective, especially when compared to residential fires.

Passive fire protection products, including intumescent sealants, can be found in nonresidential structures such as those used for educational, institutional, public assembly, stores and offices, industry, utility, defense, storage in structures, and special structures. Table 10.1 shows the number of fires, deaths, injuries, and dollar loss that occurred in nonresidential properties in the United States from 1996 to 2005.

In 2003, the estimate for fire deaths included 100 fire deaths in the Station Nightclub fire in Rhode Island, and 31 deaths in two nursing-home fires in Connecticut and Tennessee [2]. Most recently published statistics from The National Fire Protection

TABLE 10.1
Fire-Related Incidents in Nonresidential Structures

Year	Fires	Deaths	Injuries	Direct Dollar Loss in Millions
1996	150,500	140	2,575	2,971
1997	145,500	120	2,600	2,502
1998	136,000	170	2,250	2,326
1999	140,000	120	2,100	3,398
2000	126,000	90	2,200	2,827
2001	125,000	80	1,650	3,231
2002	118,000	80	1,550	2,687
2003	117,500	220	1,525	2,604
2004	115,500	80	1,350	2,366
2005	115,000	50	1,500	2,318

Association in 2006 show that 17% of all reported fires occurred in apartment-type dwellings, contributing to 13% of all civilian fire deaths [3a].

Effective fire protection encompasses not only smoke detection from smoke detectors and fire suppression from sprinkler systems but also passive fire protection. It is an integral part of fire protection engineering. In a building, there are many openings in fire-resistive construction materials through which pipes, cables, wires, duct work, and beams pass. These passive fire-stop systems protect life and building from the spread of flame, smoke, and deadly combustion by maintaining fire-rated integrity of walls, floors, and ceiling assemblies. Passive fire protection provides additional time for safe exit from buildings in the event of a fire before flame, smoke, and toxic gases can spread throughout the building. It is meant to compartmentalize the fire, keeping the fire contained in the area in which it started, for a certain period of time. There are many products used in this market segment, including fire-rated sealants for joints and top-of-wall assemblies, intumescent wrap strips for plastic pipes and large openings, cast in place solutions, intumescent collars and sleeves, etc.; but intumescent sealants are considered the workhorse of the construction passive fire protection industry. They are the product of choice in the majority of passive fire-stopping applications due to their ease of use, relative low cost, and good performance.

Their purpose is to form an insulative, foamed char upon contact with heat or fire to compartmentalize the fire, keeping it from spreading to other areas in a building thus allowing occupants time to exit safely. They are mainly used in apartment buildings, condominiums, hotels, office buildings, and hospitals, where through penetrants such as metal and plastic pipes, conduits and cables pass through walls and floors. Their use is highly regulated through third-party testing organizations such as Underwriters' Laboratories (UL) and Intertek, which employ rigorous testing methods, including ASTM E814/UL 1479 and E119.

10.2 TESTING REQUIREMENTS

Proper application and installation of passive fire protection products, including intumescent sealants, are governed by building codes such as the National Building Code (NBC), Uniform Building Code (UBC), International Building Code (IBC), Standard Building Code (SBC), and the code of the International Code Council (ICC), which comprises BOCA (Building Officials and Codes Administrators), ICBO (International Conference of Building Officials), and SBCCI (Southern Building Code Conference International). These building codes require that all fire-stop systems be rigorously tested in accordance with ASTM E814/UL1479 and ASTM E119 by third-party testing laboratories such as Underwriters Laboratories, Warnock-Hersey, Factory Mutual, Southwest Research, or Intertek. Testing is system or application specific; for example, an intumescent sealant is tested multiple times with different pipes, substrates, annular spaces, and sealant depths. A specific system might include a 5.08 cm poly(vinyl chloride) (PVC) pipe in a 10.16-cm-thick concrete floor with a 7.62 cm opening and a 1.27-cm-thick sealant surrounding the centered pipe.

Figure 10.1 illustrates a specific application using a 1.27-cm-thick intumescent sealant over backer rod (used to hold the sealant in place within the opening) with 2.54 cm PVC pipe in a 11.43-cm-thick concrete floor. The annular space is 0.95 cm to 1.91 cm in a 7.62 cm opening.

Figure 10.2 shows a different application with a 10.16 cm insulated copper pipe in a gypsum wall assembly. The pipe is centered using 1.27-cm-thick sealant with backer rod.

If the particular test system passes the ASTM E814 or UL1479 performance criteria, then it is published for use. Application by the end user must comply with the details of the listed system. Tested systems are found in the UL handbook by supplier [3b]. Often, a sealant will be tested for hundreds of different scenarios, each system specific and listed separately [4].

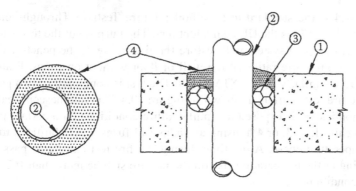

FIGURE 10.1 Plastic pipe in a concrete floor assembly with backer rod (3) and intumescent sealant (4).

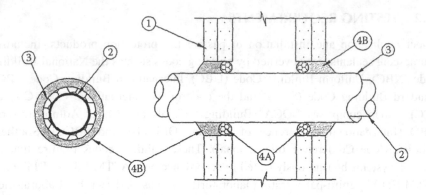

FIGURE 10.2 Insulated copper pipe with sealant through a gypsum wall assembly.

10.2.1 DESCRIPTION OF ASTM E119

ASTM E119 is the standard test method for Fire Tests of Building Construction and Materials. It is a measure of the fire-stopping endurance or fire-resistive performance of a particular barrier. Fire-stop systems in floors and walls are evaluated in terms of their ability to prevent the passage of smoke and flame as well as to prevent the temperature on the nonfire side of the floor or wall from rising high enough to ignite combustible material on the nonfire side. These are given F and T ratings. The F rating is the time in hours for which the test assembly is able to keep fire from moving to the nonfire side. The T rating is the time in hours for which the temperature of the unexposed side of the assembly remains no more than 162.8°C above the room temperature in any one location or no more than 121.1°C as an average over all thermocouple locations at the substrate surface. The furnace temperature follows a standardized time–temperature curve from which these ratings are derived. The curve is shown in Figure 10.3.

10.2.2 DESCRIPTION OF ASTM E814

ASTM E814 is the standard test method for Fire Tests of Through-Penetration Fire-Stops. UL 1479 is the UL equivalent test. The purpose of the test is to ensure that the fire-stopping system will restore the fire rating of the penetrated assembly to the original fire rating of the wall or floor without openings. F and T ratings are used similarly as in ASTM E119. The test is composed of two parts. The first is the fire test portion, as shown in Figure 10.4. The intumescent sealant in a specific application is exposed to simulated fire conditions for a certain period of time, typically 1, 2, 3, or 4 h, using an industrial furnace standardized to the UL time–temperature curve. At no time during the fire test can flame pass through the opening or the temperature rise on the nonfire side be more than 162.7°C plus ambient conditions.

The second requirement of ASTM E814/UL1479 is the hose stream test. It is an integral part of the standard. It serves two purposes. The first is to provide an indication of the physical integrity of the fire-stop once the fire test is completed. The

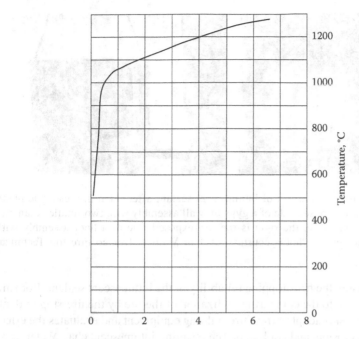

FIGURE 10.3 ASTM E119 time–temperature curve. (From UL Fire Directory.)

FIGURE 10.4 ASTM E814 fire testing of an intumescent sealant with PVC pipe in a gypsum wall. (Photo courtesy of Tim Mattox, Tremco Firestop Technical Services Manager.)

FIGURE 10.5 Hose stream test of intumescent sealant after a 1 h fire test. The photo on the left is the fire-exposed side of a gypsum wall assembly with two smaller-diameter plastic pipes. The photo on the right is the fire-exposed side of a floor assembly with one large-diameter pipe. (Photos courtesy of Tim Mattox, Tremco Firestop Technical Services Manager.)

second is to measure the fire-stopping reliability of the intumescent sealant. Both are extremely important to the compartmentalization of the fire by the fire-stop sealant. The hose stream test is adapted from fire fighting equipment and evaluates the effect of water impact, erosion, and cooling on the resulting intumescent char. Water pressure is typically 207 kPa (30 psi) for a duration of 6–16 s, depending on the surface area of the test assembly. Figure 10.5 shows pictures of the hose stream test on a wall assembly (left) and a floor assembly (right) with plastic pipe through penetrations. In each case, the pipes have burned away leaving the charred intumescent sealant as an insulative plug in each opening. To pass, these plugs must remain intact after the hose stream exposure.

While the code bodies at this time consider ASTM E814 and UL1479 equivalent, there is one major difference between them. UL1479 includes a recently adopted accelerated aging requirement for intumescent fire-stopping materials. Some intumescent sealants, based on silicate chemistry, were discovered to exhibit decreased volume expansion upon aging after a few years, thus limiting fire-stop performance and service life. Typically, the service life of intumescent sealants is considered to be equivalent to ~ 50 years, the average lifetime of a building. These materials must retain their original performance after accelerated aging. Aged volume expansion must be within 90% of the original volume expansion, and the onset of expansion, as measured by expansion pressure, must be within 3 s of the original value. Aging conditions are shown in Table 10.2.

10.3 INTUMESCENT SEALANT PERFORMANCE CRITERIA

As with any sealant, intumescent sealants must also exhibit typical application properties including acceptable viscosity, nonsag, gunnability, and toolability to the end user. The difference between these and other sealants used in construction applications lies in their ability to expand to several times their original volume when exposed to excessive temperature, as well as their ability to form a strong char to

TABLE 10.2
UL1479 Accelerated Aging Conditions

Type	Environmental Condition	Duration in Days
1	37.8°C/100% RH	180
2	70°C	270
3	80°C	140
4	90°C	70

withstand the hose stream test. With these performance criteria in mind, formulating intumescent sealants to meet the rigors of fire testing as well as end-user performance and application requirements is challenging.

10.3.1 Intumescent Sealant Formulation Considerations

The starting point for any sealant formulation is the choice of binder resin. Intumescent sealants are generally organic-based systems. Most intumescent sealants are water based and to a lesser extent are silicone based. Silicone sealants are formulated using silicone polymers of varying molecular weights or viscosities, and they are generally difunctional. Neutral cure mechanisms using oxime crosslinkers are most common. Their use as flame-retardant insulative sealants dates back to as far as 1969, as discussed in U.S. patent 3425967 assigned to General Electric. More recently, intumescent silicone sealants are described in U.S. patents 5246974 and 5508321.

More choices are available for water-based systems. The binder resin choices are numerous. Almost any thermoplastic acrylic latex or dispersion can work as long as they can be formulated to maintain some flexibility to allow optimal expansion. Under the influence of heat, the binder resin must exhibit an optimal melt viscosity to allow intumescence to occur. Most common are the acrylic latices with a low glass transition temperature, below 0°C. Most patents list latex polymers from the group consisting of acrylate, methacrylate, vinyl acetate, and combinations thereof. Also included are latex polymers composed of rubber, styrene, butadiene, copolymers of styrene and butadiene, butadiene–acrylonitrile copolymers, polyisoprene, polybutadiene, polychloroprene, and PVC [5–10].

Typical latex sealant additives can also be incorporated into the intumescent sealant formulation, including typical processing aids such as surfactants and dispersing aids for stability, plasticizers for added flexibility, pigments, flame retardants, and fillers. Thixotropes and rheology modifiers are an important part of sealant formulation, which gives the nonsag characteristics to the sealant. Most commonly used thixotropes include fumed silica and hydroxyethylcellulose derivatives. A comprehensive list of these additives is described in References 5–10.

10.3.2 BASIC INTUMESCENT CHEMISTRY

Intumescent chemistry has been around for over 100 years. The basic chemistry can still be found today in intumescent sealants. There are three critical ingredients that are required for intumescence to occur: a catalyst, a carbonific, and a spumific. The catalyst is a material that contains a high percentage of active phosphorous and decomposes at 150°C with heat to yield phosphoric acid. Common catalysts are ammonium polyphosphate (APP), urea phosphates, melamine phosphates, and diammonium phosphates. A carbonific is a material that when reacted with the acid decomposition by-product from the catalyst forms a carbonaceous foamed char at higher temperatures. This foamable carbon yields a noncombustible insulative barrier. Typical carbonifics are mono-, di-, and tri-pentaerythritols; sugar; starches; and polyols. The third necessary ingredient is a spumific, which serves as the blowing agent. On decomposition, a spumific generally releases significant quantities of gas by-product causing foaming. Melamine is an example of a spumific that gives off ammonia at approximately 300°C [11]. The general reaction scheme is depicted in Figure 10.6. Several reactions between ammonium polyphosphate, melamine, and pentaerythritol can occur. These are depicted in Figures 10.7–10.10 [12].

On decomposing, APP loses ammonia and water. Figure 10.7 illustrates two possible crosslinking reactions. Figures 10.8 and 10.9 illustrate two possible crosslinking reactions between polyphosphoric acid and pentaerythritol, followed by cyclization reactions in Figure 10.10. It has been shown that melamine not only acts as a spumific releasing ammonia on decomposition, but also forms crosslinked carbon-nitrogen polymers by a thermal deamination mechanism. These crosslinked structures contribute to char formation [13,14].

When formulating this type of intumescent chemistry into an intumescent sealant, a good starting composition for the APP/melamine/pentaerythritol system is 6:1:1 based on 100 part formulation. Depending on the desired volume expansion and resulting char strength of the carbonaceous foam left from the burned intumescent sealant, optimization of the concentration of each ingredient would be required.

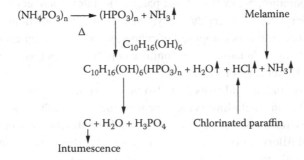

FIGURE 10.6 General intumescent reaction scheme using ammonium polyphosphate (APP) catalyst.

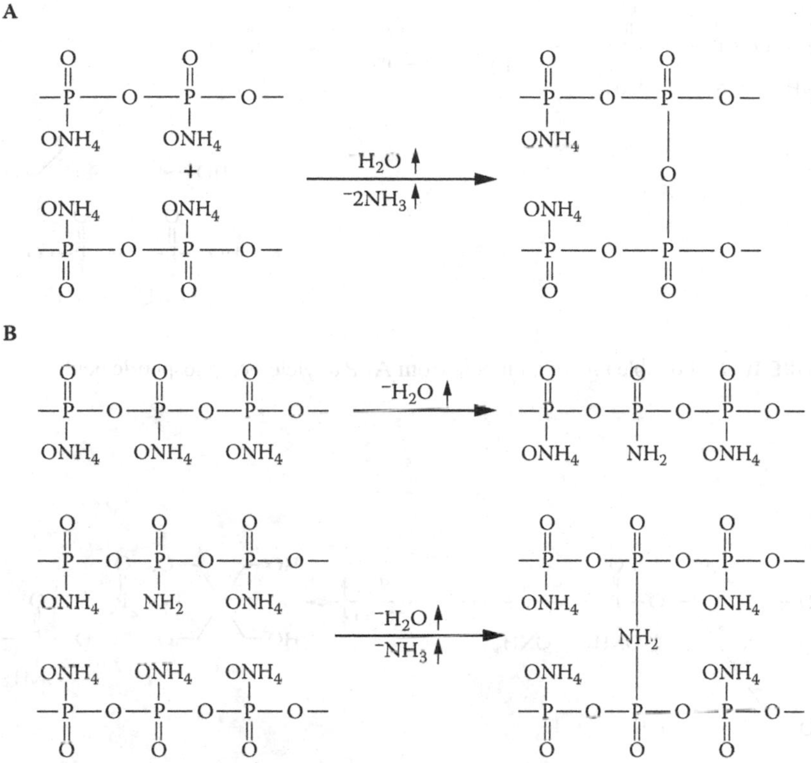

A

B

FIGURE 10.7 Possible crosslinking reactions of APP.

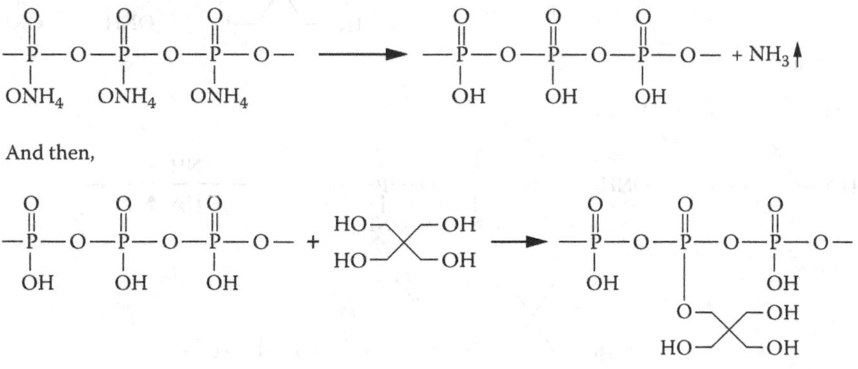

And then,

FIGURE 10.8 Possible direct phosphorylation of pentaerythritol.

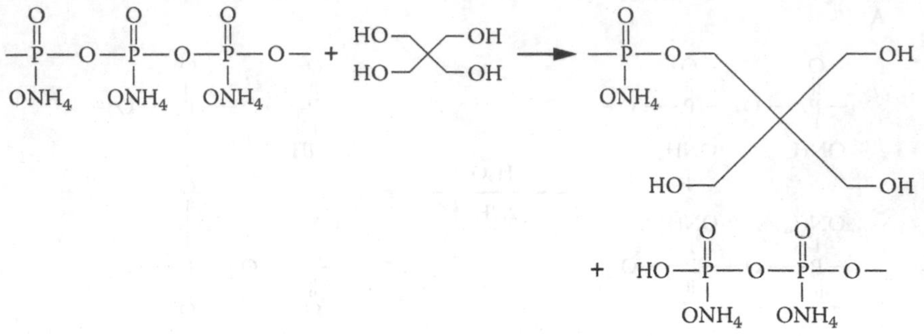

FIGURE 10.9 Possible loss of ammonia from APP to yield polyphosphoric acid.

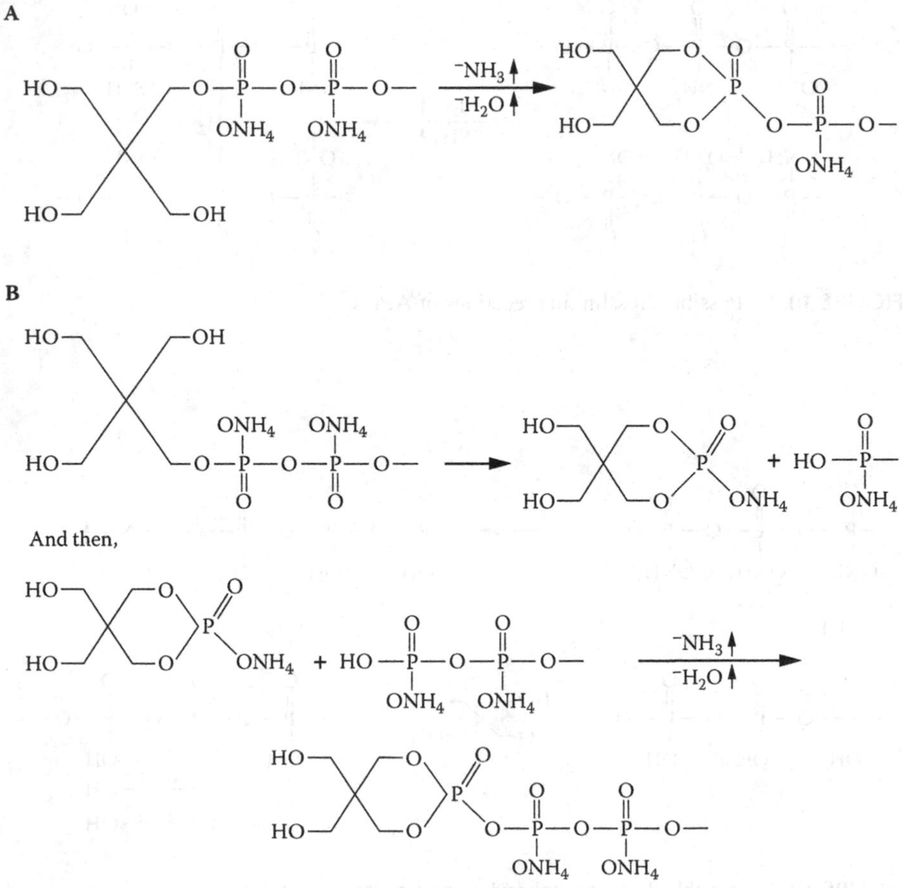

A

B

And then,

FIGURE 10.10 Possible cyclization mechanisms for char formation.

10.3.3 DESCRIPTION OF OTHER INTUMESCING AGENTS

Intumescent compounds should be chosen on the basis of their ability to expand at least 1.5 times their original volume in order to make the sealant material more thermally insulating and to adequately fill openings and voids. This volume change is activated by increasing the temperature. Intumescent additives must also exhibit limited solubility. Additives that exhibit higher water solubility tend to sublime out of the sealant as water evaporates. This phenomenon reduces the ability to intumesce. Most commonly used intumescent additives include expandable graphite, hydrated alkali metal silicates, vermiculite, perlite, borosilicates, gas-filled microspheres, mica, and combinations thereof.

Since the 1990s, the use of expandable graphite has increased dramatically in the area of intumescent sealants. Expandable graphite is flaked, interlaminar graphite that has been intercalated with an acid such as nitric acid or sulfuric acid. The graphite flakes exhibit a rapid increase in volume in the temperature range 150–300°C as the acid is released. Volume expansion can range from 50:1 to as high as 300:1 [15]. Figure 10.11 shows scanning electron micrographs of an exfoliated graphite flake that has expanded to several hundred times its original size. Figure 10.11A magnifies 20× the structure of an exfoliated expandable graphite flake. Figure 10.11B magnifies the flake 191×, illustrating the lamellar structure in the expanded state.

Suppliers of commercially available grades of expandable graphite include Asbury Graphite (Asbury, New Jersey), Timcal (Division of Imerys, Westlake, Ohio), and UCAR Carbon (Cleveland, Ohio), as well as several Chinese suppliers. In general, 2–40 wt% can be incorporated into a sealant formulation, depending on the desired volume expansion as described in the patent literature [5–10].

Alkali metal silicates can also be used as intumescent additives in sealants. These additives are endothermic in nature, releasing water at temperatures above 100°C or on contact with flame. The resulting solid is foamed hydrated silica. Silicates impart strong, rigid physical properties to the intumesced sealant due to their glassy nature [16]. Counterions include sodium, calcium, and magnesium. Silicates are generally incorporated at concentrations from 10–60 wt% of the total formulation. Generally, the concentration is such as to provide volume expansion of at least five times the original sealant volume [17]. A disadvantage of alkali metal silicate intumescent additives is their inherent sensitivity to high humidity. Under these conditions, the alkali metal tends to leach out from the silicate, thus reducing the effective volume expansion. Borate-stabilized and fatty-acid-stabilized silicates have been developed that exhibit increased resistance to humidity [18]. 3M (Minneapolis, Minnesota) has developed their fire protection business around alkali-metal-silicate-based intumescent sealants [19].

Both vermiculite and perlite are inorganic intumescent additives that expand by releasing of moisture. They are inherently heat and fire resistant and are lightweight. Vermiculite is a type of mica, whereas perlite is derived from silica-based volcanic glass [20]. Grace Construction Products (Cambridge, Massachusetts) supplies these types of additives as well as fire-stopping intumescent sealants based on this chemistry.

The last intumescent additive of interest is gas-filled microspheres. Specified Technologies, Inc. (Somerville, New Jersey) patented a two-stage intumescent sealant

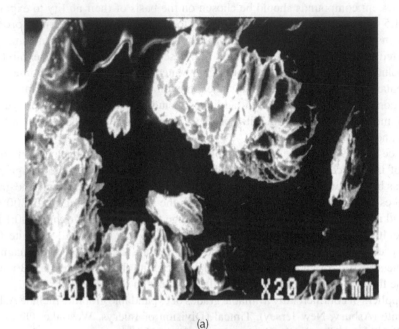

(a)

(b)

FIGURE 10.11 Scanning electron micrographs of an exfoliated expandable graphite flake: (a) 20× magnification and (b) 191× magnification. (Photo courtesy of A. Tamashausky, Asbury Graphite, Inc.)

using gas-filled microspheres as a secondary expansion [21]. These isobutane-filled microspheres used in conjunction with a primary source of intumescence such as expandable graphite, vermiculite, or alkali metal silicate provide good dimensional stability and enhanced volume expansion of the resulting foam char. These microspheres can be obtained from Expancel, Inc. (Duluth, Georgia), or H.M. Royal, Inc. (Trenton, New Jersey). These intumescent additives can be used by themselves or in various combinations to yield the desired volume expansion and resulting foamed insulative char.

10.3.4 Description of Endothermic Additives

Endothermic compounds are those that absorb heat by releasing water of hydration. These additives aid in the dissipation of heat by releasing small molecules such as ammonia, carbon dioxide, or water, which transport the heat away from the area protected by the intumescent sealant. The endothermic reaction should occur in the temperature range 90°C–1500°C. Suitable endothermic additives are typically inorganic. Examples include alumina trihydrate (ATH), zinc borate, calcium sulfate (gypsum), magnesium hydroxide, and magnesium ammonium phosphate. These are usually used in the range 14–95 wt% [22]. Certain combinations of endothermic additives have been shown to exhibit synergistic effects. US Borax, Inc. (Valencia, California) has shown that the combination of zinc borate and ATH yields superior flame retardance and char formation. Zinc borate dehydrates at approximately 290°C to yield zinc hydroxide and boric acid, while ATH decomposes into aluminum oxide and water. The zinc hydroxide and aluminum oxide react to yield ceramic compounds that form a stronger, more flame-resistant char layer than either zinc borate or ATH individually [23].

10.3.5 Optional Additives

There are other compounds that can be incorporated into the intumescent sealant formulation to enhance the fire performance, such as fillers, pigments, fire retardants, and processing aids. Fillers can be added for reinforcement, to adjust viscosity and toolability/gunnability, as well as to optimize cost. Glass fibers, powder, and frit can be added to improve strength of the sealant as well as the resulting char. Glass melts well above 500°C and upon cooling forms a rigid glassy continuous phase that aids in increasing the char strength of the intumesced sealant. Several patents describe the use of glass frit in intumescent, fire barrier compositions [24]. Flame-retardant additives include several compounds already mentioned earlier. These include boron-containing additives for smoke suppression and flame retardance, phosphorous-containing compounds, and metal oxides and hydrates. Concentrations to be used depend on the desired performance. Starting concentrations of 10% based on 100 parts formulation is a general rule of thumb. Processing aids include surfactants, dispersing aids, defoamers, fungicides/mildewcides, and adhesion promoters. Typical concentrations range from 0.1 to 1 wt%. Pigments are generally added for color identification. Many intumescent sealants are pigmented with iron oxide pigments to different shades of red. Iron oxide possesses flame-retardant properties by itself [25].

10.4   EXAMPLES OF INTUMESCENT SEALANT FORMULATIONS

A generic formulation of an intumescent sealant is shown as follows:

    25–45 wt% latex
    0–0.5 wt% fungicide
    0.2–1 wt% dispersing agent and wetting agent
    1–5 wt% plasticizer
    0–2 wt% propylene glycol
    20–40 wt% fillers
    0–3 wt% pigments
    4–6 wt% expandable graphite
    16–25 wt% zinc borate, glass fiber, and ammonium polyphosphate
    0.1–1.5 wt% cellulose thickener

More specific examples are as follows:

Example 1

    169.5 parts Neoprene 654 latex (binder resin, Dupont, Wilmington, DE)
    7.68 parts Foamaster 111 (defoamer, Cognis, Monheim, Germany)
    4 parts Triton X405 (surfactant, Rohm and Haas, Philadelphia, PA)
    4 parts Tamol 850 (dispersing aid, Rohm and Haas, Philadelphia, PA)
    2 parts Agerite Stalite S (antioxidant, R.T. Vanderbilt, New York)
    5 parts zinc oxide (ionic crosslinker/scavenger, U.S. Zinc, Inc., Pittsburgh, PA)
    5 parts iron oxide (pigment, DayGlo Color Corporation, Cleveland, OH)
    85 parts Expantrol 4 (intumescent additive, 3M, Minneapolis, MN)
    30 parts Techfill A-212 (ATH, Great Lakes Chemical, West Lafayette, IN)
    10 parts 1/16″ Fiberglass 731ED (glass fiber, Owens Corning)

Example 2

    250 parts HA-8 latex (Rohm and Haas, Philadelphia, PA)
    4.6 parts Triton X405 (Rohm and Haas, Philadelphia, PA)
    4.6 parts Tamol 850 (Rohm and Haas, Philadelphia, PA)
    162 parts Expantrol 4B (intumescent additive, 3M, Minneapolis, MN)
    28.8 parts Firebrake ZB (zinc borate, US Borax, Inc., Valencia, CA)
    28.8 parts Techfill A-212 (ATH, Great Lakes Chemical, West Lafayette, IN)
    14.4 parts Ceepree C200 (glass frit, ICI Chemicals, London)
    43.1 parts Phoschek P-30 (APP, Astaris, St. Louis, MO)
    2 parts Foamaster 111 (Cognis, Monheim, Germany)

10.5   SUMMARY

Intumescent sealants stop the spread of fire by forming an insulative foamed char upon exposure to a fire or heat source, using key chemical ingredients such as expandable graphite or APP. Intumescent sealants are used mostly to seal openings in floors and walls, preventing the spread of fire from room to room or floor to floor in commercial and nonresidential buildings. These sealants must perform to rigorous testing conditions by third-party testing facilities according to ASTM E814 or UL 1479. Their use is highly regulated by these facilities as well as by various building code officials. For optimal sealant performance in the desired applications, the intumescent and fire-retardant properties must be balanced with the rheological (flow, viscosity, and sag) properties by proper choice of sealant ingredients.

ACKNOWLEDGMENT

The author would like to thank Muneer Mohammed for help in the preparation of this manuscript.

REFERENCES

1. www.nfpa.com.
2. National Fire Protection Association, "Fire Loss Report in US during 2005," Abridged Report.
3a. NFPA.org/assets/files//PDF/os.firelosstrends.pdf.
3b. UL Fire Directory.
4. www.ifc.com.
5. U.S. Patent 07045079 (2006).
6. U.S. Patent 06153668 (2000).
7. U.S. Patent 06153674 (2000).
8. U.S. Patent 04945015 (1990).
9. U.S. Patent 06747074 (2004).
10. U.S. Patent 06238594 (2001).
11. S. C. Upadhya, *Paint India, 50*(11), 45–52 (2000).
12. H. L. Vandersall, *J. Fire Flammability*, 2, 97 (1971).
13. H. May, *J. Appl. Chem.,* 9, 340 (1959).
14. C. Van der Plaats, *Proceedings of the Second European Symposium on Thermal Analysis*, D. Dollimore (Ed.), p. 215, Heyden, London (1981).
15. K. Shen and B. Schilling, Recent Advances with Expandable Graphite in Intumescent Flame Retardant Technology, www.nyacol.com/exgraphadv.htm.
16. www.pqcorp.com/technicalservice/understanding_silicatechemistry.asp.
17. U.S. Patent 06153668 (2000).
18. U.S. Patent 04776355 (1988), U.S. Patent 04218502 (1980).
19. www.3M.com.
20. www.na.graceconstruction.com.
21. U.S. Patent 05132054 (1992).
22. U.S. Patent 06153674 (2000).
23. www.boraxfr.com/techmain.html.
24. U.S. Patent 05532292 (1996), U.S. Patent 05643661 (1997), U.S. Patent 06616866 (2003).
25. A. P. Mouritz and A. G. Gibson, *Fire Properties of Polymer Composite Materials Series: Solid Mechanics and Its Applications*, Vol. 143, 237–286, Springer Publishing, Netherlands (2006).

11 Urethane Waterproofing Membrane Systems for Sealing Concrete Structures

James Dunaway

CONTENTS

11.1 INTRODUCTION

Concrete is by far the most widely used construction material in the world. Its wide use is due both to its versatility as well as its durability [1]. There are structures that were built in Roman times using materials similar to modern concrete that are still standing today. Two major events in the development of concrete occurred in the eighteenth century: the invention of Portland cement and the use of steel bars to reinforce concrete. In the twentieth century, most of the research in using concrete focused on getting the most utility out of every pound of concrete.

As a result of the evolution of concrete and building techniques using concrete, a need arose for waterproofing membrane systems for use on concrete structures. These waterproofing membrane systems served two major purposes. The first was to keep water out of occupied spaces. In the past, concrete walls were so massive that the sheer thickness of the wall served as a barrier to water ingress. As developments in concrete technology allowed for the design of thinner walls that could still bear the same amount of load, these thinner walls did not offer as much protection from water infiltration. Steel reinforcing bars were one of the major advances that allowed for the higher-strength concrete construction and thinner concrete walls and floors. Unfortunately, when steel comes in contact with water, particularly water containing dissolved ions such as road de-icing salts or airborne salt spray from the ocean, the steel rebar can corrode and put stress on the concrete, leading to spalling and delamination [2]. Therefore, the second major purpose of concrete waterproofing membrane systems is to keep water and dissolved salts from causing corrosion of steel in steel-reinforced concrete structures.

In this chapter, we will review the use of polyurethane technology in waterproofing membrane systems for concrete. This review will include the chemistry of polyurethane waterproofing membranes, the unique properties of polyurethanes that make them ideal for use in waterproofing concrete structures, various types of waterproofing membrane systems, and the performance specifications commonly used in the industry.

11.2   TYPES OF WATERPROOFING MEMBRANE SYSTEMS

The two broad categories of waterproofing membrane systems that will be discussed here are below-grade waterproofing systems, such as those applied to the exterior surfaces of the foundations of concrete structures, and traffic-bearing membrane systems, which are applied to concrete surfaces designed to support either pedestrian or vehicular traffic. *Below Grade* is a term typically used to refer to parts of a structure below ground level. A parking garage would be an example of a concrete structure that supports vehicular traffic, which could require a waterproofing membrane system. A pedestrian plaza or walking bridge between two buildings would be examples of concrete structures that support pedestrian traffic, which could require a waterproofing membrane system. An example of a traffic-bearing membrane system being applied to a parking deck is shown in Figure 11.1. An example of a below-grade waterproofing system being applied to a foundation wall is shown in Figure 11.2. These membranes can be applied using a variety of application methods, but the most commonly used methods are a combination of squeegees, rollers, and spraying.

The performance requirements for the two different applications (below grade and traffic bearing) have some similarities, but there are also some key differences. In a below-grade application, the waterproofing system will consist of one or two layers of a thick build (15–20 mm), and highly elastomeric coating that is applied to the exterior surface of the foundation of the structure and then covered with dirt once it is cured. The below-grade membrane system is not exposed to direct solar radiation or precipitation once it is in place and covered. One key requirement for the below-grade waterproofing membrane is that it be able to bridge the cracks that might exist in the concrete. To do this, the membrane must have a high degree of elongation, typically >400% as measured by ASTM D-412. The industry-standard crack-bridging test is outlined in ASTM C836. There are also new test procedures being developed to better understand the crack-bridging ability of membranes by combining visual inspection with stress–strain measurements [3]. Another key

FIGURE 11.1 Traffic-bearing waterproofing system being applied to the top deck of a parking garage.

FIGURE 11.2 Below-grade waterproofing system being applied to a foundation wall.

requirement is that the cured membrane should not be degraded by the conditions it will be subjected to in a buried application, especially constant contact with water.

In the traffic-bearing application, the waterproofing membrane system is applied directly to the traffic-bearing surface of the structure. For example, in a parking garage, the waterproofing membrane system would be applied to the driving surfaces of the garage. The types of coatings used in traffic-bearing waterproofing systems have changed over the years [4,5]. The systems have evolved from simple asphalt overlays to more sophisticated and durable multilayer polymeric membranes. Traffic-bearing membrane systems typically need to be made up of at least two different types of materials. The first membrane, which is applied directly to the concrete surface, is the primary waterproofing layer of the system; typically, it is referred to as the *base coat*. The base coat needs to be elastomeric, with a high degree of elongation to be able to bridge small cracks in the concrete surface without rupturing and allowing water to penetrate into the concrete. Since the base coat needs to be elastomeric, it needs to be protected by another layer that is tough and durable. This other layer is typically referred to as the *top coat*. The top coat is the part of the system that must be able to withstand the abrasion from either pedestrian or vehicular traffic. Also, if the system is applied to a surface that is exposed to ultraviolet (UV; sun), the top coat must be UV-stable in order to maintain its properties after extended periods of UV exposure.

11.3   URETHANE CHEMISTRY

The terms *urethane* and *polyurethane* tend to be used interchangeably to describe a broad class of polymers that contain urethane linkage in their backbone [6]. The chemical structure of a urethane linkage is shown in Figure 11.3.

The urethane linkage imparts unique features to the polymer chain as a result of its high chemical polarity. Two of the most important features that the urethane linkage contribute to a polymer that are useful in formulating a coating to be used in a concrete waterproofing membrane system are high strength and good adhesion. The urethane linkage contributes to high strength via intermolecular attractions between polymer chains. Higher intermolecular attractions between polymer chains make it more difficult for the chains to slide past each other and, therefore, make the bulk polymer tougher and more durable. The polarity of the urethane linkage contributes to adhesion by increasing the attractive forces between the polymer and the substrate, particularly high-surface-energy substrates.

Another important attribute of urethane polymers that makes them useful in formulating the types of high solids coatings typically used as concrete waterproofing membranes is that these polymers can be designed to be low-viscosity reactive liquid prepolymers. This is useful since low-viscosity prepolymers will require much less solvent or other diluent to achieve an acceptable application viscosity in the

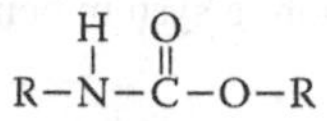

$$\text{R}-\overset{\text{H}}{\underset{\text{}}{\text{N}}}-\overset{\text{O}}{\underset{\text{}}{\text{C}}}-\text{O}-\text{R}$$

FIGURE 11.3   Chemical structure of urethane linkage.

formulated coating. As volatile organic content (VOC) limits become more and more restrictive, there is added pressure on coatings formulators to reduce the amount of solvents in their products. Reactive urethane prepolymers are one way to address this issue. Depending on how the reactive liquid prepolymers are designed, they can be formulated into a single-component coating that reacts with moisture from the air to cure or crosslink, or they can be formulated into a two-component coating into which a curative is mixed, causing the coating to cure.

Urethane polymers are typically synthesized by reacting materials containing hydroxyl groups, referred to as *polyols*, with isocyanates as shown in Figure 11.4. There are a wide variety of commercially available polyols, but the two types most often used to make urethane polymers are propylene oxide-based polyols and polyester-based polyols. These polyols usually have either two (difunctional) or three (trifunctional) hydroxyl groups per molecule and vary in molecular weight from about 300 g/mol up to as high as 8,000 to 10,000 g/mol.

Figure 11.5 shows the synthesis and structure of a typical propylene-oxide-based polyol. Figure 11.6 shows the structure of a caprolactone-based polyester polyol.

There are two basic types of diisocyanates used in the synthesis of urethane prepolymers employed in waterproofing membrane systems for concrete. The first and most widely used are the aromatic diisocyanates. Examples of these are 2,4 toluene diisocyanate (2,4 TDI) and diphenylmethane 4,4′ diisocyanate (MDI). The aromatic diisocyanates are characterized by having an aromatic ring in their structure. This aromatic structure contributes strength and durability to coatings made using these types of polymers, but it also contributes to yellowing of the coating after UV exposure. The second class of diisocyanates is referred to as *aliphatic diisocyanates* because their chemical structures do not include any aromatic rings. As a result of

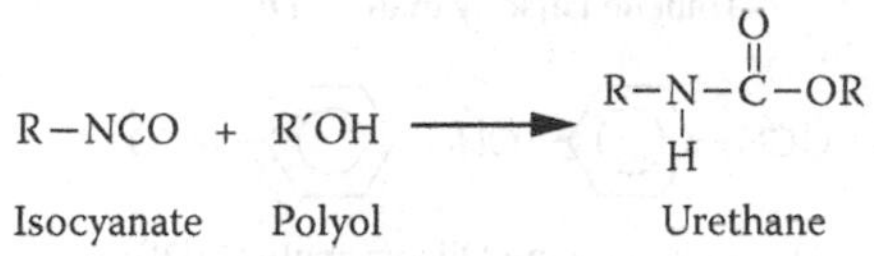

$$R-NCO \ + \ R'OH \ \longrightarrow \ R-N-\overset{\overset{\displaystyle O}{\|}}{C}-OR$$

Isocyanate   Polyol   Urethane

FIGURE 11.4   Reaction of an isocyanate with polyol.

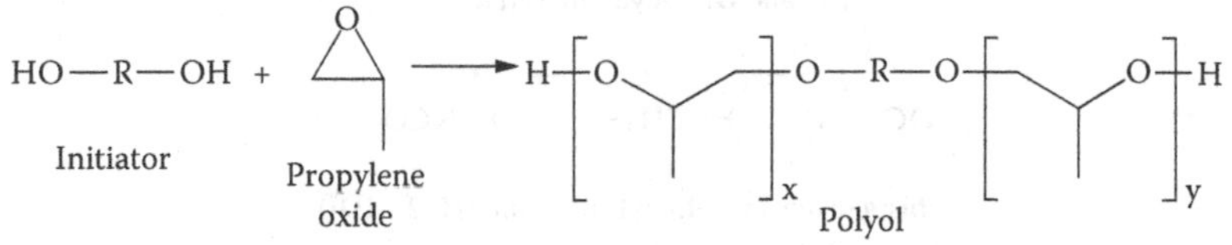

FIGURE 11.5   Synthesis and structure of propylene-oxide-based polyol.

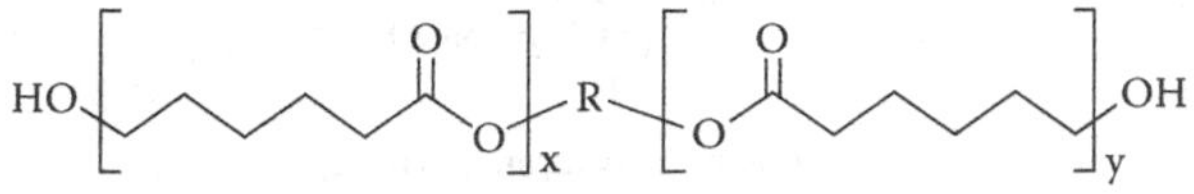

FIGURE 11.6   Structure of caprolactone-based polyester polyol.

this, urethane polymers based on aliphatic diisocyanates yield coatings that are less prone to yellowing than polymers based on aromatic diisocyanates [7]. Examples of the more common aliphatic diisocyanates are Hexane Diisocyanate (HDI), bis(4-Isocyanatocyclohexyl)methane (commonly referred to as *H-12 MDI* because of its similarity in structure to MDI), and Isophorone Diisocyanate (IPDI). The chemical structures of these common diisocyanates are shown in Figure 11.7.

As can be seen from Figure 11.7, the chemical structures of the common, commercially available diisocyanates are very different. However, some of these, such as MDI, HDI, and H12-MDI, have very symmetrical structures. This symmetry leads to more efficient hydrogen bonding between the urethane polymers made from these diisocyanates. Because of this, these diisocyanates are usually the ones used when higher-strength polymers are required. Another result of the symmetrical structure of these diisocyanates is that both the isocyanate groups on the molecule tend to have very similar reactivities [8]. This makes it more difficult to use these diisocyanates to synthesize low-molecular-weight low-viscosity prepolymers. Both TDI and IPDI are asymmetric molecules, and their isocyanate groups differ in reactivity. The isocyanate in the 2-position on TDI tends to be less reactive than the isocyanate group in the 4-position, while the secondary isocyanate in IPDI tends to be more reactive [9,10]. This differential reactivity makes TDI and IPDI better suited for use in synthesizing low-molecular-weight low-viscosity prepolymers.

Toluene Diisocyanate (TDI)

Diphenylmethane Diisocyanate (MDI)

$$OCN—(CH_2)_6—NCO$$

Hexane Diisocyanate (HDI)

bis(4-Isocyanatohexyl)methane (H12-MDI)

Isophorone Diisocyanate (IPDI)

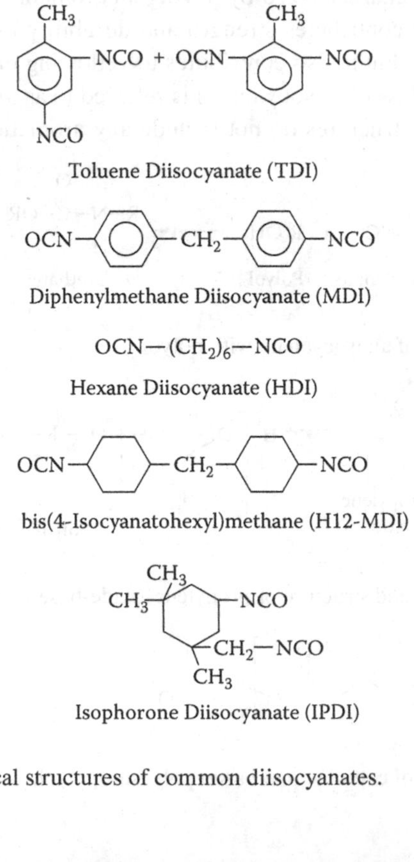

FIGURE 11.7 Chemical structures of common diisocyanates.

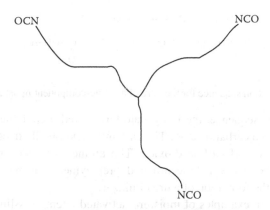

FIGURE 11.8 Idealized structure of an isocyanate-capped prepolymer.

By varying the polyol and diisocyanate used to make the urethane prepolymers, the properties of the cured coating can be varied over a wide range. It is even possible to use mixtures of polyols and diisocyanates to arrive at an even wider range of properties. An example of a urethane prepolymer that might be used in a below-grade waterproofing membrane formulation or the base coat of a traffic-bearing waterproofing membrane system would be the reaction product of a 6000-molecular-weight propylene oxide triol and MDI. An idealized structure of this type of prepolymer is shown in Figure 11.8. Coatings or membranes based on urethane prepolymers can be designed to have tensile strengths that range from 3 MPa up to over 45 MPa and ultimate elongations that range from 5 to 10%, up to well over 1000%.

The high molecular weight of the propylene oxide polyol gives the polymer its low modulus and high elongation, while the end capping of the prepolymer with isocyanate provides a reactive site for crosslinking. Also, the propylene oxide backbone has excellent hydrolytic stability.

11.3.1 URETHANE CURE CHEMISTRY

For the type of high solids urethane coatings (based on reactive liquid polymers) typically used in concrete waterproofing membrane systems, *curing* is the term used to describe the various reactions that take place when a prepolymer changes from a liquid, flowable material, to a solid. As mentioned earlier, urethane coatings used as concrete waterproofing membranes can be formulated to be either single-component coatings, which react with the moisture in the atmosphere to effect the cure, or two-component materials, which comprise a base material and a curative.

11.3.1.1 Single-Component Urethane Cure Chemistry

In a single-component or moisture-activated urethane coating, ambient moisture either reacts directly with the urethane prepolymer to cause crosslinking, or the water reacts with a moisture-activated latent curative, which then acts to crosslink the urethane prepolymer [11]. The reaction sequence for a standard moisture-cured urethane cure reaction is shown in Figure 11.9.

$$R-NCO + H_2O \longrightarrow R-NH_2 + CO_2$$

$$R-NH_2 + R'-NCO \longrightarrow R-NH-\underset{\underset{O}{\|}}{C}-NH-R'$$

FIGURE 11.9 Reaction sequence for the cure of a single-component moisture-cured urethane.

In this reaction sequence, the isocyanate-terminated prepolymer initially reacts with water to form a carbamic acid. The carbamic acid rapidly decomposes to form an amine and gives off carbon dioxide. The amine-terminated prepolymer then reacts with another isocyanate-terminated prepolymer, causing the crosslinking reaction through the formation of a urea linkage.

Oxazolidines are examples of moisture-activated latent crosslinkers that can be used in certain types of single-component urethane coatings [12]. When an oxazolidine reacts with moisture, it splits off an unreactive (toward isocyanates) aldehyde and generates a secondary amine group and a hydroxyl group [13]. The secondary amine and hydroxyl group can then react with the isocyanate-terminated prepolymer. A bis-oxazolidine is a molecule with two oxazolidine groups. When a bis-oxazolidine reacts with two molecules of water, it generates two secondary amine groups and two hydroxyl groups.

The bis-oxazolidine molecule is a much more efficient crosslinker than a molecule containing only one oxazolidine group. The reason for this is that the reaction rate of the secondary amine group with an isocyanate is about one order of magnitude faster than the reaction rate of the hydroxyl group. As a result, when bis-oxazolidine reacts fully with water to liberate two secondary amines, these amines will react almost instantaneously with two isocyanate groups, causing crosslinking to occur. In a molecule with only one oxazolidine group, after the reaction with water, the secondary amine will react with any available isocyanate very quickly, but the hydroxyl group is much slower to react. Since crosslinking occurs when isocyanate groups from two different prepolymers are reacted, the slower reaction rate of the hydroxyl group reduces the rate of crosslinking. By choosing the proper level of bis-oxazolidine added to the coating, the formulation chemist can cause all of the crosslinking reactions to occur through the reaction of the secondary amine and an isocyanate or through both the secondary amine and hydroxyl groups. Figure 11.10 shows the structure of a bis-oxazolidine and its reaction with water. Oxazolidines do not give package-stable single-component coatings when used with aromatic isocyanate prepolymers. Oxazolidines give the best package stability when used with prepolymers based on the less reactive aliphatic diisocyanates such as IPDI.

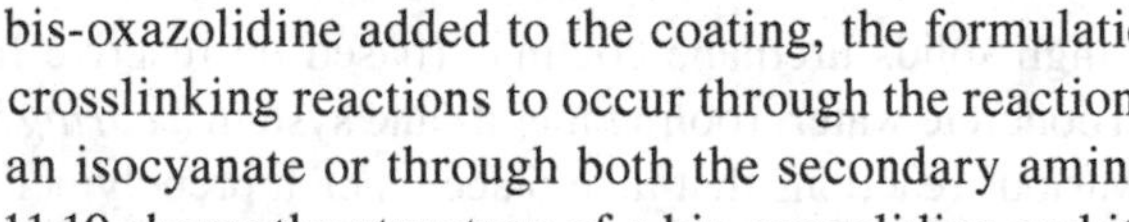

FIGURE 11.10 Structure of a bis-oxazolidine.

11.3.1.2 Two-Component Urethane Cure Chemistry

The basic chemistry of two-component urethane coatings used as concrete water-proofing membranes is similar to that of single-component urethane coatings. In a two-component system though, the isocyanate functional material, instead of reacting with water from the atmosphere to cure, reacts with a curative that is mixed in at the job site just prior to application. Two-component urethane coatings are sometimes referred to as *chemical cure* coatings since they do not rely solely on moisture from the atmosphere for their cure. One major advantage of two-component curing systems is that no carbon dioxide is generated while they are curing, and this greatly reduces the tendency of these types of coating to gas or blister after application.

11.4 FORMULATING SINGLE-COMPONENT URETHANE COATINGS

The compounding or formulating of moisture-curing urethane coatings used in waterproofing membrane systems is similar to that for other types of coatings in that the same types of additives, fillers, pigments, UV stabilizers, solvents, mildewcides, etc. are used [14]. The one major difference is that all compounding ingredients must be substantially free from water contamination. This is necessary for package stability reasons. Moisture levels even as low as 0.1% can be enough to cause an unacceptable rise in viscosity of a single-component moisture-curing urethane coating with time. Moisture scavengers can be added to the formulation but they are effective only up to a point. Therefore, the raw materials used in formulating a moisture-curing single-component urethane coating must either come from the supplier with an acceptably low moisture level or be dried prior to use. An example of a raw material that can be easily dried with the proper equipment is ground calcium carbonate, which can be dried to <hr 0.05% moisture content in a rotary-kiln-type drier. The following is a brief description of the various types of raw materials typically used in formulating single-component moisture-curing urethane coatings:

1. Prepolymer/binder: As mentioned earlier, there is a wide choice of polymers to use as the binder in a moisture-curing urethane coating. The choice of polymer is the primary factor determining the final properties of the coating.
2. Crosslinker/catalyst: Oxazolidines can be used as latent crosslinkers in some aliphatic single-component moisture-curing urethane coatings. For standard moisture-curing urethane coatings, a catalyst for the isocyanate–water reaction is generally required. Tin catalysts such as dibutyl tin dilaurate (DBTDL) are often used, as are amine catalysts.
3. Solvents/diluents: Volatile solvents such as xylene and mineral spirits can be used to reduce the viscosity of the coating. Since these solvents do evaporate, they will have a minimal effect on the properties of the cured coating but they are counted as being volatile organic content (VOC). VOC-exempt solvents such as parachlorobenzotrifluoride (PCBTF) or *t*-butyl acetate can be used if VOC limits are a concern. Plasticizers such as phthalate and terephthalate

esters can also be used to reduce the viscosity of the coatings, but since these materials remain in the coating after it has cured, they have a more pronounced effect on the physical properties of the coating than solvents.

4. Fillers: Commonly used fillers include ground calcium carbonate, clay, precipitated silica, and talc. These fillers are primarily used as extenders but some, such as clay and talc, do contribute to thixotropy of the wet coating and reinforcement of the cured coating. Another type of filler that is commonly used in membranes for below-grade waterproofing is asphalt or tar. The primary purpose of the asphalt and tar is to increase the hydrophobicity of the membrane to make it even more water resistant.

5. Adhesion promoters: Silanes are the most common adhesion promoters found in urethane coatings used as concrete waterproofing membranes. Silanes such as γ-glycidoxypropyltrimethoxysilane are particularly useful in improving the adhesion of base coats to concrete.

6. UV stabilizers: UV stabilizers are primarily used in top coats. Most of the time, a combination of a UV absorber and a UV stabilizer is used. Care must be taken to ensure that UV stabilizers do not contain reactive groups that will adversely affect the package stability of the coating.

7. Pigments: Light-stable pigments that are unreactive toward isocyanate groups such as iron oxides, carbon black, and titanium dioxide are commonly used.

8. Mildewcides/fungicides: There is a wide range of mildewcides and fungicides that can be used in moisture-curing urethane coatings such as the 3-iodo-2-propynyl butyl carbamate (IPBC) type. Again, the active ingredients of the fungicide as well as any carrier solvent used must be compatible with the moisture-curing urethane coating.

9. Thixotropes: If the coating needs to be applied to a vertical surface, a thixotrope or nonsag agent may be added to prevent running or dripping. Common thixotropes include fumed silica and carbon black.

10. Miscellaneous additives: Other additives that can be found in these types of coatings include flow control agents, bubble release agents, and dispersing aids.

A typical formulation for a single-component moisture-curing below-grade waterproofing membrane is given in Table 11.1. This formulation uses an MDI-based, isocyanate-capped prepolymer made with a propylene oxide polyol. The petroleum tar is used both to increase the hydrophobicity of the membrane as well as to reduce the volume percentage of prepolymer needed in the formulation. Ground calcium carbonate is used as a filler, and carbon black is used both for pigmentation and as a thixotropic agent. DBTDL is a commonly used catalyst for this type of system.

11.5 FORMULATING TWO-COMPONENT URETHANE COATINGS

The requirements for formulating two-component urethane coatings used in concrete waterproofing systems are similar to those for single-component moisture-curing

TABLE 11.1

Formulation for a Single-Component Moisture-Curing Below-Grade Waterproofing Membrane

Raw Material	Weight%
1. MDI/propylene-oxide-polyol-based polymer, isocyanate capped, 1.6–2.0% NCO	40
2. Petroleum tar	30
3. Ground calcium carbonate of mean particle size 12 μm	20
4. Carbon black	3
5. Solvent	6
6. Moisture scavenger	0.5
7. γ-Glycidoxypropyltrimethoxysilane (adhesion promoter)	0.3
8. DBTDL[a] (catalyst)	0.2

[a] Dibutyl tin dilaurate.

urethane coatings. Two approaches can be used in designing the polymer and cure system. The first is to simply use a polyol as the resin in the "A" component, and use an isocyanate functional resin as the curative or "B" component. When the "A" and "B" components are mixed, the isocyanate reacts with the polyol and the coating cures. The advantage of using this approach is that the polyol is not reactive toward moisture, so if compounding additives are added to the polyol containing "A" component, less care needs to be taken to use "ultra" dry raw materials. However, the "A" component cannot contain too much moisture, or it will cause gassing or blistering when it is mixed with the isocyanate containing the "B" component.

The other approach to designing a two-component urethane coating is to use an isocyanate-terminated prepolymer as the bulk of the resin and the binder in the "A" component of the formulation. The curative could then be a small molecule hydroxyl functional material such as a glycol or a less reactive diamine such as a sterically hindered aromatic diamine. The advantage of this approach is that some of the isocyanate is already reacted into the polymer, so less of the reaction has to take place in the field after mixing. As a result of this, the cure rate of these types of two-component coatings tends to be somewhat less temperature sensitive.

A generic formulation for a gray two-component aromatic urethane coating suitable for use in a traffic-bearing system as a top coat in applications in which the top coat is not exposed to UV is given in Table 11.2. This formulation uses a low-molecular-weight liquid polyester in the "A" component combined with a polymeric MDI curative in the "B" component as the polymer or resin for the coating. This combination of polyol and polymeric MDI will give a tough coating with high tensile strength and modulus, but with relatively low elongation (typically, under 100%). Since the curative uses an aromatic MDI-based isocyanate, it will have a tendency to yellow more than a coating that uses an aliphatic isocyanate as the curative. This coating uses talc both as filler and reinforcing agent. The carbon black and titanium dioxide make up the pigment system, and the fumed silica acts as the primary thixotropic agent.

TABLE 11.2

Generic Formulation for a Two-Component Aromatic Urethane Coating

Raw Material	Weight%
Component A	
1. Liquid, branched polyester polyol (equivalent weight = 330–360 g/eq)	74.4
2. Dispersing aid	0.3
3. Talc	15
4. Carbon black	0.5
5. Titanium dioxide	9
6. Fumed silica	0.5
7. Silicone (flow modifier)	0.3
Component B	
1. Aromatic polyisocyanate, MDI based, 30%–32% isocyanate content	100
Mix ratio = 3.4:1 (A:B) by volume or 3.2:1 by weight	

11.6 SPECIFICATIONS

The two major performance specifications used most commonly for concrete waterproofing membrane systems in North America are ASTM C836 "Standard Specification for High Solids Content, Cold Liquid-Applied Elastomeric Waterproofing Membrane for Use with Separate Wearing Course" for below-grade systems and ASTM C957 "Standard Specification for High-Solids Content, Cold Liquid-Applied Elastomeric Waterproofing Membrane with Integral Wearing Surface" for traffic-bearing systems.

ASTM C836 specifies a minimum Shore OO hardness as well as a minimum non-volatile (or solids) content. It also requires testing of flexibility and crack-bridging capability at −26°C. In the crack-bridging test, the coating is applied to two pieces of concrete that have been joined together to form a simulated crack between the pieces. The coating is applied at 1.5 mm and cured for 14 days at 23°C/50% relative humidity followed by 1 week at 70°C. After the full cure, the concrete block assembly is put in an automatic compression and extension machine conditioned at −26°C, and the samples are cycled at 3.2 mm per hour through 10 cycles of opening and closing to a 3.2 mm gap. Three samples are run, and the passing criterion is that no cracks form within the coating membrane. Other aspects of this specification include nonsag test, extensibility after heat aging, and peel adhesion. Full details of the specification, including test procedures, can be found in Volume 4.07 of the Annual Book of ASTM Standards.

ASTM C957 specifies a low-temperature flexibility and crack-bridging test carried out in a similar manner as ASTM C836 with the exception that the crack bridging is carried out on the entire system of base coat and top coat; the top coat is allowed to crack, but the base coat should not exhibit any cracking. The test is run at −26°C, but the samples are only extended to 1.6 mm instead of 3.2 mm for C836. There is also a minimum nonvolatile test for the base coat and a peel adhesion test. Three test requirements found in ASTM C957 are related to the fact that these types of systems are used in exposed traffic-bearing applications. The first is a chemical resistance test in which the cured base coat and top coat are immersed in water, ethylene glycol, and mineral spirits for 14 days each, and their tensile strength is measured and compared to results measured without the chemical immersion. The other test is a weathering resistance test using an accelerated weather machine, and the third test is an abrasion resistance test. Again, the full details of the specification, including test procedures, can be found in Volume 4.07 of the *Annual Book of ASTM Standards*.

REFERENCES

1. S.G. Babcock and T.B. Battles, *Architectural Precast Concrete*, Chapter 1, Prestressed Concrete Institute, Chicago (1973).
2. M.M. Al-Zahrani, S.U. Al-Dulaijan, M. Ibrahim, H. Saricimen and F.M. Sharif, *Cement Concrete Composites*, 24, 127–137 (2002).
3. M. Delucchi and G. Cerisola, *J. Protective Coatings Linings*, 12(8), 44–48 (2001).
4. M. Steele, *J. Protective Coatings Linings*, 23(6), 52–55 (2006).
5. A.P. Chrest, M.S. Smith and S. Bhuyan, *Parking Structures*, Chapter 6, Van Nostrand Reinhold, New York (1989).
6. D. Dieterich, E. Grigat, W. Hahn, H. Hesper and H.G. Schmeitzer, in *Polyurethane Handbook: Chemistry, Raw Materials, Processing, Application, Properties*, G. Oertel (Ed.), Chapter 2, Hanser Publishers, New York (1985).
7. K.C. Frisch, in *Advances in Urethane Science and Technogy*, K.C. Frisch and S.L. Reegen (Eds.), vol. 2, pp. 1–28, Technomic Publishing, Lancaster, PA (1973).
8. J.H. Saunders and K.C. Frisch, *Polyurethanes Chemistry and Technology Part I*, Chapter IV, Robert E. Krieger Publishing Co., Huntington, New York (1978).
9. R. Lombolder, F. Plogmann and P. Speier, *J. Coatings Technol.* 69, 51–57 (May 1997).
10. W. Sultan and J.P. Busnel, *J. Thermal Anal. Calorimetry*, 83, 355–359 (2006).
11. Z.W. Wicks, F.N. Jones and S.P. Pappas, *Organic Coatings: Science and Technology*, Chapter 12, John Wiley & Sons, New York (1992).
12. T. Bartels, G.J. Maters and G.J.K. Hemke, *Advances in Organic Coatings Science and Technology Series*, 10, 18–22 (1986).
13. E.D. Bergman, *Chem. Rev. 53*, 309–352 (1953).
14. C.H. Hare, *Protective Coatings*, Chapter 1, Technology Publishing Co., Pittsburgh, PA (1994).

12 Selection of Foam Backup Material for Use in Sealed Joints in Cladding and Curtain Walling

A. R. Hutchinson

CONTENTS

12.1 INTRODUCTION

This chapter is concerned with the successful sealing of relatively high-movement joints in building cladding and curtain walling using wet-applied sealant systems. The general form of such joints is well understood, and one of the crucial elements is the foam or backup material used to control the joint depth. The selection, specification, and installation of this element are very important in joints that are subject to movement.

The relatively high incidence of premature sealant failure in curtain-walled structures can be variously ascribed to inadequate specification, design and detailing, cyclic movement, and poor workmanship [1]. The seals in joints in curtain walling systems may be subjected to movement immediately after installation and may, therefore, need to accommodate continuous cyclic movement during the curing period. This movement, coupled with slow sealant curing rates, can lead to permanent and detrimental changes in sealant bead geometry, sealant adhesion, and joint performance characteristics [1–3]. Other commonly cited technical problems include poor sealant adhesion to substrates, adhesion of sealant to the backup material resulting in three-sided tack or adhesion [4], and cyclic movement in service [5–7].

Aspects of joint construction and geometry are reviewed before dealing with the considerations surrounding foam selection. The available foam materials are then discussed and characterized, and the adhesion of primers and sealants to backup materials is highlighted. The performance of sealed joint systems is then considered, and recommendations are made for testing and evaluation as part of the selection process. The chapter closes with a summary of foam backup material selection considerations for movement joints.

12.2 JOINT CONSTRUCTION AND SEALED JOINT SYSTEMS

12.2.1 GENERAL

Joints represent discontinuities in the external envelope of a building, and they are designed to accommodate movements between components or parts of the building. Such movements may be brought about by settlement, shrinkage, creep, thermal expansion and contraction, loading and wind, and may be accommodated in the joint by deforming the seal by compression and extension or by deformation in shear (see Figure 12.1a,b,c).

Sealed joints comprise a substrate (usually aluminum or concrete), a primer (e.g., a silane or an isocyanate dissolved in an organic solvent), the sealant itself (a polymer base with additives such as pigments, fillers, and catalysts), and the backup material, which is often an expanded polymeric material (polyethylene or polyurethane foam). This combination of materials may be termed a *sealed joint system*. The backup material is utilized as a joint filler to control the depth of seal, to support the sealant during bead formation, and as a bond breaker to prevent the sealant from adhering to the back of the joint. The collective term used for all aspects of sealant bead formation (application, consolidation, and smoothing of the profile) is *tooling*.

A number of seal shapes are commonly encountered in joints and these, together with typical failure initiation sites for joints subject to movement, are illustrated in Figure 12.2. In order that the seal can deform to accommodate these movements, it is essential that the seal be bonded to the two opposing faces of the joint, leaving the remaining faces of the seal free to move. This is to enable the mass of sealant to deform without inducing localized stress that would cause premature failure.

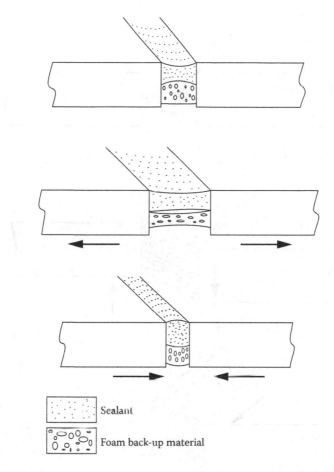

Sealant

Foam back-up material

FIGURE 12.1a   Butt joint (top) in tension (middle) and compression (bottom).

12.2.2   JOINT DEPTH

Optimum joint depth depends on the type of movement affecting the joint and the elastic character of the sealant. In joints subject to shear movement, the seal depth (in the direction of movement) should be at least equal to the width. In joints subject to tension and compression movement, the joint depth depends on the elastic nature of the sealant [1,8,9].

The depth of sealant in the joint is important, both in the interests of performance and economy. Fully elastic sealants confer optimum movement accommodation when used in thin sections, whereas plastic sealants perform best when the seal depth (D) is at least equal to the joint width (W). Elastoplastic materials tend to give the best results when the depth is half the width.

A minimum depth of sealant is necessary both for durability and for ease of application. Sealants are applied under a wide variety of conditions and the joints in which they are used are formed in various ways and under different cir-

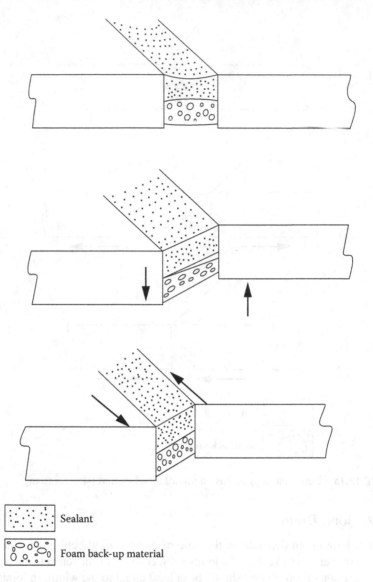

Sealant

Foam back-up material

FIGURE 12.1b   Butt joint (top) in out-of-plane shear (middle) and in-plane shear (bottom).

cumstances. There must, therefore, be a reasonable degree of tolerance on width and depth of joint seal to accommodate these circumstances.

For normal applications in cladding and curtain walling, the recommended seal depths (Figure 12.3) are shown in Table 12.1.

Where the depth of sealant is increased to give better resistance to pressure on the joint, the stress in the seal will be increased if the joint is subjected to movement. This results in some reduction in the movement accommodation capability of the joint. In such joints, the movement accommodation factor (MAF) may be reduced by

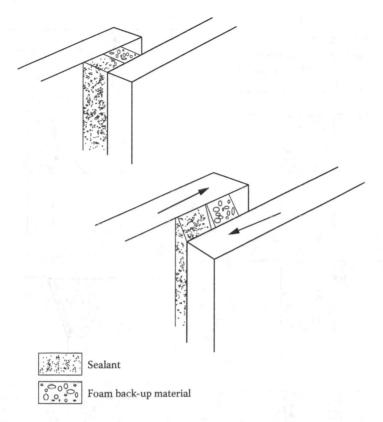

| Sealant |
| Foam back-up material |

FIGURE 12.1c   Lap joint (top) subjected to shear (bottom).

10% or more. The joint width required to accommodate the joint movement should, therefore, be increased.

If the joint is deep, a compressible backup material should be used to fill the inner part of the joint and to control the depth of sealant. If the sealant is likely to stick to the surface of the backup material, then the backup should be covered with a bond-breaker tape to prevent the sealant from adhering to the back of the joint and creating third surface (three-sided) adhesion.

The backup material assists in the application of the sealant into the joint by controlling the depth of sealant and providing resistance to sealant flow, encouraging the sealant to spread to the sides of the joint to wet the joint faces, and to achieve good adhesion.

In the absence of backup material, sealant tends to flow through the joint, barely wetting the sides and resulting in poor adhesion. Further, because of the lack of resistance at the back of the joint, it is more difficult to tool the sealant to achieve a smooth surface. When sealing joints in thin claddings, such as glass assemblies, it is necessary to either apply a temporary backing to the joint that is removed after cure or, in some cases, it may be possible to apply sealant and tool it from both sides simultaneously. This is a difficult, specialized technique, requiring considerable skill and practice.

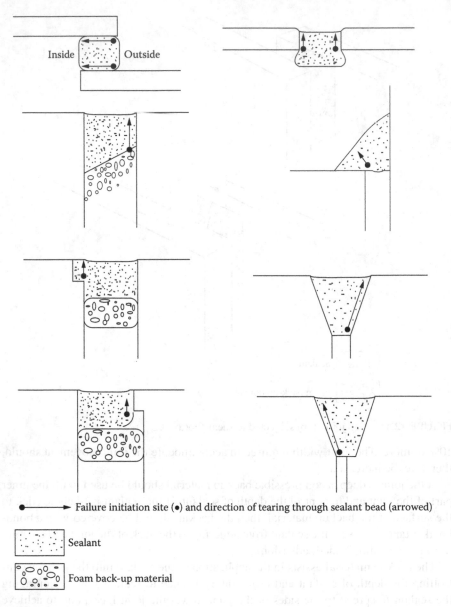

● ——→ Failure initiation site (●) and direction of tearing through sealant bead (arrowed)

Sealant

Foam back-up material

FIGURE 12.2 Joint geometries, seal shapes, and failure initiation sites in joints subject to movement.

12.2.3 BACKUP MATERIALS

A backup material should be such that the sealant does not adhere to it when cured. The sealant should adhere to the opposing faces of the joint only, allowing the sealant bead to stretch or compress without impediment. Cellular foam materials are preferred because they can be compressed in the joint initially, and they will expand

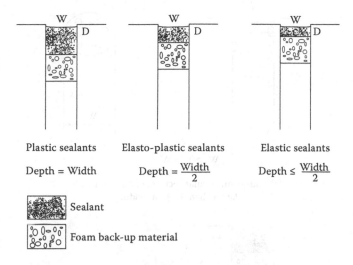

FIGURE 12.3   Recommended seal depths for different sealant types in butt joints.

TABLE 12.1
Recommended Sealant Bead Depths for Butt Joints

Sealant Type	Minimum Depth (mm)	Normal Guideline	Maximum Depth (mm)
Plastic sealants	6	$D = W$	25
Elastoplastic sealants	6	$D = W/2$	50
Elastic sealants	6	$D = W/2$ or $W/3$	20

when the joint opens and compress when it closes, thus following the joint movement and providing a continuous support for the sealant. The latter is particularly important during the cure period when the seal is easily deformed or distorted.

The most widely used material is closed-cell polyethylene foam, which is available in various grades and forms. It is supplied as lengths cut from a sheet or a circular rod (Figure 12.4). Open-cell polyurethane foam can also be used as a backup material. It is surprising that the perceived disadvantages of using an open-cell material, such as adhesion to the seal and the material being able to act like a wick to draw water through it, are not borne out by experimental evidence. This aspect is discussed in Sections 12.5 to 12.7.

Rod-form polyethylene foam has been preferred historically because it is difficult to twist or insert it the wrong way. Provided it is used at least 25% oversize, it results in the sealant bead having a large bonding surface with a slightly concave (hourglass) section, giving good stress distribution when deformed by movement (see Figure 12.4a). Rod-form backup material has disadvantages in that if used in sections less than 20% oversize, the difference in seal depth across the joint can cause excessive strain in the seal;

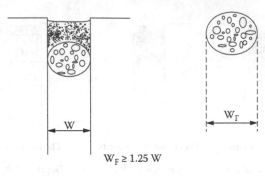

$W_F \geq 1.25\,W$

(a) Rod-form, circular section, of diameter 25% larger than the joint width

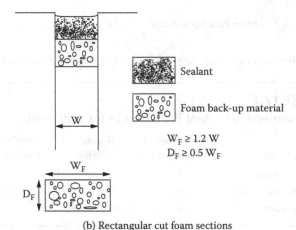

Sealant

Foam back-up material

$W_F \geq 1.2\,W$
$D_F \geq 0.5\,W_F$

(b) Rectangular cut foam sections

FIGURE 12.4 Recommended foam backup material shapes, at least 20% wider than the joint width and having a thickness of at least half the joint depth.

also, the large round section of backup material requires a deep joint to accommodate it. Rod form also tends to be expensive.

Cut foam, either as sheet or section, should be at least 20% wider than the joint to ensure that the foam is under sufficient compression to expand and follow joint movement during the cure period (see Figure 12.4b). The section should be at least half as deep as the width to minimize buckling when it is compressed in the joint. Cut sections of foam can easily be distorted or twisted as they are inserted in the joint, causing uneven seal sections and consequential weaknesses that may cause premature failure (see Figure 12.5). The use of thin sections of foam can cause buckling and consequential bulges or hollows in the seal, giving an untidy appearance.

Insufficient compression of the foam can result in a loose backing, which does not support the sealant as it is applied, resulting in poor compaction and bulging or slump after tooling. Insufficient compression of the foam may also cause the backup material to become loose as the joint opens, placing stress on the corners and causing

Back-up material folded or laminated

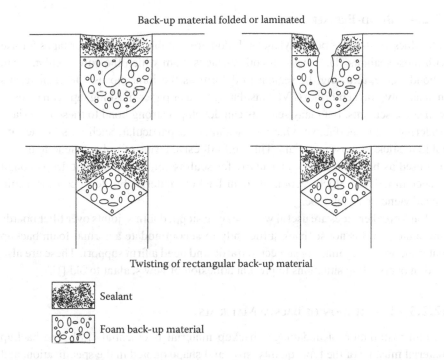

Twisting of rectangular back-up material

Sealant

Foam back-up material

FIGURE 12.5 Cohesive failures of sealant caused by incorrect placement of backup material.

splits or tears in the back of the seal that may progress to cause premature failure. Cut polyethylene foam is relatively cheap and readily available in any required size.

Sponge or expanded rubber (EPDM or butyl) can also be used as a joint backup material. It is particularly suitable for load-bearing joints because it can provide physical support to the joint seal while not impairing movement accommodation. Care should be taken to ensure that the sealant does not adhere to the rubber, which should be covered by a bond-breaker tape if necessary. Rubbers, both natural and synthetic, contain curing agents, antioxidants, and extenders that can migrate into sealants, causing staining or discoloration [1]. Where light-colored sealants are used, it may be necessary to separate the seal from the rubber by a section of polyethylene foam or backing strip.

Joint filler materials, such as polystyrene foam or fiberboard, are not good backup materials for movement joints because, although they will compress to allow closing movement, they do not expand to follow opening movement of the joint and, therefore, leave the seal unsupported. Many sealants adhere to these materials unless the surface is covered by a bond-breaker tape. Polystyrene foam is attacked by many solvents that are used in sealants and primers. Fiberboard joint fillers may be impregnated with bitumen or other materials that are easily extracted by solvents, spreading contamination and sometimes also causing staining of sealants if not isolated adequately.

12.2.4 BOND-BREAKER TAPES

Self-adhesive strips of polyethylene or Teflon are suitable bond-breaker tapes for use with most sealants. They have smooth surfaces from which the cured sealants can debond easily, allowing the sealant to deform as the joint moves. Other materials such as poly(vinyl chloride) (PVC) insulating tape or paper-masking tape can be used with some sealants, but many sealants can develop a strong bond to these materials, rendering them ineffective. One-part sealants, in particular, such as silicone, are liable to stick to most materials. Thin self-adhesive strips of polyethylene foam are also used as bond breakers, particularly for sealing joints that are not deep enough to accommodate a thick section of foam backup material but too deep for a thin polyethylene or Teflon tape.

Bond-breaker tapes are useful when sealing stepped joints, joints over filler boards where the board is not set back sufficiently to accommodate a normal foam backup material, or where joints are subject to traffic and need a firm support. These are also used in oversealing situations to prevent adhesion of new sealant to old [1].

12.2.5 INSTALLATION OF BACKUP MATERIALS

Careful installation of undamaged backup material is essential [1,8]. The backup material must be of the type, quality, size, and shape quoted in the specification, and it should be correctly located in the joint without damage. Cellular backing should be tight in the joint to give firm support to the sealant. It should be compressed such that it will expand to follow the movement of the joint during cure and be thick enough to prevent buckling as the joint closes.

Rod-form backup material requires a deep joint; if the joint is too shallow, the sealant section may be too thin and may, therefore, be prone to cohesive failure. Rectangular-section backup material should be at least half the uncompressed width to prevent buckling during joint movement.

Bond-breaker tapes should adhere to the back of the joint, should cover the moving joint, and should be at least as wide as the minimum joint width. The tape must not be allowed to cover up the sides of the joint, because it would reduce the bonding area of the sealant and could impose additional stress on the adhesive bond to the sides of the joint.

In cavity construction, care should be taken to ensure that the backup foam material does not fall into the cavity or form a bridge.

12.3 FOAM MATERIALS

12.3.1 FOAM PRODUCTION PROCESSES

Foams can be made by several different production processes [10]. Generally, gas bubbles are generated in a molten polymer, and the expanded structure is then stabilized. The general production procedure is illustrated in Figure 12.6.

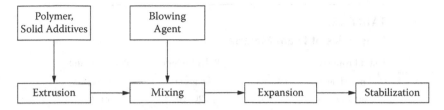

FIGURE 12.6 The foam production process.

The polymer (usually high-density polyethylene [HDPE]) is mixed with additives, extruded, and then mixed with blowing agent. There are two kinds of blowing agents: physical foaming agents and chemical foaming agents [11].

Physical blowing agents are usually dissolved in the polymer under high-pressure conditions (e.g., with N_2, air). During decompression and heating the gas expands, forming the cellular structure of the foam. CO_2 is used in a low-pressure absorption method. Cyclohexane, trichloroethylene, 1,2-dichloroethane, acetone, and 1,1,2-trichlorotrifluoroethane are also used as physical blowing agents.

Chemical blowing agents liberate large volumes of gas as a result of thermal decomposition at elevated temperatures. They are divided into inorganic and organic blowing agents. Examples of inorganic agents are $(NH_4)HCO_3$ (releases NH_3, H_2O, and CO_2), $NaHCO_3$ (releases CO_2 and H_2O), and $NaBH_4$ (releases H_2). Organic blowing agents are compounds with azo, N-nitroso, sulfohydrazo, or azido groups; they all release N_2 [10–12].

After the polymer expansion, the cellular structure must be stabilized rapidly or else it would collapse. There are two methods of stabilization: physical stabilization and chemical stabilization.

Physical stabilization is used if the macromolecular compound is a thermoplastic. The expansion of the polymer is carried out above the melting point and is then cooled down rapidly to below the melting temperature.

Chemical stabilization may take the form of crosslinking of the polymer through radiation or chemical reactions immediately following the expansion step. Chemical crosslinking reactions start through thermal homolysis of peroxide, which releases two oxy radicals. The oxy radicals abstract hydrogen atoms from the polymer and generate radicals, which react to form crosslinks in a way analogous to the polymer radicals generated through radiation [10]. Polyethylene foams are generally crosslinked by chemical reaction with peroxide.

Polyurethane (PU) foams can be generated through the reaction of excess isocyanate with water, or organic carboxylic acids in the reaction mixture, which releases carbon dioxide. The carbon dioxide creates the bubbles in the foam.

12.3.2 SURFACE AND MATERIAL PROPERTIES OF POLYMER FOAMS

The manufacturing process determines the surface chemistry of the resultant foam. The cooling conditions of the melt between injection and mould expansion determine the thickness of the skin on the foam, and the cell structure generally determines the mechanical properties of a foam [10].

TABLE 12.2

Properties of Foam Materials

Foam Property	Polyethylene	Polyurethane
Cell morphology	Closed	Open
Density (kg/m^3)	20–70	30–32
Water absorption (kg/m^3)	4–49	437–674
Compressive strength at 50% (kPa)	30–160	0–12
Tensile strength (kPa)	140–390	100–150
Elongation at break (%)	30–100	124–144
Tensile modulus (kPa)	30–260	30–100
Shore A hardness (no skin)	4–14	0
Shore A hardness (skin)	9–25	N/A

Source: Adapted from S. Iglauer, A.R. Hutchinson and T.C.P. Lee, *Cellular Polymers*, 23, 77 (2004).

TABLE 12.3

Foam Surface Free Energies

		Surface Free Energy (mJ/m^2)	
Foam Type and Surface		Polar Component, γ^p	Total Energy, γ
	No skin	0.3–6	25–48
Polyethylene	Skin	0.3–10	40–50
	Solid sheet	6	44
Polyurethane	No skin	16	39–41

Source: From S. Iglauer, A.R. Hutchinson and T.C.P. Lee, *Cellular Polymers*, 23, 77 (2004).

The surface and material properties of a range of typical open- and closed-cell foams, used as backup material in cladding and curtain walling, were reported by Iglauer et al. [13]. Some of the foams had "cut" (no skin) surfaces, and some had a skin. Six closed-cell PE and two open-cell PU foam materials were used. A summary of their findings is given in Table 12.2.

It is evident that the polyurethane materials were "softer" and more flexible, but with the capacity to absorb a large amount of water.

Iglauer et al. [13] measured the surface free energies of the polyethylene materials, both unskinned and skinned, and compared them with the values obtained from a solid sheet of polyethylene. The surface free energy of PU foam was also measured. Video contact angle analysis was utilized, using at least three different liquids. A summary of the data is given in Table 12.3.

The skinned PE foams all had higher surface free energies than their unskinned counterparts; some were significantly higher.

Infrared spectroscopy and x-ray photoelectron spectroscopy (XPS) were used by the authors to investigate the main chemical constituents of both the bulk and the surface phases of the different materials. Oxygen, nitrogen, and silicon were found on the surfaces, and these elements were probably responsible for the polar surface free energy components, and for the relatively high total surface free energies of up to 50 mJ/m^2 for one of the skinned foams. It should be noted that this value for total surface free energy is similar to that for anodized aluminum in air [13]. The literature suggests a total surface free energy value of 33 mJ/m^2 for pure PE [14].

All bulk phases of the PE foams, determined by cutting through the structure, exhibited smaller polar surface free energies and total surface free energies than the corresponding foam skins and surfaces. This suggests that the polar constituents were concentrated in the surface of all of the materials. In other words, all of the foam surfaces were heavily oxidized. To a lesser extent, traces of the blowing agents used in the foam manufacture were present and accounted for the levels of nitrogen and silicon.

Amide groups in PU are polar and create a polar surface free energy component. In this case, a value of 16 mJ/m^2 was measured. Carbon, oxygen, nitrogen, and silicon were found on the surfaces. The total surface free energies measured were between 39 and 41 mJ/m^2. These are similar to literature values reported for pure PU of 39 mJ/m^2 [14].

The clear implication of this research is that sealants will stick to both PE and PU foams. In joints subject to movement, this may lead to reduced performance. It will also be a significant issue for the case of seals that are subject to movement during cure.

Finally, Iglauer et al. [13] conducted detailed cell structure analysis of the foam materials in order to determine cell sizes and wall thicknesses. From this it was concluded that a cut surface presented a planar surface area of between 15% and 26% of a comparable solid surface. It was also noted that smaller cell sizes corresponded with higher foam densities and reduced water absorption.

12.4 ADHESION OF SEALANTS TO POLYMER FOAM BACKUP MATERIALS

Quantitative data on adhesion, as measured in any mechanical test, are impossible to obtain. This is because of the complex interaction of stresses and strains that are present in the materials that comprise any sort of joint when under load. However, semiquantitative data can be derived from certain tests that subject the interface to direct tension or peel. The other major caveat is the effect of sealant modulus on measured joint performance.

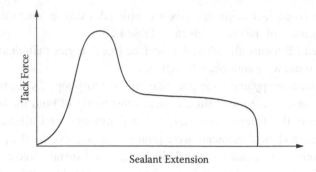

FIGURE 12.7   Typical tack force–extension curve for fresh sealant pressed against a surface.

12.4.1   TACK

Tack experiments can provide an indication of the instant, short-term adhesion between sealants and backing foams. However, tack is a complex property because it depends on the bond formation, the substrate preparation, and the nature of the separation of the materials involved when under load [15]. Pressure-sensitive tack is defined as the property of a material that enables it to form a bond of measurable strength immediately upon contact with another surface [16].

Hutchinson and Iglauer [17] devised a procedure to measure the tack-free time of a number of foam–sealant and anodized aluminum–sealant combinations. Essentially a sealant layer that had been allowed to cure for particular lengths of time was pressed against various substrates. The two materials were then separated immediately. The force and failure mode required to detach the substrate from the tacky sealant surface were recorded. The tack-free time was reached when the tensile load reduced to approximately zero and the failure mode became 100% interfacial. A typical tack force–extension curve is shown in Figure 12.7.

The authors found different tack-free times for different sealants: 3–6 h for one-part sealants, 6 h for polysulfide, and 24 h for modified silicone polyether. A maximum tack force point (critical point) was observed for every sealant. The peak tack force of one-part modified silicone was about 40% higher on an aluminum substrate than on a PE foam. It was assumed that different wetting characteristics or different surface chemistries were responsible for this difference. Overall, the results of the investigation did not correlate with other information relating wetting, adhesion, and experimental joint performance.

12.4.2   PEEL TEST DATA

180° peel testing can provide semiquantitative adhesion data regarding the foam–sealant interface [18]. The peel force delivers information about the joint strength and the work of adhesion at the foam–sealant interface. The mode of failure of the joint has to be taken into account, because it influences peel force strongly and is important in the data interpretation. Only 100% interfacial failure at the sealant–foam interface provides a reasonable indication of adhesion.

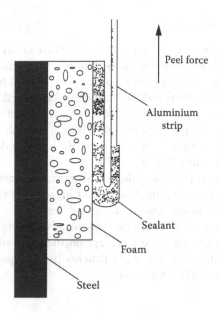

FIGURE 12.8   180° peel test configuration.

Hutchinson and Iglauer [17] undertook 180° peel testing on a variety of foam–sealant combinations using a joint configuration of the form depicted in Figure 12.8. A rigid foam substrate was created initially by bonding pieces of candidate foam materials to lengths of steel. Sealant and aluminum foil were then applied to this foam surface. Some of the test specimens employed foam that had been primed to simulate the overspill when priming the sides of a cladding panel prior to sealant application. Load–extension curves were recorded for all of the combinations, and the locus of failure was determined visually. The reproducibility of the peel test data was low, but a general qualitative trend in "adhesion" was obtained.

Peel forces generally lay between 2.5 and 16 N, with peaks of up to 50 N. Even a small peel force is significant in terms of the performance of sealed joints [4]. Peak forces of 50 N are clearly very significant because this value equates to an average peel force between a sealant and a substrate such as anodized aluminum. The cured mechanical properties of sealants such as modulus (between 320 and 900 kPa at 25% extension for the particular materials studied) exert a significant influence on joint performance.

The peel test data were very revealing. The authors found that the sealant and the foam surface had a strong influence on the peel force, but the locus of failure was influenced mainly by the sealant. One-part PU joints failed predominantly in adhesion, while one-part silicone and one-part modified silicone joints showed a mixed (adhesion, and cohesive in foam and sealant) failure. Three-part polysulfide and modified silicone polyether joints showed mainly adhesion failure, or cohesive failure within the sealant very close to the bonded interface.

The presence of primer generally increased the peel force, and higher peel forces were generally found on skinned foams.

12.4.3 ADHESION MECHANISMS

The major contribution to adhesion is primarily from chemical interactions between the foam surface and the uncured sealant, with mechanical interlocking contributing only marginally. Chemical interaction requires close contact between the surfaces, which implies an influence of the surface free energy, and this determines the wetting behavior at the interfaces. High foam surface free energy (with a particular sealant) leads to good sealant wetting and increased adhesion through chemical interaction. This is apparent from the correlations between 180° peel test data and foam surface free energy, shown in Figures 12.9 and 12.10.

Hutchinson and Iglauer [17] found that skinned foams generally had higher total surface free energies than unskinned ones, and higher peel forces were generally associated with skinned foams. One-part sealants gave rise to a higher peel force than multipart sealants, probably because their reactive functional groups are more prone to react with functional groups present on the foam surfaces and to create chemical bonds at the interface. There is a roughly linear dependency between peel force and the total foam surface free energy.

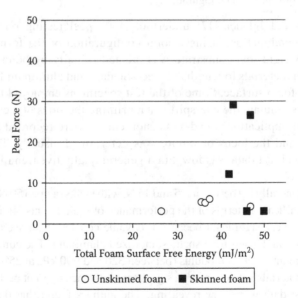

1-part silicone sealant:
Total surface free energy = 29.6 mJ/m^2
Dispersion component of surface free energy = 29.2 mJ/m^2
Polar component of surface free energy = 0.4 mJ/m^2

FIGURE 12.9 Relationship between peel force for joints made with one-part silicone sealant and foam surface free energy. (From A.R. Hutchinson and S. Iglauer, *Int. J. Adhesion Adhesives*, 26, 555, 2006.)

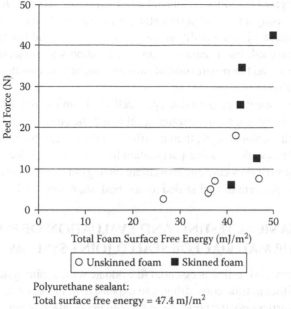

Polyurethane sealant:
Total surface free energy = 47.4 mJ/m^2
Dispersion component of surface free energy = 41.5 mJ/m^2
Polar component of surface free energy = 5.9 mJ/m^2

FIGURE 12.10 Relationship between peel force for joints made with one-part polyurethane sealant and foam surface free energy. (From A.R. Hutchinson and S. Iglauer, *Int. J. Adhesion Adhesives*, 26, 555, 2006.)

12.5 EFFECT OF FOAM BACKUP MATERIAL ON CURE OF ONE-PART SEALANTS

One-part sealant materials generally require the presence of moisture for their cure. This moisture is from the air and, therefore, the onset of cure is the moment the sealant is expelled from the sealant cartridge. Initial moisture contact is with the outer section of the extruded sealant, and a skin is rapidly formed as a result of crosslinking of this outer layer. For the inner section of the sealant bead to cure, the moisture must diffuse through the skin to regions of uncured sealant material [19].

As the cure of this outer region progresses, the thickness of the skin increases and the rate of diffusion of moisture to the inner section of the sealant bead slows down. One-part sealant materials, therefore, cure from the outside in, in a relatively slow manner, especially under conditions of low temperature and humidity. One-part sealants usually skin after 1 h of exposure to ambient conditions and are completely cured generally after 2–4 weeks. As a result of this anisotropic cure, an inhomogeneously cured bead is formed that, especially in the initial stage, may lead to stress concentrations and premature failure. PE backup material stops the progression of moisture to the back face of a joint [20].

It has been observed (e.g., Reference 5) that cohesive and adhesion failures occur at the *back* of joints, where the sealant bead is in contact with backing foam. Iglauer et al. [20] investigated the influence of backing foam on the *skin development* of one-part sealants by using a novel experimental configuration. Three different types of sealants (a silicone, a silicone-modified polyether, and a PU) and two types of foams, PE and PU, were used. Each sealant–foam combination was subjected to different curing conditions, and the progression of *skin formation* was monitored at different cure intervals of up to 14 days.

The cure experiments revealed that open-cell PU foam did not affect the cure of one-part sealants, but wet foam or water accelerated the cure speed. Closed-cell PE foam inhibited the sealants' cure, thereby affecting the development of the mechanical and physical properties of a one-part sealant bead. It was, therefore, predicted that a combination of very slow cure and movement during cure could result in significant reductions in the performance of sealed joints, both short term and long term.

12.6 MECHANICAL TESTING AND EVALUATION OF FOAM BACKUP MATERIAL FOR SEALED JOINT SYSTEMS

Laboratory mechanical testing is essential to evaluate sealed joint systems. However, it can be very difficult, time consuming, and expensive to incorporate large numbers of parameters, particularly the presence of backup materials, movement during cure, and fatigue cycling following cure.

There exists considerable experimental evidence that two-part sealant systems perform significantly better than one-part systems for large-movement applications [2,5]. It is also clear that the presence of backup materials in laboratory joints plays an important role in joint performance [4,21], and that this has very important implications for the first few days that a newly sealed joint is subjected to cyclic movement [22,23].

Iglauer et al. [4] selected a range of foams and sealants in order to provide a comprehensive investigation of the influence of different foam–sealant combinations on sealed joint performance. All combinations were evaluated in terms of their static and fatigue performances using standard ISO 8339 aluminum tensile adhesion joints. The width of the substrates was, however, doubled to support a compressed piece of foam backup material. The freshly made test specimens were allowed to cure both in a static condition and under conditions of movement using a purpose-built rig [3,23]. The joint performance results were compared with peel test data from sealant-on-foam specimens. One vital attribute of a sealant is its capacity to stretch to maintain a seal. Thus, the tensile strain at failure for the tensile adhesion joints was used as an important performance parameter. Figure 12.11 shows data for the relative tensile strain at break versus the 180° peel test information discussed earlier, for joints made with a one-part silicone sealant. Figure 12.12 summarizes the fatigue test data for joints made with the same sealant. Performance trends for joints made with other types of sealants are reported in the paper [4].

Static tensile testing was found not to be very sensitive to the presence of foam, but it did demonstrate the influence of dynamic cure and the use of primers. The fatigue tests displayed considerable sensitivity to the presence of foam backup

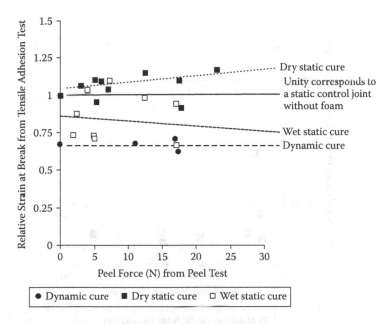

FIGURE 12.11 Correlation of relative strain at break with peel force for joints made with one-part silicone sealant. (From S. Iglauer, A.R. Hutchinson, and T.C.P. Lee, in *Durability of Building and Construction Sealants and Adhesives,* A.T. Wolf (Ed.), STP 1453, pp. 184–205. ASTM International, West Conshohocken, PA, 2004.) Linear regression lines are shown in dotted form.

material, reducing the performance of the test joints significantly, as shown in Figure 12.12.

One-part sealant joints, which were subjected to movement during cure and had a closed-cell PE foam attached, exhibited a dramatic reduction in their fatigue performance. The PU foam adhered strongly to all sealants, but its presence reduced the fatigue performance of joints only slightly. This is shown as the outlier in Figure 12.12. The low compressive strength and hardness of the PU foam compared with the PE foam was probably the main reason for this behavior, together with the fact that it did not affect sealant cure. While dry foam reduced fatigue life, wet foam introduced early adhesion failure that also led to a significant reduction in fatigue performance. Correlations were found between measured peel forces and fatigue performance for all systems except those associated with polysulfide sealants. The 180° peel test procedure described in Section 12.4.2 was, therefore, a fairly good predictor of performance reduction trends.

12.7 SUMMARY

There are a number of considerations involved in foam backup material selection for movement joints in cladding and curtain walling. Perhaps the most important consideration relates to the frequency and magnitude of predicted joint movement during the period following sealant application. This is when three-sided adhesion

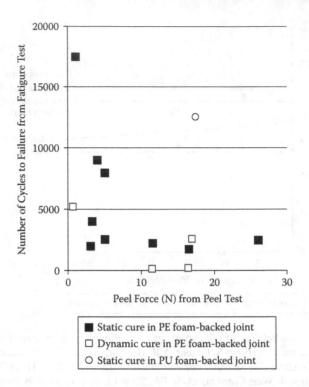

FIGURE 12.12 Correlation of cycles to failure with peel force for joints made with one-part silicone sealant. (From S. Iglauer, A.R. Hutchinson, and T.C.P. Lee, in *Durability of Building and Construction Sealants and Adhesives,* A.T. Wolf (Ed.), STP 1453, pp. 184–205. ASTM International, West Conshohocken, PA, 2004.) High peel forces correspond with significantly reduced performance of joints incorporating PE foam backup material.

can be very detrimental to the integrity of the seal achieved. If significant movement is anticipated, for example, in joints between dark-colored insulated aluminum cladding panels, then an open-cell PU foam is recommended. Other significant aspects relate to joint geometry, foam type and condition, the nature of the foam surface, the nature of the sealant, and the presence or absence of primer. If closed-cell PE foam backup material is used, then unskinned foam is preferable to skinned foam and a relatively soft, low-modulus foam is desirable.

Some general joint performance trends are collected in Table 12.4. It is not possible to provide more specific information because every sealant/primer/substrate/backup material combination is unique. A simple comparative 180° peel test should, however, provide a reliable indicator of the relative long-term performance of sealed joints subject to movement. Such a test is much more cost-effective to perform than using elaborate tensile adhesion joints subjected to movement during cure or fatigue test cycles.

TABLE 12.4

General Performance Trends Associated with Foam-Backed Joints

Variable	Effect on Performance	
	Better	Worse
High foam density		✓
High foam surface free energy		✓
Wet foam		✓
Cut foam surface	✓	
Skinned foam surface		✓
Primed foam surface		✓
Joint movement during cure		✓
Joint movement following cure		✓
One-part sealant		✓
Multipart sealant	✓	

REFERENCES

1. R. Woolman and A.R. Hutchinson, *Resealing of Buildings— A Guide to Good Practice*, Butterworth-Heinemann, Oxford (1994).
2. T.G.B. Jones and M.A. Lacasse, in *Durability of Building Sealants—A State-of-the-Art Review*, A.T. Wolf (Ed.), pp. 73–105. RILEM Publications, Paris (1999).
3. A.R. Hutchinson, T.G.B. Jones, and K.E. Atkinson, in *Durability of Building Sealants*, A.T. Wolf (Ed.), Vol. 2, pp. 99–116. E & FN Spon, London (1999).
4. S. Iglauer, A.R. Hutchinson, and T.C.P. Lee, in *Durability of Building and Construction Sealants and Adhesives*, A.T. Wolf (Ed.), STP 1453, pp. 184–205. ASTM International, West Conshohocken, PA (2004).
5. Y. Matsumoto, in *Science and Technology of Building Seals, Sealants, Glazing, and Waterproofing*, C.J. Parise (Ed.), STP 1168, pp. 30–44. ASTM International, West Conshohocken, PA (1992).
6. J. Margeson, in *Science and Technology of Building Seals, Sealants, Glazing, and Waterproofing*, C.J. Parise (Ed.), STP 1168, pp. 22–29. ASTM International, West Conshohocken, PA (1992).
7. J.R. Brower, in *Science and Technology of Building Seals, Sealants, Glazing and Waterproofing*, C.J. Parise (Ed.), STP 1168, pp. 5–8. ASTM International, West Conshohocken, PA (1992).
8. J.R. Panek and J.P. Cook, *Construction and Sealant Adhesives*, 3rd edition, John Wiley, New York (1991).
9. S.R. Ledbetter, S. Hurley, and A. Sheehan, *Sealant Joints in the External Envelope of Buildings*, Report 178. Construction Industry Research and Information Association, London (1998).
10. D. Klempner and K.C. Frisch, *Handbook of Polymeric Foams and Foam Technology*. Carl Hanser Publishers, Munich (1991).
11. J.N. Methven, *Foams and Blowing Agents*. RAPRA Review Report No 25, RAPRA Technology, Shrewsbury, U.K. (1990).

12. D. Feldman, *Polymeric Building Materials*. Elsevier Science, London (1989).
13. S. Iglauer, A.R. Hutchinson and T.C.P. Lee, *Cellular Polymers*, *23*, 77 (2004).
14. S. Wu, in *Polymer Handbook*, 4th edition, pp. 521–524. John Wiley & Sons, New York (1999).
15. R. Bates, *J. Appl. Polym. Sci.*, *20*, 2941 (1976).
16. ASTM D1878-61T, Tentative method for pressure-sensitive tack of adhesives, ASTM International, West Conshohocken, PA (1961).
17. A.R. Hutchinson and S. Iglauer, *Int. J. Adhesion Adhesives*, *26*, 555 (2006).
18. N.E. Shephard and J.P. Wightman, in *Science and Technology of Building Seals, Sealants, Glazing and Weatherproofing*, M.A. Lacasse (Ed.), STP 1271, pp. 226–238. ASTM International, West Conshohocken, PA (1996).
19. K.W. Allen, A.R. Hutchinson, and A. Pagliuca, *Int. J. Adhesion Adhesives*, *14*, 117 (1994).
20. S. Iglauer, A.R. Hutchinson, and T.C.P. Lee, in *Durability of Building and Construction Sealants and Adhesives*, A.T. Wolf (Ed.), STP 1453, pp. 171–183. ASTM International, West Conshohocken, PA (2004).
21. M. Koike, K. Tanaka, and Y. Munakata, Movement capability of sealants cured in moving joints, Report of the Research Laboratory of Engineering Materials. Tokyo Institute of Technology, Vol. 4, pp. 173–180 (1979).
22. T.G.B. Jones, A.R. Hutchinson, and M.A. Lacasse, in *Durability of Building and Construction Sealants*, A.T. Wolf (Ed.), RILEM Proceedings Vol. 10, pp. 211–227. RILEM Publications, Paris (1999).
23. T.C.P. Lee, T.G.B. Jones, and A.R. Hutchinson, in *Durability of Building and Construction Sealants*, A.T. Wolf (Ed.), RILEM Proceedings Vol. 10, pp. 297–313. RILEM Publications, Paris (1999).

13 Construction Sealants

Andreas T. Wolf

CONTENTS

13.1   INTRODUCTION

Construction sealants are widely used in a variety of industrial, commercial, and residential applications. Applications range from low-end, unspecified uses, such as crack filling in driveways, to high-end, specified, and regulated uses, such as structural glazing. Construction sealants are utilized in a multitude of applications, for example, in glazing and perimeter joints on windows and doors; roofing terminations; perimeter joints in bathroom and kitchen appliances; weather-sealing joints in facades and curtain walls; expansion joints in airport runways, highways, bridges, plazas, and parking decks; joints in water and wastewater treatment facilities (including submerged environments); and in fire-stop sealing of joints and penetrations in public facilities (fire partition wall sealing), such as in hospitals, schools, or power plants. Due to the wide use of construction sealants, the global market for these materials now exceeds 800,000 tons in volume and $3 billion in sales value.

A construction sealant is a material that is installed into an opening (gap) or joint between two or more substrates to prevent air, water, chemicals, dirt, or other environmental elements from passing through the joint or gap, while permitting limited movement of the substrates. The sealant must be able to maintain the seal while accommodating variations in joint size due to manufacturing and construction deviations and repeated building movements induced by externally applied loads and changes in environmental conditions. In addition to the basic functions of sealing and movement accommodation, certain applications place additional demands on the sealant, such as vibration damping; fire-stopping; and electrical, acoustic, or thermal insulation capabilities.

Construction sealants can be categorized by form, ability to cure or harden, chemical binder, performance, or end-use market. The major distinction in product form is between preformed sealants (gaskets and strips) and in situ cured or hardened (wet-applied) sealants. This chapter focuses primarily on in situ cured sealants, and much of the detailed discussion is limited to the modern high-performance elastic sealant types. In situ cured sealants are applied to a joint in paste-like or liquid form, and they harden or cure in place. Similar to adhesives, in situ cured sealants rely on their adhesion properties to maintain a durable seal and perform their function. However, in situ cured sealant formulations typically possess higher elasticity and flexibility than adhesives and generally display lower cohesive strength. Some sealants, termed *structural sealants,* are used in the same manner as adhesives to transfer loads between building components; however, these sealants still maintain a higher level of movement capability than conventional adhesives.

13.2 CLASSIFICATION OF CONSTRUCTION SEALANTS

Various sealant formulations have been developed over the years that meet performance specifications mandated by building codes as well as the specific and unique needs of end users. The large number and diversity of sealant formulations make their classification difficult. Often, there is considerable overlap between the various classifications. In general, construction sealants are segmented into classes based on the following characteristics:

- Product form
- Ability to harden or cure
- Chemical nature of binder
- Cure type
- Performance
- End use (markets and applications)

13.2.1 PRODUCT FORMS

Figure 13.1 shows a classification scheme for product forms. Sealing strips and gaskets are flexible materials that are manufactured in a range of sizes and cross sections. These products are preformed in the factory by extrusion, molding, or foaming and cutting processes. There are generally three types of preformed sealants:

- Mastic strips
- Cellular strips
- Elastomeric strips

Mastic strips are usually manufactured from blends of relatively soft and tacky synthetic rubber and polymers. The strips are supplied on a release liner, which is removed prior to their application. A typical example of this sealant category is butyl tape.

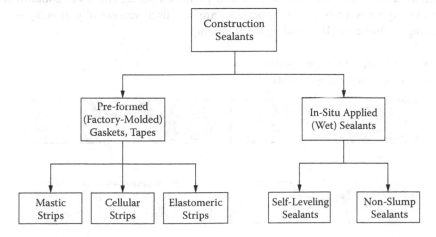

FIGURE 13.1 Classification of construction sealants by product form.

Cellular strips are cut from foamed synthetic polymers, which may also be edge-coated with an adhesive layer. Typical of this type of sealant are open-cell polyurethane foam strips that have been impregnated with a sticky material. The sticky material may be bitumen, poly(vinyl alcohol), or acrylic polymer. The cellular sealing strips are supplied precompressed to about 30%–50% of their normal thickness, and expand after placing. The strips can either be inserted into an existing joint or fixed to one side of the joint before placing the component forming the other side of the joint. These two categories of preformed strips and gaskets mainly rely on compression to achieve a tight seal, although some adhesion to the joint faces may take place. While preformed seals are only effective in joints in which the extent of expansion and compression movement is low, they can compensate considerable lateral shear movements. This feature makes them ideal for joints that are exposed to large shear movements, such as settlement joints between different buildings or building sections.

The third category, elastomeric strips and gaskets, are made from cured rubber materials, such as ethylene-propylene-diene terpolymer (EPDM), chloroprene, silicone, polysulfide, or polyurethane. EPDM and chloroprene rubber seals may also be manufactured as cellular strips and gaskets, which are typically used as compression seals. Silicone, polysulfide, or polyurethane tape seals are commonly extruded or molded and cured in a controlled manufacturing environment. These products are packaged in cured form (often rolls) and are attached to the substrate via a thin, adhesive layer of wet sealant applied on the outer edges, which allows the nonbonded center section of the tape to move freely during expansion or contraction of the joint.

While tape sealants are traditionally considered less appealing in appearance than their wet counterpart products, innovative designs are changing this perception. Figure 13.2 shows (clockwise from the upper left) schematic examples of a U-joint configuration, a concave bridge joint, a recessed bridge joint, and a recessed concave bridge joint.

Besides remedial work over a failed sealant in expansion joints, butt joint applications, undersized expansion and window perimeter joints, and leaking aluminum window systems, preformed tape sealants are also used successfully in many other sealing applications [1], including the following:

- Tilt-up concrete panel joints
- Roofing and parapet joints

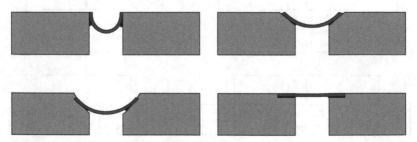

FIGURE 13.2   Schematic examples of tape sealant joint designs.

- Curtain wall joints
- Skylight seals
- Heating, ventilation, and air-conditioning (HVAC) system seals
- Fillet beads for showers and bathtubs

In situ applied sealants fall into two classes: sealants that flow under their own weight and nonslump sealants. Flowable sealants are generally self-leveling and are used in horizontal joints, for example, on parking decks or highways, while non-slump sealants are used in vertical joints. Nonslump sealants are also termed *nonsag* or *thixotropic*; the latter term reflects their rheological behavior of thinning under shear while building viscosity at rest. Nonslump sealants are typically applied as cartridges by using a sealant gun ("gun grade"), or from pails or drums by using suitable pumping and dosing equipment.

13.2.2 Ability to Harden or Cure

In situ (wet) applied sealants are often classified as "hardening" or "nonhardening" (see Figure 13.3). Nonhardening sealants are typically heavily bodied (highly filled) viscous fluids that are rendered immobile by the addition of fillers and fibers. These sealants do not "set," that is, harden or cure; rather, they stay wet or flowable after application and have physical properties similar to those before their application. Often, the term *mastic* is used to describe this type of sealant. Mastics are formulated based on "drying" or "nondrying" oils. Drying mastics form a skin on the surface exposed to the atmosphere due to an oxidative cure process. Since this cure proceeds over time into the bulk of the material, these drying mastics are not truly nonhardening sealants. Only mastics formulated based on nondrying oils, such as polybutene or

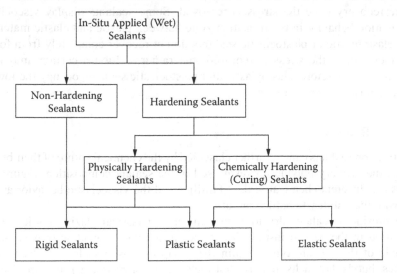

FIGURE 13.3 Classification of sealants by hardening (setting) process.

polyisobutylene or combinations of these binders, may be considered nonhardening, although even with these mastics properties drift gradually with aging.

Hardening sealants are also formulated based on heavily bodied viscous fluids; however, they set to form rigid, semiflexible, or flexible seals. The setting process may be either physical or chemical in nature, depending on whether primarily physical (i.e., weak) or chemical (i.e., strong) interactions are involved. Physically setting sealants are based on thermoplastic polymers that can be melted by temperature increase or be dispersed in an organic medium or in water. Chemically curing sealants contain functional polymers that react via polycondensation or polyaddition reactions. Examples of physically setting sealants are hot melts and solvent- or waterborne acrylates (acrylics).

Examples of chemically curing sealants are silicones, polyurethanes, polysulfides, and silicon-modified organics (such as silyl-functional polyethers, etc.). Some sealants set by both physical and chemical means. An example of such a sealant type is a chemically curing hot melt, such as a polyurethane hot melt. With these "dual" set sealants, physical setting typically dominates the handling properties during application, while chemical curing influences the performance properties. Dual-set sealants are generally classified by the setting process that has the strongest influence on their performance properties. Polyurethane hot melts, mentioned earlier as an example, are therefore classed as chemically curing sealants. The author suggests that, following the same logic, silicone latex sealants should also be considered chemically curing sealants.

Hardening sealants can be further subdivided into rigid, plastic, and elastic sealants. Rigid sealants have little or no resiliency and are characterized by their inability to flex and accommodate movement. Elastic sealants are based on functional polymers that cure to an elastomeric network. These sealants show high elastic recovery and return to their original dimensions after compression and elongation. Plastic sealants flow when placed under stress and show little or no elastic recovery once the stress is removed. Some sealants display viscoelastic deformation behavior in the transition zone between plastic and elastic materials. Such elastoplastic or plastoelastic sealants do not recover completely from forced movements once the stress is removed, but rather, exhibit a certain amount of viscous flow; therefore, elastoplastic and plastoelastic sealants occupy the low-to-medium elastic recovery range.

13.2.3 BINDERS

Construction sealants can be further classified by the chemical nature of their binder, that is, the primary polymer or resin used in the sealant composition. Figure 13.4 shows the different chemical sealant families and the viscoelastic behavior associated with their ability to harden or cure.

Oleoresinous sealants dry to a rigid material. Classical glazing putties, which were used in the past to install glass panes into wooden window frames, are an example of this sealant type. Within the group of plastic sealants, solvent-borne acrylics, butyls, hot melts, reactive hot melts, and waterborne acrylics are listed in the sequence of increasing flexibility and elastic recovery. Some waterborne acrylic

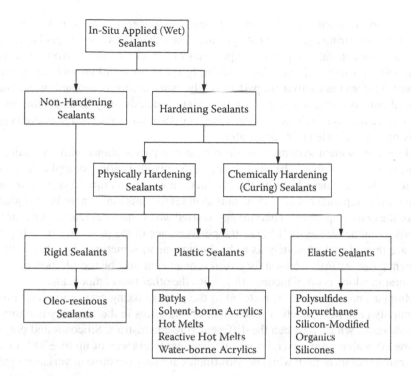

FIGURE 13.4 Chemical sealant families and the viscoelastic behavior associated with their ability to harden or cure.

formulations can be grouped with elastic sealants since they display high elastic recovery. Polysulfides, polyurethanes, telechelic silicon-modified organics (silicon hybrids), and silicones qualify as elastic sealants.

13.2.4 CURE TYPES

Chemically curing sealants may be characterized by their method of cure. The most popular classification is into single-component and multicomponent systems. However, this is only a distinction between product forms. Other classifications exist that relate to the type of cure (e.g., moisture-induced versus oxidative cure) or the specific type of crosslinker employed (e.g., acetoxy, oxime, or alkoxy cures) in silicones.

In one-part sealants, all formulation ingredients are contained in a single package. Depending on the cure chemistry, the mixed composition must be packaged in a water- or oxygen-impermeable container to prevent crosslinking or cure deactivation within the package. One-part construction sealants generally have a shelf life of 6–24 months. For moisture-induced cure systems, a water scavenger is required in the formulation. On compounding, the scavenger chemically dries the filler by reacting with any surface-adsorbed water. During storage of the finished product, the scavenger maintains the shelf stability of the unopened container by reacting with any water diffusing into the package.

In two-part systems, the key formulations components, that is, polymer, filler, crosslinker or curing agent, and catalyst, are distributed between two packages. The two packages can either separate the polymer and filler (in part A) from the curing agent and the catalyst (in part B), or separate the polymer, filler, and curing agent (in part A) from the catalyst (in part B). If the latter course is adopted in moisture-induced cure systems, it is important to maintain strictly anhydrous conditions for the mixture of polymer, filler, and curing agent. Two part construction sealants generally have a shelf life of 12–24 months.

Three-part sealant systems may simply be two-part sealants with an additional pigment paste container. The pigment paste is mixed in prior to application, thus providing field pigmentability to the product. However, multipart systems are also often used to separate out certain formulation ingredients that cannot be included in one of the two components. This strategy is used when chemical reactions occur that negatively influence the shelf life or the performance of the product. These ingredients are then placed separately as a third component, sometimes together with the pigment paste. A further reason for a third component may be the addition of a cure accelerator, which is often incompatible with the other two components.

Multipart products must be metered to the proper mixing proportions and mixed thoroughly prior to application. The effect of tolerances in the metering and mixing process varies widely between the different cure chemistries. Silicone- and polysulfide-based sealants are generally quite forgiving; tolerances of up to ±20% in mix ratio can be compensated, while polyurethanes are very sensitive to variations in the mix ratio.

One-part systems rely on an environmental trigger to initiate the cure process. Most commonly, moisture is used to trigger the sequence of reactions that lead to crosslink formation and, thus, the cure of the polymer to a three-dimensional elastomeric network. Some oxidative cure chemistries also involve atmospheric oxygen in the cure process. Because the cure of one-part products is dependent on the diffusion of moisture or oxygen into the material and on the diffusion of cure by-products from the material, all one-part sealants cure from the outside inward. For low-permeable sealants, such as polysulfides and polyurethanes, this is a relatively slow process, and some of these products may cure only 1–2 mm per week. Such slow-curing one-part products are not suitable for joints undergoing any substantial movement during cure. Silicones, because of their higher permeability at ambient conditions, cure much faster, typically at a rate of 2–4 mm per day.

Since all relevant components for the completion of the cure are provided in multipart sealants, these materials cure more rapidly and more homogeneously than one-part products. Multipart products are often used in the assembly of prefabricated units, since their rapid cure allows a reduction in the time required for the manufacturing cycle, the storage prior to shipment, or both. Faster cure is also important in on-site installations when the sealant is used outdoors. The time from application of the sealant until its first exposure to abrasion, joint movement, weather conditions (such as rainfall), etc., may be critical. Bridge joints are an excellent example of an application in which cure time is of paramount importance. Rerouting traffic because of bridge closure causes inconvenience and loss of money. Therefore, fast-curing multipart sealants are used in horizontal traffic joints on bridges.

13.2.5 Performance Characteristics

In situ cured sealants rely on their adhesion properties to maintain a durable seal and perform the required functions during their service life. They must accommodate joint movements without failing interfacially or cohesively or causing failure to substrates. On the basis of these requirements, the most important performance properties in construction sealants are movement capability, modulus, adhesion, and life expectancy (durability).

13.2.5.1 Movement Capability

In most sealant selection processes, movement capability is a key criterion. Movement capability is the amount of repetitive displacement a sealant can endure continuously over its service life without failing. Existing test standards allow only an assessment of the initial movement capability of a sealant; they do not account for changes in the movement capability as a result of aging. Various test standards that subject sealed joint specimens to cyclic movement exist, often combined with temperature variations, or water immersion, or both. Examples of such standards are ASTM C719 "Standard Test Method for Adhesion and Cohesion of Elastomeric Joint Sealants Under Cyclic Movement" [2] and ISO 9047 "Building Construction—Jointing Products—Determination of Adhesion/Cohesion Properties of Sealants at Variable Temperatures" [3] for elastic sealants, as well as ISO 9046 "Building Construction—Jointing Products—Determination of Adhesion/Cohesion Properties of Sealants at Constant Temperature" [4] for plastic sealants.

Movement capability is typically reported as a percentage of the original joint width, that is, ±hr x% or +x%/−y%, where the positive value refers to extension and the negative value to compression of the seal. A sealant with a +50%/−25% rating can be repeatedly extended by 50% or compressed to 25% of the original joint width after initial cure. Sealants can be broadly divided into classes according to the amount of movement they can successfully handle (see Figure 13.5). High-performance sealants such as silicones, polyurethanes, polysulfides, and silicon hybrids can typically handle movements of 25% or higher. Medium-performance sealants such as certain acrylics can handle movements of 10%–25%. Low-performance sealants such as butyls, putties, and caulks accommodate movements of less than 5%–10%.

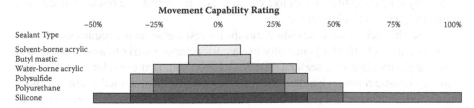

FIGURE 13.5 Movement capability ratings according to ASTM C719 of typical products within sealant families. (*Note:* The darker shaded colors within a material class indicate movement capability ranges of typical mainstream products, while lighter colors indicate movement capability ranges for higher-performance products.)

Sealant manufacturers commonly report movement capability ratings of their products. However, these numbers should be interpreted with care since the actual ratings depend on the substrate, cure time, rate of movement, etc. Some sealant manufacturers deviate from the standard test protocol and, for instance, assess the movement capability only at room temperature and under dry conditions. If the application in question involves a working joint that will be exposed to high and low temperatures as well as water, the sealant to be specified should be tested to the full rigors of the official standard test, such as ASTM C719. Such testing should be carried out by an independent laboratory recognized by an external validating organization, for example, the Sealants, Weatherproofing, and Restoration Institute (SWRI) in the United States, the national standards compliance and quality certification institutes in Europe, or similar organizations.

Movement capability tests such as ASTM C719 are useful in that they expose the sealant to conditions close to actual field conditions. However, these tests are typically carried out on sealants that have been almost fully cured without being exposed to any stress. Sealants installed in working joints in the field cure while joint movement occurs. Several recent studies have evaluated the performance of sealants exposed to movement during cure (see Reference 5 and literature cited therein). Results indicate that movement during cure may induce premature failure in a sealed joint. This failure may occur by one of several possible modes; however, the most serious and frequent mode of failure is the cracking or splitting of the sealant surface. This failure is related to the length of time needed for a sealant surface to develop an elastic skin; sealants that develop elastic surfaces rapidly seem to resist cracking better. However, even with one-part sealants that form elastic surfaces rapidly, continuous movements of greater than ±15% during cure can cause failures. If large movements during cure are expected, two-part sealants that cure more uniformly in depth may provide better resistance to failure, because two-part sealants generally cure faster than corresponding one-part sealants.

Furthermore, as mentioned earlier, tests such as ASTM C719 assess only the initial movement capability of a sealant. They do not take the effects of sealant degradation resulting from movement fatigue or long-term weathering into account. Sealant manufacturers should consider these durability aspects when providing movement capability ratings, either by establishing reliable performance histories for a new sealant product via long-term outdoor testing and extensive in-service field evaluations or by subjecting the product to a credible durability test (see relevant discussion in this chapter and in Chapter 6).

Due to the lack of an established durability test for sealants, requirement standards such as ISO 11600 [6] currently use a performance matrix to assess the movement capability class of a sealant. The ISO 11600 standard considers the sealant's elasticity (elastic recovery); adhesion and cohesion after maintained extension, both before and after water immersion; and—for glazing sealants—adhesion and cohesion properties after exposure to heat, artificial light, and water in addition to the movement capability rating obtained in the ISO 9047 or ISO 9046 test in the assessment of a sealant's movement capability class.

While there are commercial sealants capable of obtaining a higher rating (e.g., ±37.5% or ±50%) in the ISO 9047 test standard, ISO 11600 takes a conservative approach by limiting the highest classification for movement capability an elastic sealant can claim to 25%. The reason for this decision by the ISO committee TC59/SC8 was one of caution. If sealants with higher movement capability are specified to seal narrow joints, then the margin for error is reduced and undue pressure is put on the builder to meet accurate tolerances in building components and construction. Furthermore, unless the sealant supplier can claim with confidence that the movement capability for the specified sealant will remain more or less unchanged throughout the claimed life expectancy of the sealant, it is better to build a "safety" margin for property drift into the sealed joint design. In future revisions, as experience and confidence in the ISO sealant classification scheme grows, it is anticipated that higher-movement capability classes will be formally recognized [7].

13.2.5.2   Modulus of Elasticity

The value of the stress of a sealant at a specific strain is referred to as the *modulus of elasticity,* sometimes called the *secant modulus.* This important sealant property describes the force exerted per unit area by a sealant under stress. Because a primary function of a sealant is to adhere to the substrates it is in contact with, the forces generated by a joint opening or closing are transmitted by the sealant to the substrate–sealant interface. For this reason, it is important to know the modulus of the sealant and also the strength of the substrate. The use of a high-modulus sealant on a weak substrate, such as asphalt, concrete, or sandstone, can result in stresses higher than those that the substrate can tolerate. Whenever the sealant induces stress into the substrate that exceeds its strength, there is a high potential for failure, primarily cohesive failure (spalling) of the substrate. For example, concrete can be considered a high-strength substrate when in compression. However, by comparison, it is quite weak in tension. Sometimes, on closer inspection, perceived interfacial failures on these weak substrates are actually cohesive failures of the substrate near the substrate–sealant interface induced by a high-modulus sealant.

Although the sealant manufacturers' literature commonly reports modulus values, these values must be interpreted carefully. Specimen dimensions, test rate, cure conditions, and the time a sealant has been allowed to cure when tested can all have a significant effect on modulus. Therefore, for a true comparison, sealants should be evaluated by a standard test that examines all sealants by the same procedure. In general, the longer a sealant has been allowed to cure, the more realistically the modulus data match the sealant's long-term field behavior. This is because in most sealant types, cure initially proceeds rapidly, while it later takes considerable time to reach completion. In the field, it is often difficult to differentiate this so-called "postcure" from the initial stages of sealant aging induced by environmental factors.

In the ISO 11600 standard, a sealant with a room-temperature tensile modulus greater than 0.4 N/mm^2 is classed as high modulus (HM). Sealants with a modulus of 0.4 N/mm^2 or less are classed as low modulus (LM). This subclassification is used only with the high-movement classes of elastic sealants (classes 20 and 25), since modulus is a more important property in sealants used to seal joints undergoing large or rapid movements. The modulus of a sealant can increase considerably at lower temperatures. Hence, the relevant ISO 8339 [8] test standard also measures the tensile modulus at −20°C. At this temperature, a sealant with a tensile modulus greater than 0.6 N/mm² is classed as HM, even if it tested as LM at room temperature.

13.2.5.3 Adhesion of Sealants to Substrates

In situ applied sealants need to develop adhesion to substrates for them to perform their function. Three aspects of adhesion are important: the rate at which adhesion is developed after the sealant has been applied to the joint, the level and quality of the adhesion gained, and the durability of the adhesion. In many applications, construction sealants are exposed to movement during cure. Therefore, a sealant that is capable of developing adhesion rapidly to substrates is less likely to fail in the initial phase of its service life. The strength of the bond that develops between the sealant and the substrate depends on the sealant formulation, surface chemistry of the substrate, environmental conditions during cure and service of the sealant, duration of the sealant's cure, and various other, but usually minor, factors.

Early on, during the commercialization of the first elastic sealants, adhesion tests that had been initially developed for adhesives were utilized. While these tests provide some assessment of the interfacial bond strength, they often ignore the dissipative effects of the elastomeric network, which, because of the larger bulk involved in sealed versus adhesively bonded joints, play a substantial role in the practical adhesion of sealants [9,10]. The experimenter needs a solid understanding of the physics behind the specific adhesion test in order to assess its usefulness in addressing certain aspects of sealant adhesion [11]. Since energy dissipation depends on the specific adhesion test used, one cannot directly compare the results of two adhesion experiments carried out with methods involving different levels of energy dissipation, for example, peel and tensile adhesion tests.

In the assessment of the quality of adhesion gained to a specific substrate, it is important to choose an adhesion test method that reflects the prevalent type of mechanical and environmental stresses experienced by the sealant in actual field applications. Different test methods should be selected depending on whether the sealant is primarily exposed to tensile, shear, or peel forces. Furthermore, the test regime should reflect the environmental exposure conditions experienced in service.

Frequently, the bond strength between the sealant and the substrate can be substantially improved by the use of a primer. While sealant manufacturers generally do not consider recommending a primer a highly marketable aspect of the value proposition, it is often a wise choice, since it can help avoid adhesion problems in the long run. A sealant specifier or user should, therefore, seek information on whether the adhesion of a sealant to a specific substrate can be further enhanced by use of a primer. On porous substrates, the primer penetrates into the substrate and may help avoid spalling problems by reinforcing the weak top layer of the substrate. On the other hand, using a primer on a substrate for which it has not been developed and recommended can have detrimental effects. In terms of the service life of sealed joints, the durability of adhesion is often the most important aspect for consideration and will be addressed in the following subsection.

13.2.5.4 Durability

Once installed, construction sealants are exposed to a variety of environmental and service influences. One of the most important characteristics of a sealed joint is its tolerance of the operating environment. This endurance is also referred to as the joint's *permanence* or *durability*.

A primary factor in sealant durability is its ability to resist degradation induced by environmental elements. Outdoor degradation processes in construction sealants are primarily caused by weathering factors such as solar radiation (especially, the ultraviolet [UV] portion), water, oxygen, and temperature and its cycles. Aggressive atmospheric contaminants or components may also contribute to the degradation of sealed joints in highly polluted areas or industrial environments. In certain applications, specific environmental degradation factors may override the importance of the aforementioned weathering factors. In this regard, microbiological attack is the most important environmental (biological) degradation factor. For instance, in sealant applications below grade (underground), wastewater treatment plants, or food processing plants, the primary degradation factor is often microbiological attack.

Service factors augment and modulate environmental factors. Mechanical stress, for instance, is a degradation factor by itself; however, it also acts in synergy with various environmental factors. Any durability test method for sealants specified for applications involving joint movement should include a realistic reflection of joint movement exposure (during and after cure) in combination with other relevant stress factors that are part of the expected operating environment (for a more detailed discussion, see Chapter 6).

Environmental and service factors may cause degradation of the surface, the bulk, or the interface to the substrate during the service life of a sealant. Surface changes that may occur are degradation in color and gloss, wrinkles, crack formation (crazing, mud cracking, etc.), and chalking (whitening). Bulk properties may degrade and cause hardening or softening of the sealant. Typically, hardening is caused by a postcure mechanism. Often, this postcure mechanism is simply an extension of the initial cure mechanism induced by a catalyst retaining its activity in the cured sealant. Frequently, softening is caused by degradation of crosslinks or by melting of thermoplastic components of the sealant. Hydrolysis of crosslinks is the most common factor in elastic sealant softening. For certain types of sealants, hydrolysis in warm, humid environments may become so severe as to transform the cured sealant back into a fluid. This type of degradation is termed *reversion*.

The service life of a sealant is determined by both the durability of the bulk material and the durability of the adhesion to the substrates. A sealant may have excellent resistance to sunlight; however, if its adhesion degrades with aging and the sealant fails interfacially within a short period of time, it is of little use. The same can be said of a sealant with superior adhesion characteristics but poor resistance to UV light. Either situation results in a short service life.

Certain interactions between the physical properties and adhesion of a sealant as it is aging may also affect its service life. For instance, changes in the modulus of elasticity can strongly influence the adhesion life (the time period until interfacial failure occurs). Some sealants may harden with aging as a result of plasticizer loss or continued crosslinking. As a sealant hardens, the modulus increases, and more stress is placed on the sealant–substrate interface. If the tensile forces exceed the adhesion strength of the sealant–substrate interface, the seal fails interfacially. If the tensile forces exceed the cohesive strength of the substrate, the substrate fails cohesively. Therefore, credible durability tests for sealants must assess performance degradations resulting from the interplay between physical properties and adhesion.

A sealant's durability is often predicted on the basis of its performance in accelerated weathering tests that expose the material to high temperatures, high humidity, water spray or immersion, and UV or simulated sunlight. Because there is much debate on how to predict actual service-life performance on the basis of accelerated testing, the duration of accelerated test exposure varies widely with the test method and may range from hundreds to thousands of test hours. Previous studies of weathering effects on the surface and bulk properties of sealants indicate that at least 1000–2000 h of accelerated exposure are required to induce changes similar to 1 year of outdoor exposure in moderate-to-subtropical climates (see Reference 12 and literature cited therein). In contrast, many standardized weathering tests for sealants still require only 250 or 500 h in an accelerated weathering machine.

Claims by sealant manufacturers that their products have passed such standardized accelerated weathering tests may have little meaning in the real world. In fact, these tests are misleading and provide the user or specifier with a wrong sense of security. Durability studies of construction sealants specified for outdoor applications should be based on 5,000 to 20,000 h of artificial weathering and ideally include joint movement during cure as well as during the artificial weathering of the specimens.

Another way of determining the durability of a sealant is to monitor the history of field installations. Sealant manufacturers often have case histories of successful installations. For the user or specifier of sealants, often it is more efficient to determine the suitability of a specific sealant product by inspecting job sites where the sealant has already been installed for several years, rather than waiting for long-term accelerated weathering tests. However, when using this method of field performance assessment, the sealant user or specifier should ask the manufacturer to certify that the sealant formulation has not changed over time since the early projects were completed.

13.2.6 APPLICATIONS OF CONSTRUCTION SEALANTS

Construction sealants are often classified according to their function or end use. Typically, interior sealants act to contain a medium, while exterior sealants exclude contaminants. Some sealants perform both functions. Some of the major application segments for construction sealants are as follows:

Low- and high-rise building construction:
- Residential windows and doors (glazing and perimeter seals, insulating glass (IG)-edge seals, etc.)
- Commercial glazing (IG edge-seal, glazing seal, weatherproofing seal, structural glazing seal, etc.)

- Exterior weatherproofing (weatherproofing seals in light- and heavy-weight façades, expansion and control joints, connection joints, etc.)
- Interior sealing (perimeter seals at interior windows and doors joints; baseboards; moldings; sanitary perimeter seals at bathtubs, showers; etc.)
- Roofing (seam seals, seals at roofing terminations, etc.)
- Tilt-up concrete exterior walls (weatherproofing of expansion and control joints)

Public works and civil engineering:
- Plazas, parking decks, pavements, sidewalks, and stadiums (waterproofing of expansion and control joints)
- Bridges and highways (waterproofing of expansion and perimeter joints)
- Airport runways and aprons (waterproofing of pavement joints)
- Water and wastewater treatment facilities (sealing of all kinds of joints, including those in submerged environments)
- Institutional buildings (prisons, schools, hospitals, etc.; all kinds of joints, some of which require special features, e.g., ability to decontaminate seal, radiation-resistant seals, security seals, etc.)
- Power plants (fire-stop materials in joints and penetrations)
- Gasoline stations, chemical plants, and hazardous material containment structures (seals that resist chemicals and gasoline)

13.3 SEALANT TYPES: PROPERTIES, CHEMISTRY, FORMULATIONS, AND APPLICATIONS

13.3.1 SILICONES

The first silicone construction sealants were introduced in the mid-1960s, and many of the sealants applied in the early days are still functional today. Products are available in a variety of forms, from paste-like nonslump materials to flowable self-leveling sealants. Both single-component and multicomponent versions are available. Silicone sealants can be formulated based on different cure chemistries; however, the vast majority of commercial products sold for construction applications are based on condensation cure chemistry. Most of these products are applied and cured at ambient conditions and are therefore termed *room-temperature-vulcanizing* or *RTV* products. Specialty sealants that chemically cure hot melts or waterborne dispersion products exist; however, even these sealants utilize condensation cure chemistry.

13.3.1.1 Properties of Silicones

Silicones occupy the regime between inorganic silicates and organic polymers, and exhibit some of the properties of each material type because of the combination of the partially ionic siloxane (Si-O-Si) bond and the organic substituent groups. The commercial importance of silicone construction sealants is due to their unique combination of properties, as shown in Table 13.1.

Silicone sealants show extraordinary weatherability. The lack of chromophores (light-absorbing groups) along the polymeric backbone and the high Si-O bond energy

TABLE 13.1

General Benefits and Limitations of Silicone Sealants

Benefits	Limitations
• Excellent weather and thermal stability (no property drift with aging; durable, that is, UV-resistant, adhesion)	• High gas permeability (lack of barrier property)
• Ozone and oxidation resistance	• Nonpaintable
• Nonyellowing clear and transparent formulations feasible	• Somewhat higher cost
• Good electrical and fire properties	• Odor issue with some cure systems
• High gas permeability (desired permeation property, resulting in fast-curing one-part sealants)	• Dirt pickup and staining with certain formulations
• Ability to wet difficult substrates (due to low surface tension), good adhesion to a variety of substrates	• Should not be applied to wet or damp surfaces (poor adhesion) with the exception of silicone latex sealants
• High movement capability (up to +100%/−50%), high elastic recovery (85%–98%)	• Poor tear resistance with transparent formulations
• Low temperature dependency of uncured and cured properties; low-temperature application feasible; extreme low-temperature flexibility	
• Fast-curing one-part sealants feasible	

provide inherent resistance to sunlight. Because of this stability, silicones, unlike organic sealants, do not require addition of UV absorbers to their formulations and can be used in highly demanding UV-exposed applications such as in structural glazing or solar collector systems. Their excellent heat stability allows exposure to high service temperatures, for example, in solar collector systems; heating, ventilation, and air-conditioning (HVAC) systems; curtain wall and roofing applications; etc.

Silicone sealants display good resistance to oxidation by ozone and oxygen. The oxidation of the hydrocarbon side groups results in the formation of carbonyl groups [13]. Since carbonyl groups do not interact strongly with other chemical groups in silicone polymers, the oxidation has little effect on the mechanical properties of the sealant. This is consistent with the fact that even after 20 years of outdoor weathering in sunny climates, silicone sealants show comparatively little changes in physical properties [14,15].

Silicone sealants display extreme low-temperature flexibility; standard formulations remain flexible down to −60°C and special formulations down to −90°C. They are able to withstand the temperature extremes that are experienced in climates such as in the polar regions, and are able to maintain a good degree of flexibility at very low temperatures where other materials would stiffen and crack.

In general, silicone sealants are good insulators and are characterized by high dielectric strength, high volume resistivity, low dielectric constant, and low dissipation factor. The low dissipation factor is a desirable property because low dissipation minimizes the waste of electrical energy as heat. However, by changing the type and amount of compounding ingredients, the formulator is able to influence the electrical properties over a wide range.

In contrast to organic-polymer-based sealants, silicones display very favorable fire performance characteristics, for example, a very low heat release rate, a unique insensitivity of burn rate to fire severity, and minimal nonthermal hazard of key combustion products for both humans and property. The combustion of silicones results in the formation of carbon dioxide, water, amorphous silica, and low yields of carbon monoxide, but no other polymer-specific toxic gases are generated. The combustion products demonstrate minimal potential for corrosive damage unless the silicone sealant is filled with a halogen-containing compound. A substantial portion of the amorphous silica generated during the combustion process is deposited on the fuel-generating burning surface, resulting in the formation of silica char. These surface silica deposits are believed to play an important role in determining the unique burning characteristics of silicones, that is, the relatively low heat release rate and the minimal dependence of heat release rate on applied external heat flux (fire severity).

The large free volume and mobility of silicone polymer chains give rise to a high diffusion coefficient and high permeability of gases or vapors. For many applications, high gas permeability is a desirable property. For instance, the higher moisture transmission rate of silicone at ambient and lower temperatures allows the formulation of fast-curing one-part systems that cure at temperatures as low as −40°C. For applications, in which high permeability is not desired, special silicones with reduced permeability have been developed or system solutions are available. An example of an ideal system solution is a dual-sealed insulating glass unit, where the primary polyisobutylene-based seal provides the barrier function and the silicone secondary seal ensures the structural integrity of the unit under all relevant loads.

Because of their low surface tension, silicone sealants wet most substrates even under difficult conditions and, when formulated with suitable adhesion promoters, exhibit very good adhesion.

Silicone polymers also exhibit much less change in physical properties with change in temperature compared to organic systems. This results in sealants that are almost as easy to extrude in cold weather as in warm weather. It also means that the cured sealant shows less change in properties, for example, modulus, as a function of temperature when compared to organic sealants.

Because of their low surface tension (ca. 21–22 mN/m), silicone sealants are not easy to adhere to. Silicones have a critical surface tension of wetting (24 mN/m) that is higher than their own surface tension. This enables liquid silicones to spread on cured silicones, a process that promotes film formation and surface coverage as a first step to good adhesion. However, organic materials, such as paints, have much higher surface tensions and, therefore, do not spread well on and adhere to silicone sealants. This is why, unless special formulation efforts are made, silicone sealants in general are considered nonpaintable. In reality, for most applications, this is a blessing in disguise. Sealants in movement joints should not be painted or coated with a material of lesser movement capability than the sealant. Most paints, with very few exceptions, have much less movement capability than silicone sealants, typically only in the 5%–10% range. When such paints are used on sealed joints undergoing higher movements, cracks forming in the paint film can propagate into the sealant and cause premature joint failure.

Silicone sealants are somewhat higher priced than polyurethane sealants. Some silicone sealants, especially the acetoxy, amine, and oxime types, exhibit a strong odor upon cure. Silicone sealants formulated with a nonreactive silicone plasticizer have the potential to stain substrates. However, this staining potential can be minimized or completely eliminated with proper formulation techniques. Because of their excellent durability, silicone sealants tend to retain the dirt deposited on their surface, while degradation of the surface of organic sealants causes dirt particles to be flushed off by rain. With the exception of waterborne silicone latex sealants, silicone sealants should not be applied to wet or damp surfaces as water contamination of the substrate surface interferes with adhesion buildup. Finally, most clear or transparent silicone sealants show poor tear resistance; however, this is seldom an issue in the specific applications in which these products are used.

13.3.1.2 Chemistry of Silicone Sealants

As mentioned earlier, the vast majority of commercial construction silicone sealants are based on condensation cure chemistry. Because of their ease of application, one-part products dominate the market, especially for consumer applications. For a more detailed discussion of silicone cure chemistries, see Reference 16.

13.3.1.2.1 One-Component Cure Systems

In one-part condensation cure sealants, an excess of crosslinker is used as water scavenger to maintain the shelf stability of the unopened container. Upon mixing the reactive (silanol functional) silicone polymer and the typically trifunctional hydrolyzable silane crosslinker during manufacture of the sealant, the condensation (endblocking) reaction shown in Figure 13.6 rapidly occurs in the presence of a condensation catalyst. Some crosslinkers are self-catalytic and do not require an external condensation catalyst.

Once the one-part package is opened, atmospheric moisture reacts with the hydrolyzable groups attached to the polymer chain ends. This hydrolysis reaction is followed by condensation of the silanol and another hydrolyzable group (X) attached to silicon at the polymer end. The net result of these two reactions is shown in Figure 13.7. This reaction continues until a three-dimensional elastomeric network is formed.

$$\sim SiR_2OH + X - \underset{\underset{R}{|}}{\overset{\overset{X}{|}}{Si}} - X \overset{(catalyst)}{\rightleftharpoons} \sim SiR_2 - O - \underset{\underset{R}{|}}{\overset{\overset{X}{|}}{Si}} - X + HX$$

FIGURE 13.6 Reaction of crosslinker (trifunctional silane, $RSiX_3$, where X represents the hydrolyzable group) with silanol group at siloxane-polymer end (only one silanol end is shown).

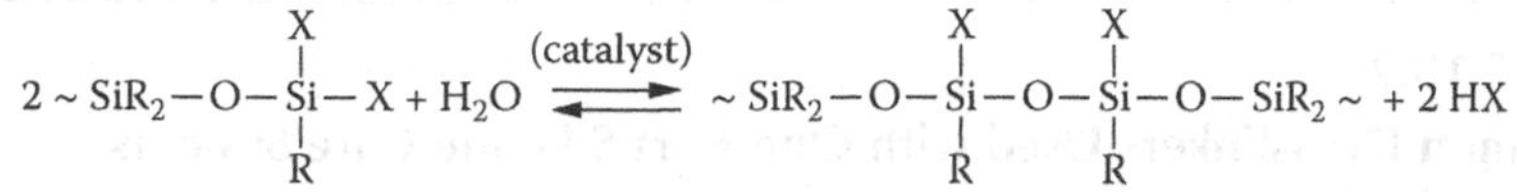

$$2 \sim SiR_2-O-\underset{\underset{R}{|}}{\overset{\overset{X}{|}}{Si}}-X + H_2O \underset{\text{(catalyst)}}{\overset{\text{(catalyst)}}{\rightleftharpoons}} \sim SiR_2-O-\underset{\underset{R}{|}}{\overset{\overset{X}{|}}{Si}}-O-\underset{\underset{R}{|}}{\overset{\overset{X}{|}}{Si}}-O-SiR_2 \sim + 2HX$$

FIGURE 13.7 Hydrolysis and condensation reactions as first steps in the crosslinking mechanism.

One-part cure systems depend on atmospheric moisture to initiate cure. The cure of one-part silicones is considerably faster than that of organic one-part sealant formulations, which generally show a lower permeability toward water. One-part silicone sealants typically cure at a rate of about 3–4 mm per day. The cure by-products, which diffuse out of the room-temperature-vulcanizing (RTV) silicone and often cause a characteristic smell during cure, include alcohols [17–19], ketoximes [20], carboxylic acids [21–26], amides [27,28], hydroxylamines [29–32], ketones [33,34], and amines [35–38]. Some of the more common crosslinkers that are used with one-part silicone cure systems are shown in Table 13.2.

Alkoxysilane, oximosilane, and enoxysilane crosslinkers eliminate by-products and display a pH-neutral chemical reaction; that is, they are neither acidic nor basic and, therefore, are the key components of neutral noncorrosive cure chemistries. Amidosilane-based cure chemistries are often also considered neutral, although they may cause corrosion on sensitive substrates (for instance, stress corrosion has been observed on brass in contact with benzamide-cured silicone). The acetoxysilane-based cure system releases acetic acid; aminoxysilane and aminosilane-based cures form alkaline by-products. Acidic cure silicone sealants should not be used on alkaline substrates such as concrete, as this leads to the formation of salts, resulting in very poor adhesion. Furthermore, both acidic and alkaline cure products should not be allowed to contact corrodible substrates.

Ultralow modulus silicone products can be formulated using chain extension technology. Chain extension is based on the concept that the polymer chain is first extended by reaction with a difunctional silane before crosslinking occurs with a trifunctional silane. As the two reactions compete with each other, the difunctional silane must have a substantially higher reactivity than the trifunctional one. Since the reactivities of silanol groups vary with the number of electron-withdrawing substituents on the silicon atom, difunctional silanes, R_2SiX_2, are much less reactive than trifunctional ones, $RSiX_3$, for the same organic substituent R. This requires careful selection of the molecular design of the chain extender as compared that of the crosslinker.

During the 1980s, a formulation based on methylvinyldi-(N-methylacetamido) silane chain extender and a polyfunctional aminoxysilane crosslinker [39–41] set the standard for ultra-low-modulus silicone sealants and became very successful commercially. This chain-extension technology achieves about 1.5–2.5 extensions of the polymer chain length before crosslinking occurs, resulting in an increased average degree of polymerization D_p between chain links. A further outstanding feature of the cured elastomers of this invention is their "knotty tear." Other chain-extension chemistries have been patented, for instance, use of methylvinylbis(N,N-diethylaminoxy)

TABLE 13.2

Common Drosslinkers Used with One-Part Silicone Cure Systems

Cure System	Typical Silane	Chemical Name
Acetoxy	$CH_3Si\left[-O-\overset{\overset{\displaystyle O}{\|\|}}{C}-CH_3\right]_3$	Methyltriacetoxysilane
Alkoxy (Methoxy)	$CH_3Si(-O-CH_3)_3$	Methyltrimethoxysilane
Oxime	$CH_3Si\left[-O-N=C\overset{\displaystyle CH_3}{\underset{\displaystyle C_2H_5}{\big\langle}}\right]_3$	Methyl tris(methylethylketoximo) silane
Amine	$CH_3Si\left[-\overset{\overset{\displaystyle H}{\|}}{N}-\langle S\rangle\right]_3$	Methyl tris(cyclohexylamino) silane
Amide	$CH_3Si\left[-\underset{\overset{\displaystyle \|}{CH_3}}{\overset{\overset{\displaystyle O}{\|\|}}{N}}-C-CH_3\right]_3$	Methyl tris(N-methylacetamido) silane
Amide	$CH_3Si\overset{\overset{\displaystyle OC_2H_5}{\|}}{}\left[-\underset{\overset{\displaystyle \|}{CH_3}}{N}-\overset{\overset{\displaystyle O}{\|\|}}{C}-\langle\bigcirc\rangle\right]_2$	Methyl (ethoxy) bis(N-methylbenzamido) silane
Enoxy	$CH_3Si\left[-O-\overset{\overset{\displaystyle CH_2}{\|\|}}{C}\underset{\displaystyle CH_3}{}\right]_3$	Methyl tris(isopropenoxy) silane

silane as chain extender in combination with trimethoxymethylsilane, MeSi(OMe)$_3$, as crosslinker [42] and use of methylvinyl di(ϵ-caprolactamo) silane chain extender in an aminoxysilane-cured formulation [43].

In many cases, the cure by-products are sufficiently acidic or basic to act as condensation catalysts; these chemistries are referred to as *self-catalytic cure systems*. While these systems provide good in-depth cure, an additional condensation catalyst is generally required to achieve a tack-free surface. This condensation catalyst may be a tin compound [44], such as stannous octoate and dibutyltindilaurate (DBTDL), or an organotitanate [18–19]. Titanates are employed primarily in neutral cure (alkoxy, amide, or oxime) systems, while tin catalysts are added to acetoxy-, oxime-, and amine-cure formulations. There is also the option of formulating alkoxy-cure systems with Sn (IV) catalysts; however, to achieve good shelf stability, this requires a composition free of water and silanol [45–48]. Amine compounds, such as lauryl amine, as well as amino-functional adhesion promoters, such as γ-aminopropyltrialkoxysilanes, are known to act as hydrolysis and condensation catalysts. The catalytic nature of amines results from their alkalinity and, in the case of γ-aminoorganoalkoxysilanes, from their ability to coordinate with silicon in a pentavalent configuration.

Typical titanate catalysts used are tetraalkoxy titanates and chelated titanates. The tetraalkoxy titanates are the more catalytically active species. A possible reaction mechanism of organotitanate compounds in silicone crosslinking chemistry involves the reaction of water with the titanate ester group ($\equiv$TiOR) in the catalyst, leading to the formation of $\equiv$TiOH groups. The $\equiv$TiOH groups subsequently react with $\equiv$SiOR′ to form $\equiv$Si-O-Ti$\equiv$ bonds, which further react with $\equiv$SiOH groups to form $\equiv$Si-O-Si$\equiv$ bonds and $\equiv$TiOH [49].

13.3.1.2.2 Two-Component Cure Systems

In two-part condensation cure systems, the more common practice is to package the crosslinker and catalyst together as one reactive component, leaving the mixture of polymer and filler as the unreactive component. The decision as to whether the crosslinker should be incorporated with the polymer and filler mixture depends on the tendency of the specific organosilane crosslinker to react with particular formulation ingredients even in the absence of a condensation catalyst.

Two-part condensation cure sealants must be mixed shortly before application. Once mixed, the crosslinking reaction proceeds without the need for external atmospheric moisture. However, it should be noted that a very small amount of moisture is needed to initiate the condensation reaction. This moisture generally is present in the form of residual moisture on the filler. Many two-part condensation systems also contain excess crosslinker, requiring additional moisture to initiate the cure. Furthermore, to achieve successful completion of cure, it is important to ensure that the cure by-products completely leave the curing sealant.

Two-part condensation cure systems generally have shelf lives of 6–12 months. The key benefit of two-part condensation cure systems is their ability to achieve deep section cure and high "green strength" within a few hours. For full development of physical properties, several days may be required, because the cure by-products must still leave the bulk of the cured product.

As in one-part systems, condensation cure involves a functional silicone chain-end, typically silanol, and a polyfunctional silane. Suitable crosslinkers are trifunctional, tetrafunctional, or multifunctional materials, such as alkyl orthosilicate esters, and esters of orthosilicic [50] or metasilicic acid (polyalkylsilicates) [51] (see Table 13.3). N-propyl orthosilicate and tetraethyl orthosilicate are the most commonly used crosslinkers in two-part systems. Generally, the same condensation cure catalysts that apply to one-part systems are also employed in two-part systems.

As in one-part cure systems, chain extension may be utilized to achieve ultra-low-modulus products. Adaptation of the acetamidosilane chain extender and aminoxysilane crosslinker chemistry to two-part sealants has been successfully demonstrated [52,53].

13.3.1.3 Formulations

Table 13.4 provides an overview of the different formulation ingredients and their functions in silicone condensation cure sealants [54].

TABLE 13.3

Common Crosslinkers Used with Two-Part Silicone Cure Systems (Note: All are Alkoxy-Cure Systems)

Cure System	Typical Silane	Chemical Name
Ethoxy	$Si(OC_2H_5)_4$	Tetraethyl orthosilicate
n-Propoxy	$Si(O-n-C_3H_7)_4$	Tetra-n-propyl orthosilicate
i-Propoxy	$Si(O-i-C_3H_7)_4$	Tetra-i-propyl orthosilicate
Methoxyethoxy	$Si(OCH_2CH_2OCH_3)_4$	Tetrakis(2-methoxyethyl) orthosilicate
Ethoxy	$[OSi(OC_2H_5)_2]_{4-5}$	Polyethyl silicate

13.3.1.3.1 Polymers

RTV condensation cure silicone sealants are commonly formulated from α,ω-silanol-endblocked polymers (see Figure 13.8) with a molecular weight of about 20,000 to 200,000 Da, corresponding to a viscosity of about 1,000 cP to 300,000 cP (mPa·s). The majority of silicone sealants are based on poly(dimethylsiloxane) (PDMS) polymer. However, polymerization versatility in the preparation of silicone polymers allows the easy inclusion of sufficient amounts of different organic groups to extend the range of achievable properties. For example, 3,3,3-trifluoropropyl groups in poly [(trifluoropropylmethyl)(dimethyl)siloxane] contribute excellent oil and fuel resistance. Inclusion of phenyl groups in poly[(phenylmethyl)(dimethyl) siloxane] results in improved heat and radiation resistance, and lowers the gas permeability of the cured elastomer.

13.3.1.3.2 Plasticizers

Trimethylsiloxy-terminated PDMS polymer, that is, bis(trimethylsiloxy)poly (dimethylsiloxane), often referred to as *silicone fluid*, is widely used as plasticizer in silicone sealants because of its excellent compatibility both with the liquid siloxane polymer as well as the cured network. The plasticizer can be selected so as to lower either the viscosity or the modulus of the formulation, or both variables simultaneously. Optimization of plasticizer viscosity and level will generally also allow formulations with increased elongation at break and tensile strength. Silicone plasticizers are available over a very wide range of molecular weights. Commercially, most important for silicone sealants and adhesives are silicone fluids with viscosities of 100 to 60,000 cP. These plasticizers have the same durability characteristics as the PDMS network, which explains the excellent environmental stability of silicone sealants compounded with silicone fluid plasticizers.

Silicone polymer plasticizers have historically been used in many formulations. Their nonreactivity means that, if improperly used, they can migrate from

TABLE 13.4

Formulation Components and Their Functions in Silicone Condensation Cure Sealants

Component	Typical Chemicals	Function
Polymer	PDOS diols HOSi~SiOH	Backbone required to form the elastomeric network
Plasticizer	Trimethylsiloxy-endblocked poly(dimethylsiloxane) (PDMS)	Adjustment of mechanical properties such as hardness, viscoelasticity, and rheology
	Paraffin oil of high boiling point, polyisobutylene, and alkylbenzenes	Organic plasticizers (diluents) may reduce formulation cost as well as provide some special ease-of-use properties
Reinforcing fillers (active)	Fumed silica (SiO_2), precipitated calcium carbonate ($CaCO_3$), and carbon black	Thixotropic reinforcing agents (nonslump), and adjustment of mechanical properties (cohesion); provide toughness to the elastomer as opposed to brittle materials
Nonreinforcing fillers (passive)	Ground calcium carbonate ($CaCO_3$)	Reduce formulation cost; adjust rheology, and mechanical properties
Crosslinkers	Acetoxy ~Si(OOCCH$_3$)$_3$ alkoxy ~Si(OR)$_3$ oxime ~Si(ON=CRR')$_3$ amine ~Si(NHR)$_3$	Crosslink polymeric component; provide network structure
Specific additives	Catalysts: organo-Sn, organo-Ti, organo-Pt, organo-Zn, and organo-Rh	Control of the rate of the curing process
	Adhesion promoters: X-CH$_2$CH$_2$-Si(OR)$_3$	Enhance the adhesion to substrates
	Water scavengers	Prolong shelf life
	Pigments	Offer a wide range of colors
	Rheology additives	Adjust ease-of-use characteristics and features
	Biocides	Provide fungus growth resistance, for example, sealing of sanitary devices and equipments

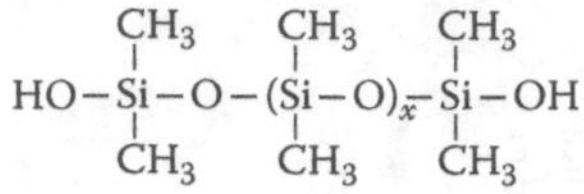

$$
\begin{array}{ccc}
CH_3 & CH_3 & CH_3 \\
| & | & | \\
HO-Si-O-(Si-O)_x-Si-OH \\
| & | & | \\
CH_3 & CH_3 & CH_3
\end{array}
$$

FIGURE 13.8   Silanol (SiOH)-terminated dimethylsiloxane polymer ($x = 300–1600$).

the sealant and stain certain substrates. Staining has been a widely publicized flaw of silicone sealants; however, the potential of a formulation to stain a substrate can be minimized or even eliminated with proper formulation. In general, this is accomplished by using nonplasticized formulations for stain-sensitive substrates. So-called reactive plasticizers may also be used to minimize the staining potential of a silicone sealant. In these plasticizers, a substantial fraction of the PDMS molecules are trimethylsiloxy-terminated on one end and silanol-terminated on the other end. These reactive fluids tie into the network during cure, forming loose polymer ends. By doing so, they internally plasticize the elastomeric network. The benefit of these plasticizers is that they cannot migrate (bleed) from the cured formulation.

13.3.1.3.3   Fillers

Silicone polymers when cured into elastomers by themselves are weak gel-like materials. For this reason, fillers must be used to provide reinforcement. The type of fillers used in silicone sealants varies widely; two of the most common fillers are fumed silica and calcium carbonate. Fumed silica, a highly reinforcing filler, is usually added in amounts ranging from 6% to 20%. Silica is usually used when a high-strength sealant is desired. Several silicas having different surface areas are available, and surface treatment with silanes may be used as well.

For lower-modulus sealants, either ground or precipitated calcium carbonate or combinations of ground and precipitated calcium carbonate are commonly used. These fillers are much less reinforcing than silica and may represent 50% by weight or more of a formulation. The surface of calcium carbonate is often treated with fatty acids, for example, stearates, to provide improved compatibility and dispersion into the silicone. By using various filler types, loadings, and filler treatments, the rheological nature of the sealant can vary from nonsag to self-leveling with low or high extrusion rates, and tensile strengths can vary from 0.35 to 7 MPa.

13.3.1.3.4   Crosslinkers

As mentioned earlier, the moisture-reactive crosslinkers used in silicone sealants are of the form $R_nSi(OR')_{4-n}$, where $n = 0$ or 1 and R may be any organic group, for example, methyl, ethyl, or vinyl. Common crosslinkers for one- and two-part systems are shown in Tables 13.2 and 13.3, respectively.

13.3.1.3.5   Adhesion Promoters

Many moisture-curing silicones have good inherent adhesion to a variety of substrates; this is especially true for the amino- and amidosilane-cured sealants. However, adhesion can be markedly improved with different combinations of silanes

[55]. The more common silane adhesion promoters are categorized as amino-functional, for example, 3-aminopropyltrimethoxysilane, 3-aminopropyltriethoxysilane, or *N*-(2-aminoethyl)-3-aminopropyl trimethoxysilane; as epoxy-functional, for example, 3-glycidoxypropyltrimethoxysilane and 2-3 (3,4-epoxycyclohexyl) ethyltrimethoxysilane; mercapto-functional, for example, 3-mercaptopropyltrimethoxysilane; or methacrylate-functional, for example, 3-methacryloxypropyltrimethoxysilane.

13.3.1.3.6   Formulation Examples

Table 13.5 provides an overview of typical silicone sealant formulations. The actual addition levels of the various ingredients depend on the cure chemistry (type of crosslinker, catalyst, and polymer) and the desired properties for the specific end use. As an example, Table 13.6 shows the formulation of a translucent, oxime-cured, LM construction sealant [56].

13.3.1.4   Applications of Silicone Sealants

Silicone sealants, primarily as one-part room-temperature-vulcanizable (RTV) products, are widely used by the construction industry for applications such as sealing expansion joints in buildings and highways; general weatherproofing joints in porous and nonporous substrates; sanitary joints around bathroom and kitchen fixtures; as well as fire-rated joints around pipes, electrical conduits, ducts, and electrical wiring within building walls and ceilings. In a variety of applications, silicone sealants also perform the functions of an adhesive, that is, they act as structural sealants. For example, silicones are used in structural glazing, where the cured sealant becomes part of the overall load-bearing design, or in insulating glass secondary seals, which structurally bond two panes of glass together.

Structural silicone glazing (SSG) is the method of bonding glass, ceramic, metal, stone, or composite panels to the frame of a building by utilizing the bond strength,

TABLE 13.5

Typical Formulations for Clear (Silica-Filled) and Pigmented (Calcium-Carbonate-Filled) One-Part Silicone Sealants (Addition Levels Are Shown in Percent by Weight)

Ingredient	Clear (SiO_2 filled)		Pigmented ($CaCO_3$ filled)	
	High Modulus	Low Modulus	High Modulus	Low Modulus
Polymer	80%–85%	55%–65%	40%–65%	30%–50%
Plasticizer		20%–25%		10%–20%
Crosslinker	5%–8%	5%–8%	5%–8%	5%–8%
Adhesion promoter	0.5%–3%	0.5%–3%	0.5%–3%	0.5%–3%
Fumed silica filler	10%–15%	8%–10%	3%–10%	0%–5%
$CaCO_3$ filler			30%–50%	30%–50%
Catalyst	0.005%–0.3%	0.005%–0.3%	0.005%–0.3%	0.005%–0.3%

TABLE 13.6
Formulation of an Oxime-Cured LM Construction Sealant

Ingredient	Chemical	Addition Level (weight%)
Polymer	SiOH-endblocked PDMS (50,000 cP)	65.9
Plasticizer	SiMe$_3$-endblocked PDMS (1,000 cP)	20
Crosslinker	Methyltris(methylethylketoximo)silane	5
Adhesion promoter	Aminopropyltriethoxysilane	1
Fumed silica filler	Fumed silica (SiO$_2$) filler (150 g/m^2 surface area)	8
Catalyst	Dibutyltin dilaurate (DBTDL)	0.1

movement capability, and durability of a silicone structural sealant. The SSG technique provides the architect with almost limitless design possibilities, while promising improved quality and performance to the building owner. A recent study, published in 1998 in France, revealed a significantly lower failure rate with SSG façades than with conventional cladding.

A remarkable extension of the structural glazing technology is the use of wet-applied RTV silicone structural sealants in the manufacture of windows. This technology, which has been recently developed in Europe for poly(vinyl chloride) (PVC) windows, provides several key benefits to window manufacturers. With the structural sealant transferring loads between the frame and the insulating glass, the strength of the IG unit contributes to the overall stability of the window. Structural window glazing thus enables a reduction in the PVC profile stiffness by using thinner PVC extrusions or elimination of steel reinforcement inserts, resulting in raw material cost savings for the window manufacturer. The simple assembly technique involving adhesive fixation of the insulating glass unit to the vinyl frame results in substantial productivity gains.

Knowledge of the performance of SSG systems has led naturally to the expansion of the use of silicone glazing in new high-performance façade systems, especially in protective, safety, and security glazing systems for either hurricane glazing or bomb-blast-resistant façades [57]. Bomb blasts are generally characterized by extremely high loads acting on the building over a short impact time; typical bomb blasts may generate impulses of up to 550 kPa · ms. Silicone sealants display a very low glass transition temperature (T_g) due to the flexibility of their siloxane polymer backbone. Because of the time-temperature-equivalence principle developed by Williams, Landel, and Ferry, a low T_g implies high flexibility of the polymer chains at high temperatures within very short time scales. The theory thus predicts the particularly good performance of silicone sealants when submitted to a sudden load condition typical of a bomb blast, hurricane, or earthquake.

The very favorable fire performance characteristics of silicones, such as very low heat release rate, insensitivity of burn rate to fire severity, low toxicity and low corrosivity of combustion products, as well as silica char formation that results in higher fire endurance, when combined with their excellent durability and longevity make them ideal elastomeric sealants for fire protection applications.

13.3.2 POLYURETHANES

Polyurethane construction sealants saw a rather slow market introduction hampered by technical problems in the late 1960s (for some of the early technologies utilized in sealant products, see References 58–60). However, with most of these problems having been resolved during the 1970s, the use of polyurethane sealants showed a dramatic increase from the mid-1980s onward. Polyurethane sealants today are among the most widely used sealant types in both new and remedial construction, second only to silicone sealants. The versatility of urethane chemistry allows the formulation of a wide range of products, which are available as flowable and nonslump materials in both single-component and multicomponent versions. Furthermore, reactive hot melt and waterborne (dispersion) products exist that utilize similar cure mechanisms as the RTV polyurethane materials. Reactive polyurethane hot melts were introduced to the market in the early 1990s, while waterborne systems, the newest type of polyurethane sealants, became available in the early 2000s.

13.3.2.1 Properties of Polyurethane Sealants

The properties of polyurethane sealants are determined by both their cure mechanism and their polymeric backbone. Unlike silicone sealants, which all utilize the same type of polymer, that is, poly(diorgano-siloxane), polyurethane sealants can be formulated with a vast variety of polymers. However, the majority of commercial polyurethane sealants are based on poly(propylene oxide) polyols, and a substantial number of specialty polyurethane sealants, for example, for insulating glass (IG) [61] use, are based on polybutadiene polyols. Polyurethane hot melts are often based on polyester polyols, while waterborne polyurethane sealants may contain dispersions of acrylic polyols.

The standard cure mechanism is via isocyanates; however, other types of urethane polymers have been developed that are endblocked with mercapto or alkoxysilyl groups and that cure via different mechanisms. Of these so-called "modified" urethane polymers, only the silicon-curable type will be discussed in detail in the separate section dealing with silicon hybrid sealants. Finally, polyurethane sealant formulations can be modified by use of substantial portions of additives, such as tar or bitumen, which have a strong influence on the overall properties. The commercial importance of polyurethane construction sealants is due to their unique combination of properties, as shown in Table 13.7.

Because of their formulation versatility, polyurethanes can be formulated with a wide range of properties that permit them to satisfy important needs in a broad variety of applications. Their most prominent features are excellent toughness, abrasion resistance, good flexibility (especially at low temperatures), and good adhesion to a wide variety of substrates. Some of the shortcomings of polyurethane sealants are poor temperature resistance (limited to about 85°C), sensitivity to moisture and UV radiation, and accurate metering and mixing requirements (two-component systems only).

TABLE 13.7
General Benefits and Limitations of Polyurethane Sealants

Benefits	Limitations
• High movement capability (up to +50%/−35%) and high elastic recovery (75%–85%)	• Low gas permeability (resulting in slow-curing one-part formulations)
• High puncture, tear, and abrasion resistance	• Poor low-temperature application properties
• Paintable	• Discoloration (yellowing) for standard (aromatic-isocyanate-based) formulations
• Low cost (especially multipart formulations)	• Should not be applied to wet or damp surfaces (poor adhesion and bubbling)
• Low gas permeability (desired barrier property)	• Limited elevated-temperature resistance
• Excellent hydrolytic stability and resistance to microbiological attack (for tar and other specially modified versions)	• Insufficient resistance of glass adhesion to UV light (not allowing stopless or structural glazing)
• Low odor	• Possible reversion (for certain formulations)
• Good cold-temperature flexibility (although modulus increases)	• Variety of formulations can cause wide differences in performance
• Good adhesion to a variety of substrates	• Retardation of alkyd paint-drying process
	• Durable products available only in pigmented versions

As mentioned earlier, the properties of polyurethane sealants depend strongly on their polymeric backbone. Poly(propylene glycol), the most commonly used polymer in polyurethane sealants, offers good hydrolysis resistance, low-temperature flexibility, and lower cost when compared to alternative polyols. Acrylate-based polyurethanes show excellent weatherability and good movement capability at ambient temperatures, but they have poor cold-temperature flexibility and poor solvent resistance. Polyester-based sealants have good fuel and weather resistance, but poor low-temperature flexibility and poor resistance to water. An exception to this is the class of polyester-polyols, which are prepared by the ring opening and subsequent polymerization of lactone rings, such as polycaprolactone, polybutyrolactone, or polyvalerolactone, and which have relatively good hydrolysis resistance as well as low viscosity and yield polyurethane sealants and adhesives of high tensile strength and elongation. In contrast to the polyester-based variety, polybutadiene-based polyurethanes have good resistance to water, but have moderate weather resistance and a limited service temperature range. Finally, polythioethers provide chemical resistance in elastomeric materials that cannot be achieved in polyether-polyurethanes.

Polyetherurethane sealants, especially the multipart versions, show high elastic recovery (75%–85%). This feature accounts for their good performance in high-movement joints, such as in building façades. Many of the commercial two-part polyetherurethane sealants are particularly noted for their puncture, tear, and abrasion resistance, which makes these products especially suitable for horizontal joints in traffic-bearing surfaces. It is this combination of high movement capability and good tear resistance that causes polyurethane sealants to be perceived as having higher performance than polysulfide sealants, although their weather resistance is still often inferior to that of polysulfides.

Polyurethane sealants are paintable with most paint types. Some retardation in the drying of alkyd paints may occur when these paints are applied on one-part polyurethane sealants within their first week of cure. Polyurethane sealants, especially the polybutadiene-based sealants, display very low moisture and low gas permeabilities, which is advantageous for some applications; however, this also causes the one-part formulations to cure rather slowly. Tar- or bitumen-modified polyetherurethane sealants show excellent hydrolytic stability and resistance to chemicals and microbiological attack. Unless formulated with a solvent, polyurethane sealants have very little odor. Polyetherurethane sealants have a low brittle point and therefore display good cold-temperature flexibility although their modulus increases substantially at lower temperatures. One of the key advantages of multipart polyetherurethane sealants is their lower cost when compared to silicone and polysulfide sealants, which largely contributed to their commercial success. Isocyanates react readily with a variety of other functional groups, including those present on most substrates, thus obtaining strong adhesion bonds.

Most polyurethane sealants are difficult to extrude at lower temperatures; however, this property is somewhat formulation dependent. The majority of polyurethane sealants used in construction are based on aromatic isocyanates, which make them prone to yellowing when exposed to sunlight. As with most other sealants, polyurethane sealants should not be applied on wet or humid substrates, since residual water negatively affects adhesion buildup to the substrate. However, with polyurethane sealants, water also reacts with the isocyanate curing agent, causing formation of carbon dioxide and bubbling of the sealant.

Polyurethane sealants are far less thermally stable than silicones; their stability is comparable to that of polysulfide sealants. Some polyetherurethane sealants are claimed to have heat stabilities up to 135°C, but typical polyurethane sealants have a maximum stated use temperature of 70°C–85°C. Tremendous improvements in the resistance of their glass adhesion to UV light have been made since the commercialization of the first polyurethane sealant formulations, allowing their use in less demanding residential IG applications. However, the glass adhesion is not sufficiently stable for polyurethanes to be used in stopless or structural glazing applications.

With the use of proper stabilizing additives, polyether-based polyurethanes generally provide good weathering properties. However, some polyurethane sealant formulations have reverted to liquid materials in the field when exposed to high levels of heat and humidity [62]. Exposure to UV light has been shown to further accelerate this degradation process [63]. As mentioned earlier, the formulation versatility associated with polyurethane chemistry has resulted in a vast variety of formulations with different performances. While this has allowed polyurethane sealants to penetrate a wide variety of market segments, it has also caused some uncertainty in meeting the expectations of sealant users and building owners with regard to the performance of generically specified polyurethane sealants [64].

$$R-N=C=O + R'-OH \rightleftharpoons R-\overset{H}{\underset{|}{N}}-\overset{O}{\underset{||}{C}}-O-R'$$

FIGURE 13.9 Reaction of isocyanate with alcohol to form urethane.

13.3.2.2 Chemistry of Polyurethane Sealants

The key reaction on which polyurethane chemistry is based was discovered more than 150 years ago. In 1849, Wurtz reacted isocyanates with alcohols to give esters of carbamic acid, which were named *urethanes* (see Figure 13.9).

Then, in 1937, Otto Bayer discovered the polyaddition reaction of polyisocyanates with polyols that led to the production of high-molecular-weight polyurethanes. All polyurethane sealants cure by the reaction of a hydrogen donor with a polyisocyanate, generally in the presence of a catalyst. In two-part systems, the hydrogen donor is a polyol; in one-part systems, it is water or, for blocked one-part systems, an amine. Since the reaction of a diol with a diisocyanate leads only to an extension of the polymer chain, formulations must contain triols or even higher functional polyols in order to cure into an elastomeric material.

The toughness of cured polyurethane sealants is a result of the separation of the polymer matrix into two phases: a continuous phase of soft segments contributed by the polyol and a dispersed phase of hard segments formed by the isocyanate blocks. These two phases are segregated into domains, the dispersed hard-segment blocks being largely held together by covalent and hydrogen bonds. The segregation into distinctive domains provides an internal reinforcement to polyurethane polymers.

13.3.2.2.1 One-Component Cure Systems

In one-part cure systems, an isocyanate-functional prepolymer is prepared by reacting a polyol and a polyisocyanate (see Figure 13.10). In the reaction shown, the molar ratio of isocyanate (NCO) to hydroxyl (OH) is 2:1 in order to produce a fully isocyanate-endblocked polymer. If the molar NCO to OH ratio had been less than 1:1, a fraction of the polymers would have been hydroxyl terminated.

In most prepolymers, the polyols will consist of both diols and triols, the latter to produce the crosslinks in the cured elastomer. The ratio of diols to triols (D/T) has a substantial effect on both the rheology of the wet product and the cured properties of the elastomer.

Once the one-part sealant is exposed to ambient moisture, the free isocyanate reacts in a two-step reaction process. The first reaction of the isocyanate with water forms an amine and carbon dioxide. Since amines readily react with isocyanates, the intermediate amine quickly reacts further with available free isocyanate groups, extending the chain via urea bonds (see Figure 13.11). If the average functionality of the polyol mix is >2, an elastomer is formed.

$$O=C=N{\sim}R{\sim}N=C=O + HO{\sim}R'{\sim}OH \rightleftharpoons O=C=N{\sim}R{\sim}\overset{H}{\underset{|}{N}}-\overset{O}{\underset{||}{C}}-O{\sim}R'{\sim}O-\overset{O}{\underset{||}{C}}-\overset{H}{\underset{|}{N}}{\sim}R{\sim}N=C=O$$

FIGURE 13.10 Endblocking reaction in the preparation of the isocyanate-functional polymer.

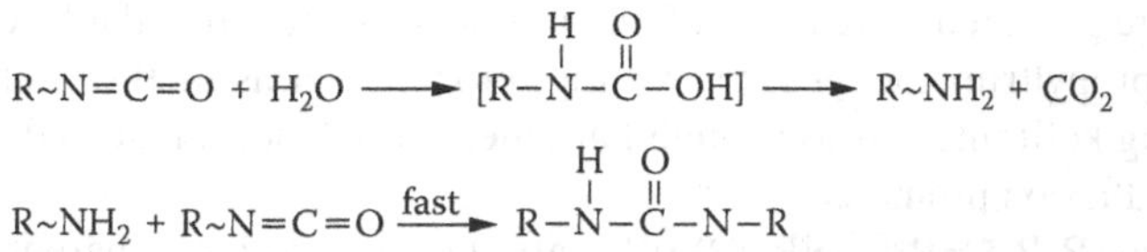

$$R\sim N=C=O + H_2O \longrightarrow [R-\overset{\overset{\displaystyle H}{|}}{N}-\overset{\overset{\displaystyle O}{\|}}{C}-OH] \longrightarrow R\sim NH_2 + CO_2$$

$$R\sim NH_2 + R\sim N=C=O \xrightarrow{\text{fast}} R-\overset{\overset{\displaystyle H}{|}}{N}-\overset{\overset{\displaystyle O}{\|}}{C}-N-R$$

FIGURE 13.11 One-part polyurethane cure reaction.

An obvious disadvantage of the one-part cure chemistry shown in Figure 13.11 is the formation of carbon dioxide (CO_2) gas. In a one-part sealant based on moisture-curing NCO-terminated prepolymer, it is impossible to have a liquid medium free of CO_2 gas. Much of the CO_2 gas is solubilized in the medium, and as long as the rate of diffusion of the gas from the sealant to the ambient air exceeds its rate of formation, no bubbles are formed. However, since most sealants are filled with some particulate fillers, these filler particles provide nucleation centers where bubble formation may occur. Since all fillers carry some level of adsorbed moisture on the surface of the particles, almost invariably certain environmental conditions exist where bubble formation occurs, particularly at the site of the reaction between isocyanate and adsorbed water on the filler particles. Frequently, bubbling occurs when the freshly applied, supersaturated sealant is struck by sunlight. The subsequent temperature rise results in an increase in the partial pressure of the dissolved CO_2, starting the bubbling process.

Thus, additional mechanisms need to be employed to prevent bubble formation. In general, the approach pursued is to provide an alternative reaction partner for the isocyanate that is liberated by ambient moisture, but it must react faster with the isocyanate than water does. Such reaction partners are called *latent hardeners*. The majority of commercial one-part polyurethane products today utilize a cure chemistry based on the reaction of isocyanates with amines, which are released when the latent hardeners are hydrolyzed. As shown in Figure 13.11, amines react very readily with isocyanates.

Ketimines are one type of latent hardener, which releases polyamines upon exposure to moisture (see Figure 13.12). Ketimines are prepared according to the reverse reaction by refluxing ketones and diamines in an azeotroping solvent.

However, mixtures of prepolymers with ketimines generally show poor shelf stability; therefore, further blocking mechanisms are often employed. One such method is to block the prepolymer with phenol, resulting in the formation of urethane groups. Upon exposures to moisture, the diamine produced by hydrolysis of the diketimine displaces the phenol and cures the product. An interesting variation of this approach, which also provides one-part polyurethane compositions with improved adhesion, is the use of alkoxysilyl-ketimines as latent hardeners. These

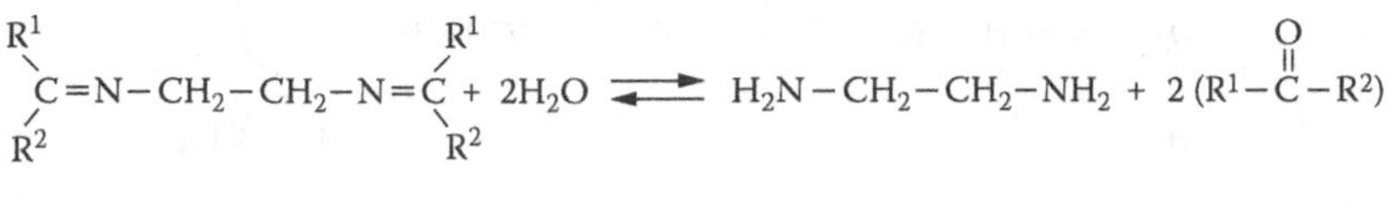

$$\overset{R^1}{\underset{R^2}{\diagdown}}C=N-CH_2-CH_2-N=\overset{R^1}{\underset{R^2}{\diagup}}C + 2H_2O \rightleftharpoons H_2N-CH_2-CH_2-NH_2 + 2\,(R^1-\overset{\overset{\displaystyle O}{\|}}{C}-R^2)$$

FIGURE 13.12 Formation of diamine by hydrolysis of diketimine.

hardeners are prepared by reaction of an aminoorgano-functional alkoxysilane, such as γ-aminopropyltrimethoxysilane, with a ketone, such as methyl isobutyl ketone. The resulting ketimine acts as a latent hardener for the isocyanate and, additionally, also as an adhesion promoter.

Enamines ($R^1R^2C=R^3C–NR^4R^5$) offer another path to latent hardeners for polyurethane sealants. Reaction of an enamine with water yields a ketone (usually cyclohexanone) and a secondary amine, which cures the isocyanate prepolymer. Preparation of the enamine is by reverse reaction using azeotropic distillation.

Another one-part cure chemistry, which has seen widespread use, is based on oxazolidines as latent hardeners. Upon exposure to water, the oxazolidine produces an aminoalcohol, which cures the isocyanate-functional prepolymer. As with ketimines, the hydrolysis reaction is the reverse of the reaction by which oxazolidines are formed (see Figure 13.13).

The most commonly utilized oxazolidine-cure system is based on a bisoxazolidinyl material [65]. It is formed by first reacting a ketone or aldehyde with a bis(hydroxyalkyl) amine and then reacting the N-hydroxyalkyl-oxazolidine with a diisocyanate, typically the prepolymer or hexamethylenediisocyanate, to form a urethane-bisoxazolidine derivative (Figure 13.14).

Attempts have been made to accelerate the speed of cure by increasing the rate of opening (activation) of the oxazolidine ring by changing the substituents. An interesting approach is based on the reaction of the N-hydroxyalkyl-oxazolidine with an alkoxysilane or organoalkoxysilane to form oxazolidinesilyl ethers [66]. Use of these materials allows better control of the ring-opening rate, results in better shelf stability, and reduces the risk of foaming during shelf life and cure, since the alkoxysilane also functions as a chemical drier (water scavenger) for the formulation.

Both the urethane endblocking and cure reactions can be catalyzed by suitable additives. Generally, tertiary amines, Sn (II), Sn (IV), Bi (III), and Sb (III)

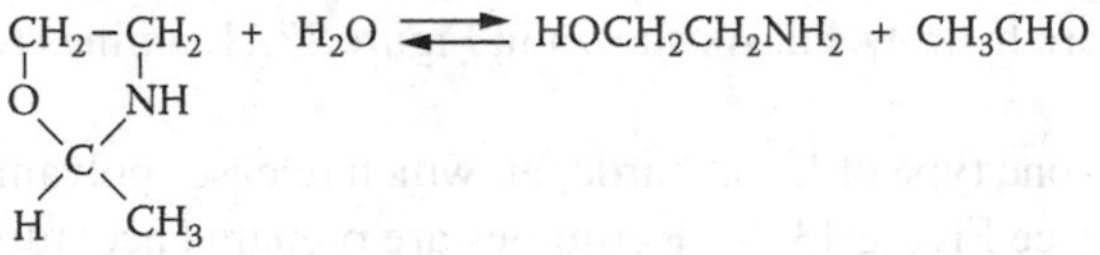

$$CH_2-CH_2 + H_2O \rightleftharpoons HOCH_2CH_2NH_2 + CH_3CHO$$

FIGURE 13.13   Formation of aminoalcohol by hydrolysis of diketimine.

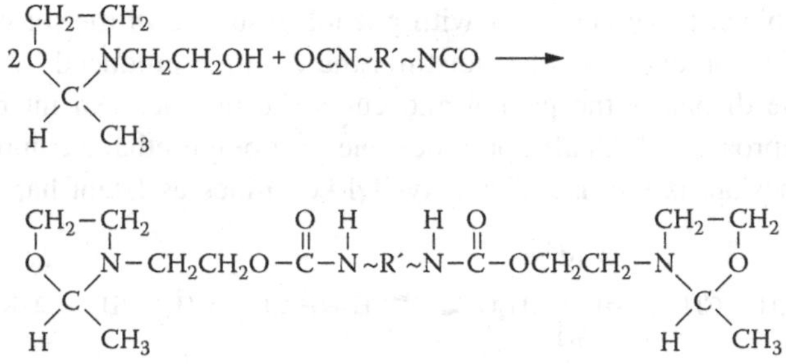

FIGURE 13.14   Formation of bisoxazolidine.

compounds are used as catalysts. Dibutyltin dilaurate (DBTDL) is often the catalyst of choice, since it provides a good balance between shelf life and catalysis of both the prepolymer reaction and the moisture cure reaction.

13.3.2.2.2 Two-Component Cure Systems

In a two-component formulation, the diisocyanate and additives are packaged separately from the polyols and other additives. Immediately before application, the two parts are mixed; for optimal properties, the two components must be kept dry, and the stoichiometric ratios must be precisely controlled.

The cure reaction of a two-part polyurethane is between a polyisocyanate and a polyol. Mixtures of different polyols and isocyanates are often used to achieve the physical properties needed in the final cured form. The polyisocyanate can be in the form of the original monomer, as adduct or oligomer of the monomer, or as a prepolymer. Generally, use of the prepolymer is preferred for handling and toxicity reasons. Since no carbon dioxide is generated during the cure reaction, bubbling is not an issue with two-part polyurethanes unless they are applied to damp or wet substrates. Tertiary amines, especially Sn (IV) (DBTDL) compounds, are used to catalyze the two-part urethane reaction. In addition to the primary reaction of the polyisocyanate with a polyol, up to four side reactions may be occurring, and some of them may be desirable for product performance:

1. Reaction of isocyanate with an amine to form a urea: This reaction occurs mostly as a side reaction unless amine compounds are added for a specific function.
2. Reaction of three isocyanates to form an isocyanurate: This reaction increases the chemical resistance of the cured material.
3. Reaction of isocyanate with a urethane to form an allophanate: This reaction increases the cure speed, increases the thixotropy, and improves adhesion to some substrates.
4. Reaction of isocyanate with a urea to form a biuret: This is primarily an undesirable side reaction.

13.3.2.3 Polyurethane Formulations

Table 13.8 provides an overview of the different formulation ingredients and their functions in polyurethane sealants.

13.3.2.3.1 Polymers

Generally, poly(oxypropylene) polyols are used because of their low viscosity, low cost, versatility, and acceptable properties. Poly(tetramethylene ether glycol) (PTMEG) polyols are premium types of polyether polyols that display excellent physical properties. Polyethers are available in a wide range of molecular weights, from about 400 to greater than 6000 Da. Diols and triols are most common, but higher functionalities are also available. Hydrophobic butadiene-based polyols are used when the formulator wishes to extend the sealant with other hydrophobic materials such as mineral oil and petroleum residues (hydrocarbons) or when low

TABLE 13.8
Formulation Components and Their Functions in Polyurethane Sealants

Component	Typical Chemicals	Function
Polymer	Organic diols and triols with ~R_2COH endblocks that react with the isocyanate to form prepolymers	Backbone required to form the elastomeric network
Endblocker	Isocyanates (MDI, PMDI, TDI, XDI, and HDI)	Chain extension and crosslinking of the polymer
Latent hardener (in one-part systems)	Ketamines, enamines, oxazolidines, and silyl-modified versions of the latent hardeners	Blocked reaction partner and catalyst (released by reaction with ambient moisture) required to achieve good shelf life and reaction rate in one-part cure systems
Physical or chemical drier	Zeolites, silica gel, para-toluene sulfonyl isocyanate, and alkoxysilanes	Elimination of moisture by adsorption or chemical reaction (to improve shelf life and reduce potential for bubbling)
Solvent (mainly in one-part systems)	Xylene, heptane, and acetone	Viscosity reduction to increase extrusion rate
Plasticizer	Phthalates (DIDP, DINP, BBP, TCP, etc.), adipates (DOA), aromatic sulfonic esters, and N-(2-hydroxy propyl) benzene sulfonamide	Adjustment of mechanical properties such as hardness, viscoelasticity, and rheology
Reinforcing fillers (active)	Treated fumed silica (SiO_2), talc, precipitated calcium carbonate, and carbon black	Thixotropic reinforcing agents (nonslump), adjustment of mechanical properties (cohesion); provide toughness to the elastomer as opposed to brittle materials
Nonreinforcing fillers (passive)	Ground calcium carbonate	Reduce formulation cost; adjust rheology and mechanical properties
Stabilizers	Antioxidants, UV absorbers, and hindered amine light stabilizer (HALS)	Protect polymeric backbone from degradation by photooxidation caused by visible and UV components of solar radiation
Specific additives	Catalysts: tertiary amines, organo-Sn (IV), and Bi (III) compounds	Control of the rate of the curing process
	Adhesion promoter: $X-CH_2CH_2CH_2-Si(OR)_3$	Enhance adhesion to substrates
	Pigments	Offer a wide range of colors
	Rheology additive	Adjust ease-of-use characteristics and features
	Biocides	Provide fungus growth resistance, for example, sealing of sanitary devices and equipments

moisture and gas permeabilities are desired. Polyester polyols are used when higher thermal stability, solvent resistance, abrasion resistance, and better appearance are required; the disadvantages are medium-to-high cost and high viscosity. The newly developed reactive urethane hot melt adhesives are usually based on polyester polyols because of their high crystallinity or high glass transition temperature in the case of amorphous polyester.

13.3.2.3.2 Endblockers

Common diisocyanates employed are liquid toluenediisocyanate (TDI), which is commercially available as a mixture of 2,4 and 2,6 isomers, and 4,4-methylenediphenyldiisocyanate (MDI). While TDI has the lowest cost of any polyisocyanate and provides good physical properties to the cured sealant, toxicity concerns are leading toward increased use of MDI. MDI is available in three broad types of compositions:

* Monomeric MDI (MMDI)
* Modified MDI
* Polymeric MDI (PMDI)

MMDI is a purified material distilled from a polymeric MDI mixture. It consists of over 97% 4,4'-MDI with small amounts of 2,4'-MDI and traces of the 2,2'-isomer. It is a solid with a melting point of about 38.5°C. MDI is relatively expensive, tends to crystallize and dimerize upon storage [67] and, as a solid, is more difficult to handle. Modified MDI is a liquid, made by converting some of the isocyanate groups into carbodiimide groups, which then react with the excess isocyanate. Other types of liquid MDI (also called *prepolymers*) are made by the reaction of the diisocyanate with small amounts of glycols (see Reference 68 and the prior art discussed in this patent). PMDI, also called *polymethylene polyphenyl isocyanate* (PMPPI), is a brownish liquid. PMDI is becoming increasingly important since it is a liquid, is easy to handle, and has a lower cost than the pure MDI product. PMDI is a crude product whose exact composition varies. The main constituents are 40%–60% 4,4'-MDI, the remainder being other isomers of MDI (2,4'-, 2,2'-), trimeric species, and higher-molecular-weight oligomers. Sealants made from MDI prepolymers usually display lower performance qualities than those made from TDI prepolymers. The aromatic diisocyanates are most commonly used (chiefly TDI and MDI), since the aromatic types give more rapid cures and higher tensile strengths compared with the aliphatic diisocyanates. Aliphatic isocyanates—hexamethylene-1,6-diisocyanate (HDI) or isophorone diisocyanate (IPDI)—offer better light stability, but they are substantially more expensive. HDI is highly toxic and, therefore, is being increasingly phased out.

13.3.2.3.3 Latent Hardeners

As mentioned earlier, ketamines, enamines, and oxazolidines are the most commonly used latent hardeners in one-part polyurethane formulations.

13.3.2.3.4 Physical or Chemical Driers (Dehydrating Agents, Water Scavengers)

Water is a big problem with all polyurethane sealants. Polyols are hydroscopic and often require drying prior to prepolymer preparation. Fillers that are added to the sealant formulation contain adsorbed water. Much of this water in the sealant formulation can be removed during compounding by mixing the heated sealant composition under high vacuum. During this process, referred to in the industry as *cooking,* free isocyanate reacts with free water to form carbon dioxide, which can be stripped off with the vacuum. However, any water that remains in the formulation has a negative effect on shelf life and increases the potential for bubbling upon application of the sealant. Therefore, physical or chemical driers are added to the formulation. Physical driers, such as molecular sieves or silica gel, work by adsorbing the water. Chemical driers, such as para-toluene sulfonyl isocyanate (PTSI) or alkoxysilanes, act as water scavengers. The PTSI readily reacts with water to form carbon dioxide that can be removed during compounding. Alkoxysilanes convert water to alcohols, which are captured by further reactions with isocyanates.

13.3.2.3.5 Solvents

Solvents, such as xylene, heptane, or acetone, are used primarily in one-part products to adjust viscosity, improve flow characteristics, and increase extrusion rate. While in the past polyurethane sealant formulations contained up to 15% (by weight) solvents, nowadays the solvent content is minimized to reduce the amount of volatile organic content (VOC). The trend is toward solvent-free products. High-solid or nonsolvent-type formulations can be attained by using high levels of plasticizers or by applying sealants at elevated temperatures to reduce viscosity. High levels of plasticizers, however, can cause staining of the adjacent substrates.

13.3.2.3.6 Plasticizers

A wide variety of plasticizers exist for polyurethane sealant formulations that are selected on the basis of their solubility parameter to maximize compatibility with the specific urethane polymeric backbone.

13.3.2.3.7 Fillers

The fillers used in polyurethane formulations are similar to those used in silicones; calcium carbonate, talc, clays, and silica are among the most common. Because of the undesirable reaction of isocyanates with water, fillers used in polyurethane formulations must be dry.

13.3.2.3.8 Adhesion Promoters

Polyurethane sealants have good inherent adhesion to most substrates, but silane adhesion promoters are often used to improve this adhesion [55]. Epoxy-, amino-, and mercapto-functional silanes are the most commonly used ones because of their dual reactive nature. The alkoxysilane end can react with surface hydroxyls; the epoxy, amino, or mercapto end reacts with the isocyanate.

13.3.2.3.9 UV Stabilizers

A wide range of commercial antioxidants, UV absorbers, and hindered amine light stabilizers (HALS) are used in both one- and two-part polyurethane sealant formulations to improve the resistance against heat, UV radiation, and visible light. However, in one-part formulations, these additives are often ineffective since these stabilizers contain active hydrogen in their molecules, which is likely to react with the isocyanate-functional prepolymer. This reaction may be slow, and sometimes it takes place only over extended periods during the shelf life of the sealant. However, when stabilizers get attached to the main polymer matrix, they become immobilized and no longer provide the required protection. Consequently, some one-part polyurethane sealant formulations lose their stabilization properties during extended shelf life; that is, the same sealant product may display different degrees of environmental resistance at the beginning and end of its shelf life.

13.3.2.3.10 Formulation Examples

Table 13.9 provides an overview of typical formulation ranges for plasticized polyurethane construction sealant formulations [69]. Table 13.10 provides a specific example of an unplasticized, fast-curing, single-component, gun-grade polyurethane sealant formulation suitable for use in construction applications [70].

13.3.2.4 Applications of Polyurethane Sealants

Because of their tear and abrasion resistance, polyurethane sealants are widely used to seal horizontal joints in traffic-bearing surfaces. Typical examples include sealing of expansion, control, and perimeter joints in parking decks, pavements, plazas, malls, patios, driveways, interior wood or concrete floors in factory or institutional buildings, or any other areas subject to foot and light vehicle traffic. Special grades that are designed for heavy traffic areas such as seals in highway joints, industrial

TABLE 13.9

Formulation Ranges for Plasticized Polyurethane Construction Sealants

Ingredient	Addition Level (weight%)
Diisocyanate prepolymer	30–35
Filler	30–45
Plasticizer	15–35
UV absorber	1–3
Antioxidant	1–2
Dehydrating agent	1–3
Adhesion promoters	1–3
Thixotrope	2–3
Solvent	3–5
Pigment paste	2–3

TABLE 13.10

Unplasticized, Fast-curing, Single-Component, Gun-Grade Polyurethane Construction Sealant Formulation

Ingredient	Chemical	Addition Level (weight%)
Prepolymer	NCO prepolymer, equivalent weight 1355, fast curing	53.6
Reinforcing filler	Carbon black	3.6
UV stabilizer	Titanium dioxide	2.4
Filler	Calcium carbonate	33.3
Reinforcing filler	Silica	6.0
Solvent	Toluene	1.2

building floors, or warehouse floors exposed to abrasion by cars, trucks, or forklifts exist. Polyurethane grades that resist exposure to solvents, cleaning detergents, lubricants, and a broad range of other chemicals also exist. These sealants may be used in floor joints in chemical laboratories, show rooms, airplane hangars, and textile and other industrial plants.

Many polyurethane sealants are readily paintable after completion of their cure and are therefore used in applications that require frequent repainting of the substrate. Examples are seals at house sidings, baseboards, trims, log cabins, mobile homes, as well as at certain interior joints such as the top of non-load-bearing walls, etc.

Polyurethane sealants are also frequently used for sealing movement and control joints in concrete and natural stone façades. Note that on these substrates, polyurethane sealants containing high levels of plasticizers can sometimes cause bleeding and staining, just as with some silicone sealants. As with silicones, suitability for a given application must be tested with each specific sealant product. However, with polyurethane sealants, each color of each sealant should be tested since considerable differences exist between formulations of differently pigmented products. Testing each color is especially necessary in assessing the weatherability of polyurethanes, since some pigments, typically titanium dioxide, zinc oxide, and carbon black, have good UV-blocking qualities. Therefore, sealants of other colors not containing these pigments may have widely different durabilities. For applications requiring a nonyellowing quality, an aliphatic-isocyanate-based polyurethane sealant must be specified.

Two-part polybutadiene-based polyurethane sealants are increasingly used in the less demanding residential insulating glass (IG) production segment, replacing the higher-priced polysulfide sealants. However, IG units made with these sealants cannot be used in structural glazing or other applications where the edge seal of the unit is openly exposed to sunlight. Furthermore, since polyurethane sealants are based on a stoichiometric addition cure system, the mix ratio between the two components must be closely controlled in order to maintain good cured properties of the sealant.

13.3.3 POLYSULFIDES

Building sealants based on liquid polysulfide polymers came to the market in the late 1940s. They were the first high-performance elastic sealants to be commercialized. Their market introduction was due to the advent of curtain wall construction in the United States, which created the need for high-performance sealants. The previously used mastics were incapable of accommodating movements between the building elements. The use of polysulfide sealant in 1954 to reglaze Lever House office building in New York signaled the acceptance of this sealant type for the modern curtain wall. Being the right material available at the right time, polysulfide sealants enjoyed a tremendous rise in popularity during the 1950s and 1960s. However, with the commercialization of acrylic, silicone, and polyurethane sealants, they started to lose market share in the 1970s. Today, polysulfides are still widely used in IG manufacture and in various high-value civil engineering applications. The unique properties resulting from polysulfide chemistry allow the formulation of highly specialized products, which are available as flowable and nonslump materials in both single-component and multicomponent versions.

13.3.3.1 Properties of Polysulfide Sealants

The properties of polysulfide sealants are determined primarily by their polymeric backbone, then by the type of curing agent, and finally by their formulation. Similar to silicone sealants, the majority of commercially available polysulfide sealants are based on one common mercapto-endblocked poly(bis(ethyleneoxy)methanedisulfide) backbone. Various versions of hybrid polysulfides have been developed over the years, involving nonaliphatic polymeric backbone structures and nonmercaptan end groups on the polymer; however, these types of sealants represent a relatively small share of the construction market and will not be discussed here. The standard cure mechanism for polysulfide sealants is oxidation of the mercapto end groups resulting in the formation of disulfide links. The commercial importance of polyurethane construction sealants is based on their unique combination of properties, as shown in Table 13.11.

The unique backbone structure of the polysulfide polymers accounts for most of the performance characteristics of the formulated sealant products. Sulfur is the obvious distinguishing feature of commercial liquid polysulfide polymers, since it accounts for about 37% of the total molecular weight of the polymer. The high sulfur content determines the polymer's solubility parameter and imparts resistance to swell in hydrocarbon fuels and organic solvents and good resistance to many chemicals, such as dilute acids and bases. The solubility parameter of bis(ethyleneoxy)-methanedisulfide-based polysulfide sealants is between 10 and 11 $MPa^{1/2}$ and, thus, their resistance to aliphatic hydrocarbons, aviation fuels, and motor oils is not surprising. Resistance to swell in aromatic hydrocarbons, alcohols, esters, ethers, and ketones is very good, especially when compared to other elastic sealants.

Because of their relatively high surface energy, polysulfides are paintable with most paints. However, when painted with waterborne acrylic paints, plasticizer migration from the sealants to the paint often occurs, making the paint film tacky.

TABLE 13.11

General Benefits and Limitations of Polysulfide Dealants

Benefits	Limitations
• Excellent resistance to swell in hydrocarbon fuels and solvents	• Low gas permeability (resulting in slow-curing one-part formulations)
• High resistance to many chemicals	• High stress relaxation (compression set)
• Paintable	• Limited elevated temperature resistance
• Reasonable low-temperature flexibility (although modulus increases)	• Should not be applied to wet or damp surfaces (poor adhesion)
• Low gas permeability (desired barrier property)	• Cure is highly temperature dependent
• Excellent hydrolytic stability and resistance to microbiological attack	• Insufficient resistance of glass adhesion to UV light (not allowing stopless or structural glazing)
• Good weatherability	• Persistent odor
• High tolerance to mix ratio variations due to nonstoichiometric cure chemistry (two part)	• Potential paint compatibility issue (plasticizer migration) with acrylic paints
• Good adhesion to a variety of substrates	• Available only in pigmented versions
• High stress relaxation	

However, this effect is formulation dependent and can be overcome by a suitable choice of plasticizer.

The absence of tertiary carbon atoms in the polymer backbone reduces the susceptibility of polysulfide sealants to attacks by oxygen and UV radiation. Well-formulated polysulfide sealants, therefore, show good weatherability. Many buildings sealed 25–30 years ago still have the same polysulfide sealant in place and are performing well. However, over the past 50 years of their use, the average polymer content of polysulfide sealants has fallen significantly. A sample of sealant taken from a building sealed in 1952 was shown to have a polymer content of 60% (by weight). In contrast, polysulfide sealants used today in the manufacture of IG units have polymer content in the range of 22%–32%, and for building sealants the range is 19%–30% by weight. The durability of these products will almost certainly be less as a result, despite the fact that properly formulated compounds with low polymer content pass standardized certification tests [71].

The formal ($-O-CH_2-O-$) group in the bis(ethyleneoxy)methanedisulfide segment gives the polymer low-temperature flexibility. The glass transition temperature (T_g) of the pure polymer is $-59°C$. When formulated into a sealant product, the T_g value increases depending on the polymer content and the compounding ingredients used, but generally falls in the range of $-20°C$ to $-30°C$ for most commercial polysulfide sealants.

Polysulfide sealants have low permeability to water vapor, solvent and fuel vapors, and gases, once again a consequence of the polymer structure. While these are beneficial properties for many applications, the low moisture permeability also causes one-part polysulfide sealants to cure very slowly in depth once a skin has formed on their surface. The thin skin developed at the start of the moisture-induced cure in one-part sealants can tear if extended. Furthermore, cyclic movements can strongly deform slow-curing one-part sealants, causing cohesive failure.

The polysulfide polymer structure is resistant to hydrolysis; due to this reason, polysulfide sealants exhibit excellent durability in applications involving long-term water immersion. Polysulfide sealants with high polymer content can be formulated for excellent microbiological resistance using certain curatives (MnO_2) and plasticizers (chlorinated paraffin) [72].

Since the cure induced in two-part polysulfide sealants by most of the inorganic oxides and peroxides can be classed as heterogeneous, maintaining an exact stoichiometric ratio between the two parts while mixing is unnecessary. To achieve the required cure rate, the curative is always used in excess and small variations on either side of this ratio are inconsequential to the rate of cure or the development of properties. This is a great advantage over the addition cure system utilized in polyurethanes, which requires close maintenance of the mix ratio.

With the use of adhesion promoters, polysulfide sealants acquire strong adhesion to many substrates. However, the UV stability of glass adhesion, while generally being better than that of polyurethane sealants, is not sufficient to allow use of polysulfides in applications such as structural glazing, in which the glass–sealant interface is directly exposed to sunlight.

At temperatures above 50°C, cured polysulfide sealants do not fully recover from applied stress, exhibiting tensile or compression set. This effect is due to the thermal lability of the disulfide groups in the polymer backbone, leading to stress relaxation and the observed set. Whether chemical stress relaxation is considered an advantage or disadvantage depends on the specific sealant application. If a joint undergoes essentially a single major movement, for instance due to building settlement, a sealant incapable of chemical stress relaxation is placed under permanent stress, generally leading to joint failure. Polysulfide sealants perform rather well in these conditions, since the stress rapidly declines due to relaxation. However, in other joints undergoing rapid cyclic movements, such as in curtain walls, a polysulfide sealant creases and folds owing to its elastoplastic nature. This can be aesthetically displeasing and may lead to cohesive or interfacial failure. The same rearrangement of disulfide bonds that cause poor elastic recovery also results in poor creep resistance. Because of their lack of creep resistance and insufficient UV stability, polysulfides cannot be used as structural sealants.

The upper service temperature for polysulfide sealants is about 90°C, but this is influenced by the formulation and, in particular, the curing system. As with most other RTV sealants, polysulfide sealants should not be applied on wet or damp surfaces, since poor adhesion to the substrate may result. The cure rate of polysulfide sealants is highly temperature dependent. At temperatures below –25°C, cure will almost come to a standstill. This effect has been utilized in the past by shipping premixed two-component polysulfide sealants to building sites in refrigerated trucks. Once warmed up, these sealants were applied like a one-part material, but cured rapidly because of their two-part chemistry. At elevated temperatures, typically above 40°C, the cure rate of polysulfide sealants may become so high as to cause problems, especially with automatic application equipment. Finally, polysulfide sealants have a persistent odor caused by cyclic sulfur-containing products formed in the

$$4 \text{ HS}\sim\text{R}\sim\text{SH} + O_2 \longrightarrow 2 \text{ HS}\sim\text{R}\sim\text{SS}\sim\text{R}\sim\text{SH} + 2 H_2O$$

FIGURE 13.15 Oxidative cure of polysulfide sealants (shown here for the mercapto-functional chain ends).

condensation polymerization process used in the manufacture of the polymers. The odor is noticeable both in the uncured sealant as well as the cured elastomer, a fact that is often disliked.

13.3.3.2 Chemistry of Polysulfide Sealants

Most commercial sealants based on liquid polysulfide polymers cure by oxidation of the terminal mercapto groups, as shown in Figure 13.15, to form disulfide bonds. Early workers had examined the cure of polysulfides with manganese dioxide (MnO_2) [73] and nonmetallic oxidizing agents [74], and it had been proposed that cure proceeded via a free radical mechanism [75]. A more recent work [76] suggested a complicated multistage cure mechanism for polysulfides via the formation of thioyl radicals, thiolate anions, and radical anions, forming disulfide linkages. Isocyanates and epoxides readily react with mercapto groups and, therefore, form the basis of alternative cure chemistries; however, these will not be discussed in this chapter.

Since the oxidation of mercapto groups leads only to an extension of the polymer chain, polymers do contain a certain controlled amount of branching (functionality > 2) to ensure proper cure of an elastomeric material. Table 13.12 lists the most commonly used oxidants for polysulfide construction sealants.

The source of oxygen in the oxidative cure of mercaptan-terminated polymers is either an inorganic oxide or peroxide, or an organic hydroperoxide. The first one is the

TABLE 13.12
Commonly Used Curatives (Oxidants) for Polysulfide Construction Sealants

Curative (Oxidant)	Accelerator	Retarder
Lead peroxide (PbO_2)	Sulfur	Stearic acid and lead stearate
Manganese dioxide (MnO_2)	Tetramethylthiuram disulfide (TMTD), tetramethyl guanidine (TMG), and water	
Sodium perborate ($Na BO_3 \cdot H_2O$)	TMG, di-o-tolyl guanidine (DOTG), barium oxide (BaO), and sodium hydroxide (NaOH)	
Potassium permanganate ($KMnO_4$)	Water and bases	
Calcium peroxide (CaO_2)	BaO	
Barium peroxide (BaO_2)	Calcium oxide (CaO)	
Cumene hydroperoxide	Bases	

main choice in the construction sealants industry. All oxidative cures of polysulfide polymers require water for initiation and are base-catalyzed. Inorganic oxides and peroxides are insoluble in the polysulfide sealant compound and, hence, the cure can be classed as heterogeneous. Robustness of the sealant cure, despite variations in the mix ratio, is the key technological advantage resulting from heterogeneous cure chemistry. The disadvantage is that the activity of the oxide depends on particle size, surface area, activation, and the degree of dispersion. While sealant manufacturers conduct analytical characterization of the oxides, analytical parameters obtained on the oxide powder do not always equate to reproducible activity in the polysulfide sealant.

The choice of the curing system controls a number of factors, such as cure rate, heat resistance, resistance to bacterial attack, or simply color, as some of the oxides and peroxides are strongly coloring. The use of manganese dioxide in two-part polysulfide sealants provides improved high-temperature stability along with better resistance to UV light under glass; these two improvements were the reason for selecting this system for all IG sealants. However, manganese-dioxide-cured sealants do absorb water and tend to swell. Lead peroxide has been widely used in the past for two-part sealants as it gives fast cure and its pot life can be controlled by stearic acid or stearate retarders. The use of lead peroxide is becoming less common nowadays because of environmental concerns; its use has been phased out entirely in Western Europe (during the 1990s) and Japan (at the end of 2002).

During the transition phase, two-part polysulfide sealants, especially those designed for IG applications, often utilized a combined cure system based primarily on manganese oxide with small amounts of lead peroxide as a "starter" to initiate rapid cure after a relatively flat induction period. The use of potassium permanganate is limited by its strong inherent color. Although curing with organic hydroperoxides (and peroxides) is possible, it has to be controlled to avoid chain scission and reversion. Cumene hydroperoxide has been used commercially to produce white two-part building sealants. Sodium perborate is used in both one-and two-part formulations. Unlike the cure induced by other oxide curatives, the reaction of mercapto groups with sodium perborate is stoichiometric, indicating that the perborate is active in an aqueous solution and the mercaptan oxidation occurs either in the aqueous phase or at the interface between the aqueous phase and sealant medium. By chance, it was noticed that sealants cured with sodium perborate were not susceptible to fungal attack. Cumene hydroperoxide, calcium peroxide, barium peroxide, and sodium perborate are white and, therefore, they are used for the formulation of white or light-colored sealants.

13.3.3.2.1 One-Component Cure Systems

Because water is required for the initiation of all oxidative cures, it is possible to halt the cure by removing the water from a composition through desiccation. This concept forms the basis for the design of all one-component polysulfide sealants. On extrusion, the sealant absorbs water, initiating the oxidative cure. Often, desiccants such as calcium oxide or barium oxide are used, which become highly basic by reaction with water, thus further accelerating the cure. Calcium peroxide [77] and sodium perborate [78] are the preferred curatives for one-part sealants since they are very shelf stable in an anhydrous environment. Bases that can be introduced into the formulation as an anhydrous powder, typically dispersed in a dry plasticizer, are used as accelerators for one-part cure systems. The preferred accelerators are barium oxide and calcium oxide; sodium hydroxide is also often used with the perborate-cure system.

13.3.3.2.2 Two-Component Cure Systems

Manganese dioxide [79] and sodium perborate are the preferred curatives for two-part formulations. Organic bases, such as tetramethylthiuram disulfide (TMTD), di-o-tolyl guanidine (DOTG), and tetramethyl guanidine (TMG), are used as accelerators. To further accelerate cure, water is added to the base component of two-part products. To ensure uniform distribution during mixing, water is sometimes first adsorbed on a superabsorber fiber prior to its incorporation into the sealant formulation.

13.3.3.3 Polysulfide Formulations

Table 13.13 provides an overview of the different formulation ingredients and their functions in polysulfide sealants.

13.3.3.3.1 Polymers

Commercial polysulfide polymers are produced in aqueous dispersion by the reaction between alkyl chlorides and sodium polysulfide [80]. Over 95% of all commercial liquid polysulfide polymers are now made with 2,2′-dichlorodiethoxymethane (also known as *bis-2-chloroethylformal*) as chain extender and 1,2,3-trichloropropane as initiator of chain branching [81]. The chemistry underlying the manufacture of liquid polysulfides is uniquely versatile, allowing the manufacture of high-molecular-weight polymers with narrow, near-Gaussian molecular-weight distribution (low polydispersity). However, the technology also has some disadvantages. First, it is difficult to produce polymers with terminal groups other than mercaptan. Second, the process uses large amounts of water, and the by-products, carried in the wastewater, are toxic and environmentally unpleasant. Thus, considerable cost goes into the treatment of the wastewater and the proper disposal of the hazardous by-products. The liquid, low-molecular-weight polymers used in polysulfide sealants are then produced by reductively cleaving the high-molecular-weight polymer using a mixture of sodium hydrogen sulfide (NaSH) and sodium sulfite (Na_2SO_3) [82].

The commercial liquid polysulfide polymers range in molecular weight from 1000 to 8000 Da. The degree of branching is expressed as the mole percentage of trichloropropane used in the polymerization process. This ranges from 0% to 2%, although the actual level of incorporation of trichloropropane is lower due to its hydrolytic instability. The level of branching influences primarily the modulus of the cured product. Figure 13.16 shows a simplified representation of the structure of the liquid polysulfide polymers.

In reality, the structure of a polysulfide polymer is somewhat random in that the C_2H_4O and CH_2O units are not evenly spaced as suggested by the aforementioned idealized molecular formula. In fact, Mahon [83] in her work describes 11 variants with 4 subvariants.

Polysulfide polymers can be easily modified by introduction of other hydrocarbon segments. For example, hydrocarbon segments can be introduced by the use of

TABLE 13.13

Formulation Components and Their Functions in Polysulfide Sealants

Component	Typical Chemicals	Function
Polymer	Partially branched, mercapto-endblocked bis(ethyleneoxy)-methanedisulfide-based polymer	Backbone required to form the elastomeric network
Curatives	PbO_2, MnO_2, $Na\ BO_3 \cdot H_2O$, $KMnO_4$, CaO_2, BaO_2, and cumene hydroperoxide	Oxidation agents that cure the polymer by oxidation of the terminal mercapto groups to form disulfide bonds
Physical or chemical drier (one-part systems)	Zeolites, CaO, and BaO	Elimination of moisture by adsorption or chemical reaction (to halt cure during the shelf life of one-part products)
Solvent (pourable and self-leveling products)	Xylene and heptane	Viscosity reduction to increase flowability
Plasticizer	Chlorinated paraffins and phthalate esters	Adjustment of mechanical properties such as hardness, viscoelasticity, and rheology
Reinforcing fillers (active)	Precipitated calcium carbonate, silica, and carbon black	Act as thixotropic reinforcing agents (nonslump), adjustment of mechanical properties (cohesion), providing toughness to the elastomer as opposed to brittle materials
Nonreinforcing fillers (passive)	Ground calcium carbonate	Reduce formulation cost; adjust rheology and mechanical properties
Specific additives	Catalysts: bases, that is, inorganic oxides (convertible to hydroxides), hydroxides, TMTD, and guanidines	Control of the rate of the curing process
	Adhesion promoter: $X\text{-}CH_2CH_2CH_2\text{-}Si(OR)_3$	Enhance adhesion to substrates
	Pigments	Offer a wide range of colors
	Rheology additive	Adjust ease-of-use characteristics and features
	Biocides	Provide microbial resistance

$$HS-(CH_2CH_2OCH_2OCH_2CH_2SS)_x-CH_2CH-CH_2-(SS-CH_2CH_2OCH_2OCH_2CH_2)_y-SH$$
$$(SS-CH_2CH_2OCH_2OCH_2CH_2)_z-SH$$

FIGURE 13.16   Structure of liquid polysulfide polymers.

dibromohexane, dibromodecane, and dichloro-*p*-xylene as comonomers. All polymers based on these comonomers exhibit improved thermal resistance. Elastomers based on the dibromodecane-modified polymer also have a moisture transmission rate that is reduced by 40% compared to standard polysulfide elastomers [84].

For some applications, a higher moisture vapor transmission rate is advantageous, for example, in one-part sealants with moisture-initiated cure. This can be achieved by incorporation of polar monomers into the polymer backbone. A polysulfide based exclusively on the monomer triglycol dichloride has a water permeability 25% higher than conventional commercial polysulfides [84].

13.3.3.3.2 Curatives

Calcium peroxide and sodium perborate are the preferred curatives for one-part sealants, while manganese dioxide and sodium perborate are the most important curatives for two-part formulations.

13.3.3.3.3 Physical or Chemical Driers (Dehydrating Agents, Water Scavengers)

Zeolites, calcium oxide, and barium oxide are used as dehydrating agents in one-part formulations to halt cure during the product's shelf life.

13.3.3.3.4 Solvents

As the liquid polysulfide polymers are quite viscous and their viscosity increases even further by the use of fillers in the formulation, it is difficult to make pourable or self-leveling sealants. It is not unusual for the formulator to resort to the use of solvents, such as xylene, toluene, and heptane; however, then shrinkage of the cured product may become a problem.

13.3.3.3.5 Plasticizers

Only a limited range of plasticizers is compatible with polysulfides because the solubility parameter of an aliphatic polymer is high. Nearly all the early polysulfide sealants used chlorinated biphenyls. These plasticizers were cheap, miscible up to high levels, and imparted good durability to the sealant. However, in the 1970s, they were banned as carcinogens. Because of their good durability, polysulfide sealants containing polychlorinated biphenyls (PCBs) are still installed on joints in buildings today. The resealing of joints containing PCB-plasticized polysulfide sealants requires special handling and environmental cleanup methods (see References 85 and 86).

Chlorinated paraffins with chlorine content greater than about 40% were found to be quite good alternatives and were also relatively low priced [87]. In recent times, however, the chlorinated paraffins have also been subjected to scrutiny as possible carcinogens, but the current European guidelines continue to permit the use of most types of short-chain chlorinated paraffins in polysulfide sealants [88,89]. Alternative plasticizers that have been found to be suitable include di-isoundecyl phthalate (DIUP), polymeric plasticizers, certain phosphate plasticizers, and butylbenzyl phthalate (BBP), although the last is reported to be less efficacious. The key problem encountered with alternative plasticizers is that they are more prone to bleeding from the sealant product, thus affecting both the sealant and the substrate. This is indeed the case with BBP in certain sealant products. Low-volatility phthalate esters are used particularly in sealants used for IG production. To avoid phthalate plasticizers, which may possibly cause endocrinal effects in humans and animals, alternative plasticizers are currently being sought. For example, the use of benzoate plasticizers in two-part polysulfide sealants has been disclosed [90].

13.3.3.3.6 Fillers
The fillers used in polysulfide formulations are similar to those used in silicones; calcium carbonate, talc, clays, and silica are among the most commonly used ones. Fumed silicas give good reinforcement, but they also cause a marked increase in the viscosity of the formulation. Small amounts of silica are used for rheology control. Fillers used in one-part formulations must have low water content. The pH of the ingredients used in the formulation can impact the cure rate of polysulfide sealants either positively or negatively. Bases accelerate the cure, while acids decrease the cure rate. Small-particle-size calcium carbonate is often coated with organic surface-treating agents, such as stearic acid, to improve the dispersing properties, making the incorporation of the filler into the formulation easier during compounding. However, such an acidic treating agent can slow down the cure rate.

13.3.3.3.7 Adhesion Promoters
As polysulfide sealants do not have good inherent adhesion, suitable adhesion promoters must be used in the formulation. The most widely used industrial adhesion promoters are organoalkoxysilanes, which are very effective in polysulfide sealants, especially when adhesion to glass is important. As is the case with urethane sealants, silanes with a dual-reactive nature are typically used. Examples of such silanes are mercapto- and epoxy-functional silanes. Organic titanates may also be used [91].

13.3.3.3.8 Formulation Examples
Table 13.14 provides an overview of typical formulations of plasticized polysulfide construction sealants [92]. Table 13.15 provides a specific formulation of a one-part gun-grade polysulfide sealant [93].

13.3.3.4 Applications of Polysulfide Sealants
The low moisture and vapor permeability of polysulfide sealants at ambient conditions was the key factor in the initial selection of this sealant type by the IG industry in the late 1960s. Today, polysulfide sealants are still widely used for both residential and commercial IG applications. Polysulfide sealants, however, cannot be used for structural glazing, butt jointing, and other applications in which the edge seal of the IG unit is openly exposed to sunlight. IG manufacture is still, by far, the largest application for polysulfide sealants in construction, despite the turmoil that was created in 2001 by the sole U.S. manufacturer and key global supplier of polysulfide polymers exiting the market because of concerns over environmental cleanup costs and the poor economic performance of these products. The scarcity of polysulfide polymers then provided an opportunity for other sealants, such as polyurethanes and silicones, to enter the supply-critical IG market.

TABLE 13.14

Overview of Typical Plasticized Polysulfide Construction Sealant Formulations

Component	One-Part Sealant (weight%)	Building Sealant (weight%)	Insulating Glass Sealant (weight%)
Polysulfide polymer	20	35	30
Fillers	50	40	50
Plasticizers	25	20	15
Adhesion promoter	2	2	2
Curing agents	3	3	3

TABLE 13.15

One-Part Gun-Grade Polysulfide Sealant Formulation

Ingredient	Weight%
Polysulfide liquid polymer	50
Epoxidized soy oil	4
Pyrogenic (fumed) silica	2
Calcium carbonate filler (surface treated)	5
Titanium dioxide (white pigment and filler)	22
Dehydrated lime	2
Synthetic zeolite (moisture scavenger)	2
Calcium peroxide (curing agent)	4
Phthalate plasticizer	2
Aminopropyltriethoxysilane	1
Toluene	5

Over the years, polysulfide sealants have maintained and expanded their share in various segments of the civil engineering market. Because of their excellent water resistance, polysulfides are widely used in the construction of canals, reservoirs, dams, and other water-retaining civil engineering structures. A good example is the irrigation canal system in California, which was built in the 1960s and runs for several hundred miles. These canals were sealed with polysulfide sealants and have remained virtually maintenance free over the years.

The excellent resistance of some polysulfide sealant formulations to water and microbial attack are key factors in their use in wastewater treatment plants. A study showed that sealants with high polymer content (>40%) formulated with manganese dioxide as curative and chlorinated paraffin as plasticizer remained unattacked in the activated sludge channel of a sewage treatment plant for 11 years [72]. The type of plasticizer appears to be critical for bacterial resistance, as other polysulfide products included in the same study degraded rapidly by reversion. The authors postulated that other plasticizers were degraded by bacterial attack, producing acidic by-products capable of cleaving the polysulfide backbone.

In recent years, the protection of groundwater and other environmental concerns have sparked the emergence of a new class of polysulfide sealants that combine durability, strength, and chemical resistance to gasoline, diesel, and engine oils. These products were first specified in the Netherlands, where the water table is very near to the surface. Dutch, and in the meantime also German, legislation requires that all gasoline service stations be sealed to prevent fuel, particularly diesel, from seeping into the groundwater. Several technologies were evaluated, but the combination of heavy-duty impermeable paving with a special grade of polysulfide sealant performed best. Currently, only specially formulated polysulfide sealants meet the Dutch Environmental Authority's (KIWI) specification. Similar work has been initiated in the United Kingdom, and it is likely that the European Union will adopt a pan-European specification based on the original KIWI standard. Furthermore, the technology has been extended to all applications where the water table is under threat from chemicals, such as chemical storage tanks, fuel spill secondary containment areas in airports and tank farms, chemical plants, refineries, etc.

13.3.4 POLYACRYLATES

There are two principal classes of polyacrylate (acrylic) sealants: waterborne latex acrylics and solvent-borne acrylics. The solvent-borne products are based on medium-molecular-weight polymers dissolved in an organic solvent; in waterborne acrylics, high-molecular-weight polymers are dispersed in water. Cure proceeds nonchemically in both cases by evaporation of the solvent or water as the dispersing medium present in the compound.

The first solvent-borne acrylic caulks appeared in the construction market in the late 1950s. The mid-1960s saw the commercialization of the first acrylic latex polymer for waterborne caulks. These early products were low-performance plasticized pigmented caulks used primarily as gap fillers in interior applications. The acrylic latex products were valued both by professionals and do-it-yourselfers for their low price, low toxicity, and easy application.

During the 1970s, sales of latex acrylics grew rapidly, while solvent-borne acrylics experienced only modest sales growth. Over the same time period, the performance of acrylic latex sealants, especially their water resistance, continued to improve. This allowed the use of these sealants in certain exterior applications that placed low demands on their movement capability. The "internally plasticized" caulks, based on acrylic polymers with low T_g, became available during the same decade [94]. Despite the substantial earlier technological advances, by the early 1980s the performance of acrylic latex sealants had reached a plateau and some people already felt that the end of their life cycle was in sight.

A number of key innovations occurred during the 1980s that drastically changed this perception. First, highly plasticized products were introduced in the market in the mid-1980s. These so-called "plasticized" acrylic latex sealants enjoyed a dramatic rise in sales. The term *plasticized* acrylic sealant was somewhat misleading, since the first-generation acrylic latex sealants were already plasticized products. The key difference between the old and the new generations of acrylic latex sealants was the level of plasticizer used. While the old generation had a polymer-to-plasticizer ratio from 3:1 to 4:1 based on solid acrylic polymer content, the new generation was formulated with a much higher plasticizer ratio, close to 2:1. This resulted in very flexible sealants with movement capability of ±25% and the ability to meet most of the performance requirements of high-performance sealant specifications, such as ASTM C 920. The second important innovation was based on the use of silane adhesion promoters [95] or small amounts of silicone fluids [96] to impart better water resistance to the adhesion of these sealants. The silane or silicone portion typically represented far less than 2% of the total formulation; however, in a clever marketing ploy, these sealants were advertised as "siliconized" acrylics, projecting the image of silicone performance combined with the ease-of-use attributes of latex acrylics. Finally, transparent (clear) and translucent acrylic latex products were introduced [97], first based on polymer blends and later on single-polymer formulations, which became an important part of the market. With the increased exterior use of acrylic sealants, resistance of the uncured sealant to rainfall (wash out) became an important consideration. To meet this need, by the end of the 1980s, the first wash-out resistant formulations were launched for exterior use [98].

Today, cutting-edge formulations of clear, transparent, and pigmented acrylic latex sealants are capable of meeting the stringent performance requirements of architectural sealant specifications. Siliconized acrylic latex formulations now represent the vast majority of acrylic sealant sales. Internally plasticized acrylic sealant formulations are available that meet or exceed the requirements of ASTM C 920 without the need for a high level of external plasticizer that could cause staining. Solvent-borne acrylics are still used in a limited number of niche applications; however, because of concerns over their flammability and toxicity, they are increasingly being replaced by other product types.

13.3.4.1 Properties of Acrylic Sealants

The properties of acrylic sealants are determined by their polymeric backbone, the type of vehicle (i.e., water- versus solvent-borne systems), and their formulation. Acrylic polymers used for commercial sealants are tailor-made copolymers of acrylic and/or methacrylic esters and other monomers. Generally, the alkyl acrylate or methacrylate represents the predominant component of the copolymer, with the comonomers being present in small amounts. However, a rather large choice of monomeric modifiers exists, and the properties of the final acrylic copolymer critically depend on the type of modifier as well as its level and means of incorporation during the copolymerization process.

Acrylic monomers are extremely versatile building blocks, and polymers having a variety of properties that can be adjusted to meet specific market needs are available. Key polymer properties that influence the performance of the formulated sealant are T_g, molecular weight, and—for latex polymers—minimum film formation temperature, solids content, and surfactant level and type. The vehicle employed to deliver the acrylic polymer—solvent or water—places certain restrictions on the acrylic polymer and determines some types of formulation additives. Finally, the basic type of the formulation—unfilled (clear sealants) or filled (translucent or pigmented sealants)—and the specifics of the formulation, especially the types and levels of plasticizer and filler, have a dramatic effect on the performance of the formulated sealant. In low-end caulks, the acrylic latex polymer may be blended with lower-cost poly(vinyl acetate) polymer; the blend ratio then has a strong effect on the overall performance.

The ultimate performance of acrylic sealants depends on their actual formulation. Sealant products can be formulated over a broad spectrum of qualities ranging from low- to high-end performance. The variation in the performance of the various acrylic sealants, especially for the latex sealant type, is vast and, thus, there are only a few typical properties. Table 13.16 shows typical characteristics of solvent-borne thermoplastic acrylic construction sealants.

Solvent-borne acrylic sealants develop excellent unprimed adhesion to a wide variety of surfaces. They place very few demands on substrate preparation and cleaning. Often, good adhesion is achieved even on difficult surfaces, such as oil-contaminated metals, because the solvent contained in the product allows the sealant

---

TABLE 13.16

General Benefits and Limitations of Solvent-Borne Thermoplastic Acrylic Sealants

Benefits	Limitations
• Excellent unprimed adhesion to almost all substrates and ability to gain adhesion on oil- or bitumen-contaminated substrates	• Noxious odor
	• High volatile organic content (VOC)
	• Flammability
• Excellent UV resistance and weatherability	• High shrinkage (solvent evaporation)
• One-part product	• Toxicity (solvent and acrylic monomer content)
• Paintable	
• Higher polymer solids content (~80%) than latex acrylics	• Extrusion rate is highly temperature dependent
• High stress relaxation	• Limited elevated-temperature resistance (thermoplasticity)
• Available in translucent and pigmented versions	• Very poor elastic recovery, poor movement capability, and poor low-temperature flexibility
	• Residual surface tack (thermoplasticity)
	• Cannot be used as a structural sealant (thermoplasticity)
	• Potentially poor mold resistance (plasticizer dependent)
	• Limited solvent resistance
	• Should not be applied to wet or damp surfaces (poor adhesion) or on sensitive plastic substrates that are attacked by the solvent

to wet out the surface by dissolving the contamination and incorporating it into the sealant. Solvent-borne acrylics have excellent weathering characteristics, resist UV light and ozone deterioration, and are color stable. Typically, longevity of the sealed joint is not limited by the weatherability of the product but by its limited movement capability. Excessive joint movement then results in creases and folds and, often, in cohesive failure. Solvent-borne acrylics are one-part products and do not require any mixing on the job site. Generally, paintability with alkyd paints is good. Plasticizer migration into acrylic paints has been observed with some sealant products, depending on the nature of the plasticizer used. Solvent-polymerized acrylic polymer has a solid content of approximately 80%, which is higher than that for acrylic latex polymer. However, the solvent content of the sealant still gives rise to considerable shrinkage (25%–30%) upon evaporation. Due to their plastic deformation behavior, solvent-borne sealants display very high stress relaxation, which makes them ideally suited for joints undergoing only unidirectional movement.

Solvent-borne acrylic sealants have a strong odor resulting from the solvent and monomer content of the formulation and, therefore, they are generally not used in closed rooms or occupied buildings. Toluene, xylene, and ethyl benzene, typically found in solvent-borne acrylics, pose flammability, environmental (VOC), and health risks associated with chronic exposure. Flash points may be as low as 4°C–12°C. VOC typically ranges from 300 to 400 g/L, the value being highest for clear formulations. Solvent-borne acrylics may emit trace amounts of ethyl acrylate, acrylic acid, and acrylonitrile, which are suspected carcinogens. Some companies ban or restrict the use of products containing acrylic acid monomer.

A number of limitations result from the thermoplasticity of solvent-borne acrylics. For instance, the extrusion rate is highly temperature dependent. Solvent-borne acrylics cannot be applied below 5°C–10°C and must be heated at the job site for application in cold climates. Manufacturers typically quote the upper service temperature limit as being 80°C, although considerable softening occurs already at temperatures above 40°C. Solvent-borne thermoplastic acrylics have poor elastic recovery and cold-temperature flexibility and, therefore, very low movement capability; a typical value quoted is ±10%, although this may be rather optimistic, considering the substantial increase in modulus experienced by many solvent-borne acrylics at low temperatures. Surface skinning of the sealant after application may take up to 36 h and, typically, some residual surface tack remains. Because of thermoplasticity, this surface tack increases with higher service temperatures, resulting in considerable dirt pickup and discoloration of lighter-colored sealants. Obviously, with their high stress relaxation and thermoplasticity, solvent-borne acrylics cannot be used as load-bearing structural sealants. Finally, depending on the plasticizer used, solvent-borne acrylics may show a high susceptibility to mold growth. In contrast, the typical characteristics of waterborne latex acrylic construction sealants are shown in Table 13.17.

Acrylic latex sealants are one-part products and thus do not require any mixing prior to application. The success of latex acrylic sealants has been built primarily on their waterborne nature, which results in such attractive characteristics as easy application and tooling, water cleanup, and adhesion to damp substrates. The last property allows exterior application of the sealant after rain or under

TABLE 13.17
General Benefits and Limitations of Waterborne Latex Acrylic Sealants

Benefits	Limitations
• One-part product	• Requires protection from freezing during
• Ease of application	storage
• Water cleanup	• High shrinkage (water evaporation)
• Can be applied on damp substrates	• Cure is highly climate dependent
• Low toxicity	• Should not be used below 5°C
• Low VOC	• Sensitive to early joint movement
• Low odor	• Limited elevated-temperature resistance
• Paintable	(thermoplasticity)
• Excellent UV resistance and weatherability	• Poor mold resistance (plasticizer dependent)
• Fair-to-good adhesion to commonly used	• Limited water resistance (redispersible)
construction substrates	• Hardening with aging (formulation
• High-end products meet architectural	dependent)
performance specifications based on	• Poor-to-fair low-temperature flexibility
excellent flexibility, high elastic recovery,	(formulation dependent)
and movement capability of up to ±25%	
• Moderate-to-high stress relaxation	
• Clear and translucent formulations, when	
applied, are opaque and whitish	
• Relatively low cost	

damp conditions. As waterborne sealants show a lack of wet tack, that is, they are not messy or sticky when applied, they are easy to handle and tool. Acrylic latex polymers have inherently low toxicity, because the high-molecular-weight polymer contains only very small amounts of monomers, and the cure is nonchemical in nature. For the same reasons, and because of their solvent-free nature, these products have low VOC and low odor. When applied under moderate climatic conditions, acrylic latex sealants are typically tack free in less than 1 h, and acrylic paints may be applied shortly afterward. Coating the sealant with an alkyd paint film restricts the evaporation of water. Generally, it is recommended to wait for 3 days before painting the sealant with an alkyd paint. Regardless of the paint type, the paint films are readily accepted and held. However, with some plasticizers, exudation and migration of the plasticizer into acrylic paints may occur, resulting in glossy surfaces with increased dirt pickup.

Excellent weatherability is another key strength of acrylic latex sealants, and it is utilized in exterior applications [99]. Acrylic latex sealants adhere well to the most commonly used porous (absorbing) building materials, often without the use of a primer. They also show a fairly good adhesion to metals, poly(vinyl chloride), and painted substrates. High-end products meet architectural performance specifications because of excellent flexibility, high elastic recovery, and movement capability of up to ±25%. Sealants can be formulated to show moderate-to-high stress relaxation. When applied, clear and translucent formulations are opaque and whitish. Depending on joint dimensions and climatic conditions, it takes 1–2 weeks of cure for them to attain their final appearance. This initial opaqueness is generally viewed as an advantage. As the material whitens, it is easy to see whether the sealant has been correctly applied and to clean it up if necessary. Finally, compared with other sealant types, acrylic latex sealants are relatively less expensive.

While many acrylic latex products are claimed to have freeze/thaw resistance, typically for 3–5 cycles down to −18°C, most manufacturers recommend protection of the sealant from freezing during storage. The waterborne nature of this sealant type also results in its greatest limitation: shrinkage. It is virtually impossible to formulate high-performance acrylic latex sealants with shrinkage lower than 20%–25% for pigmented versions and 35%–40% for transparent and clear versions. Low-performance pigmented products may show shrinkage as low as 15%–20% due to higher filler content. However, transparent and clear low-end products may shrink as much as 50%–55%. Shrinkage is generally viewed as negative, as it places a greater stress on the sealed joint. However, at least to some extent, this stress is relieved by the relaxation occurring in the cured sealant. Products with up to 40% shrinkage perform well in the field; if the formulation allows sufficient stress relaxation, no, or only limited, hardening occurs with aging, and the joint experiences limited movement. High-end (pigmented) products with 20%–30% shrinkage are expected to perform better, even with higher exposure to movement.

As the cure proceeds by evaporation of water, these sealants are highly sensitive to the climatic conditions, especially temperature and humidity. Acrylic latex products should not be applied at temperatures below 5°C. In cold, foggy climates, cure proceeds extremely slowly, making the product vulnerable to external influences such as joint movement, rainfall, dirt deposition, etc. Some acrylic latex products appear to be sensitive to early joint movements in a "semigelled" state, when some, but not all, water has evaporated and the viscosity of the paste has substantially increased. The upper service temperature of acrylic latex sealants is generally quoted as 85°C–95°C. As with the solvent-borne type, acrylic latex sealants may show a high susceptibility to mold growth, depending on the plasticizer, thickener, and other formulation components used. Because the cure is physical in nature, acrylic latex sealants are to a certain extent redispersible and should not be immersed in water for prolonged periods. Finally, field experience shows that many acrylic latex sealants tend to harden with aging. Sometimes, aging is dramatic, and cohesive or interfacial joint failure occurs within a few years. This hardening is caused almost exclusively by plasticizer loss and can be avoided either by using internally plasticized polymers or by skillful formulation techniques based on compatible low-volatile plasticizers.

13.3.4.2 Chemistry of Polyacrylate Sealants

With thermoplastic acrylic polymers, cure proceeds nonchemically by evaporation of the solvent or water as the dispersing medium present in the compound. The physical setting process in solvent-borne acrylics is determined simply by the speed of evaporation of the solvent. In contrast, physical setting of acrylic latex sealants is more complex; it first involves the evaporation of water as the dispersing medium, followed by the coalescence of the acrylic polymer micelles. The physical properties of such sealants depend on the physical entanglements and secondary bonding forces between long polymer chains.

13.3.4.3 Polyacrylate Formulations

Table 13.18 provides an overview of the different formulation ingredients and their functions in acrylic sealants. Polymer, plasticizer, filler, thixotropic additive or thickener, pigment, adhesion promoter, solvent or dispersing medium, and surface-drying regulator are typically common to both solvent-borne and latex acrylic formulations. Further additives are required to achieve a stable waterborne latex acrylic dispersion, such as surfactant, freeze/thaw stabilizer, primary and secondary dispersants, defoamer, and an aqueous base.

13.3.4.3.1 Polymers

The ability to manufacture a high-molecular-weight acrylic polymer by the solution polymerization process is limited by the polymer solution viscosity, which increases rapidly with increasing molecular weight. In contrast, the viscosity of an acrylic latex is not influenced by molecular weight. Thus, very-high-molecular-weight polymers can be attained by an emulsion polymerization process. The molecular weights of typical solvent-polymerized acrylic polymers are in the 80,000–120,000 Da range, while acrylic latex polymers generally have molecular weights greater than 1,000,000 Da. High-molecular-weight latex acrylic polymers are prepared by emulsion polymerization of alkyl esters of acrylic acid. Monomer, water, surfactants, and an initiator are mixed and polymerized until the acrylic monomer is depleted. The maximum level of solids for the acrylic latex polymer is approximately 63%, while in solution polymerization a solids content of approximately 83% can be attained. Table 13.19 provides a comparison between the characteristics of acrylic latex and solution polymers, while Table 13.20 shows the glass transition temperature of typical acrylic homopolymers.

Commercial acrylic polymers are made from a variety of monomers, primarily those having ethyl, butyl, and 2-ethylhexyl ester groups. In addition to these "nonfunctional" monomers, comonomers are used to modify the polymers. The two most common methods by which acrylic polymers are modified are by using high glass-transition-temperature (T_g) monomers or monomers with adhesion-enhancing moieties or both. Copolymerization with high-T_g monomers, listed in Table 13.21, will not only raise surface hardness and increase the modulus, but can also dramatically improve the hydrophobicity and elasticity of the polymer chain. Copolymers of monomers, such as acrylonitrile, that impart an undesired yellowing to the sealant are not useful in a clear product, but they are used in pigmented products.

The use of a number of polar monomers to improve adhesion is widely known and practiced. Table 13.22 shows such monomers. The acidic monomers perform an additional function in that they help to stabilize the emulsion particles during and after polymerization. Since acidic monomers have such high T_g values, even small amounts can substantially change the polymer T_g, which, in turn, will change the low-temperature performance of the sealant and may also increase its modulus.

Typically, the T_g of acrylic polymers useful for sealant formulations falls within the range from about −50°C to about +10°C. The correct polymer choice in terms of its T_g is critical to obtaining sealants having the desired flexibility and limited

TABLE 13.18
Formulation Components and Their Functions in Acrylic Sealants

Component	Typical Chemicals	Function
Polymer	Copolymers of acrylate esters and other comonomers	Backbone required to form the plastomeric (solvent-borne acrylics) or elastomeric (latex acrylics) network
Plasticizer	Benzylbutyl phthalate, dipropylene glycol dibenzoate, propylene glycol alkyl phenyl ether, and N-n-butylbenzene sulfonamide	Enhances flexibility of acrylic sealant, lowers glass transition temperature, and improves mechanical properties
Filler (active)	Calcium carbonate, talc, and silica	Aids in adjustment of rheology (body, slump control), surface tack, and mechanical properties (cohesion); provides toughness to the polymer; and increases solids content of formulation (reduces shrinkage)
Thixotropic additive and thickener	Silica, vitreous aluminosilicate fibers, alkali-swellable poly(acrylic acid), hydroxyethyl cellulose, and poly(vinyl alcohol)	Aids in attaining slump resistance and improves "body" (resistance felt during tooling) of the formulation, resulting in better handling and tooling properties
Pigment	Rutile TiO_2 (whitening agent) and other pigments	Allows adjustment of sealant color
Adhesion promoters	Epoxy-, amino-, and ureidoorgano alkoxy silanes, X-$CH_2CH_2CH_2$-$Si(OR)_3$; preferred are γ-glycidoxypropyl trimethoxysilane (waterborne acrylics) and γ-methacryloxypropyl trimethoxy silane (solvent-borne acrylics)	Enhance the wet adhesion on nonporous substrates (metals, glass, and glazed ceramic tile)
Solvent/dispersing medium	Xylene, toluene (solvent-borne acrylic), and water (latex acrylic)	Allows adjustment of solids content of formulation (rheology)

Surface-drying regulator	Mineral spirit (acrylic latex), and methanol (solvent-borne acrylic)	Retards skin formation of acrylic latex to allow adequate tooling properties, and accelerates skin formation of solvent-borne acrylics to reduce dirt pickup and allow early paintability
Surfactant (latex acrylic)	Anionic or nonionic surfactants, for example, sodium lauryl sulfate and alkylaryl polyethers	Acts as wetting agent as well as emulsifier in aqueous media, which stabilizes the polymer and filler dispersion, improves mechanical stability of dispersion, lowers sealant consistency, and enhances shelf stability
Freeze/thaw stabilizer (latex acrylic)	Ethylene glycol and propylene glycol	Provides freeze/thaw stability to the dispersion
Dispersant (latex acrylic)	Potassium tripoly-phosphates (primary dispersant) and acrylic polymer (secondary dispersant)	Deflocculates filler and pigment particles, which is essential in the preparation of a homogeneous and creamy low-consistency latex sealant with good package stability
Defoamer (latex acrylic)	Mixtures of mineral oil, silica derivatives, and organic particles; mixtures of siloxanes and organic particles or silica or both	Prevents excessive air entrapment during compounding
Aqueous base (latex acrylic)	NH_4OH	Allows pH adjustment— pH optimization is especially important in silane-modified formulations to ensure good shelf stability
Other additives	Biocides, for example, isothiazoline derivatives	Provide microbial resistance (in can as well as in use)

TABLE 13.19

Characteristics of Latex and Solution Acrylic Polymers Used in Sealant Formulations

Characteristic	Latex Polymer	Solution Polymer
Appearance	Opaque	Clear
Typical molecular weight (Da)	1,000,000	100,000
Viscosity (cP)	100–500	50,000–200,000
Weight-solids (%)	55–63	83
Particle size (μm)	0.1–0.5	Homogeneous solution
Mechanism of film formation	Coalescence/sintering of closely packed latex particles	Increased concentration of polymer by solvent evaporation

TABLE 13.20

Glass Transition Temperatures of Various Acrylic Homopolymers

Monomer	Glass Transition Temperature (°C)
Methyl acrylate	+9
Ethyl acrylate	−22
i-Butyl acrylate	−43
n-Butyl acrylate	−54
2-Ethylhexyl acrylate	−68
Lauryl acrylate	−74

TABLE 13.21

High T_g Monomeric Modifiers

Monomer	T_g of Homopolymer (°C)	Characteristics
Acrylonitrile	97	Solvent resistance and toughness
Methyl methacrylate	105	Durability
Styrene	100	Hydrophobicity
t-Butyl acrylate	45	Durability and hydrophobicity
Butyl methacrylate	20	Good general performance
Vinyl chloride	81	Inexpensive

TABLE 13.22

Monomers with Polar Functionality to Aid Adhesion

Monomer	T_g of Homopolymer (°C)
Hydroxyethyl acrylate	−15
Hydroxyethyl methacrylate	55
Acrylamide	165
N-methylol acrylamide	—
Acrylic acid	106
Methacrylic acid	228
Itaconic acid	—
Allyl glycidyl ether	—

surface tack, and this choice depends on the formulation with regard to filler loading as well as plasticizer type and level. Plasticizer-free pigmented sealants are formulated from very low T_g polymers. Higher-T_g polymers are appropriate for clear sealants as well as for pigmented sealants that contain a plasticizer.

13.3.4.3.2 Plasticizer

Acrylic polymers, regardless of whether they are latex or solvent borne, can be formulated into clear or pigmented sealants with or without external plasticizers, as needed to meet performance requirements. Most pigmented acrylic latex sealants on the market today are formulated with a supplementary external plasticizer. This is because the use of external plasticizers is a more cost-effective method of achieving good flexibility of the sealant than using a very low T_g internally plasticized acrylic polymer. A number of compatible plasticizers are available for acrylic polymers that lower the T_g of the composition very effectively. Benzylbutyl phthalate, dipropylene glycol dibenzoate, diethylene glycol dibenzoate, dipropylene glycol monobenzoate, propylene glycol alkyl phenyl ether, and mixtures of these plasticizers are used widely (see description in Reference 100). In silica-containing transparent (clear) formulations, benzylbutyl phthalate and N-n-butylbenzene sulfonamide or other N-alkylaryl sulfonamide plasticizers are preferred for product clarity [98].

Since plasticizers also greatly reduce the tensile strength at moderate-to-high loadings, one approach frequently pursued is to use a strong, higher-T_g polymer in combination with an effective plasticizer. However, this option involves some long-term risk if the plasticizer migrates from the sealant with time. Aging then results in a hard, brittle seal; regrettably, this effect can be frequently observed in the field. This risk is minimized when the correct type and level of plasticizer is used and when the polymer is not so hard that a small loss of plasticizer would render the sealant unfit. Suitable plasticizers, therefore, have low volatility even at elevated temperatures of up to 80°C. Plasticizer level and compatibility are also important aspects for paint compatibility. Visual examination of acrylic latex sealants installed in the field often shows that the highly plasticized formulations have glossy surfaces. When these sealants are later painted with exterior flat latex paints, they form "shiners," that is, they cause glossy and tacky paint surfaces. Both phenomena result from an insufficient compatibility of the plasticizer at the chosen addition level.

13.3.4.3.3　Filler

The choice of fillers as well as the amount used can affect the stability, crack resistance, adhesion, and rheological properties of the sealant. Calcium carbonate and talc are widely used in pigmented acrylic sealants, while silica may be used in clear and translucent formulations. In latex systems, wetting agents and dispersants are usually required for efficient thickening and reinforcement by the fillers.

13.3.4.3.4　Thixotropic Additive and Thickener

Silica and vitreous aluminosilicate fibers are frequently used as thixotropic additives to attain slump resistance in solvent-borne acrylics. Silica may also be used for the same purpose in acrylic latex formulations. Alkali-swellable poly(acrylic acid), hydroxyethyl cellulose, and poly(vinyl alcohol) or other water-soluble or water-dispersible thickeners are added to acrylic latex sealants to improve the handling and tooling properties.

13.3.4.3.5　Adhesion Promoter

Organofunctional alkoxy silanes having the formula X-$CH_2CH_2CH_2$-$Si(OR)_n$ are added to solvent-borne or latex acrylic formulations to impart improved "wet adhesion" on nonporous substrates, which is generally defined as *cohesive failure* after 1 week of water immersion [55]. In solvent-borne acrylics, mercapto-, acrylic-, and vinyl-functional silanes are frequently employed, while amino-, ureido-, and, increasingly, epoxy-functional silanes are added to acrylic latex formulations. In polymers used with solvent-acrylic sealants, the acrylic- or vinyl-functional adhesion promoter may be copolymerized into the polymeric backbone. Recently, epoxysilanes have attracted much attention as adhesion promoters for acrylic latex sealants [101], because they do not cause yellowing and have substantially less impact on viscosity changes during shelf life than aminosilanes.

Use of dialkoxyorganosilanes, such as glycidylpropylmethyldiethoxysilane, overcomes the negative impact that trialkoxyorganosilane addition has on maximum elongation while maintaining the beneficial effect on adhesion durability. The silane may be added neat, dissolved in an organic medium (protected addition), or as an aqueous emulsion. In waterborne systems, the alkoxysilane adhesion promoter is likely to undergo hydrolysis and condensation reactions. If the formulation contains silica, condensation reactions may also occur between silanols on the silica and the hydrolyzed silane. In any case, the net result is a loss of effectiveness of the adhesion promoter during the shelf life of the product. Since hydrolysis and condensation reactions are catalyzed by acids and bases, the best shelf stability of the silane in an aqueous system is generally achieved at neutral pH.

13.3.4.3.6 Solvent and Dispersing Medium

To achieve the desired rheology of the acrylic sealant, its solids content is adjusted during compounding. For solvent-borne sealants, usually xylene, toluene, or other alkylbenzenes are added, while for latex acrylics, solids content is adjusted with distilled or deionized water. The maximum level of solids for pigmented solvent-borne and acrylic latex sealants is approximately 75%–80%. Above this point, the viscosity increases rapidly and the product stability is poor. Solids content in clear versions is considerably lower and typically ranges from 55% to 65% in acrylic latex sealants and from 65% to 70% in solvent-borne acrylic sealants.

13.3.4.3.7 Surface-Drying Regulator

The rate of skin formation in acrylic sealants is adjusted with the help of a surface-drying regulator. Skin formation in acrylic latex sealants is retarded with mineral spirit, while methanol is added to solvent-borne formulations to accelerate skin formation.

13.3.4.3.8 Surfactant

The surfactants used are of special concern to sealant formulations because they can interfere with adhesion if improperly used. One approach to solving this problem is to incorporate the surfactant into the polymer backbone during polymerization (copolymerizable surfactant). This approach, which places the surfactant in an ideal location to stabilize the dispersion, does not allow the surfactant to migrate through the aqueous phase and interfere with adhesion, because the surfactant is connected to the polymer backbone.

13.3.4.3.9 Dispersant

A dispersant or wetting agent provides efficient thickening and reinforcement by the filler. Without a dispersant, the filler would steal surfactant from the polymer latex, and coagulation would result. Thus, dispersing agents provide viscosity and package stability. Sodium polymetaphosphate and low-molecular-weight poly(carboxylic acid) salts are good dispersants in many acrylic latex formulations.

13.3.4.3.10 Formulation Examples

Tables 13.23 and 13.24 provide information on typical formulations of high-performance plasticized latex acrylic [102] and solvent-borne acrylic [103] construction sealants.

13.3.4.4 Applications of Acrylic Sealants

Solvent-borne acrylic sealants are still more widely used in North America than in Europe; however, they are increasingly limited to niche applications because of concerns over health and worker safety issues. Their excellent inherent adhesion makes solvent-borne acrylics the ideal choice for difficult-to-seal exterior joints that do not undergo much movement. Examples of typical uses are sealing of exterior exposed joints on metal sidings, skylights, metal-flanged window frames, rivet seams, roof rails, roof curbs, perimeter door joints, corner molding, fabricated roof-lap seam assemblies, as well as use as primary seal for "drop in" glazing in wood, vinyl, and aluminum windows.

TABLE 13.23

High-Performance Plasticized Latex Acrylic
Construction Sealant Formulation

Ingredient	Addition Level (weight %)
Acrylic polymer dispersion (62.5% solids)	39.7
Propylene glycol	1.2
Nonionic surfactant	1.1
Inorganic dispersants	0.1
Organic anionic dispersants	0.1
Titanium dioxide pigment	1.1
Calcium carbonate	44.0
Precipitated silica	0.2
Plasticizer	9.3
Organosilane	0.1
Mineral spirits	1.9
Preservatives	0.2
Anionic thickener	0.8
Ammonium hydroxide (14%)	0.2

TABLE 13.24

Solvent-Borne Acrylic Construction
Sealant Formulation

Ingredient	Addition Level (weight%)
Acrylic polymer solution (68% solids)	80
Fumed silica	6.3
Toluene	13.6
Organosilane	0.1

Because acrylic latex caulks and sealants are spread over such a wide quality range, they also cover a rather large variety of applications. In general, these caulks and sealants can be classified into three distinctive groups, each associated with specific applications:

- Highly filled gap fillers
- Economic caulks with medium performance targeted at the do-it-yourself (DIY) market
- Exterior and interior high-performance sealants

Gap fillers are used for filling cracks and voids undergoing limited movement in similar and between dissimilar materials in interior as well as exterior applications, such as in wood, masonry, or concrete driveways. They are also applied as caulking materials between masonry, wood, and metals prior to application of conventional finishes.

The medium-performance caulks sold in the DIY market are designed as "one-for-all" products, and often they can be used primerless on a variety of common building materials such as wood, concrete, metal, glass, stone, brick, stucco, or marble. While widely used for exterior applications in low-rise buildings, most of these caulks are not specially formulated for exterior use and still carry a "Do not apply where joint will be subject to heavy rainfall within 8 hours of application" warning. Interior uses include sealing against air, moisture, and noise infiltration, such as on heating, ventilation, and air-conditioning (HVAC) ducts; extractor fans; chimneys; perimeter seals on windows and doors; penetration seals in walls; top-of-wall seals; etc. Exterior applications include perimeter seals on window and door frames; seals in aluminum, wood, or PVC sidings; skylights; and vents.

The high-performance sealants sold to professional waterproofers essentially cover the same range of applications as the medium-performance materials used by the do-it-yourselfers; however, they can accommodate a higher movement exposure and are often formulated for exterior use, that is, early rain wash-out resistance. Some specially formulated acrylic latex sealants are used in fire-stopping applications. These intumescent materials expand when exposed to heat and fill voids created by combustible materials burned during a fire.

13.3.5 SILICON-CURABLE ORGANIC SEALANTS

Some newer sealant types are based on so-called hybrid polymer technologies. In simplistic terms, hybrids combine the backbone of one sealant family with the reactive groups—typically positioned at the polymer terminals—of another sealant polymer type. For example, the polymercaptan polymers have either a hydrocarbon or a polyether (polyurethane) backbone coupled with mercaptan end groups. This allows cure to be carried out with conventional polysulfide-curing agents, while imparting the thermal stability and other durability characteristics of the backbone to the sealant network.

Over the past 20 years, a rather large number of hybrid polymer systems have been produced by mixing and matching the cure chemistry of one conventional polymer type with the backbone of another polymer type. Many of these hybrid technologies provide unique features that are valued in high-performance markets such as aerospace or electronics. Within the construction industry, the majority of commercially available hybrid polymers combine the cure chemistry of silicones with the backbone of a still rather limited number of polymers. This class of hybrids is often referred to as *silane modified, silyl modified, silicon modified, modified silicone*, or *silicon(e) hybrids*. The author does not consider any of these terms to accurately describe this class of polymers and prefers to use the term *silicon-curable organic* or *SCO*.

SCO polymers based on polyether, polyurethane, polyisobutylene, and polyacrylate backbones are commercially available. Sealants based on these polymers have a similar cure chemistry as standard silicone-based sealants; the vast majority of commercially available SCO sealants are based on tin-catalyzed alkoxy condensation cure chemistry. Because of this similarity with silicone sealants, the cure chemistry of SCO sealants will not be discussed in detail; however, some important features will be pointed out.

13.3.5.1   SCO Polyethers

Sealants based on SCO polyethers having dimethoxymethylsilyl groups at the polymer terminals—the so-called *telechelic alkoxysilyl polyethers*—have been commercialized in Japan since the early 1980s. They are available as both one- and two-part formulations; however, the majority of commercial SCO polyether sealants sold globally are one-part products. After their introduction, their market share in Japan increased gradually compared to mainly polyurethane and polysulfide sealants. Since 1997, they have been enjoying a dominant share of the Japanese sealants market (35%), followed by silicone (25%) and polyurethane (25%) sealants. The market share of SCO polyethers in Japan has stabilized since then. In Europe and North America, silicones continue to dominate the market, but sealants based on SCO polyethers are penetrating some market segments, especially the DIY segment, since the late 1990s with a higher growth rate than that of conventional sealants. With increased competition in Europe and North America from other SCO sealant types, especially SCO polyurethanes, it is likely that this growth rate will decline over the next few years.

13.3.5.1.1   Properties of SCO Polyether Sealants

The properties of SCO polyether sealants are determined primarily by their poly(propylene oxide) backbone. First-generation products were based solely on this polymeric binder [104]; second- and third-generation products introduced during the early-to-mid-1990s contained a blend of a polyacrylate [105] or SCO polyacrylate [106] and an SCO polyether to improve the durability of the cured sealant. The commercial importance of SCO polyether construction sealants is based on their unique combination of properties, as shown in Table 13.25.

Many of the most customer-visible benefits of SCO polyether sealants are derived from their silicon-cure chemistry. When first introduced in Japan, they were recognized for their environment-friendly cure chemistry, free of heavy metal oxides. At that time, the majority of two-part polysulfide sealants still employed a curative based on PbO_2 (in Japan, PbO_2 was phased out as late as by the end of 2002). Even today, SCO polyether sealants are advertised as being isocyanate free. This beneficial property has helped them substantially in gaining share against conventional one-part polyurethane sealants in Europe, especially in the DIY market segment, which has more stringent regulations on consumer health protection. Since SCO polyether sealants employ an alkoxy-cure chemistry that yields alcohol as a leaving group, they are essentially noncorrosive (depending on the specific formulation) and release very little odor during cure. Because SCO polyether sealants do not contain any isocyanate, they do not form bubbles or blisters during application, which is, again, a key advantage over conventional polyurethanes.

TABLE 13.25
General Benefits and Limitations of SCO Polyether Sealants

Benefits	Limitations
• Environment-friendly cure chemistry—free of heavy metals (when compared to PbO$_2$-cured polysulfides)	• Medium gas permeability (slower-curing one-part formulations)
• Isocyanate-free cure (when compared to polyurethanes)	• Surface tack
• Low odor, noncorrosive	• Cannot be applied to wet or damp surfaces (poor adhesion)
• No gassing or bubbling (when compared to polyurethanes)	• Limited elevated-temperature resistance
• Paintable	• Insufficient UV stability of glass adhesion to allow stopless or structural glazing
• Reasonable low-temperature flexibility (although modulus increases)	• Durable products available only in pigmented versions
• Medium gas permeability	
• Good weatherability (for generation II and III products)	
• Easy to extrude at low temperatures (when compared to polyurethanes)	
• Can tolerate high filler addition levels	
• Low-VOC, solvent-free formulations feasible	
• High elastic recovery	

A number of benefits result from the polyether backbone and the associated formulation latitude in terms of plasticizer selection or ability to blend the polymer with other polymers. First-generation SCO polyether sealants showed reasonable paintability with alkyd paints; however, waterborne acrylic paints had only poor adhesion to the cured sealant. Blending the polymer with a polyacrylate as in the second-generation SCO sealants substantially improved the overall paintability with all types of paints. The poly(propylene oxide) backbone in SCO polyether sealants results in reasonable low-temperature flexibility. While modulus increases with decreasing temperature, it does so to a lesser extent than for polyurethane sealants. Since the gas permeability of SCO polyether sealants is in the medium range, one-part SCO polyether products cure faster than do one-part polysulfide products, but still substantially slower than one-part silicone sealants.

Generation II and III SCO polyether products show good weatherability. Since the organic backbone of SCO polyethers does not contain hydrolyzable groups, these sealants do not undergo reversion in climates characterized by high humidity and high temperatures. However, the resistance of glass adhesion to degradation by sunlight is not sufficient to support applications such as structural or stopless glazing. SCO polyether sealants usually show a lower dependency of extrusion rate on temperature than polyurethane sealants, but they are still more temperature sensitive than silicone sealants.

SCO polyether sealants have better low-temperature extrudability than polyurethane sealants since they lack the polar urethane groups, which increase interaction between polymer chains by hydrogen bonding. This lack of urethane groups has two further beneficial effects: SCO polyethers, when compared to polyurethane polymers of the same molecular weight, have lower viscosity and therefore, they tolerate a higher filler addition level and generally do not require solvent addition to adjust viscosity and extrusion rate. The first benefit is primarily of interest to the formulator who is attempting to reduce formulation cost. The second benefit results in very low VOC, which is important for the environment and also reduces the risk to workers. Finally, since the poly(propylene oxide) backbone is not undergoing rearrangement reactions, SCO sealants can be formulated to display high elastic recovery combined with low modulus.

Probably the most noticeable disadvantage of SCO polyether sealants is their high surface tack when compared to polyurethane or silicone sealants. Various attempts have been made to overcome this issue; however, it is still a problem even with Generation II and III products. Depending on the environment and exposure conditions, high surface tack may result in substantial dirt pickup and retention.

Generally, manufacturers of SCO polyether sealants claim an upper service temperature in the range of 90°C–120°C.

13.3.5.1.2 Chemistry of SCO Polyethers

Telechelic SCO polyethers are obtained from allyloxy-endblocked poly(propylene oxide) polymers [107] that are reacted with dimethoxysilane by hydrosilylation reaction [108]. In SCO polyether sealant formulations, the hydrolysis and condensation reactions of the dimethoxymethylsilyl-endblocked poly(propylene oxide) polymer are commonly catalyzed with a tin compound. Sn (II) compounds are used in some two-part products, while one-part products generally use Sn (IV) compounds such as dibutyltindilaurate or dibutyltinbisacetylacetonate, often in combination with an amine such as laurylamine as cocatalyst. Since one-part Sn (IV)-catalyzed alkoxy-cure products must be free of moisture and silanol in order to achieve good shelf life, fillers and other moisture-containing formulation components must be predried. Generally, a silane water scavenger (dehydration agent) having a higher hydrolysis rate than the alkoxysilyl-endblocked polyether is added to the formulation to maintain shelf life [109].

13.3.5.2.3 Formulations for SCO Polyether Sealants

Products are formulated based on dimethoxymethylsilyl-endblocked poly(propylene oxide) polymers with different degrees of branching, phthalates or aliphatic oils as plasticizers, fillers—typically calcium carbonate, silica, or talc—and other additives, for example, catalysts, adhesion promoters, drying agents, UV stabilizers (HALS), and antioxidants to obtain sealants that cure in the presence of moisture.

Table 13.26 provides an overview of some recommended starting formulations for plasticized SCO polyether construction sealants.

TABLE 13.26
Recommended Starting Formulations for Plasticized SCO Polyether Construction Sealants

Ingredient	Chemicals	Addition Level (weight%)	
		Medium Filled [110]	Highly Filled [111]
Polymer	Dimethoxymethylsilyl-endblocked poly(propylene oxide) polymers (linear and branched blended at a 60:40 ratio)	32.8	25
Plasticizer	Phthalate, for example, di-iso-undecyl phthalate (DIUP)	18	17
Extending filler	Stearate-coated ground calcium carbonate, specific surface area 5 m²/g; mean particle size 2.0 μm	39.3	50
UV stabilizer	Titanium dioxide	6.6	3
Reinforcing filler	Fumed silica	0.7	
Thixotropic agent	Micronized amide wax		3.5
Antioxidant		0.3	
Dehydration agent	Vinyltrimethoxysilane	0.7	0.7
Adhesion promoter	N-(β-aminoethyl)-3-aminopropyltrimethoxysilane	1.0	0.5
Condensation catalyst	Dibutyl tin dilaurate, di-(n-butyl) tin bisketonate	0.6	0.3
Cocatalyst	Laurylamine	0.2	

13.3.5.1.4 Applications of SCO Polyether Sealants

SCO polyether sealants are very versatile. They are isocyanate free, easy to extrude and apply, paintable, and place low demands on substrate preparation, making them very suitable for the DIY markets. SCO polyether sealants can be formulated over a wide range of properties, from soft to reasonably hard and tough materials, allowing a broad range of applications. In professional applications, their lack of UV-stable glass adhesion typically limits their use in glazing applications; however, they are finding increased use in curtain wall and masonry wall applications. Because of their excellent paintability, high-modulus SCO polyether sealants are also used in sealing hardwood parquet floors.

13.3.5.2 SCO Polyurethanes

SCO polyurethanes were one of the first SCO polymers and sealants to be invented during the early 1970s. These sealants build on the strengths of standard urethane chemistry, such as adhesion and toughness, as well as the versatility of silicon-cure chemistry. SCO polyurethane polymers may be obtained either by reaction of an isocyanate-terminated prepolymer with an organofunctional alkoxysilane having a Zerewitinoff-active hydrogen atom, for example, secondary-amino-, ureido-, or mercapto-organoalkoxysilanes [112,113], or by reaction of a hydroxyl-terminated prepolymer [114] with an isocyanatoorganoalkoxysilane. Both routes lead to alkoxysilyl-terminated polyurethane polymers. In the reaction with isocyanate-functional prepolymers, secondary-aminoorganoalkoxysilanes are preferred since they do not impart an unpleasant odor to the final product and perform more favorably. A simple and low-cost industrial route to secondary-aminoorganoalkoxysilane-endblockers (N-alkoxysilylalkyl-aspartic acid esters) is described in Reference 115. One of the early findings, which considerably helped the industrial utilization of these sealants, was that the cure rate of SCO polyurethanes could be substantially accelerated by the use of aminosilane adhesion promoters as cocatalysts [116].

Sealants initially formulated with these polymers were generally high-viscosity, high-modulus, and low-elongation materials, used primarily in transportation and industrial applications. During the mid-1980s, efforts were increasingly focused on developing SCO polyurethane sealants for construction applications. Various approaches were pursued to achieve low-modulus, high-elongation materials with good adhesion to a broad variety of substrates. One of the first successful attempts was that of Berger and coworkers, who, using secondary amino bis(alkoxysilane) endblockers, were able to improve flexibility and wet adhesion [117]. Pohl and Osterholtz described the use of dialkoxysilane endblockers [118] as a means of reducing modulus. Another approach to reducing the crosslink density of the cured elastomer is to use secondary aminosilanes with a bulky substituent, such as a phenyl group, on the nitrogen as endblockers. However, the use of dialkoxysilane or sterically hindered silanes as endblockers suffers from several drawbacks. The secondary amine containing silanes are slow to react with the urethane prepolymer, while polymers endcapped with dialkoxy-functional silanes are typically very slow to cure. Furthermore, the formation of urea groups, which is experienced when using aminosilane endblockers, leads to a substantial increase in viscosity of the prepolymer, potentially resulting in processing or application problems.

Today, various approaches to low-modulus SCO polyurethane sealants that are based on the specific properties of the silane endblocker and the isocyanate prepolymer or polyetherdiol polymer exist. Increasing the molecular weight of the prepolymer while simultaneously increasing its functionality creates a softer material without compromising its cohesive strength [119]. Higher maximum elongation and lower modulus can be achieved by using a silane endblocker that has a branched midsegment [120], such as N-ethyl-3-amino-isobutyl trimethoxy silane. This is because such an endblocker will result in less efficient packing and, thus, higher free volume near the crosslinking sites, facilitating faster motion and disentanglement when the sealant is strained. Use of high-molecular-weight polyols, for example, those >8000 Da, with low unsaturation allows preparation of low-modulus prepolymers when using special N-(3-trimethoxysilylpropyl) aspartic acid diethyl ester endblocker [121]. Low-modulus sealants can also be obtained by reaction of a high-molecular-weight hydroxyl-endblocked urethane prepolymer with an isocyanate-functional alkoxysilane [122]. These polymers have fewer polyurethane hard segments, which reduces the possibility of hydrogen bonding, leading to lower polymer viscosities at equivalent molecular weights and, therefore, lower-modulus sealants. Contrary to the standard SCO polyurethane technology, these polymers can be produced in the absence of plasticizers, which gives producers more latitude to optimize the sealant formulation.

These new-generation products are expected to rapidly gain share in construction markets against traditional polyurethane and polysulfide as well as SCO polyether sealants. Growth of this sealant type will be enhanced by the fact that SCO polyurethane polymers are nonproprietary and can be manufactured by any polyurethane formulator without additional equipment requirements—a key difference when compared to other SCO polymer technologies.

13.3.5.2.1 Properties of SCO Polyurethane Sealants

The properties of SCO polyurethane sealants are determined by the nature of their polymeric backbone, the specific endblockers used, and their silicon curability. A wide range of polyols (polyether, polyester, or polybutadiene) and isocyanates (aliphatic or aromatic) can be chosen and reacted at varying molar ratios to create prepolymers having different molecular weights. Such prepolymers may then be reacted with a silane selected from a variety of endblockers to yield SCO polyurethane materials that can be formulated into low- or high-modulus sealants. As with the previously discussed SCO polyether sealants, many of the customer-visible properties of SCO polyurethane sealants are derived from the environmentally friendly silicon-cure chemistry. However, the typical polyurethane structure of soft and hard segments can be used to dramatically vary physical properties such as modulus, flexibility, and strength, which makes this class of sealants unique. Table 13.27 shows a compilation of the characteristics of SCO polyurethane sealants.

SCO polyurethane sealants are frequently compared with their nonsilicon-modified counterparts. In SCO polyurethane sealants, any isocyanate functionality

TABLE 13.27

General Benefits and Limitations of SCO Polyurethane Sealants

Benefits	Limitations
• Environment-friendly isocyanate-free cure chemistry (when compared to polyurethanes)	• Low gas permeability (resulting in slow-curing one-part formulations)
• No gassing or bubbling (when compared to polyurethanes)	• Should not be applied to wet or damp surfaces (poor adhesion)
• Low odor, noncorrosive	• Limited elevated-temperature resistance
• Good weatherability	• Insufficient UV stability of glass adhesion to allow stopless or structural glazing
• Adhesion to a broad range of substrates	• Available only in pigmented versions
• Paintable	
• Good low-temperature flexibility (although modulus increases)	
• Low gas permeability (beneficial barrier property)	
• High elastic recovery	

originally present on the prepolymers or on smaller molecules, such as on the iso-cyanate chain extenders, completely reacts with an excess of the silane endblocker. While isocyanates still need to be handled during manufacture, the sealants are isocyanate-free in the application, leading to less environmental impact and fewer worker's health concerns. Because no isocyanates are present, bubbling during cure is not a concern, as it often is in traditional polyurethane sealants. Additionally, the absence of isocyanates broadens the range of formulation additives available to the formulator, allowing the use of aminosilane adhesion promoters and hindered-amine light stabilizers (HALS), which cannot be employed in one-part polyurethane seal-ants. Due to the addition of suitable stabilizers and adhesion promoters, SCO poly-urethanes display substantially better weather resistance and adhesion to a broader range of substrates than conventional polyurethane sealants. SCO polyurethane seal-ants are also generally immediately paintable after application with both solvent- and waterborne paints, while traditional polyurethanes can only sometimes be painted. Note, however, that coating a moisture-curable sealant with paint immediately after application will slow down the cure of the sealant.

The remaining beneficial properties, such as low-temperature flexibility, low gas permeability, and high elastic recovery, primarily result from the urethane polymer backbone and are essentially unaffected by the silicon curability of the polymer. The same statement applies for all of the listed limitations. However, note that SCO polyurethane sealants generally cure faster than SCO polyether sealants because the polymer is trimethoxysilyl rather than dimethoxysilyl endblocked, resulting in higher reactivity.

13.3.5.2.2 Chemistry of SCO Polyurethane

Telechelic SCO polyurethanes are prepared in a two-step reaction sequence. First, a urethane prepolymer is prepared by the reaction of a diisocyanate and a high-molecular-weight polyol in the conventional manner. Low-viscosity urethane prepolymers suit-able for the production of low-modulus construction sealants can be prepared using diisocyanates having two isocyanate groups of differing reactivity, such as toluene dii-socyanate (TDI) or isophorone diisocyanate (IPDI), to control the molecular weight dis-tribution of the prepolymer. The diisocyanate is reacted with a polyol component having a high molecular weight (>8000 Da) by selecting an NCO/OH ratio either between 1.3 and 1.5, resulting in an isocyanate-endblocked prepolymer, or between 0.65 and 0.75, resulting in a hydroxyl-endblocked polymer. The prepolymer is then silylated by cap-ping with a suitable organosilane. Isocyanate-functional prepolymers are capped with secondary aminoorgano alkoxy silanes, for example, phenylaminopropyl trimethoxy silane or N-ethyl-3-aminoisobutyl trimethoxy silane; hydroxyl-functional prepolymers are reacted with 3-isocyanatopropyl trimethoxysilane. This two-step reaction sequence is often carried out in a continuous reactor; however, it may also be completed in the same kettle mixer in which the sealant is to be compounded.

In SCO polyurethane sealant formulations, the hydrolysis and condensation reac-tions of the trimethoxysilyl-endblocked polymer are commonly catalyzed using a tin compound. Frequently, Sn (IV) compounds such as dibutyltindilaurate or dibutyltin-bisacetylacetonate are used often in combination with an amine such as laurylamine as cocatalyst. Since one-part Sn (IV)-catalyzed alkoxy-cure products must be free of moisture and silanols in order to achieve good shelf life, fillers and other moisture-containing formulation components must be predried. Generally, a silane water scavenger (dehydration agent) having a higher hydrolysis rate than the alkoxysilyl-endblocked polymer is added to the formulation to maintain shelf life.

13.3.5.2.3 Formulations for SCO Polyurethane Sealants

Products are formulated based on trimethoxysilyl-endblocked polymers, phthalates or aliphatic oils as plasticizers, fillers (typically, calcium carbonate, silica, or talc), and other additives, for example, catalysts, adhesion promoters, drying agents, UV stabilizers (HALS), and antioxidants to obtain sealants that cure in the presence of moisture. Note that "standard" aminoorgano-functional adhesion promoters are widely used with SCO polyurethanes and are known to offer good adhesion to a variety of substrates. However, recently, branched silanes, such as 4-(trimethoxysilyl)-2,2-dimethyl butane amine, have been demonstrated to yield a substantially higher elongation at break and lower modulus; these properties are highly desirable in weatherproofing construction sealants [123]. Table 13.28 provides a recommended starting formulation for a low-modulus plasticized SCO polyurethane construction sealant [123].

13.3.5.2.4 Applications of SCO Polyurethane Sealants

SCO polyurethane sealants have only recently entered the construction markets; it is therefore difficult to predict the specific applications for which these materials will be preferred. However, being even more versatile than SCO polyether sealants, SCO polyurethanes ranging from low-modulus sealants to tough structural sealants or adhesives can be expected to cover a wide variety of applications. In professional applications, their lack of UV-stable glass adhesion will typically limit their use in glazing applications; however, similar to SCO polyethers, they can be expected to find increasing use in curtain wall and masonry wall applications.

TABLE 13.28

Starting Formulation For a Low-Modulus Plasticized SCO Polyurethane Construction Sealant

Ingredient	Chemical	Addition Level (weight%)
SCO urethane polymer	Trimethoxysilyl-endblocked EO/PO polyether polymer	22.9
Plasticizer	Diisodecyl phthalate	18.3
Calcium carbonate	20/80 mixture of precipitated and ground $CaCO_3$	55.0
Titanium oxide		1.2
Thixotropic agent	Treated fumed silica	1.2
UV absorber		0.5
Dehydration agent	Vinyltrimethoxysilane (VTM)	0.3
Adhesion promoter	4-(Trimethoxysilyl)-2,2-dimethyl butane amine	0.5
Condensation catalyst	Dibutyltin dilaurate	0.05

13.3.5.3 SCO Polyisobutylene

Alkoxysilyl-endblocked polyisobutylene constitutes another class of SCO polymer. Although this polymer type was launched in the mid-1990s, formulated sealants based on this polymer are currently commercially available only in Japan. Because of the extremely low vapor permeability of polyisobutylene, sealants based on SCO polyisobutylene can only be formulated as two-part sealants. Due to the high price of the SCO polyisobutylene polymer, most of the sealants commercially available in Japan are being sold for high-value nonconstruction applications. However, it can be speculated that a possible future decline in polymer price will justify the polymer's entering construction markets on a larger scale.

13.3.5.3.1 Properties of SCO Polyisobutylene Sealants

The properties of SCO polyisobutylene sealants are determined by their isobutylene polymeric backbone and silicon curability. Table 13.29 compiles the characteristics of SCO polyisobutylene sealants, as far as they are known today.

As for the whole class of SCO sealants, the key end-use benefits are derived from the environment-friendly and worker-health-friendly silicon curability. Sealants based on SCO polyisobutylene show very low vapor and gas permeabilities. This property makes them excellent candidates for sealing in or sealing out gaseous environments, which is important in electric, electronic, or IG applications; however, it also implies that one-part moisture-curing products can only be formulated by giving up this key benefit [124]. Thus, commercial products that build on the low permeability strength of SCO polyisobutylene are two-part formulations. Due to their lack of unsaturation and the formulator's ability to use a wide range of stabilizers, SCO polyisobutylene sealants display very good weatherability. The sealants also show reasonable-to-good adhesion to a variety of substrates. Their low T_g, and therefore good low-temperature flexibility, results from their isobutylene backbone.

TABLE 13.29
General Benefits and Limitations of SCO Polyisobutylene Sealants

Benefits	Limitations
• Environment-friendly, isocyanate-free cure chemistry (when compared to polyurethanes)	• Should not be applied on wet or damp surfaces (poor adhesion)
• No gassing or bubbling (when compared to polyurethanes)	• Limited elevated-temperature resistance
• Low odor, noncorrosive	• Probably insufficient UV stability of glass adhesion to allow stopless or structural glazing
• Very good weatherability	• High polymer cost
• Paintable	• Surface tack
• Reasonable-to-good adhesion to various substrates	
• Good low-temperature flexibility	
• Very low gas permeability	
• Reasonable elastic recovery	

The key disadvantage of this sealant type is the (still) high polymer cost. Currently, only a single company manufactures the polymer. As the demand for this type of polymer grows, it is likely that cost reduction resulting from economy of scale will allow price reduction. Surface tack appears to be an inherent problem with this polymer technology according to the coverage of this issue in the patent literature. However, it appears that this problem can be overcome by addition of photo- or oxidation-curable substances to the sealant formulation.

13.3.5.3.2 Chemistry of SCO Isobutylene Polymers

SCO isobutylene polymers are obtained by hydrosilylation of telechelic allyl-terminated polyisobutylene. Allyl-functional telechelic polyisobutylene is accessible via "inifer" polymerization. During the early 1980s, a novel process of making α,ω-di(t-chloro)polyisobutylene was invented. In this method, a multifunctional compound capable of simultaneously initiating polymerization and acting as a transfer agent is employed. For this compound, the term *inifer* was chosen, derived from the words *initiator* and *transfer*. Using p-dicumyl chloride and BCl_3 as a fully chlorinated Friedel–Crafts coinitiator, isobutylene can be cationically polymerized to α,ω-di(t-chloro)polyisobutylene [125]. The tertiary chloro-termination can be converted to iso-propenyl ($-C(CH_3)=CH_2$) ends by selective elimination of HCl with a strong base, such as potassium t-butanoxide or, more conveniently, by electrophilic substitution (allylation) of tertiary chloro-capped poly(isobutylene) with allyltrimethylsilane [126]. Alkoxysilyl functionality then can be introduced by hydrosilylation of the unsaturated end groups with methyldichlorosilane followed by alcoholysis [127].

Limited public information is available at this stage on the product chemistry and formulation of SCO polyisobutylene sealants. As with other SCO chemistries, hydrolysis and condensation reactions of the dimethoxysilyl-endblocked polymer are commonly catalyzed with tin compounds. Because of their low moisture permeability, condensation cure cannot be initiated by ambient moisture diffusing into the sealant. Two-part products therefore contain water or water-generating substances in one of the two parts of the formulation. However, to achieve good shelf life, the "curable" part of the formulation must still be sufficiently free of moisture. Sn (II) or Sn (IV) compounds may be used in two-part products either on their own or in combination with an amine, such as laurylamine, as cocatalyst.

13.3.5.3.3 Formulations for SCO Isobutylene Sealants

Products are formulated based on dimethoxysilyl-endblocked polymers, plasticizers, fillers (typically, calcium carbonate, silica, or talc), and other additives, for example, thixotropic agents, catalysts, adhesion promoters, UV stabilizers (HALS), photo- or oxidation-curable substances, physical property modifiers, and antioxidants. Table 13.30 presents the formulation for a plasticized SCO polyisobutylene sealant [128].

13.3.5.3.4 Applications of SCO Polyisobutylene Sealants

SCO polyisobutylene sealants have been recently launched in the construction markets in Japan. Because of the high cost of the polymer, these sealants are expected

TABLE 13.30

Formulation for a Plasticized SCO Polyisobutylene Sealant

Part	Ingredient	Chemical	Addition Level (weight%)
A	Polymer	Dimethoxymethyl-endblocked PIB polymer	19.0
	Plasticizer	Paraffinic process oil	17.1
	Filler (extender)	Ground CaCO$_3$	34.3
	Filler (extender)	Talc	19.0
	Light stabilizer	Nickel dimethyldithiocarbamate	0.6
	Thixotropic additive	Hydrogenated castor oil	1.0
	Antioxidant	Hindered phenol	0.2
	UV stabilizer	Salicylate UV absorber	0.2
	Light stabilizer (HALS)	Hindered amine	0.2
	Photocurable substance	Dipentaerythritol penta/hexaacrylate	0.6
	Adhesion promoter	γ-glycidoxypropyltrimethoxysilane	0.4
		γ-isocyanatopropyltriethoxysilane	0.8
	Modifier	Diphenyldimethoxysilane	0.1
B	Plasticizer	Paraffinic process oil	3.0
	Filler (extender)	Ground CaCO$_3$	1.9
	Pigment	Carbon black	0.5
	Cure initiator	Water	0.4
	Silanol condensation catalyst	Dibutyl tin dimethoxide	0.8

to address specialized needs of only high-value market segments. As polymer cost decreases in the future, it is expected that insulating glass (IG) will be the main application for SCO polyisobutylene sealants.

13.3.5.4 SCO Acrylates

SCO acrylates were among the first SCO polymers to be invented during the late 1960s. The initial polymers were prepared by radical polymerization of alkylacrylate or alkylmethacrylate monomers using a radical initiator, for example, azobisisobutyronitrile (AIBN), and a chain transfer agent, typically a mercaptan, for controlling the molecular weight of the final polymer. The alkoxysilyl functionality was introduced by one or more of the following three methods: (1) copolymerization of an unsaturated group containing silane, for example, γ-methacryloxypropyltrimethoxysilane; (2) use of an alkoxysilyl-functional radical initiator, for example, α,α'-azobis-5-trimethoxysilyl-2-methyl-valeronitrile; or (3) use of an alkoxysilyl-functional chain transfer agent, for example, γ-mercapto-propyltrimethoxysilane. The alkoxysilyl functionality of the initial SCO acrylate polymers was either pendant or pendant and end-capped on the main chain, depending on the nature of the chain transfer agent used [129–131].

Sealants formulated with these initial polymers were either very brittle, when low-molecular-weight polymers were used, or required large amounts of solvents to allow use of high-molecular-weight polymers. They also suffered from a substantial increase in viscosity during their shelf life. One attempt at overcoming these problems focused on the use of mono-silanol compounds in the formulation so that upon moisture curing, a part of the alkoxysilyl groups does not act as crosslinking sites but rather as chain propagation sites [132]. Kozakiewicz and coworkers found that acrylate copolymers with lower levels of alkoxysilyl functionality gave sealants with lower modulus and higher elongation at break and that the viscosity drift during shelf life could be overcome by use of suitable water scavengers [133]. This approach led to formulation of the first commercially viable SCO acrylate construction sealant at the end of the 1990s.

During the mid-1990s, new methods of living radical polymerization for telechelic acrylates were developed. Kusakabe and Kitano realized that atom transfer radical polymerization (ATRP) could be utilized to manufacture polyacrylate polymers with narrow molecular weight distribution having alkenyl groups or curable silyl groups at the chain ends in a high functionality ratio [134]. Telechelic polymers, having silicon-curable groups at the polymer-chain ends, impart to the crosslinked sealant better mechanical properties than are obtainable from polymers with pendant functional groups. Telechelic SCO acrylate polymers became commercially available in 2003, and the first sealants based on this polymer type are currently being launched.

13.3.5.4.1 Properties of SCO Acrylates

The following discussion will be limited to telechelic SCO acrylates since the author expects this polymer type to set the standard for this future sealant family. The properties of SCO acrylate sealants are determined by their acrylic polymer backbone and silicon curability. Table 13.31 shows a compilation of the properties of SCO acrylate sealants, as far as they are known today.

As for the whole class of SCO sealants, the key end-use benefits are derived from the environment-friendly and worker-health-friendly silicon curability. This curability also differentiates SCO acrylates from conventional acrylates by eliminating the thermoplastic nature of acrylates via the formation of a cured network. The higher moisture vapor transmission of acrylates, when compared to SCO polyethers, SCO polyurethanes, or SCO polyisobutylenes, allows formulation of reasonably fast-curing one-part sealants. The high UV and hydrolysis resistance of the acrylate polymer backbone contributes to the excellent weatherability of SCO acrylate sealants. In a recent study, adhesion to glass was shown to be unaffected after 10,000 h of exposure to UV irradiation in a Super Xenon Weather Meter (irradiance: 180 W/m^2 (300–400 nm), black panel temperature: 63°C, water spray for 18 min and UV exposure for 102 min within weathering cycles of 2 h duration) [135]. The combination of excellent durability and nonstaining characteristics makes SCO acrylate sealants an ideal choice for glazing applications involving adhesion to photocatalytic self-cleaning glasses. Due to their organic backbone, SCO sealants exhibit very good paintability with a wide range of different paints. Experimental formulations of sealants based on this polymer are claimed to pass the ISO 11600 25LM/HM standard for glazing applications.

TABLE 13.31

General Benefits and Limitations of SCO Acrylate Sealants

Benefits	Limitations
• Environment-friendly, isocyanate-free cure chemistry (when compared to polyurethanes)	• Should not be applied to wet or damp surfaces (poor adhesion)
• No gassing or bubbling (when compared to polyurethanes)	• Limited elevated-temperature resistance
• Low odor, noncorrosive	• Probably insufficient UV stability of glass adhesion to allow stopless or structural glazing
• Excellent weatherability	• Surface tack
• Paintable	• High polymer cost
• Good adhesion to a wide range of substrates	
• Reasonable low-temperature flexibility	
• Medium gas permeability	
• High elastic recovery	
• Potential for both transparent and pigmented formulations	

The key disadvantage of this sealant type is the (still) high polymer cost. Presently, only a single company manufactures the polymer. As the demand for this type of polymer grows, it is likely that cost reduction resulting from the economy of scale will allow price reduction. Surface tack appears to be an inherent problem with this polymer technology according to the coverage of this issue in the patent literature. However, it appears that this problem can be overcome by addition of photo- or oxidation-curable substances to the sealant formulation.

13.3.5.4.2 Chemistry of SCO Acrylate Sealants

Very little public information is available at this stage on the product chemistry and formulation of SCO acrylate sealants. Based on the patent literature [136], it is obvious that hydrolysis and condensation reactions of the $Si(OMe)_n$-endblocked acrylate polymer are commonly catalyzed with a tin compound. Sn (II) compounds may be used in two-part products either on their own or in combination with an amine, such as laurylamine, as cocatalyst. For one-part products, Sn (IV) catalysts are used, for example, dibutyltinbisacetylacetonate. As one-part Sn (IV)-catalyzed alkoxy-cure products must be free of moisture and silanols in order to achieve good shelf life, fillers and other moisture-containing formulation components have to be predried. It is likely that effective chemical drying can also be achieved by adding a silane water scavenger (dehydration agent) to the formulation.

13.3.5.4.3 Formulations for SCO Acrylate Sealants

Products are formulated based on dimethoxysilyl-endblocked polymers, regular or polymeric plasticizers, fillers (typically, calcium carbonate, silica, or talc), and other additives, for example, thixotropic agents (such as polyamide waxes), catalysts, adhesion promoters, drying agents, photo- or oxidation-curable additives to reduce surface tack, UV stabilizers (HALS), and antioxidants to obtain sealants that cure in the presence of moisture. While regular plasticizers (such as phthalates) can be used, polymeric plasticizers with a number-average molecular weight, M_n, greater than 1000 are especially preferred to minimize staining on sensitive substrates, such as marble and granite, and improve paintability with alkyd paints. Suitable high-molecular-weight plasticizers are polyadipate, polybutene, and alkylbenzenes. Acrylic polymers made by living polymerization and having a narrow molecular weight distribution are especially preferred as plasticizers because of their excellent compatibility with the SCO acrylate polymers. Tung oil or liquid diene polymers, such as polybutadiene, may be used as oxidation-curable substances. Enhanced surface drying may sometimes be obtained by use of a metallic dryer, that is, a catalyst that promotes the oxidation-curing reaction. Pentaerythritol triacrylate and trimethylol propane triacrylate are described as photo-curable additives providing lower surface tack to the cured sealant.

Table 13.32 provides a formulation for a plasticized SCO acrylate construction sealant [135]. Note that to a commercial formulation, UV absorbers and heat and light stabilizers would also be added [136].

13.3.5.4.4 Applications of SCO Acrylate Sealant

SCO acrylate sealants are currently being launched in the construction markets. Because of the high cost of the polymer, these sealants are expected to address specialized needs of only high-value market segments. One such application that is being targeted is the glazing of photocatalytic self-cleaning glass. Glazing sealants for self-cleaning glass must be nonstaining; they also should have very high durability to resist the photocatalytic degradation of the sealant–glass interface. Regular organic sealants cannot be used for this application as they fail interfacially on photocatalytic glass after rather short exposure to sunlight. The polymeric backbones of most other SCO sealant types also do not have sufficient stability to provide an acceptable long-term solution. Whether SCO acrylates will find use in regular glazing and façade applications will depend on future polymer pricing.

TABLE 13.32

Formulation for a Plasticized SCO Acrylate Sealant

Ingredient	Chemicals	Addition Level (weight%)
SCO acrylate polymer	Dimethoxysilyl-endblocked acrylate polymer	39.1
Plasticizer	Acrylic polymer	19.5
Calcium carbonate	Precipitated $CaCO_3$	39.1
Surface modifier	Tung oil, pentaerythritol triacrylate, or trimethylol propane triacrylate	1.2
Adhesion promoter	γ-glycidoxypropyltrimethoxysilane	0.8
Condensation catalyst	Dibutyltinbisacetylacetonate	0.4

13.4 SUMMARY

Sealants are materials installed in gaps or joints to prevent water, wind, dirt, or other contaminants from passing through the joint or gap. They are commonly rated by their movement ability. High-movement sealants such as silicones, polyurethanes, and polysulfides typically accommodate joint movements of 25% or higher. Acrylic sealants are commonly used in joints where movement is between 5% and 15% and no higher than 25%. Butyl sealants are used in applications with movement less than 12.5%. The base polymer of each sealant type provides inherent properties that find use in specific applications. Silicones are resistant to weathering and are used in applications where longevity is a prime concern. Polyurethanes are tough and abrasion resistant, with good adhesion to many substrates, thus allowing them to be used in many applications in weatherproofing, especially in high-abuse areas. Polysulfides have superior chemical resistance and hence find wide use in civil engineering applications. Latex acrylics have good weatherability, are less expensive, and are easy to use, making them popular with consumers in nondemanding applications. Recently, a new class of sealants referred to as *silicon-curable organic* or *hybrid* sealants has been commercialized. These sealants afford improved adhesion and weatherability combined with properties such as paintability and nonstaining character that are typically associated with organic-based sealants.

REFERENCES

1. J. Bakus, *Constr. Specifier*, 58 (4), 80–86 (2005).
2. Anonymous, ASTM C719-93, Standard Test Method for Adhesion and Cohesion of Elastomeric Joint Sealants Under Cyclic Movement (Hockman Cycle), ASTM International, West Conshohocken, PA (2005).
3. Anonymous, ISO 9047, Building Construction—Jointing Products—Determination of Adhesion/Cohesion Properties of Sealants at Variable Temperatures, International Standardization Organization, Geneva (2001).
4. Anonymous, ISO 9046, Building Construction—Jointing Products—Determination of Adhesion/Cohesion Properties of Sealants at Constant Temperature, International Standardization Organization, Geneva (2002).
5. T.G.B. Jones and M.A. Lacasse, in *Durability of Building Sealants*, A.T. Wolf (Ed.), RILEM Report 21, pp. 73–105, RILEM Publications, Cachan, France (1999).
6. Anonymous, ISO 11600, Building Construction—Jointing Products—Classification and Requirements for Sealants, International Standardization Organization, Geneva (2002).
7. Anonymous, The BASA Guide to the ISO 11600 Classification of Sealants for Building Construction, BASA British Adhesives & Sealants Association, Stevenage, Herts, United Kingdom, available at www.basa.uk.com/.
8. Anonymous, ISO 8339, Building Construction—Sealants—Determination of Tensile Properties (Extension to Break), International Standardization Organization, Geneva (2005).
9. K.L. Mittal, in *Adhesion Measurement of Thin Films, Thick Films and Bulk Coatings*, K.L. Mittal (Ed.), STP No. 640, pp. 5–17, ASTM International, West Conshohocken, PA (1978).
10. K.L. Mittal, *Polym. Eng. Sci.*, 17, 467–473 (1977).
11. A.T. Wolf and N.E. Shephard, in *Proceedings of the 17th International Symposium Swiss Bonding*, E. Schindel-Bidinelli (Ed.), pp. 393–404, Swibotech, Rorbas, Switzerland (2003).

12. A.T. Wolf (Ed.), *Durability of Building Sealants*, RILEM Report 21, RILEM Publications, Cachan, France (1999).

13. Y. Israeli, J.L. Phillipart, J. Cavezzan, J. Lacoste, and J. Lemaire, *Polym. Degrad. Stab.*, *36*, 179–185 (1992); Y. Israeli, J. Cavezzan, and J. Lacoste, *Polym. Degrad. Stab.*, *37*, 201–208 (1992); Y. Israeli, J. Cavezzan, J. Lacoste, and J. Lemaire, *Polym. Degrad. Stab.*, *42*, 267–279 (1993); Y. Israeli, J. Lacoste, J. Cavezzan, and J. Lemaire, *Polym. Degrad. Stab.*, *47*, 357–362 (1995).

14. P.D. Gorman, in *Science and Technology of Building Seals, Sealants, Glazing and Waterproofing*, D.H. Nicastro (Ed.), Vol. 4, ASTM STP 1243, pp. 3–28, ASTM International, West Conshohocken, PA (1995).

15. D. Oldfield and T. Symes, *Polym. Testing*, *15*, 115–128 (1996).

16. A.T. Wolf and J.-P. Hautekeer, in *Sealants and Adhesives Handbook*, P. Cognard (Ed.), Vol. 3, Elsevier Science Ltd., Oxford, in press.

17. P.L. Brown and J.F. Hyde, Dow Corning Corporation, U.S. Patent 3,161,614 (1964).

18. D.R. Weyenberg, Dow Corning Corporation, U.S. Patent 3,334,067 (1967).

19. S.D. Smith and S.B. Hamilton, Jr., General Electric Company, U.S. Patent 3,689,454 (1972).

20. E. Sweet, Rhone-Poulenc, U.S. Patent 3,189,576 (1965).

21. L. Ceyzeriat, Rhone-Poulenc, U.S. Patent 3,133,891 (1964).

22. L.B. Bruner, Dow Corning Corporation, U.S. Patent 3,032,532 (1962).

23. L.B. Bruner, Dow Corning Corporation, U.S. Patent 3,035,016 (1962).

24. M.D. Beers, General Electric Company, U.S. Patent 3,382,205 (1968).

25. T.A. Kulpa, General Electric Company, U.S. Patent 3,296,161 (1967).

26. M.D. Beers, General Electric Company, U.S. Patent 4,257,932 (1981).

27. D. Goelitz, K. Damm, R. Mueller, and W. Noll, Bayer AG, U.S. Patent 3,364,160 (1968).

28. H. Sattlegger, W. Noll, K. Damm, and H.D. Goelitz, Bayer AG, U.S. Patent 3,378,520 (1968).

29. K.C. Pande and R.E. Ridenour, Stauffer Chemical Company, U.S. Patent 3,448,136 (1969).

30. J. Boissieras, L.F. Ceyzeriat, and M.J.C. Lefort, Rhone–Poulenc, U.S. Patent 3,359,237 (1967).

31. R.A. Murphy, General Electric Company, U.S. Patent 3,341,486 (1967).

32. J. Boissieras and L.F. Ceyzeriat, L.F., Rhone Poulenc, U.S. Patent 3,429,847 (1969).

33. T. Takago, T. Sato, and H. Aoki, Shin-Etsu Chemical Company, U.S. Patent 3,819,563 (1974).

34. T. Takago, Shin-Etsu Chemical Company, U.S. Patent 4,180,642 (1979).

35. S. Nitzsche and M. Wick, Wacker-Chemie, U.S. Patent 3,032,528 (1962).

36. M. Wick, P. Hittmair, E. Wohlfarth, and S. Nitzsche, Wacker-Chemie, U.S. Patent 3,464,951 (1969).

37. P. Hittmair, S. Nitzsche, M. Wick, and E. Wohlfahrt, Wacker-Chemie, U.S. Patent 3,408,325 (1968).

38. S. Nitzsche, W. Kaiser, E. Wohlfahrt, and P. Hittmair, Wacker-Chemie, U.S. Patent 3,674,738 (1972).

39. L. Toporcer and J.N. Clark, Dow Corning Corporation, U.S. Patent 3,776,934 (1973).

40. L. Toporcer and I. Crossan, Dow Corning Corporation, U.S. Patent 3,817,909 (1974).

41. J.M. Klosowski, Dow Corning Corporation, U.S. Patent 3,996,184 (1976).

42. K. Mine and T. Tamaki, Toray Silicone Company, U.S. Patent 4,387,177 (1983).

43. J.R. Hahn, Dow Corning Corporation, U.S. Patent 4,360,631 (1982).

44. C.A. Berridge, General Electric Company, U.S. Patent 2,843,555 (1958).

45. M.A. White, M.D. Beers, G.M. Lucas, R.A. Smith, and R.T. Swiger, General Electric Company, U.S. Patent 4,395,526 (1983).

46. J.J. Dziark, General Electric Company, U.S. Patent 4,417,042 (1983).
47. G.M. Lucas and J.J. Dziark, General Electric Company, U.S. Patent 4,483,973 (1984).
48. R.H. Chung, General Electric Company, U.S. Patent 4,515,932 (1985).
49. S.J. Clarson and J.A. Semlyen (Eds.), *Siloxane Polymers*, Prentice Hall, Englewood Cliffs, NJ (1993).
50. H.G. Brod and O. Schweitzer, Degussa, German Patent 1,035,358 (1958).
51. K.E. Polmanteer, Dow Corning Corporation, U.S. Patent 2,927,907 (1960).
52. R.A. Palmer and S. Spells, Dow Corning Corporation, U.S. Patent 5,246,980 (1993).
53. R.A. Palmer, Dow Corning Corporation, U.S. Patent 5,290,826 (1994).
54. F. De Buyl, *Int. J. Adhes. Adhes.*, *21*, 411–422 (2001).
55. K.L. Mittal (Ed.), *Silanes and Other Coupling Agents*, Vol. 3, VSP/Brill, Leiden (2004); K.L. Mittal (Ed.), *Silanes and Other Coupling Agents*, Vol. 4, VSP/Brill, Leiden (2007).
56. D. Flackett, in *Gelest 2000 Catalogue Silicon, Germanium & Tin Compounds, Metal Alkoxides, Metal Diketonates, and Silicones*, pp. 473–479, Gelest Inc., Tullytown, PA (1998).
57. J.-P. Hautekeer, in *Proceedings of Glass Processing Days 13–16 June 1999*, J. Vitkala (Ed.), pp. 214–219, Tamglass Ltd. Oy, Tampere, Finland (1999).
58. Anonymous, Standard Products Company, U.S. Patent 3,401,141 (1968).
59. Anonymous, Bostik Ltd., British Patent GB 1,143,229 (1969).
60. Anonymous, Sinclair Research Inc., British Patent GB 1,145,338 (1966).
61. F. Wilson, U.S. Patent 4,063,002 (1977).
62. D.H. Nicastro, in *Failure Mechanisms in Building Construction*, D.H. Nicastro (Ed.), pp. 48, 49, American Society of Civil Engineers (ASCE) Press, New York (1997).
63. T.J. Bridgewater and L.D. Carbary, in *Science and Technology of Building Seals, Sealants, Glazing and Waterproofing*, J.M. Klosowski (Ed.), Vol. 2, pp. 45–63, ASTM International, West Conshohocken, PA (1992).
64. R. Will, T. Kaelin, and A. Kishi, *Adhesives and Sealants*, SRI International, Menlo Park, CA (2003).
65. M. Hajek and K. Wagner, Bayer AG, U.S. Patent 4,002,601 (1977).
66. K. Kimura, T. Takeda, H. Hosoda, K. Ishikawa, and H. Okuhira, The Yokohama Rubber Company, U.S. Patent 5,747,627 (1998).
67. Anonymous, *Polyurethane MDI Handbook*, BASF AG, Ludwigshafen, Germany (2000).
68. P.H. Markusch, R.S. Pantone, R. Guether, and W.E Slack, Bayer AG, U.S. Patent 6,482,913 (2002).
69. R.P. Deltieure, in *Proceedings of Caulks and Sealants Short Course II*, The Adhesive and Sealant Council, Rosemont, IL (1992), available at www.ascouncil.org/.
70. E.M. Petrie, Moisture Curing Mechanisms for Adhesives and Sealants (2005), available at www.SpecialChem4Adhesives.com/.
71. T.C.P. Lee, *Properties and Applications of Elastomeric Polysulfides*, Rapra Review Reports, Report 106, Vol. 9, Rapra Technology Ltd., Shawbury, Shrewsbury, Shropshire, United Kingdom (1999).
72. T.C.P. Lee, T. Rees, and A. Wilford, in *Science and Technology of Building Seals, Sealants, Glazing, and Waterproofing*, C.J. Parise (Ed.), ASTM STP 1168, pp. 47–56, ASTM International, West Conshohocken, PA (1992).
73. V. Minkin, *Kauch. i Rez.*, *11*, 9–11 (1984).
74. R.H. Khan and R.C. Rastogi, *Chem. Ind.*, *55* (9), 282, 283 (1989).
75. G. Capozzi and G. Modena, in *The Chemistry of the Thiol Group, Part 2*, S. Patai (Ed.), pp. 785–839, John Wiley & Sons, Chichester, England (1974).
76. R.J. Coates, B.C. Gilbert, and T.C.P. Lee, *J. Chem. Soc. B*, 1387–1390 (1992).
77. E.A. Sheard, Thiokol Chemical Corporation, U.S. Patent 3,349,047 (1967).

78. M. Yamada and K. Watanabe, Toray Thiokol, Japanese Patent 58,134,123 (1983).
79. J.R. Panek, Thiokol Chemical Corporation, U.S. Patent 3,282,902 (1966).
80. J.C. Patrick, U.S. Patent 1,996,486 (1935) (application filed on April 25, 1927).
81. J.C. Patrick, Thiokol Chemical Corporation, U.S. Patent 2,142,144 (1939).
82. J.C. Patrick and H.R. Ferguson, Thiokol Chemical Corporation, U.S. Patent 2,466,963 (1949).
83. A. Mahon, Linear Polysulfides: Their Characterization and Degradation Pathways, Ph.D. Thesis, University of Warwick, Warwick, England (1996).
84. S.J. Hobbs, *Polym. Mater. Sci. Eng.*, *67*, 415, 416 (1992) and S.J. Hobbs, K.B. Potts, J.W. Nuber, and J.R. Gilmore, Morton International Inc., U.S. Patent 5,430,192 (1995).
85. M. Sundahl, E. Sikander, B. Ek-Olausson, A. Hjorthage, L. Rosell, and M. Tornevall, *J. Environ. Monit.*, *1*, 383–387 (1999).
86. M. Kohler, J. Tremp, M. Zennegg, C. Seiler, S. Minder-Kohler, M. Beck, P. Lienemann, L. Wegmann, and P. Schmid, *Environ. Sci. Technol.*, *39*, 1967–1973 (2005).
87. C.J. Yaggi, Jr., Neville Chemical Company, U.S. Patent 4,189,407 (1980).
88. Anonymous, Proposal for a Directive of the European Parliament and of the Council Amending for the 20th Time Council Directive 76/769/EEC Relating to Restrictions on the Marketing and Use of Certain Dangerous Substances and Preparations (Short Chain Chlorinated Paraffins), Commission of the European Communities, Brussels (2000).
89. C. Corden, Socio–Economic Impacts of the Identification of Priority Hazardous Substances under the Water Framework Directive—Final Report Prepared for European Commission Directorate–General Environment, RPA—Risk & Policy Analysts Ltd., Loddon, Norfolk, United Kingdom (2000).
90. M. Proebster and S. Grimm, Henkel–Teroson GmbH, U.S. Patent 6,919,397 (2005).
91. Anonymous, DuPont Tyzor Organic Titanates—Technical Note Adhesives & Sealants, E.I. du Pont de Nemours and Company, Wilmington, DE (2001).
92. J.R. Panek, in *Handbook of Adhesives*, I. Skeist (Ed.), 3rd ed., van Nostrand Reinhold, New York (1992).
93. P. Cognard, Sealants for Construction Part III-2—Elastomeric, High Performances Sealants (2004), available at www.SpecialChem4Adhesives.com/.
94. J.A. Lavelle and L.S. Frankel, *Adhesives Age*, *20* (11), 41–44 (1977).
95. W.T. Chang, Beecham Home Improvement Products, Inc., U.S. Patent 4,626,567 (1986).
96. S.L. Pratt and G.M. Lucas, General Electric Company, U.S. Patent 5,216,057 (1993).
97. See, for example, J.E. Goldstein, Air Products and Chemicals, Inc., U.S. Patent 5,124,384 (1992).
98. H. Loth, K. Helpenstein, T. Podola, and B. Knop, Henkel KGaA, U.S. Patent 5,118,732 (1992).
99. V.A. Demarest, J.A. Dionne, M. Lertora, and J.R. Magnotta, in *Science and Technology of Building Seals, Sealants, Glazing, and Waterproofing*, J.M. Klosowski (Ed.), Vol. 7, STP 1334, pp. 22–42, ASTM International, West Conshohocken, PA (1998).
100. B.E. Stanhope and W.D. Arendt, Velsicol Chemical Corporation, U.S. Patents 6,583,207 (2003) and 7,056,966 (2006).
101. M.W. Huang, Witco Corporation, U.S. Patent 6,001,907 (1999).
102. M.A. Sherwin, in *Proceedings of Adhesive and Sealant Council Short Course, October 21, 22, 2001*, The Adhesive and Sealant Council, Inc., Washington, DC (2001), available at www.ascouncil.org/.
103. J.S. Amstock, in *Handbook of Adhesives and Sealants in Construction*, J.S. Amstock (Ed.), pp. 3.1–3.26, McGraw-Hill, New York (2000).
104. See, for example: T. Mita, H. Nakanishi, J. Takase, K. Isayama, and N. Tani, Kaneka Corporation, U.S. Patent 4,507,469 (1985) and T. Hirose, S. Yukimoto, and K. Isayama, Kaneka Corporation, U.S. Patent 4,837,401 (1989).

105. For blends of acrylate polymers with SCO polyether polymer see, for example: T. Hirose and K. Isayama, Kaneka Corporation, U.S. Patent 4,593,068 (1986) and S. Kohmitsu, H. Wakabayashi, T. Hirose, and K. Isayama, Kaneka Corporation, European Patent 0,339,666 B1 (1995).

106. For blends of SCO acrylate polymers with SCO polyether polymers see, for example: H. Wakabayashi and K. Isayama, Kaneka Corporation, U.S. Patent 5,109,064 (1992).

107. K. Isayama and I. Hatano, Kaneka Corporation, U.S. Patent 3,951,888 (1976).

108. K. Isayama and I. Hatano, Kaneka Corporation, U.S. Patent 3,971,751 (1976).

109. J. Takase, T. Hirose, and K. Isayama, Kaneka Corporation, U.S. Patent 4,444,974 (1984).

110. Anonymous, MS Polymer Silyl–Silyl Terminated Polyethers for Sealants and Adhesives of a New Generation, Kaneka Corporation, Osaka, Japan (2000).

111. Anonymous, Technical Paper: Methoxysilane Sealants—One-Component Moisture Curing Methoxysilane Sealants, Cray Valley Ltd., Waterloo, Machen, Caerphilly, United Kingdom (2001), available at www.crayvalley.com/.

112. G.M. Seiter, 3M Company, U.S. Patent 3,627,722 (1971).

113. G.L. Brode and L.B. Conte, Jr., Union Carbide Corporation, U.S. Patent 3,632,557 (1972).

114. S.D. Rizk, H.W.S. Hsieh, and J.J. Prendergast, Essex Chemical Corporation, U.S. Patent 4,345,053 (1982).

115. C. Zwiener, L. Schmalstieg, and J. Pedain, Bayer AG, U.S. Patent 5,364,955 (1994).

116. E.R. Bryant and J.A. Weis, Inmont Corporation, U.S. Patent 3,979,344 (1976).

117. M.H. Berger, W.P. Mayer, and R.J. Ward, Union Carbide Corporation, U.S. Patent 4,374,237 (1983).

118. E.R. Pohl and F.D. Osterholtz, Union Carbide Corporation, U.S. Patent 4,645,816 (1987).

119. M. Huang, J. Nesheiwat, and N. Stasiak, Adhesives Age, 44 (6), 31–34 (2001).

120. M.W. Huang and B.A. Waldman, Witco Corporation, U.S. Patent 6,197,912 (2001).

121. L. Schmalstieg, R. Lemmerz, M.–H. Walter, and O. Wilmes, Bayer AG, U.S. Patent 6,545,087 (2003).

122. R.R. Johnston and P. Lehmann, Witco Corporation, U.S. Patent 5,990,257 (1999).

123. M. Huang, P. Lehmann, B. Waldman, and F. Osterholtz, Adhesives Age, 45 (4), 33–37 (2002).

124. M. Bahadur, T. Suzuki, and S. Toth, Dow Corning Corporation, U.S. Patent 6,258,878 (2001).

125. J.P. Kennedy, R.A. Smith, and L.R. Ross, Jr., University of Akron, U.S. Patent 4,276,394 (1981).

126. J.P. Kennedy, D.R. Weyenberg, L. Wilczek, and A.P. Wright, Dow Corning Corporation, U.S. Patent 4,758,631 (1988).

127. T. Iwahara, K. Noda, and K. Isayama, Kaneka Corporation, U.S. Patent 4,904,732 (1990).

128. T. Okamoto, M. Chiba, and J. Takase, Kaneka Corporation, U.S. Patent 6,335,412 (2002).

129. B.J. Sauntson, Scott Bader Company Ltd., U.S. Patent 4,333,867 (1982).

130. E.P. Plueddemann, Dow Corning Corporation, U.S. Patents 3,306,800 (1967) and 3,453,230 (1969).

131. K. Kohno, S. Nishikawa, Y. Hattori, and K. Kitao, Sunstar Giken K.K., U.S. Patent 4,478,990 (1984).

132. F. Kawakubo, S. Yukimoto, M. Takanoo, K. Isayama, and T. Iwahara, Kaneka Corporation, U.S. Patent 4,788,254 (1988).

133. W.E. Kozakiewicz, D.K. Potter, and S.A. Young, Tremco Incorporated, U.S. Patent 5,705,561 (1998).

134. M. Kusakabe and K. Kitano, Kaneka Corporation, European Patent Application 0,789,036 A2 (1997).

135. Y. Masaoka, Y. Nakagawa, T. Hasegawa, and H. Ando, J. ASTM Int., 3 (10), Paper JAI100450 (2006).

136. For detailed description see: M. Fujita, N. Hasegawa, and Y. Nakagawa, Kaneka Corporation, U.S. Patent Application 2004/0,029,990 A1 (2004).

14 Application of Self-Adhering Flashing Products in Building Openings

James D. Katsaros and Nyeleti S. Hudson

CONTENTS

14.1 INTRODUCTION

Moisture in buildings that accumulates on moisture-sensitive materials behind walls, roofs, and other areas that can trap water is a source of significant damage to buildings, including mold formation and structure rot. While there are many recognized sources of moisture intrusion in buildings, one of the most critical is the interface between the wall and openings in the wall associated with fenestration products (windows and doors). Figure 14.1 gives examples of the types of damage that can occur to buildings around window openings if they are improperly installed or flashed. A report by RDH Building Engineering Limited in Canada studied the occurrence of leakage for a wide variety of window types and assemblies, considering six potential leakage paths for water intrusion [1]. Although water leakage was found to some extent in all of the leakage paths, the "through window to wall interface to adjacent wall assembly" was the most prevalent leakage path for all types of windows tested and had a high risk of consequential damage to the building. Improper use of flashing and the overreliance on building sealants were consistently noted as key contributing effects in this report. Another report by the Partnership of Advancing Technology in Housing (PATH), which publishes the *Durability by Design Guideline,* noted that "most leakage problems are related to improper or insufficient flashing details or the absence of flashing" [2]. The PATH report also noted that "caulks and sealants are generally not a suitable substitute for flashing."

It is thus recognized that a durable seal at the window–wall interface is essential to help prevent damaging moisture intrusion in buildings and that building sealants alone cannot adequately perform this task. While mechanically attached flashing materials have been traditionally used in tandem with building sealants to deflect moisture away from and provide a moisture seal at the window–wall interface, self-adhering flashing products are now utilized to combine these properties into one product that provides moisture deflection as well as an "extension" of the sealant beyond the immediate location of window–wall interface. When applied properly,

APA photo

FIGURE 14.1 Examples of building damage due to improper flashing and installation. The figure on the left shows damage that occurs behind the building façade, while the picture on the right shows damage at the interior sill of the window.

self-adhering flashing products have been shown to be highly effective in protecting the window–wall interface for various window shapes and designs [3,4].

The self-adhering products serve as a bridge between the flange or mounting fin of a fenestration product and the water-resistive barrier (WRB), sealing the interface with a durable pressure-sensitive adhesive and topsheet that is typically at least 10 cm in width. Whereas a building sealant is designed to seal a 1–2 cm gap, the self-adhering flashing provides a much wider seal that is backed by a durable topsheet designed to maintain the seal integrity against building and joint movement and external exposure. Builders are now broadly utilizing these products because of their numerous advantages, which include ease of installation ("peel and stick"), robust water seal around the interface (an extension of the building sealant), and durability against environmental exposure and building joint movement. A survey of commercially available self-adhering flashing products as of 2001 is given in Reference 5. Since that time, numerous new product offerings have been launched in the market, including more butyl-based systems and extendable topsheet products to flash around nonlinear and three-dimensional (3-D) shapes, such as round-top windows and sill pans.

While the application of self-adhering flashing products is becoming widespread, there are a number of important considerations that must be addressed to achieve the full benefits of these products. For instance, since these products are installed in nonideal conditions compared to most industrial sealant applications, proper installation and surface conditions are essential for the successful use of these products. In the following sections, these considerations will be categorized into the following key elements: (1) types of self-adhering flashings, (2) performance attributes, based on a recently developed industry standard, and (3) installation methodologies unique to self-adhering flashing products.

14.2 TYPES OF SELF-ADHERING FLASHING PRODUCTS

The American Architectural Manufacturers Association (AAMA) has developed a material standard, "Voluntary Specification for Self-Adhering Flashing Used for Installation of Exterior Wall Fenestration Products," designated as AAMA 711-07, to specifically address the use of self-adhering flashing products [6]. This is the first consensus document to characterize self-adhering flashing products, specifying several physical property requirements, and will be examined in more detail in Section 14.3. Self-adhering flashing products are defined in AAMA 711-07 as "flexible facing materials coated completely or partially on at least one side with an adhesive material and which do not depend on mechanical fasteners for permanent attachment" [6]. The definition goes on to state: "They are used to bridge the joint (gap) between fenestration framing members and the adjacent water resistive barriers or sealed drainage plane material. The purpose of flashing is to drain water away from the fenestration product to the exterior."

The self-adhering flashing products are thus characterized primarily by (1) the type of topsheet used, which can be various polymer films, nonwoven material, or extendable material for specialty applications, and (2) the type of adhesive backing. Most products also contain a release liner, since the adhesive is very aggressive and would be difficult to unroll without this feature. The self-adhering flashing products are typically a lamination of the topsheet and a pressure-sensitive adhesive, and both components contribute significantly to the performance attributes. The topsheet provides the strength, tear resistance, durability, and flexibility (in some cases conformability) required for this demanding application. It is essential that the topsheet have sufficient durability to UV exposure, since the products are often exposed for several weeks or months after application and before the installation of the building façade, as well as thermal durability to withstand the thermal cycling that will occur both behind the façade and on the window flange. In addition, the topsheet must be flexible enough to conform around irregular shapes, maintain the integrity of the seal against building and joint movement, maintain physical integrity against thermal exposure, and, in some cases, be extendable to seal seamlessly on 3-D shapes such as the formation of sill pan flashing.

The pressure-sensitive adhesive backing is also a critical defining parameter in characterizing the end-use performance of the flashing product. Current offerings are categorized into two main adhesive classifications: (1) rubber-modified asphalt or bitumen adhesives and (2) polybutene or polyisobutylene-based "butyl" adhesives. Other types of adhesive systems, such as acrylic or block copolymer-based hot melt adhesives, are also available but are much less common than the two systems noted earlier. These adhesive systems are a blend of many components, including the base rubber, tackifiers, waxes, and other functional or nonfunctional fillers, that control the end-use properties of the adhesives. The right components, as well as proper mixing technology, are critical to achieve desired performance attributes. This chapter will focus on the end-use application and performance of these types of materials. More details on the chemical attributes of the pressure-sensitive adhesive systems can be found in a number of other sources [7–10].

The modified-asphalt-based materials were the first self-adhering flashing products on the market and currently have the majority market share, according to a recent Ducker study, whereas the butyl-based materials are the newer entrants and represent the fastest-growing portion of the market [11].

Several recent studies have compared the performance attributes of modified-asphalt-based self-adhering flashing products to butyl-based systems [12,13]. These studies have examined material durability and adhesion performance after exposure to UV light, accelerated heat aging, and adhesion performance to various substrates under various conditions (temperature, surface cleanliness, and moisture conditions) and have thrown up several important differentiating conclusions regarding self-adhering flashing systems.

To summarize, it is essential for the end user to consider the application conditions (surface, temperature, cleanliness, etc.) and the service life conditions (heat exposure, thermal cycles, and building movement) when selecting a self-adhering flashing system. A key challenge for self-adhering flashing products lies in the fact that while most industrial adhesives are applied in a controlled manufacturing environment, the self-adhering flashing products used in the building industry are applied in an external environment under all possible conditions, including wide temperature ranges, surface moisture, and a high degree of contamination (this is a construction site!). Thus, setting minimum performance attributes for these products is critical, and has been accomplished by the AAMA committee that completed the first version of the AAMA 711 standard for self-adhering flashing products, first published in December of 2005 and later revised in 2007 [6]. The contents of this standard will be the basis of the discussion of performance attributes in the section 14.3.

14.3 PERFORMANCE ATTRIBUTES OF SELF-ADHERING FLASHING PRODUCTS

Self-adhering flashing products play an essential role in the management of water in building openings, so it is critical that minimum performance requirements be standardized to ensure adequate performance. The AAMA published the first consensus standard for these products designated as AAMA 711-07, "Voluntary Specification for Self-Adhering Flashing Used for Installation of Exterior Wall Fenestration Products" [6]. This standard specifies nine property elements, with test method and minimum requirements noted, for self-adhering flashing products in building openings. These properties and minimum accepted values are listed in Table 14.1. The ICC Evaluation Services adopted these properties in the "Acceptance Criteria for Flexible Flashing Materials," AC-148, effective November 1, 2006 [14]. This section will review the basis and special considerations for material property attributes for self-adhering flashings, using the AAMA 711-07 test summary in Table 14.1 as a guideline. This AAMA standard also contains sections defining acceptable installation conditions and setting minimum dimensions such as flashing width, which is currently defined as "two inches (50 mm) past the critical interface, or at least four inches (100 mm) overall." The *critical interface* is defined in detail in the AAMA 711-07 standard.

14.3.1 TENSILE STRENGTH

As noted in Section 14.2, self-adhering flashing topsheets can be made from a wide variety of materials. These include polymer films, nonwoven sheets, and elastomeric films. A minimum tensile strength is specified so that the product has adequate strength to support the adhesive seal across the building joint and against movements in that joint without fracturing the seal. Since various materials can be used, three different test methods are specified in the AAMA 711-07 standard so that the appropriate method is utilized for a given material, as listed in Table 14.1.

14.3.2 SEALABILITY AROUND FASTENERS AND NAILS

Nail sealability is a key differentiating attribute for self-adhering flashing products. The ability to "self-heal" around penetrations through the flashing, such as nails, staples, and other fasteners found at building sites, is of great value to protect against water intrusion through fasteners used for siding, window trim, and other attachments at or near the window–wall interface, which exceeds the capabilities of mechanically attached flashings. The AAMA 711 standard specifies two different

TABLE 14.1
Property Requirements for Self-Adhering Flashing as per AAMA 711-07

Property	Test Method	Minimum Requirement
Section 5.1		
Tensile strength—rubber and thermoplastics	ASTM D 412, Method A, Die C	985 kPa (143 psi) minimum
Tensile strength—polymer modified bitumen	ASTM D 1970, Section 7.3	985 kPa (143 psi) minimum
Tensile strength—woven or nonwoven textile fabrics	ASTM D 5034	0.5 N/mm (2.9 lb/in.) minimum
Section 5.2		
Water penetration around nails:		
After 24 h @ 23°C ± 1°C (73°F ± 2°F) and 50% RH	ASTM D 1970, Section 7.9, modified per 5.2.1.	Must pass 31 mm (1.2 in.) of water head pressure
After thermal cycling (10 cycles)		Must pass
Water penetration around nails (alternative method):		
After 24 h @ 23°C ± 1°C (73°F ± 2°F) and 50% RH	ASTM E 331/E 547, or modified test per Annex 1	Must pass
After thermal cycling (10 cycles)		Must pass
Section 5.3		
90° Peel adhesion (initial)		
After 24 h @ 23°C ± 1°C (73°F ± 2°F) and 50% RH	ASTM D 3330, Method F	0.26 N/mm (1.5 lb/in.) minimum peel adhesion
Substrates noted in text		

Section 5.4		
Accelerated UV aging (14 days)	ASTM G 154, UVA cycle 1 ASTM D 3330, Method F	0.26 N/mm (1.5 lb/in.) minimum peel adhesion
90° Peel adhesion		
Appearance	Visual	Note change from original appearance in topsheet and/or adhesive
Section 5.5		
Elevated temperature exposure	ASTM D 3330, Method F	0.26 N/mm (1.5 lb/in.) minimum peel adhesion
Level 1—50°C (122°F) 7 days		
Level 2—65°C (149°F) 7 days		
Level 3—80°C (176°F) 7 days		
Appearance	Visual	Note change from original appearance in topsheet and/or adhesive
Section 5.6		
Thermal cycling (10 cycles)	AAMA (TBD), Section 5.5	0.26 N/mm (1.5 lb/in.) minimum peel adhesion
90° Peel adhesion (see Section 5.6 for temperatures)	ASTM D 3330, Method F, Section 16	
Appearance	Visual	Note change from original appearance in topsheet and/or adhesive
Section 5.7		
Cold temperature pliability	ASTM C 765	Must pass—18°C (0°F)
Section 5.8		
Peel adhesion after water immersion	AAMA 800, Section 2.4.1.3.1/2.4.1.4.3 Test B	0.26 N/mm (1.5 lb/in.) minimum peel adhesion
Section 5.9		
Resistance to peel	Annex 2	Report only

test methods for nail sealability: (1) a modification of ASTM D 1970, Section 7.9 and (2) a modification of ASTM E331/E547 for water resistance.

The modified ASTM D1970 test is a horizontal water column test. The height of the water column is specified at 31 mm (1.2 in.) of water to simulate 50 mph wind-driven rain. This is a realistic exposure for a material that will be either behind a façade, in the case of head and jamb flashings, or at the sill of a window, in the case of sill pans. A passing criterion is no water leakage under the flashing or the substrate around the fastener after 25 thermal cycles. As for the suitability of this test, the D1970 test has the advantage of being a laboratory scale test that is easy and inexpensive to run, but the flashing products are typically not in a horizontal state unless they are used in the sill pan, in which case a penetration through the flashing (nail, fastener, or staple) is not recommended. The alternative method for sealability around nails is a modified wall test, per ASTM E331/E547. This is a vertical test with three fasteners through the flashing substrate, as shown in a series of schematics in the AAMA 711-07 standard. This test better simulates the vertical application of self-adhering flashing at the heads and jambs of the windows, but requires a more complex setup that involves a test laboratory with a spray rack. Thus, there are advantages to both methods, and either method will meet the specifications of this standard.

14.3.3 PEEL ADHESION (INITIAL)

Peel adhesion is a quantitative measure of the bond between the self-adhering flashing and the substrate, which is also an indirect measure of the quality of the resulting watertight seal. Thus, this is the most critical physical property measurement to ensure adequate moisture seal performance. Peel adhesion in this application is also subject to a number of special challenges.

In typical industrial sealant and adhesive applications, the conditions of the substrate surface and exterior environment are specified and tightly controlled. This is essential to achieve reproducible results and a reliable bond between substrates. In the case of self-adhering flashings, this type of substrate and condition control is not possible. First, the products are applied on widely varying substrates: these include vinyl flanges of windows (which can be of various compositions), WRBs (which can range from polymeric films, nonwoven substrates to asphalt-impregnated building paper), or the exterior sheathing of the building, which is the largest source of variation. Building sheathing can include Oriented Strand Board (OSB), which itself is a highly variable substrate, plywood, fiberglass-coated gypsum board, poured concrete or concrete block, metal frame, and various other fibrous or film-coated materials.

While the substrate type is highly variable, the other main source of variability is the environmental and surface conditions in application. Self-adhering flashing products are applied in an exterior environment that is exposed to a continuous source of contamination and debris—it is a building site! In addition, the external exposure includes a full range of environmental conditions: wet, dry, hot, and cold are all expected states. Thus, substrate variability, surface contamination, and a full range of environmental conditions make the application of self-adhering flashing products a unique challenge to achieve a reliable, watertight seal over the joint between the fenestration and the wall. However, installers and flashing manufacturers have learned to overcome, or at least work around, these issues, and these products are being successfully utilized in a high proportion of building sites.

Another challenge is to quantify minimum "acceptable" adhesion for the end-use needs. This minimum acceptable adhesion must be adequate to hold the self-adhering flashing product in place through the expected life of the building, as well as enable a watertight seal over the joint–building interface. The AAMA 711-07 standard, Section 5.3, has established 2.6 N/cm (1.5 lb/in.) as the minimum peel adhesion force, as measured by ASTM D3330, method F, for a 90° peel between the flashing and the substrate. Conditions specified are 23°C and 50% RH, with a 24 h dwell time. The AAMA 711-07 standard specifies five substrates with which the product must meet this minimum peel adhesion value at room temperature conditions, which are oriented strand board (smooth side), aluminum, vinyl, plywood, and the self-adhering flashing-facing material (to account for adequate adhesion of laps). Self-adhering flashing products that are able to meet the minimum peel strength requirement on these substrates, as well as after exposures noted in the following text, without the aid of a primer, are categorized as Type A products. Those that require the use of a primer to meet these requirements are classified as Type B products.

As noted earlier, the actual external application environment will vary widely from the ideal conditions specified in ASTM D3330. Also, there are many other substrates that the self-adhering flashing products must adhere to other than the five noted in Section 5.3. To address this, Appendix A of the AAMA 711-07 standard provides an expanded list of substrates, including various WRBs for testing. Also, the lowest temperature where the minimum adequate peel adhesion (2.6 N/cm) is to be achieved is reported. Low-temperature adhesion is a primary application concern and represents the most challenging environment for the self-adhering flashing. Minimum temperature application is typically reported by flashing manufacturers as an indication of the adhesion performance—some report temperatures below 0°C, but this will also depend on surface moisture content (caution: such a surface will become icy below 0°C) as well as the integrity of the substrate surface, which depends on surface smoothness and continuity.

Substrate surface integrity is a key source of variability and low adhesion failure with self-adhering products. When an inadequate bond is realized, it is often the substrate surface rather than the pressure-sensitive adhesive on the flashing product that fails, because of the unbonded surface, where the components that make up the surface, (fibers, wood chips, fiberglass, etc.) pull away from the substrate. This is especially the case with the loosely bonded surface found on oriented strand board, where wood chips are pressed together to various degrees of bond, and fibrous sheathing material such as fiberglass-coated gypsum board. The mode of failure with these types of substrates is the delamination at the substrate surface. With substrates where the surface is loosely bonded, it is essential that these surfaces be adhesively bonded, through the use of a primer or other spray adhesive application, in order to achieve adequate bond strength with self-adhering flashing products.

To summarize, the actual adhesion performance of self-adhering flashing products thus depends on many factors, in addition to the adhesion strength of the pressure-sensitive adhesive. Performance testing of various commercially available self-adhering flashing products for a range of temperatures, substrates, and surface conditions has been reported by Katsaros [13]. This report indicates that certain substrates, such as fiberglass-coated gypsum board, need a primer to achieve adequate adhesion for all self-adhering flashing products tested. According to the report, butyl-based self-adhering products demonstrated better peel adhesion values at low temperatures and on challenging surfaces (such as wet or dusty OSB) than the modified-asphalt-based products.

To continue this assessment, and because of the increasing number of available butyl-based self-adhering flashing products, an update of the study referenced earlier [13] will be reviewed here. A total of 15 products were included in the evaluation. Six of the products are inherently extendable, which means they are able to conform to 3-D shapes, and were labeled as samples A, C, D, E, F, and N. The remaining 9 products are not extendable and were labeled as samples B, G through M, and O. An additional variable was the "chemistry" of the pressure-sensitive adhesive. Three of the samples were modified-asphalt or bitumen-based (A, B, and K) adhesives, 11 of the samples were butyl-based adhesives (C through G, I, J, L, and M through O), and 1 sample was an ethylene–propylene rubber-based adhesive (H) (see Table 14.2 for a listing of the samples tested along with their characteristics).

The adhesives were tested on representative substrates for the building industry, including OSB, concrete block (Concrete Masonry Unit—CMU), poly(vinyl chloride) [PVC], dusty OSB and dusty CMU to simulate real-world conditions, a representative WRB, and a fiberglass sheathing board. The temperatures used in the test reflected the range of reported installation and service temperatures for the

TABLE 14.2
List of Samples Tested

Flashing Product	Adhesive System	Topsheet Type
A	Rubber-modified asphalt or bitumen	Extendable film
B	Rubber-modified asphalt or bitumen	Extendable film
C	Polybutene or polyisobutylene butyl	Extendable spunbonded polyolefin
D	Polybutene or polyisobutylene butyl	Extendable film
E	Polybutene or polyisobutylene butyl	Extendable film
F	Polybutene or polyisobutylene butyl	Extendable spunbonded polyolefin
G	Polybutene or polyisobutylene butyl	Spunbonded polyolefin
H	Ethylene–propylene rubber	Polypropylene film
I	Polybutene or polyisobutylene butyl	Nonwoven fabric
J	Polybutene or polyisobutylene butyl	Foil
K	Rubber-modified asphalt or bitumen	Film
L	Polybutene or polyisobutylene butyl	Spunbond nonwoven
M	Polybutene or polyisobutylene butyl	Film
N	Polybutene or polyisobutylene Butyl	Extendable spunbonded polyolefin
O	Polybutene or polyisobutylene Butyl	Film

commercially available materials and were −18°C (0°F), −4°C (25°F), 21°C (70°F), and 38°C (100°F).

The first set of results concerns adhesion to dry OSB. The test data are shown in Figure 14.2. An important point to highlight is that at −18°C (0°F), all of the products had a very weak bond to the substrate. The adhesives were actually very stiff and brittle and, as a result, the value measured was the force required to bend the sample 90° in order to initiate the peel. In a number of cases, the adhesive cracked due to the low temperature. All of the samples failed interfacially at −18°C, meaning they peeled cleanly from the substrate, as illustrated in Figures 14.3 and 14.4. One product appeared to meet the 2.6 N/cm (1.5 lb/in.) requirement set by AAMA-711-07, but the actual adhesion to the OSB was minimal due to the stiffness of the sample.

At −4°C (25°F), the adhesive bond is somewhat stronger than it was at −18°C; the increase in adhesion to the OSB was at least 2× for the majority of the samples. Once again, the failure mode was interfacial for all of the samples, and one sample failed in a stick-slip manner, which is a discontinuous failure mode often found at low temperatures. When a sample fails in a stick-slip fashion, the peel values oscillate between high values, which are required to initiate peel from the substrate and low values, which occur when there is "slack" in the sample. This usually continues over the entire length of the peel surface. At −4°C, samples E and F were the only samples that were able to meet the 2.6 N/cm (1.5 lb/in.) peel adhesion requirement set by AAMA-711-07.

As expected, the peel strength increased as the temperature increased. As indicated in Figure 14.2, the samples tested at 38°C (100°F) had an average peel strength value of 5.07 N/cm (2.93 lb/in). The primary difference in the level of adhesion in this case was the chemistry of the different adhesives. Some of the adhesives failed interfacially, and some failed cohesively. The asphalt-based adhesives failed interfacially most of the time, whereas the butyl-based adhesives failed cohesively the majority of the time, with varying levels of adhesive left on the substrate (see Figure 14.5 for examples of typical results).

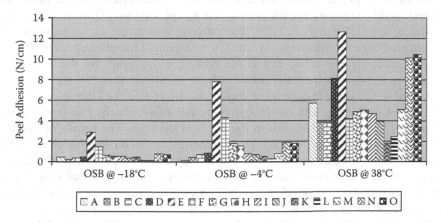

FIGURE 14.2   Peel adhesion of various self-adhering flashings to OSB at various temperatures without primer.

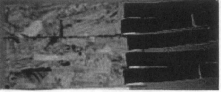

FIGURE 14.3 Representative results for extendable (left) and nonextendable (right) samples peeled from OSB at −18°C.

FIGURE 14.4 Representative results for nonextendable (left) and extendable (right) samples peeled from OSB at −4°C.

FIGURE 14.5 Representative results for asphalt-based (left) and butyl-based (right) adhesives on OSB at 38°C.

In an effort to characterize the adhesive bond to substrates that are indicative of real-world conditions, the adhesion of the samples to dusty OSB was compared to that of dry OSB at −4°C (25°F). The purpose was to select a temperature that would challenge the materials but would also highlight the differences in the surfaces. Intuitively, one would expect the peel adhesion results for the dusty OSB surface to be some factor lower than for the dry OSB samples. "Dusty" OSB was prepared in the same fashion as in Reference 13, in that the OSB samples were placed in a plastic bag with sawdust, shaken, and then the surface layer was gently blown off, leaving residual dust in the pores of the OSB. The results of the testing are shown in Figure 14.6. Of the 14 materials tested during this phase, 12 of the samples followed this trend. For the 2 samples whose peel strength actually increased, several factors may explain the results. The first is that it is impossible to ensure the same level of sawdust on each test surface, so the surfaces on which the flashings were tested may have had less sawdust than the other test boards. The second explanation is the highly variable nature of the OSB. The adhesives on the samples may have bonded to the larger particles in the board, which would require more force to peel the sample (see Figure 14.7 for pictures

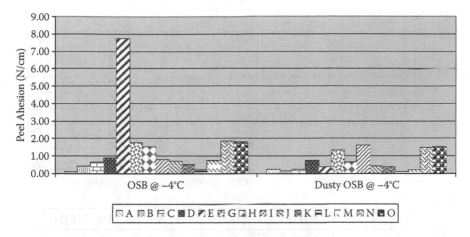

FIGURE 14.6   Peel strength of self-adhering flashing on OSB and dusty OSB.

FIGURE 14.7   Representative results for nonextendable (left) and extendable (right) samples peeled from dusty OSB at –4°C.

of the peel adhesion results for dusty OSB). The mode of failure for the materials was interfacial in all cases for the dusty OSB samples. As mentioned previously, it is difficult to replicate the same level of dust on each surface, but this in fact is the real-world situation. A consequence of this was that some samples fell off the boards prior to peeling—also found in the real world if the surface is highly contaminated. The low temperature also contributes to low adhesion, since without the sawdust the adhesion was minimal. These results are representative of what could happen in the field if surfaces are too dusty prior to applying the self-adhering flashing material.

The next set of results focus on peel adhesion for the self-adhering flashing products to a concrete block. Figure 14.8 shows the peel adhesion results for the temperatures tested. Again, the expectation is that as temperature increases, average peel adhesion to the substrate should also increase. Results for 14 of the 15 samples followed this trend. The exception was product A, which had higher adhesion at –18°C (0°F) than at –4°C (25°F). Since the recommended installation temperature for this product is 7°C (45°F), there are a number of explanations for the data. The surface of the CMU, similar to the OSB, is not flat and is porous in nature. This means that the adhesive will "flow" into the pores differently on each sample. Another interesting observation for this substrate is that the thickness and softness of the adhesive

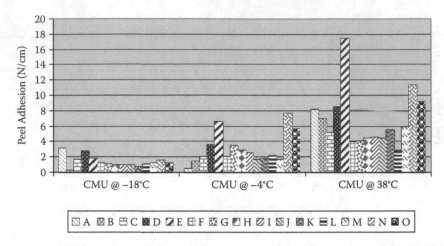

⬙A ⬛B ⊟C ◨D ⬗E ⊞F ⬚G ◪H ⬓I ⬙J ⬙K ⊟L ⬙M ⬙N ◨O

FIGURE 14.8 Peel adhesion of self-adhering flashings on CMU at various temperatures.

seemed to impact the level of adhesion to the substrate. The assumption is that this attribute allows the adhesive to flow, which would allow it to fill the pores in the concrete and form a mechanical bond to the substrate. It was noted that the pressure-sensitive adhesive systems that did not tend to "flow" had much lower average peel adhesion values at 38°C (100°F).

At −18°C (0°F), all of the materials had a very weak bond to the substrate, and failed interfacially. One notable point is that product D, which has a high thickness, had an average peel adhesion of 2.75 N/cm (1.59 lb/in.) at this temperature. Figure 14.9 is representative of the results obtained.

The adhesion performance of the self-adhering flashings improved at −4°C (25°F), but despite the stronger bond strength between the adhesive and the substrate, all of the samples still failed interfacially. Products D, E, G, H, N, and O all had adhesion greater than 2.6 N/cm (1.5 lb/in.) at this temperature. Figure 14.10 is representative of the results obtained.

All of the products had good adhesion at 38°C (100°F). The failure at this temperature was cohesive for the majority of the samples, with the exception of some of the asphalt-based adhesives, which failed interfacially. An additional observation

FIGURE 14.9 Representative results for butyl-based (left) and asphalt-based (right) samples peeled from CMU at −18°C.

FIGURE 14.10 Representative results for extendable (left) and nonextendable (right) samples peeled from CMU at −4°C.

FIGURE 14.11 Representative results for asphalt based (left) and butyl-based (right) adhesives for samples peeled from CMU at 38°C.

noted was staining on the concrete from the bitumen-based adhesives after peeling (see Figure 14.11 for the results).

Additional samples were tested on dusty concrete and primed concrete. Again, the purpose of the dusty surface was to simulate real-world conditions. Primed concrete was tested since primers are commonly used on concrete surfaces in the field. The results of the testing are shown in Figure 14.12. The expectation was that the average peel adhesion values for the dusty concrete would be less than for the dry concrete, and the values would be higher for the primed concrete. Differences in the data can be explained by the same explanations listed earlier for dusty OSB.

The performance of the self-adhering flashings on the dusty concrete was similar for the test group if it is viewed in terms of asphalt-based and butyl-based samples. All the bitumen-based samples had very little adhesion to the concrete, or fell off the concrete when the peel was initiated. The butyl-based materials showed some adhesion to the concrete, and measurable test data were obtained for all the samples. In most cases, the average peel adhesion was lower for the dusty surface, as expected. The samples for which average adhesion seemed to have increased, the data for the dry concrete and the dusty concrete were not statistically different, the assumption is that those samples had a lower level of "dust" on them (see Figure 14.13 for representative results of the tests).

As expected, all of the materials adhered well to the primed concrete. Some of the samples failed cohesively, but the majority of the samples failed interfacially

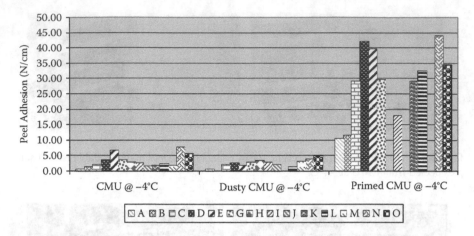

FIGURE 14.12 Peel adhesion of self-adhering flashings on CMU, dusty CMU, and primed CMU.

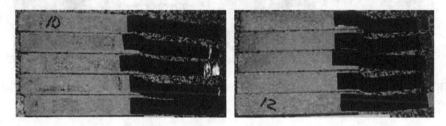

FIGURE 14.13 Representative results for butyl-based (left) and asphalt-based (right) samples peeled from dusty concrete at −4°C.

between the topsheet and the adhesive, leaving the adhesive fully bonded to the surface. Samples H, J, and M tore when the peel was initiated and as a result do not have a recorded value (see Figure 14.14 for typical examples of the peel adhesion results on primed concrete).

   To further characterize the performance of self-adhering flashings to primed concrete, additional temperatures were tested. Based on the mode of failure (interfacial between the adhesive and the topsheet) for the products at −4°C, the decision was made to test a higher temperature (21°C) instead of −18°C. The third temperature tested was 38°C. The results are shown in Figure 14.15.

   Since the performance of the samples at −4°C was discussed earlier, the discussion of the results for this set of samples will begin with those run at 21°C. The mode of failure in this case varied by sample or product type but was consistent within the set. In some cases, the failure was purely cohesive, but the failure for many samples was purely interfacial with the adhesive delaminating from the topsheet. Initially, two samples failed cohesively and, as the peel progressed, the adhesive separated from the topsheet. Once again, if the sample tore during the process, then a value was not recorded—which was an indication of excellent adhesion to the primed surface (see Figure 14.16 for representative pictures from the testing).

FIGURE 14.14 Representative results from nonextendable (left) and extendable (right) samples peeled from primed concrete at –4°C.

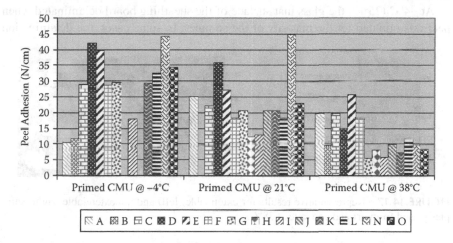

FIGURE 14.15 Peel adhesion of self-adhering flashings on primed CMU at various temperatures.

GURE 14.16 Representative results from butyl-based extendable products (left) versus asphalt-based nonextendable products (right) peeled from primed concrete at 21°C.

The performance of the self-adhering flashing products on primed concrete was consistent between product types, with one exception. The asphalt-based samples all failed interfacially between the topsheet and the adhesive. The mode of failure for sample E was the same. All of the remaining butyl-based adhesives failed cohesively with varying degrees of separation. Figure 14.17 describes the failure types observed.

The third substrate tested was a gypsum-based fiberglass sheathing board. The substrate was primed to reflect common practices in the field and to understand the adhesion bond between the sheathing material and the self-adhering flashing products. In the previous study [13], it was shown that no commercially available self-adhering flashing product had adequate adhesion to unprimed fiberglass sheathing board, due to the loosely bonded surface of the sheathing. It was shown that this surface must be primed to achieve good adhesion. Figure 14.18 shows the results of the samples tested.

At −4°C (25°F), the glass mat surface of the sheathing board delaminated when peel was initiated for the majority of the samples. In some cases, the delamination

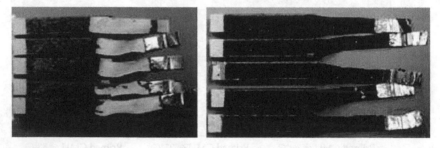

FIGURE 14.17 Representative results for extendable (left) and nonextendable (right) samples peeled from primed concrete at 38°C.

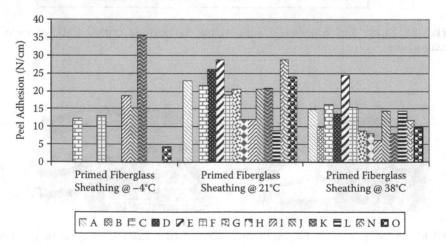

FIGURE 14.18 Peel adhesion of self-adhering flashings on primed fiberglass sheathing at various temperatures.

and interfacial failure between the adhesive and the topsheet were observed in the same set of samples. As a result, no value was recorded. It is believed that the adhesive between sheathing surface and gypsum was not meant for use in this manner at such low temperatures. All but one of the remaining samples failed interfacially between the topsheet and the adhesive; one sample (I) failed cohesively. The results are shown in Figure 14.19.

The primary mode of failure at 21°C was interfacial between the adhesive and the fiberglass sheathing or between the adhesive and the topsheet. Only one sample (B) failed in a different manner: the test was stopped when the topsheet tore during peeling. Representative results are shown in Figure 14.20.

At 38°C (100°F), all but three of the samples failed cohesively during the peel process. The remaining three samples, B, E, and K, failed interfacially between the butyl adhesive and the topsheet. Representative results are depicted in Figure 14.21.

The fourth substrate tested was a commonly used fibrous WRB. Since the surface is relatively flat and not particularly challenging, the expectation was that adhesion for all products would increase with temperature. The results for all samples are shown in Figure 14.22, and for the most part they met expectations.

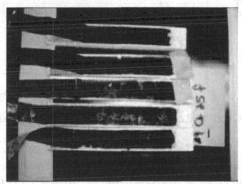

FIGURE 14.19 Representative results for extendable (left) and nonextendable (right) samples peeled from primed fiberglass sheathing at −4°C.

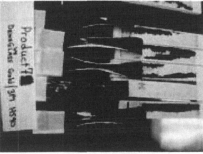

FIGURE 14.20 Representative results for nonextendable (left) and extendable (right) samples peeled from primed fiberglass sheathing at 21°C.

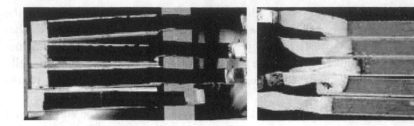

FIGURE 14.21 Representative results for nonextendable (left) and extendable (right) samples peeled from primed fiberglass sheathing at 38°C.

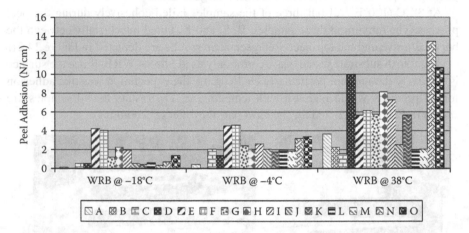

FIGURE 14.22 Peel adhesion of self-adhering flashings on a WRB at various temperatures.

At −18°C (0°F), most of the products were stiff and brittle, which led to three primary modes of failure. Several of the products failed interfacially and, therefore, had minimal adhesion to the substrate. Others failed in the stick-slip mode described earlier. Two of the products actually adhered to the substrate at this temperature, and the mode of failure was delamination (during the 90° peel the top layer of the substrate pulled away from the surface of the substrate). The products also surpassed the AAMA-711-07 requirement of 2.6 N/cm (1.5 lb/in.). Figure 14.23 depicts representative results at this temperature.

FIGURE 14.23 Representative results for two butyl-based nonextendable samples peeled from a water-resistive barrier at −18°C.

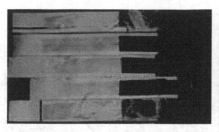

FIGURE 14.24   Representative results for nonextendable (left) and extendable (right) samples peeled from a water-resistive barrier at −4°C.

At −4°C (25°F), the adhesion for all samples improved substantially. While the types of failure were similar to the previous test, the shifts occurred in the categories of failure. For this temperature, the primary mode of failure was substrate surface delamination to varying degrees depending on the sample. Two of the samples failed in the stick-slip mode, and sample B still had minimal adhesion. Figure 14.24 shows representative results for this temperature.

As expected, the performance of all of the products improved at 38°C. Once again, the primary mode of failure was varying degrees of substrate surface delamination. In a few cases, there was a combination of substrate and cohesive failures (the adhesive splits). The outliers were one sample that failed interfacially between the butyl adhesive and the topsheet, and another that failed cohesively. Figure 14.25 depicts representative results.

The fifth substrate tested was PVC, which is representative of the flange material for many commercially available windows. The expectation was that the adhesion would be exceptional for all products since it is a relatively smooth and flat surface. The results of the test are shown in Figure 14.26.

The leading cause of failure at both −18°C (0°F) and −4°C (25°F) was the stick-slip phenomenon discussed earlier. At the lower temperature, the brittleness of some of the samples resulted in the material breaking during the peeling process. Again, stick slip is not a measure of peel adhesion; what is being measured is the force required to break the adhesion bond from the substrate. An additional mode of failure observed was the adhesive adhering to the PVC and the topsheet delaminat-

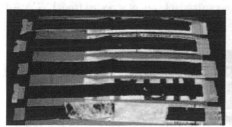

FIGURE 14.25   Representative results for asphalt-based (left) and butyl-based (right) samples peeled from a water-resistive barrier at 38°C.

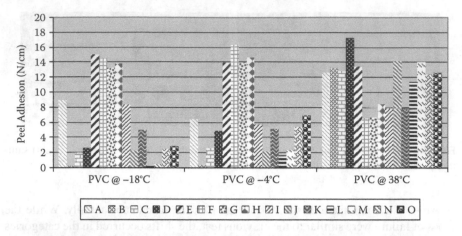

FIGURE 14.26 Peel adhesion of self-adhering flashings to PVC at various temperatures.

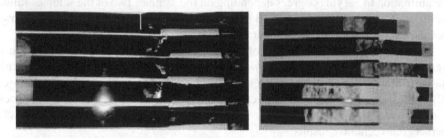

FIGURE 14.27 Representative results for asphalt-based (left) and butyl-based (right) samples peeled from PVC at −18°C.

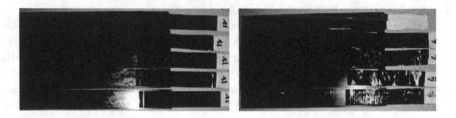

FIGURE 14.28 Representative results for nonextendable samples peeled from PVC at −4°C.

FIGURE 14.29 Representative results for butyl-based (left) and asphalt-based (right) samples peeled from PVC at 38°C.

ing from the adhesive during the peeling process. Figures 14.27 and 14.28 depict representative results for −18°C (0°F) and −4°C (25°F).

The principal mode of failure at 38°C (100°F) was cohesive for the butyl-based adhesives. All of the asphalt-based samples failed interfacially, with one sample showing mixed mode failure. It is the opinion of the authors that this is the only set of samples for which an actual peel adhesion value was obtained for this substrate. Representative results are shown in Figure 14.29.

In summary, a comparison of the products can be completed on two levels: the first being the difference between asphalt- and butyl-based products; the second being the differences in the butyl-based products. When the butyl-based and asphalt-based adhesives are compared over the range of temperatures and substrates studied, the butyl-based products outperformed the asphalt-based products. One could argue that it was not really an equitable comparison because the recommended installation temperatures for the asphalt-based products are much higher than those tested. The primary differences between the products that are butyl-based adhesives are believed to be chemistry related and were not investigated as part of this work.

The objective of this work was to compare representative commercially available flashing products by testing them under a range of conditions that they could be exposed to in the field. This in no way includes all of the possible conditions and exposures in use, nor is it a complete listing of products. The key learning from this effort is that there are a wide range of products commercially available with varying levels of performance. As a result, one should carefully consider the performance of the product versus its intended use when making selections for specific applications.

14.3.4 ACCELERATED UV AGING

As noted earlier, the self-adhering flashing products are installed during the installation of the window, which can occur several weeks or months before the exterior façade is installed. Thus, the products must have sufficient UV stability to prevent deterioration of the topsheet as well as the pressure-sensitive adhesive before being covered and protected from UV radiation. It is important that these products not be exposed indefinitely to exterior light unless specifically noted by the manufacturer.

The UV-aging performance of polymeric materials can be simulated using an accelerated UV-aging test, of which there are several versions with different light sources and exposure schemes. The AAMA 711-07 standard specifies ASTM G154, UVA cycle 1, for 14 days (336 h), whereas ICC-ES AC-148 specifies 10 h per day for 21 days (210 h), which is consistent with the acceptance criterion for WRBs, according to ICC-ES AC-38. Accelerated aging is done to simulate the "long-term" effects of aging, but it is difficult to estimate the real-time exposure with the accelerated aging tests. In general, it is best to do real-time UV exposure (typically done in Florida or Arizona) in order to correlate actual UV performance in use. The effect of long-term UV exposure on various types of self-adhering flashing products was studied in more detail and is documented [12]. This study utilized accelerated UV aging techniques and showed a dramatic difference in the performance of various commercially available self-adhering flashing products, particularly in terms of their adhesive classification (butyl based or modified asphalt or bitumen based).

14.3.5 EXPOSURE TO ELEVATED TEMPERATURE—THERMAL HISTORY

Self-adhering flashing products are exposed continuously to thermal cycling due to external exposure throughout the life of their use. These products typically are attached to the window flange and bridge the interfacial gap with the wall, so it is essential that the products maintain their integrity throughout the extremes of the thermal cycles. Temperatures behind siding that gets direct sunlight, as well as on flanges that can be made of metal (typically aluminum) or vinyl, or even reflected sunlight off the neighboring buildings can build up excessive heat behind the sidings. The photograph in Figure 14.30 was taken on an 80°F (26.5°C) day in Sacramento, California, where the surface temperature on gray-colored wood siding was measured to be 170°F (76.7°C). It can be easily imagined that on hotter days, or on metal surfaces that received direct or reflected sunlight, the temperatures can easily exceed 80°C.

While these extreme temperature exposures may last only a few hours on a given day, the products used behind these surfaces will see this exposure many times over their useful life, resulting in an accumulated thermal affect.

The effect of the thermal exposure for various types of commercially available self-adhering flashing products was studied in detail [13]. This study showed significant differential performance of the self-adhering flashings, depending primarily on the adhesive system utilized; it was found that the butyl-based systems were more thermally stable than modified-asphalt- or bitumen-based materials.

Thus, it is important that the correct self-adhering flashing products be selected based on the thermal exposure expected. To this end, the AAMA-711-07 standard has specified three levels of heat exposure that can categorize the products differentially into acceptable thermal exposure ranges: Level 1 at 50°C (122°F), Level 2 at 65°C (149°F), and Level 3 at 80°C (176°F). As detailed in Reference 12, not

Photo by Steve Easley

FIGURE 14.30 Surface temperature on a 26.5°C (80°F) day in Sacramento, California.

all commercially available self-adhering flashing products are able to withstand the Level 3 heat exposure, particularly those that feature a rubber-modified asphalt or bitumen adhesive system, as they are prone to losing integrity, causing the top-sheet to curl back. The performance criterion in AAMA 711-07 is that the products maintain minimum peel adhesion to a given surface after exposure for 7 days at the select temperature in the preceding text, along with no change (pull back) in original appearance. It is thus important to consider the heat exposure expected when select-ing a self-adhering flashing product to ensure durability throughout the expected life of the installation.

14.3.6 THERMAL CYCLING

In addition to high-temperature exposure, the self-adhering flashing products will be subject to temperature cycles in use. Thus, while a self-adhering flashing prod-uct may be exposed to temperatures above 80°C, as per the previous discussion, they may also be exposed to extremely low temperatures, such as −40°C. The adhe-sive systems for self-adhering flashing products are a blend of many components, as noted previously, including the base rubber, tackifiers, waxes, and other fillers. The adhesives are processed to ensure the correct morphology for the functional perfor-mance of the adhesive, but it is possible that this morphology may be disrupted at severe temperature exposure, causing phase separation and a shift in performance. As a result, the AAMA-711-07 standard requires thermal "shock" cycling between −40°C and 50°C for 25 cycles, followed by peel adhesion testing to ensure that mini-mum performance is maintained. This will ensure that the adhesive system will not undergo unexpected modification in morphology or physical form through expected thermal cycles in end use.

14.3.7 COLD TEMPERATURE PLIABILITY

As discussed in Section 14.3.6, self-adhering flashing products may be exposed to extreme low temperatures as well as high temperatures. In order to perform as a sealing material, the self-adhering flashing products must maintain ductile pliability and, in particular, must not form cracks or exhibit brittle failure. Thus, the AAMA 711-07 standard requires a cold temperature pliability test at −18°C (0°F), which can be done per the bend tests described in the various ASTM methods noted in the AAMA 711-07 standard.

14.3.8 ADHESION AFTER WATER IMMERSION

Self-adhering flashing products must maintain their watertight seal behind building façades, which are known to have only limited ability to hold water back. Thus, it is expected that the self-adhering flashing products will see extensive moisture expo-sure during the end-use life of these products. In fact, self-adhering flashing products can also be used at the sill below the fenestration product as sill pan flashing, in which case the products are in a horizontal position and may experience exposure to standing water for a period of time. The AAMA 711-07 standard thus requires that the self-adhering flashing products maintain integrity and minimum peel adhesion after 7-day immersion in tap water.

This test is specified to be performed on an aluminum substrate, such that the substrate itself is not affected by the water immersion (as OSB or other wood-based sheathings would be). However, in reality, these substrates will be bonded to OSB substrates during installation in wet conditions, so adhesion performance to wet surfaces is critical. The effect of moisture on the peel adhesion of various commercially available self-adhering flashing products was studied in more detail [12,13]. In one of these studies [13], the effect of dusty surfaces was examined as well.

14.4 INSTALLATION OF SELF-ADHERING FLASHING PRODUCTS

14.4.1 INSTALLATION STANDARDS

The application of correct installation procedures is essential for the successful use of self-adhering flashing products. Installation variations can be extremely complex, as many variables need to be considered. The installation details can depend on several factors, which include (1) the fenestration product integration type (integrally flanged, nonintegrally or field-applied flanged, brick mold, or box or nonflanged products), (2) the fenestration product geometry (rectangular, round top, circular or oval, or other), (3) the wall system (wood framed, concrete block, metal framed, etc.), (4) the wall system (drainage cavity, surface barrier, or concrete or mass wall), (5) the WRB system (flexible membrane, fluid or spray applied, laminated board system, etc.), and (6) the expected environmental exposure of the installation (wind, rain, groundwater, heat, etc.).

Unfortunately, no standard exists today that adequately addresses all of these factors and provides guidelines to the installer on installation details pertinent to the given situation. General sequential guidelines for the installation of windows and doors have been developed through the ASTM E2112 "Standard Practice for Installation of Exterior Windows, Doors and Skylights" [15], originally published in 2001, and which went through a major revision in 2007. The standard developed four basic methods depending on the relative sequencing of the WRB and flashing product. The sequences are often based on the local trade practices, for example, whether the framer installs the WRB before the windows are delivered, typical of the eastern United States, or whether the windows are installed first, which is more typical in the western United States. Table 14.3 summarizes these sequences for the four methods.

While all of these methods are approved for window and door installation, the self-adhering flashing products typically will use either an A or A1 method, since the benefits are not fully realized if the flashing is behind the window flange, as with B methods. This, of course, depends on the local trade practices. However, as noted earlier, the ASTM E2112 standard is a general guideline that provides basic sequential steps for installation, but does not address environmental exposure considerations, wall system variations, or different WRB systems, nor does it handle different fenestration integration systems very well other than integrally flanged windows. In addition, the original ASTM E2112 document dealt mostly with mechanically

TABLE 14.3
ASTM E2112 Installation Sequence Considerations

	Head and jamb flashings will be applied over the face of the integrally mounting flange of the window	Jamb and sill flashings will be applied behind the face of the integrally mounting flange of the window
Water-resistive barrier is to be applied after window or door installation	Method A	Method B
Water-resistive barrier is to be applied prior to window or door installation	Method A1	Method B1

attached flashing systems, specifying a minimum 22.5 cm (9 in.) width, since the self-adhering flashing products were not as established at the time of the original draft.

In addition, the codes have not yet provided much guidance on installation and the use of self-adhering flashing products. The International Code Council International Residential Code (IRC) Section 703.8 was modified in 2006 to read as follows:

IRC 703.8 Flashing. Approved corrosion-resistant flashing shall be applied shingle-fashion in such a manner as to prevent entry of water into the wall cavity or penetration of water to the building structural framing components. The flashing shall extend to the surface of the exterior wall finish. Approved corrosion-resistant flashings shall be installed at all of the following locations:

Exterior window and door openings. Flashing at exterior window and door openings shall extend to the surface of the exterior wall finish or to the water-resistive barrier for subsequent drainage.

What qualifies as "approved corrosion-resistant flashing" is not defined in the codes or referenced to any industry standard at this time. Thus, all that is required is that the products be installed shingle fashion and directed to either a drainage plane or the exterior.

14.4.2 PRINCIPLES OF INSTALLATION

Because of the many complexities involved in the installation of various types of fenestration products in all the possible wall systems and expected environmental exposures as noted in the previous section, it is impossible to prescribe installation details that cover all of the potential installation configurations. However, if the installer follows a set of guiding principles for installation, which can be applied to any configuration, the likelihood of success of the installation dramatically increases. These basic principles are summarized in the following list, followed by some supporting comments:

Principles for Installation:
1. Maintain continuity between flashing and the window–wall interface.
2. Ensure correct shingling or lapping of flashing and other water-resistive membranes and materials.
3. Provide a drainage path for leaks.
4. Test the window–wall system as a complete installation.
5. Install at conditions that promote adhesion of sealants and flashing.
6. Use materials that have adequate durability for the anticipated exposure.
7. Carry out risk assessment.

14.4.2.1 Continuity

Continuity is defined in the ASTM E2112 standard [15] as follows:

"Continuity—Continuity shall be maintained between elements in the fenestration product and the water resistive barrier that provides weather protection, air leakage control, and resistance to heat flow and vapor diffusion...."

The continuity between the fenestration product and the WRB is best provided by the flashing products, and self-adhering flashing products provide the greatest water-resistant seal over this interface if used properly. Lack of continuity at this interface is a source for leaks causing air and moisture intrusion, which is unfortunately a very common occurrence in actual field installations. Figure 14.31 provides an example of insufficient continuity in actual installations observed in the field.

14.4.2.2 Correct Shingling or Lapping of Materials

The application of flashing in correct shingle fashion is now required by the ICC Code, Section 703.8, but this is one of the most common flaws found in the field. Figure 14.32 illustrates the basic principle of applying building materials in correct shingle fashion.

Figure 14.33 shows typical shingling errors, where in one case (left) the flashing at the head of the window is installed under the flange of the window. The other picture (right) shows another common flaw: instead of the WRB being lapped over the flashing, the head flashing is applied over the WRB. In the picture on the right, the sill flashing is also reverse-shingled, as it is applied over the flange of the window at the sill rather than under the jamb. In both of these cases, water that gets behind the WRB or the flashing product (which is not self-adhering) is likely to travel behind the head of the window and enter into the wall cavity or interior of the building. This is often depicted as a "window leak," but is actually a flawed installation.

FIGURE 14.31 Example of lack of continuity between the WRB and fenestration product.

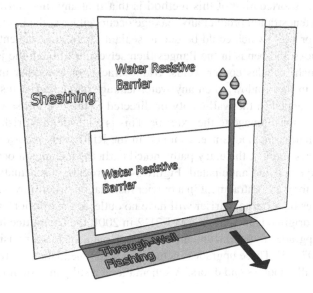

FIGURE 14.32 Application of materials in correct shingle fashion.

FIGURE 14.33 Examples of incorrect shingling of flashing at the window head. On the left, the flashing is installed behind the head flange and is not adequately attached to the wall, providing a passage for water entry into the building. On the right, the flashing is over the housewrap at the head, and the flashing at the sill is over the nailing fin, representing incorrect shingling in both cases.

14.4.2.3   Provide a Drainage Path for Leaks, "Barrier versus Drainage Installation Methods"

The original ASTM E2112 document also detailed a "barrier" installation method, in that sealant was applied around the entire perimeter of the integral flange of the fenestration product to create a full barrier at the external interface between the window and the wall system. Thus, both the air and water seals existed at the external interface. The shortcoming of this method is that if at any time during the life of the installed fenestration product any leakage occurs either at the interface between the window or wall (which could be due to sealant failure, insufficient sealant coverage, or cracks and bends in the flanges themselves) or through the joinery of the window (which includes corner welds, frame cracks, or any other mechanical or welded joint in the window), then any water leakage into a barrier installation system will be trapped in the wall cavity or directed to the interior, since there is no reliable means for drainage to the exterior. This is a key defining risk with using a barrier installation method, since, as noted in the RDH study reported in [1], leakage in windows through the entry paths noted earlier is a common occurrence that should be expected and anticipated. Figure 14.34 provides a schematic to describe the potential for water entrapment in a barrier installation, where any water intrusion that penetrates the external barrier will have no outlet to the exterior.

Since the original launch of ASTM E2112 in 2001, the committee has developed significant upgrades to the document, which was published as a revision of ASTM E2112 in 2007 [15]. These upgrades include the recommendation for the use of "sill pans" under all windows and doors. A schematic of a sill pan system is indicated in Figure 14.35.

The sill pan is a protective water-impermeable substrate at the base of the fenestration product, window or door, that acts to collect any water intrusion that may occur at the various locations noted earlier in the Canada Mortgage and Housing

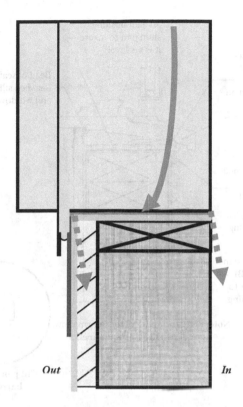

FIGURE 14.34   Side-view schematic of water entrapment potential for barrier installation methods.

Corporation (RDH) study [1], and directs this water safely to the exterior, through either the WRB in a drainage cavity wall or directly to the exterior, such as in a surface barrier wall (illustrated in the circle in Figure 14.35). The sill pans can be made up of a variety of materials, including metals, rigid or semirigid plastics, or self-adhering flashings. The self-adhering flashing option includes both extendable or conformable flashings for seamless three-dimensional shapes and nonextendable "patched" systems. The schematic shown in Figure 14.36, taken from the EEBA Water Management Guide [16], illustrates three common types of sill pan system designs, showing (from left to right) a nonextendable self-adhering flashing sill pan system, an extendable or conformable self-adhering flashing sill pan system, and a rigid or semirigid sill pan system.

The sill pan system, as explained earlier, enables the installation to manage any water intrusion that will inevitably occur through the life of the fenestration product installation. This is accomplished by leaving openings in the seal between the window flange and the sill pan flashing at the sill of the window, to allow drainage. The 2007 version of ASTM E2112 calls for at least two discontinuities in the sealant that are each at least 50 mm (2 in.) wide. Many drainage installation methods do not call for any sealant at the sill of the window. In this drainage system, a robust air and water seal at the *interior* perimeter joint between the fenestration product and the

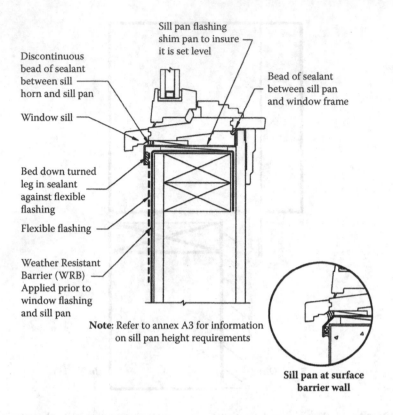

Sill pan flashing
shim pan to insure
it is set level

Discontinuous
bead of sealant
between sill
horn and sill pan

Bead of sealant
between sill pan
and window frame

Window sill

Bed down turned
leg in sealant
against flexible
flashing

Flexible flashing

Weather Resistant
Barrier (WRB)
Applied prior to
window flashing
and sill pan

Note: Refer to annex A3 for information
on sill pan height requirements

Sill pan at surface
barrier wall

FIGURE 14.35   Sill pan schematic, drawing taken from ASTM E2112 (ASTM E2112-07, "Standard Practice for Installation of Exterior Windows, Doors and Skylights").

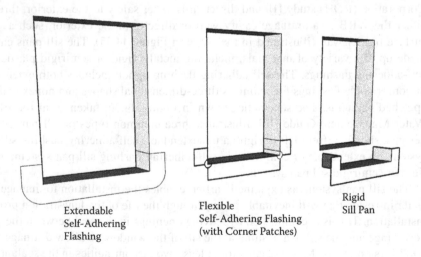

Extendable
Self-Adhering
Flashing

Flexible
Self-Adhering Flashing
(with Corner Patches)

Rigid
Sill Pan

FIGURE 14.36   Various sill pan system designs, utilizing extendable self-adhering flashing, flexible self-adhering flashing with corner patches, and a rigid sill pan.

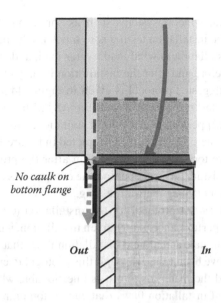

*No caulk on
bottom flange*

Out   *In*

FIGURE 14.37   Side-view schematic of drainage installation system.

wall cavity is essential. This seal must be continuous and of sufficient adhesion and durability to withstand any pressures that will be imparted through the "open" sill at the exterior interface.

A schematic of the drainage installation with a sill pan system is given in Figure 14.37. Note that the discontinuous sealant at the exterior flange is left open to enable drainage, while the interior perimeter seal now acts as the primary barrier to air and water intrusion. It is important to note that this internal seal will not be subject to the external environmental cycles that the external seals are, resulting in less joint movement over time, and thus the joint should be more durable.

Self-adhering flashing products, both extendable and nonextendable, are broadly utilized for the sill pan flashing, as well as for the more traditional application at the external jambs and heads of the fenestration product. More details on the minimum dimensions, requirements for end-dams and back-dams, and other specifics are contained in the 2007 version of the ASTM E2112 standard practice [15].

14.4.2.4   Test the Window Wall System as a Complete Installation

Window manufacturers are very diligent at testing window performance for thermal, wind, and water resistance. In addition, much testing is done on wall assemblies to measure energy efficiency and water management performance. However, there is relatively little testing required or reported on the performance of the window–wall assembly as installed. A wall testing protocol was reported along with test results on various wall configurations by Weston et al. [3], showing the advantages of the drainage installation methods. The AAMA has developed a laboratory test protocol, AAMA 504–05 [17], that is used to qualify installation methods outside the conventional standard as detailed in ASTM E2112 [15] or the InstallationMasters™ Training course [18]. Another report has reported the use of this installation test protocol for nontypical window systems such as "brick mold" windows [19]. The general principle of the installation testing is to mock up the installation in the laboratory; subject the installation to wind load, water load, and structural loading; and perform the testing before and after the installation is subjected to thermal cycling. The flowchart for a suggested protocol is given in Figure 14.38, as first reported by Weston et al. [3]. The test protocol utilizes three ASTM methods for air [20], water [21], and structural [22] performance of the installation. This is done before (Phase I) and after (Phase II) thermal exposure, as depicted in Figure 14.38. It is also critical that the installation be torn down for inspection after the protocol is completed in order to assess any moisture intrusion behind the components of the installation, that was not readily apparent during the wall testing.

While this laboratory test protocol for the installation test greatly enhances the ability to evaluate the performance of any given installation, it is very difficult to adequately predict and simulate actual field installation flaws that will occur. Purposely flawed installations have been tested through this protocol to evaluate the robustness of the installation and the materials, but it is questionable whether this adequately covers the multitude of installation flaws that can be found in the field. In addition, the thermal cycling provides some indication of durability of the installation through

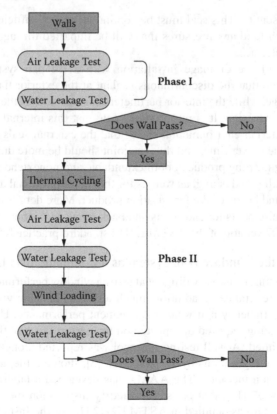

FIGURE 14.38 Suggested flowchart for installation test protocol.

joint movement and thermal history, but it is not expected that this will adequately simulate the real exposure that the materials and installation will see in actual use.

Therefore, while the laboratory wall testing is an essential component of successfu l installation performance, it cannot fully replace the specific installation guidelines as the only reliable performance measurement for a successful installation.

14.4.2.5 Installation Conditions That Promote Adhesion

The fifth principle of installation is concerned with ensuring that the self-adhering flashing products perform in the field as required based on the initial conditions for installation, that is, it should adequately adhere to the surface. There are many important surface and environmental factors to take into account to ensure proper adhesion under widely varying conditions. These elements are addressed by the AAMA-711-07 physical property standard [6], which was evaluated in great detail in Section 14.3.

14.4.2.6 Material Durability

Whereas the previous principle of installation addressed the initial conditions for installation, it is equally important that the materials used to seal the fenestration interface with the wall will perform in the field as required over the life of the installation, that is, it should maintain integrity through UV exposure, heat exposure, thermal cycling, water immersion, etc. It is possible for temperatures behind building façades or on the window flanges to reach 80°C or higher on warm sunny days, or if exposed to reflective materials. Are the materials used suitable for these exposures or will this heat exposure cause degradation and loss of moisture seal? This topic was discussed at length in Section 14.3.5, as addressed by the material property standard developed by AAMA.

14.4.2.7 Risk Assessment

While the 2007 version of ASTM E2112 "recommends" the drainage method featuring a sill pan, this system is not mandated by code or any industry standard as yet. This, along with much of the installation details, remains the decision of the design professional, installation contractor, or window manufacturer. The decision as to what installation method to use is actually an assessment of risk tolerance, bringing together a number of factors. The EEBA Water Management Guide [16] has developed a set of risk factors, including water resistance rating of the fenestration product, moisture tolerance of the assembly, exposure to the environment, rainfall amount, drying potential, and workmanship. Taking into account the preceding guiding principles, a full evaluation of all these risk factors is necessary to determine the necessary features for a successful installation. In particular, Section 14.4.2.3 describes how a drainable installation method, featuring a redundant level of water protection behind the primary barrier to the exterior, significantly reduces the overall risk of moisture damage in a building opening.

14.4.3 Sample Installation Details

While no installation guidelines exist today can address all of the installation variations and conditions that one will encounter in the field, window and flashing manufacturers have developed general installation details that can cover the bulk of the
installations in any given area or window or wall type. An excellent step-by-step
illustration for flashing a flanged window system is given by Hagstrom [23]. In addition, Hagstrom provides details for flashing a door system, featuring a brick-mold
frame rather than flanges, in a later article [24]. Figures 14.39 and 14.40 provide
sequential flashing details for a typical A1 (WRB applied before window) and A
(WRB applied after window) methods, as described in ASTM E2112 and Table 14.3.
These details utilize an extendable self-adhering flashing as the sill pan system. Also,
these details include a "round-top" window system, where extendable or conformable self-adhering flashing is used to provide a seamless installation around the arch
of the window. These details are for illustration purposes only, as there are many
other effective means available.

14.5 SUMMARY

When used properly, self-adhering flashing products are highly effective at preventing moisture intrusion in building openings. There are many physical property
requirements, installation conditions, long-term exposure conditions, and installation details that must be taken into account for proper utilization. While much progress has been made in providing these guidelines through industry standards such
as AAMA 711 and ASTM E2112, much work still needs to be done to further define
the parameters for an effective use of these and other such building products in the
wide range of conditions and systems that the installer will encounter. However, until
the standards "catch up" with the broad utility of the self-adhering flashing products,
following the Principles for Installation outlined in Section 14.4.2 of this chapter will
greatly enhance the performance attributes of the installation.

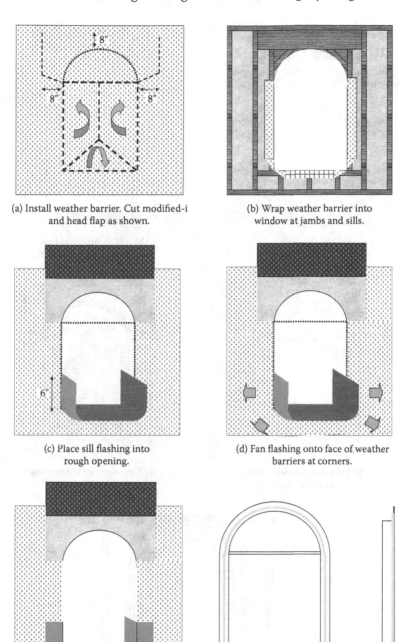

(a) Install weather barrier. Cut modified-i and head flap as shown.

(b) Wrap weather barrier into window at jambs and sills.

(c) Place sill flashing into rough opening.

(d) Fan flashing onto face of weather barriers at corners.

(e) 3-Dimensional flashing installed.

(f) Caulk jamb and head flange.

FIGURE 14.39A Installation when weather barrier is installed before the window installation.

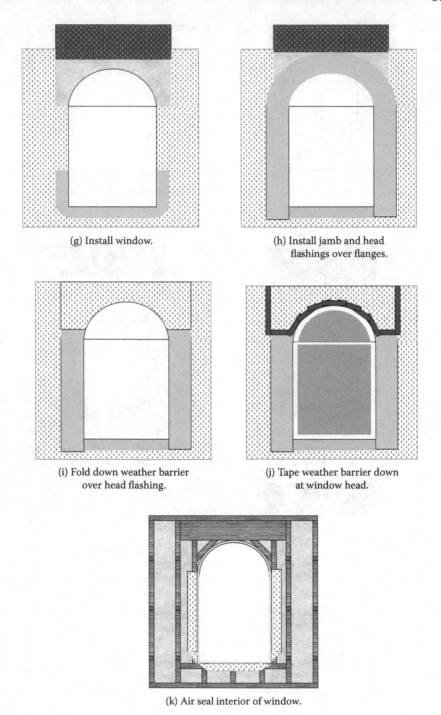

(g) Install window.

(h) Install jamb and head flashings over flanges.

(i) Fold down weather barrier over head flashing.

(j) Tape weather barrier down at window head.

(k) Air seal interior of window.

FIGURE 14.39B Installation when weather barrier is installed before the window installation.

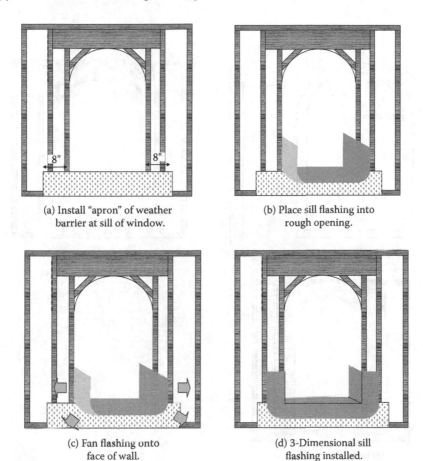

(a) Install "apron" of weather barrier at sill of window.

(b) Place sill flashing into rough opening.

(c) Fan flashing onto face of wall.

(d) 3-Dimensional sill flashing installed.

FIGURE 14.40A   Installation when water barrier is installed after the window installation.

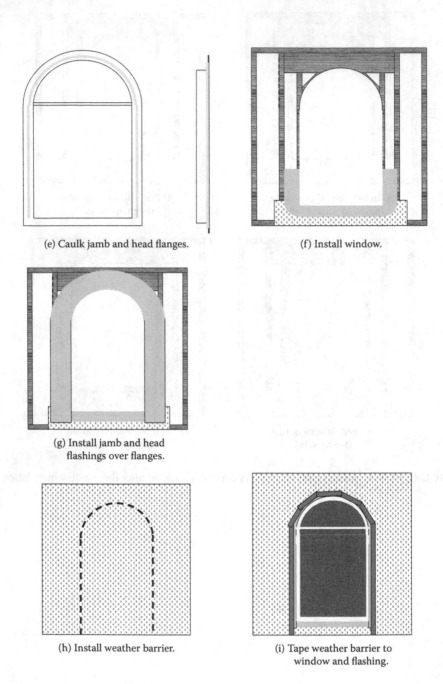

(e) Caulk jamb and head flanges.

(f) Install window.

(g) Install jamb and head
flashings over flanges.

(h) Install weather barrier.

(i) Tape weather barrier to
window and flashing.

FIGURE 14.40B   Installation when water barrier is installed after the window installation.

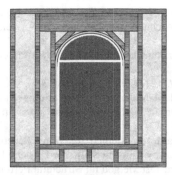

(j) Air seal window from interior.

FIGURE 14.40C   Installation when water barrier is installed after the window installation.

REFERENCES

1. Water Penetration Resistance of Windows—Study of Manufacturing, Building Design, Installation and Maintenance Factors, A Report by RDH Building Engineering Limited, Vancouver, BC, submitted to Canada Mortgage and Housing Corporation, Ottawa, Ontario (December 2002).
2. Durability by Design, A Guide for Residential Builders and Designers, Partnership for Advancing Technology in Housing (PATH), NAHB Research Center, Inc., Upper Marlboro, MD (May 2002).
3. T.A. Weston, X. Pascual, and J. Herrin, Paper presented at the Texas A&M University 13th Symposium on Improving Building Systems in Hot & Humid Climates, Houston, TX (May 2002).
4. T.A. Weston and J. D. Katsaros, *Walls and Ceilings*, pp. 34–44 (September 2003).
5. M. Holiday, *J. Light Construction*, 51–60 (June 2001).
6. Voluntary Specification for Self-Adhering Flashing Used for Installation of Exterior Wall Fenestration Products, "AAMA 711-07," American Architectural Manufacturers Association (AAMA), Schaumburg, IL, June 2007.
7. I. Skeist (Ed.), *Handbook of Adhesives,* 3rd edition, Van Norstrand Reinhold, New York (1990).
8. A. Pizzi and K. L. Mittal (Eds.), *Handbook of Adhesive Technology,* 2nd edition, Marcel Dekker, New York (2003).
9. E.M. Petrie (Ed.), *Handbook of Adhesives and Sealants,* McGraw-Hill, New York (2000).
10. J.S. Amstock (Ed.), *Handbook of Adhesives and Sealants in Construction*, McGraw-Hill, New York (2001).
11. AAMA/WDMA U.S. Industry Market Studies—Fenestration Products, Ducker Research Company, Inc., Bloomfield Hills, MI (2006).
12. A.D. Zima, T.A. Weston, J.D. Katsaros, and R. Hagood, in *Durability of Building and Construction Sealants and Adhesives, ASTM STP 1453,* A.T. Wolf (Ed.), ASTM International, West Conshohocken, PA (2004).
13. J.D. Katsaros, *J. ASTM Intl., 2*, No. 10 (2005).
14. ICC Evaluation Services, Acceptance Criteria for Flexible Flashing Materials, AC-148, Approved June 2006, Effective November 1, 2006.
15. ASTM E2112-07, "Standard Practice for Installation of Exterior Windows, Doors and Skylights."
16. EEBA Water Management Guide, Building Science Press, Inc., Westford, MA, April 2006.

17. AAMA 504-05, *Voluntary Laboratory Test Method to Qualify Fenestration Installation Procedures,* American Architectural Manufacturers Association (AAMA), Schaumburg, IL (December 2005).

18. InstallationMasters™ training course, operated by Architectural Testing Institute (ATI), New York, PA.

19. B. J. Crowder-Moore, *J. ASTM Intl., 2* No. 10 (2005).

20. ASTM E 283, "Standard Method for Determining Rate of Air Leakage Through Exterior Windows, Curtain Walls, and Doors Under Specified Pressure Differences Across the Specimen."

21. ASTM E 331, "Standard Test Method for Water Penetration of Exterior Windows, Curtain Walls, and Doors by Uniform Static Air Pressure Difference."

22. ASTM E 330, "Standard Test Method for Structural Performance of Exterior Windows, Curtain Walls, and Doors by Uniform Static Air Pressure Difference."

23. C. Hagstrom, *J. Light Construction,* 1–8 (June 2005).

24. C. Hagstrom, *J. Light Construction,* 1–8 (January 2006).

15 Sealants for Bridge Expansion Joints

A. Pizzi

CONTENTS

15.1 INTRODUCTION

Bridge sealants or, more properly, sealants needed for the expansion joints of cement bridges, need to have particular properties for their unusual requirements. They need to be flexible and elastic, they often need to bond with still wet cement and, when used for bridge repair work, they must be able to set sufficiently quickly to reopen the bridge and road to the traffic in a matter of hours. On top of all this, they need to be extremely strong to withstand the slow movements but powerful stresses induced by the whole bridge moving, expanding, and contracting with the daily temperature variations.

These are quite contrasting requirements and can be satisfied only by using sealants composed of mixtures of different resins of different characteristics. Many systems have been devised to achieve the conditions for compromise between toughness and flexibility. In general, toughness and strength are achieved by using epoxy resins, and extensive reviews of their use as bridge sealants exist [1]. Epoxide resins are distinguished by many good properties. However, cured products have flexibility and toughness that are inadequate for certain applications. A large number of additives that have a flexibilizing action have, therefore, been described in the literature, including compounds that contain thiol groups [2]. Among these, the liquid polysulfide polymers, which are known under the trade name thiokols and are in most cases used in combination with amine-curing agents, have received the most attention [3]. Several patents describe in detail a variety of these systems [4]. The U.S. Pat. No. 3,090,793 [5] describes polyesters-containing mercaptan groups, obtained by reacting mercaptoalcohols, for example, 2-mercaptoethanol, with polycarboxylic acids having at least 18 carbon atoms, especially polymeric fatty acids, and the use of these products together with basic catalysts, preferably with tertiary amines, as hardeners for epoxide resins. The hardened systems show good mechanical properties and an increased flexibility.

The U.S. Pat. No. 3,352,810 [6] also describes epoxide resins with improved flexibility that, in addition to the resin and the conventional hardeners for epoxide resins, contain an ester, having two or more SH groups, of a mercaptocarboxylic acid, with a polyol as a flexibilizer. Preferred mercaptocarboxylic acids are thioglycolic acid or 3-mercaptopropionic acid, and preferred polyols are poly(ethylene glycol)s, poly(propylene glycol)s, and poly(propylene triol)s. The hardeners used are, *inter alia*, primary, secondary, and tertiary amines.

The U.S. Pat. No. 3,472,913 [7] describes polyethers containing both hydroxyl groups and thiol groups and having a molecular weight of at least 1000, and a theoretical thiol groups/hydroxyl groups ratio greater than 1. The number of thiol groups per molecule can be varied as desired. These polymers containing thiol groups can be crosslinked with a large number of reagents, also including epoxide resins. In the case of mixtures with epoxide resins, a basic catalyst is preferably used, for example, an amine. The crosslinked products are distinguished by high toughness and extensibility.

In the U.S. Pat. No. 4,126,505 [8], mixtures of materials are described that contain (1) an epoxide resin; (2) a hardener for the epoxide resin, such as an amine having at least three H atoms that are bonded directly to an aliphatic or cycloaliphatic amine nitrogen atom, or certain tertiary amines; (3) a polymercaptan; and (4) a polyene having at least two activated ethylenic double bonds. The mixtures of materials are suitable as adhesives that, by reaction of the polymercaptan with the polyene, rapidly form a rubber-like adhesive bond and then fully crosslink as a result of the hardening of the epoxide resin by the amine. A large number of polymercaptans can be used as the component (3). Thus, (a) mercaptan-containing polyesters, (b) sulfides with mercaptan end groups, (c) polybutadienes or butadiene/acrylonitrile copolymers and poly(monomercaptancarboxylate)s having mercaptan end groups, and (d) oxyalkylene compounds with mercaptan end groups can be used.

The U.S. Pat. No. 3,914,288 [9] describes adducts obtained by reacting an epoxide resin with a polymercaptan having at least two SH groups, which are separated from one another by a chain of at least six carbon atoms or carbon atoms and oxygen atoms, and an amine. These adducts are used as hardeners for the production of flexible epoxide resins. Polymercaptans preferred for the preparation of the adducts are (a) esters of mercaptocarboxylic acids and polyols, (b) oligomers and polymers having recurring disulfide units, (c) polymercaptans containing hydroxyl groups, obtained by reacting a chlorohydrin ether of a polyhydric alcohol with a hydrosulfide in an alcoholic medium, and (c) polyesters containing thiol groups.

In spite of the many approaches suggested earlier, none of them is fully able to overcome the difficulties in the production of flexible, tough-elastic systems based on epoxide resins. For this, a different class of mixed sealants is needed. Thus, another class of compounds used for the purpose are epoxy compositions mixed with a toluene diisocyanate-poly(oxypropylene)-based polyether polyol. This contains a small amount of tert-butylphenol-blocked isocyanate (–NCO) groups. It is mixed with (a) coal tar pitch and (b) *N*-(2-aminoethyl) piperazine as a deblocker and coreagent of epoxy groups and free-NCO groups [10].

Thus, to withstand the powerful stresses involved, the resins used are two-component epoxide systems with polyamines, although these systems are still too rigid if used alone to be suitable for the flexibility needed for this application. However, the weaknesses of these epoxy resins, namely, low flexibility and poor adhesion, can be overcome by modification with blocked crosslinking polyurethane resins [11]. These introduce the necessary level of flexibility and improve adhesion to cement, even to wet cement, while a high level of mechanical resistance is obtained due to the epoxy resins.

During the curing of such three-component epoxy–diamine–polyurethane systems, the amine unblocks the polyurethane and crosslinks both epoxy resin and the long-chain polyurethane elastomers to form a cured network. Through the diamine, both epoxy to epoxy, polyurethane to polyurethane, and also epoxy to polyurethane crosslinks occur.

15.2 CHEMISTRY

Bisphenol A glycidyl ether–based epoxy resins are the most commonly used. Different types of reactive polyurethane resins can be used, that is, branched or linear with the basic structure illustrated in Figure 15.1, the characteristics of which are indicated in Table 15.1. Depending on the polypropylene polyol used, the resin can be trifunctional, as illustrated in Figure 15.1, or only difunctional. The reactive polyurethane needs to be blocked to avoid any self-hardening during storage.

The polyamine component used is generally a commercial 3,3'-dimethyl-4,4'-diaminodicyclohexyl methane (Figure 15.2) [11], but other polyamines can also be used. A three-component system comprising a low-viscosity bisphenol A epoxy resin with an equivalent weight of 190 (epoxy value 0.5–0.55), a low-viscosity polyurethane, and the polyamine in a ratio 30:70:12.5 presents, in general, a tensile strength of approximately 13 MPa and 200% elongation at break [11]. The most important characteristic of such a type of network is that it is tough yet flexible. Strength and elongation characteristics over a wide range can be obtained by varying the respective proportions of these three components.

TABLE 15.1
Characteristics of Different Types of Reactive Polyurethane (PUR) Resins

	Reactive PUR Resin 1	Reactive PUR Resin 2
Functionality	3	2
Type	Branched	Linear
Viscosity at 25°C (mPa.s)	80,000	33,000
Blocked –NCO groups (%)	2.4	1.7

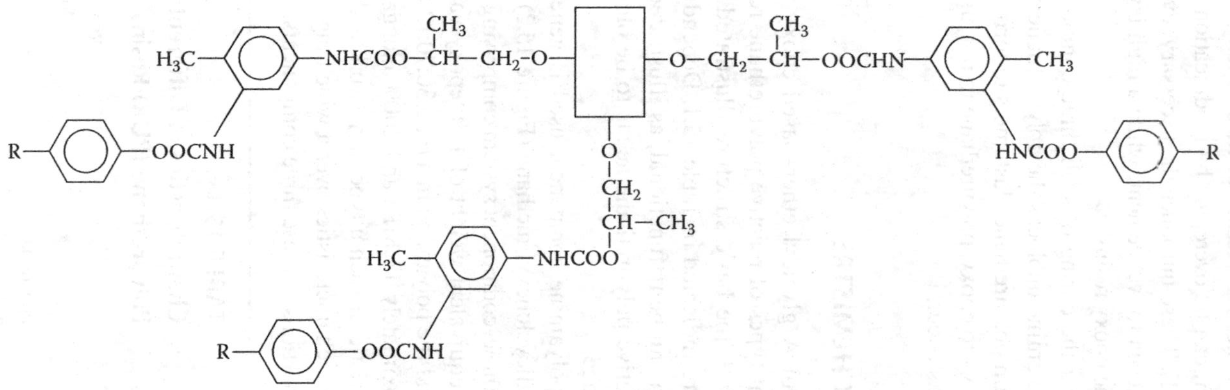

FIGURE 15.1 Schematic formula of a reactive PUR resin. Depending on the polyoxypropylene polyol used (shown as a rectangle in the center of the molecule), this resin can be branched and trifunctional, or linear and difunctional.

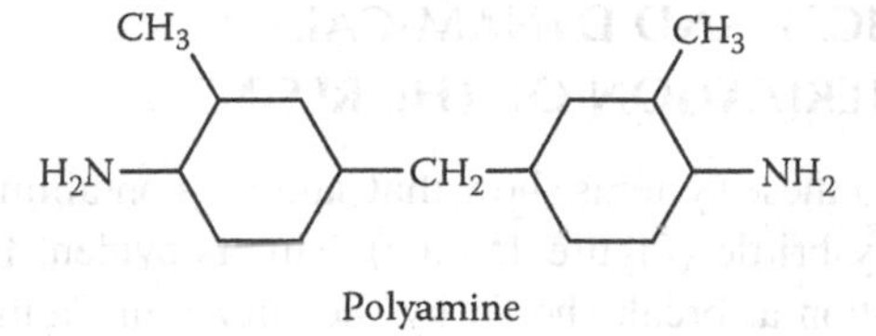

Polyamine

FIGURE 15.2 Structure of 3,3′-dimethyl-4,4′diaminodicyclohexyl methane, the polyamine-type crosslinker generally used for the epoxy–PUR–polyamine sealants.

As the main resin components are expensive, black liquid tar is added in order to reduce the cost of the product. This does not constitute a problem as the sealant will then have, on curing, the same color as the bridge road on which it is applied.

The reactions that take place during curing of a three-component epoxy-diamine-polyurethane system are shown in Figure 15.3. Two reactions between similar reagents (PUR resin with PUR resin, and epoxy resin with epoxy resin) and one reaction between two different reagents (PUR resin with epoxy resin) are possible [11].

The crosslinking reaction can be considerably accelerated by increasing the reaction temperature, and this has no significant effect on the properties of the finished sealant. At both ambient and at higher temperatures, it is possible to accelerate the curing considerably by adding a typical amine-type epoxy resin crosslinking catalyst, or a mixture of catalysts.

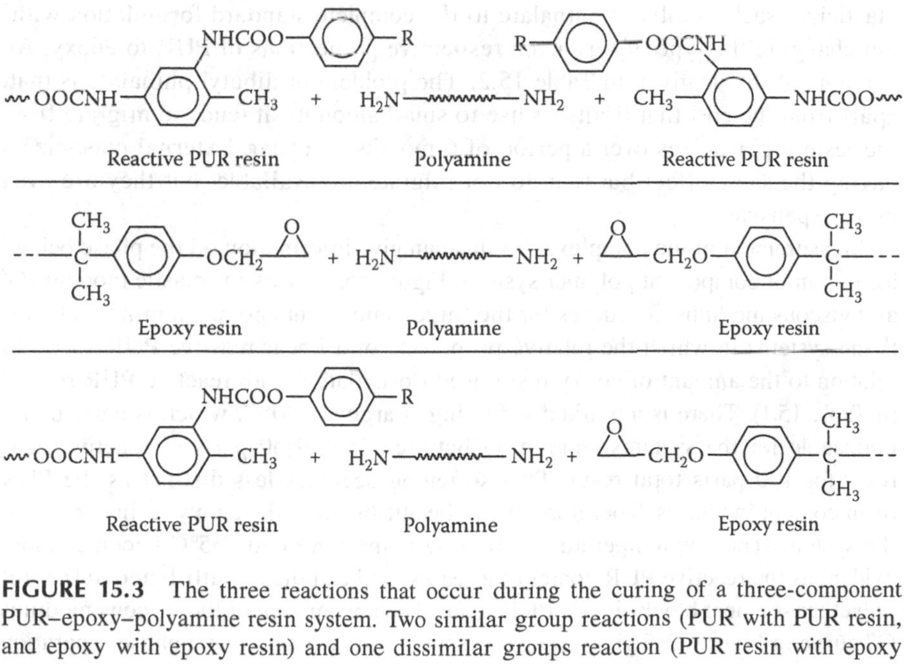

FIGURE 15.3 The three reactions that occur during the curing of a three-component PUR–epoxy–polyamine resin system. Two similar group reactions (PUR with PUR resin, and epoxy with epoxy resin) and one dissimilar groups reaction (PUR resin with epoxy resin) occur.

15.3 MECHANICAL AND DYNAMICAL CHARACTERIZATION OF THE RESINS

Stress–strain tests on these systems show that samples containing up to 25% of PUR resin 1 are still very brittle (Figure 15.4a,b). This is evident from the low tensile strength and elongation at break shown by the curves in Figure 15.4. In a sample containing around 30% reactive PUR resin, the tensile strength increases markedly to 45 MPa when tested without tar addition. It progressively decreases to below 5 MPa as the reactive PUR resin proportion is further increased. At the same time, the elongation at break rises. The maximum strength is, in general, much lower when tar is added to reduce the cost of the sealant.

As regards the use of these systems for bridge expansion joints, the ranges of elongation and strength required vary depending on the application needs. In general, the proportions of PUR to epoxy resin are around 50–60:50–40 by weight (Figure 15.4a,b) for maximum strength and an elongation up to 50%. However, in very cold climates (–20°C to –30°C) where such proportions would lead to cracking of the cured sealants, much higher proportions of reactive PUR can be used or the type of tar changed. Table 15.2 shows one example of a formulation used for very cold climates, as well of the type of latent hardener/accelerator amine used. The type of tar extender used also influences the elongation of the system. In particular, the freezing/crystallizing temperature of the liquid tar needs to be determined. If it is too high, either the type of tar should be changed or at least it should be partly combined in various proportions with tars crystallizing at much lower temperatures. An alternative to this is the addition of a plasticizer such as dibutyl phthalate to the complete standard formulation without changing the type of tar or the respective proportions of PUR to epoxy. An example of this is given in Table 15.2. The problem of dibutyl phthalate is that, apart from the cost that limits its use to small amounts, it tends to migrate from the resin it plasticizes over a period of 6 months to a year. External plasticizers having the same effect but that do not migrate are available, but they are even more expensive.

Measurement of the complex modulus can give information on the phase behavior of a multicomponent polymer system. Figure 15.5 shows the elastic modulus G′ and viscous modulus G″ curves for the three-component epoxy–diamine–polyurethane systems in which the relative proportion of a linear reactive PUR varies in relation to the amount of epoxy resin used (formulation with reactive PUR resin 2 in Table 15.1). There is a marked softening at around 100°C, which is particularly noticeable for the systems containing between 35 and 50 parts of reactive PUR resin for 100 parts total resin. This softening becomes less distinct as the PUR resin content increases. Therefore, it can be attributed to the epoxy resin content of the system. The low-temperature softening point at around –55°C becomes more evident as the reactive PUR content increases, and can thus be attributed to the soft polyether segment blocks in the PUR resin. The appearance of the viscous modulus G″ curve in Figure 15.5 gives the impression of the existance of a phase separation in the crosslinked plasticized epoxy resin.

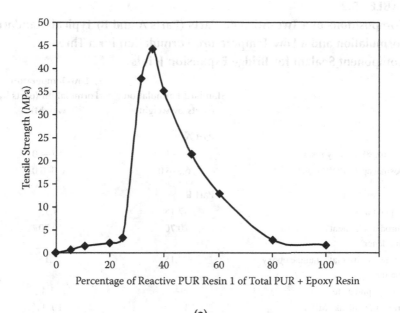

(a)

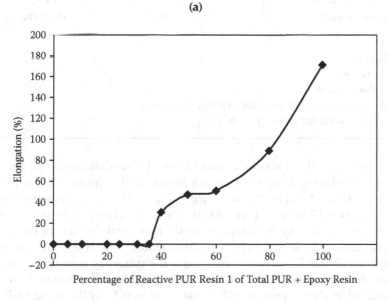

(b)

FIGURE 15.4 (a) Tensile strength and (b) elongation as a function of the resin composition (x parts PUR resin 1 and (100 − x) parts epoxy resin) cured with polyamine.

TABLE 15.2

Compositions of a Two-Mixable-Parts (Parts A and B) Typical Standard Formulation and a Low-Temperature Formulation for a Three–Component Sealant for Bridge Expansion Joints

	Standard Formulation (parts by weight)	Low-Temperature Formulation (parts by weight)
Part A		
Epikote 880 epoxy resin[a]	3.7–4.0	3.7–4.0
Desmocap 11 PUR resin[b]	6.3–6.0	6.3–6.0
Part B		
Liquid tar	7.15	7.15
Eurodur 43, latent epoxy resin hardener[c]	0.70	0.70
Anchor 1110, dialkylamine epoxy hardener[d]	2.15	2.15
Dibutyl phthalate	—	1.4
Strength at break (MPa)	11–15[e]	10–11[f]
Elongation at break	~250%[e]	150–200%[f]

[a] Shell.
[b] Bayer AG.
[c] Schering AG.
[d] Anchor Chemicals.
[e] At ambient temperature, depending on the type of tar used.
[f] At −20°C, depending on the type of tar used.

A comparison of the elastic G' and viscous G'' moduli curves for the three-component sealant systems when using a branched PUR (resin 1, Table 15.1) or a linear PUR (resin 2, Table 15.1) is shown in Figure 15.6. Each system tested contains 65 parts PUR resin. Figure 15.6 indicates the glass transition temperature T_g of the two systems by the peaks of the G'' curve and the start of the decrease of the G' curve. Although the epoxy resin softening point is less marked in these systems, the G' and G'' curves show that, in the system containing the branched reactive PUR (resin 1, Table 15.1), the glass transition temperature T_g is higher (as indicated by the G'' peak at −42°C instead of −60°C for the linear PUR resin) and the epoxy resin softening point is lower (at 82°C instead of 102°C) than in the system containing the linear reactive PUR (resin 2, Table 15.1). As the linear system has a lower T_g, it is better suited for use at lower temperatures than the branched system.

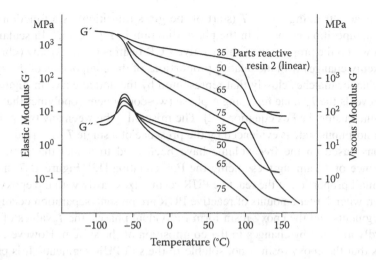

FIGURE 15.5   Elastic modulus G′ and viscous modulus G″ as a function of resin composition (PUR resin 2/epoxy resin)

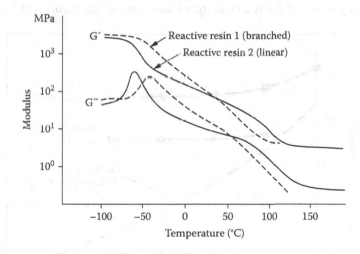

FIGURE 15.6   Direct comparison of the elastic (G′) and viscous (G″) moduli curves (65 parts reactive PUR resins 1 and 2).

15.4   THERMAL MEASUREMENTS

The thermal behavior of three-component epoxy–diamine–polyurethane sealant resins is also affected by their separation in hard and soft phases. Differential scanning calorimetry (DSC) tests on samples crosslinked at room temperature without postheating show that there are two glass transition temperatures: that of the PUR soft segments at around –50°C, and that of the hard phase at around 90°C. The values of these two T_gs vary depending on the composition of the system.

The exact "freezing point" T_f (start of the glass transition) as a function of the system composition is shown in the phase diagram in Figure 15.7. In sealant systems in which the proportion of the reactive PUR resin is relatively low (<hr 25%), T_f markedly changes with the relative proportions of the components of the system. This behavior matches closely that represented by the dotted curve in Figure 15.7, which reveals the T_g value for a single-phase two-component copolymer that would be obtained using the Fox equation [12]. The miscibility between any two polymers in the amorphous state is evidenced by the presence of a single T_g [2]. Traditionally, equations based on the free volume have been used to model the composition dependence of T_g, among these being the Fox equation [12]. Figure 15.7 indicates that a small proportion of the reactive PUR resin mixes easily with the epoxy resin; however, when larger amounts of reactive PUR are present, separation occurs. Soft PUR segments and the epoxy resin form individual phases, the T_f values of which are hardly affected by changes in the composition of the system. However, it also appears that the epoxy resin is not soluble in the soft PUR segments. It is perhaps more correct to assume that the epoxy resin forms a phase with the hard PUR segments. If after a first heating up to 250°C the sample is heated again, the freezing point of the epoxy resin phase increases. This is a well-known occurrence in cold-cured epoxy resins. There is a large difference between the theoretical T_g value of

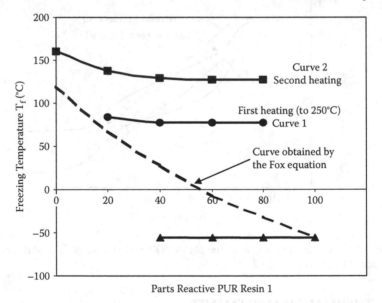

Parts Reactive PUR Resin 1

FIGURE 15.7 Freezing/glass transition temperature variation as determined by differential scanning calorimetry for a three-component PUR–epoxy–polyamine system. The curve obtained by the Fox equation indicates that T_f changes with the proportion of PUR resin. In reality, these changes are only observed at PUR ≤ 25–30 parts (curve 2 lower PUR range). The Epoxy–PUR phase separation occurs at a much higher proportion of PUR. This is shown by both the higher T_f curve (curve 2 versus curve 1) obtained with the second heating and the lack of variation of the T_f values at higher proportions of PUR, data shown by symbol ▲.

the epoxy resin when crosslinked under ideal conditions and the typically low-curing temperature needed for amine crosslinking of the same resin. As a result, the T_g of the crosslinking system increases so much after a relatively few reaction steps that the crosslinking "freezes." Only when the temperature of the system exceeds the high value of T_g, as in the case of the first DSC measurement, does the crosslinking process start and continue. This secondary reaction is also evident from an exothermic reaction that occurs at a temperature higher than the T_g in the first DSC curve.

15.5 CROSSLINKING KINETICS

The curing kinetics of the systems using a combination of blocked isocyanates as the crosslinking component can be observed particularly well by Fourier transform-infrared (FT-IR) spectroscopy and dynamic mechanical analysis [13]. By observing the percentage changes in the IR peaks at 1750 cm^{-1} (*O*-arylcarbamyl groups, urethane groups), 1730 cm^{-1} (urethane groups of polyether/TDI), and 1655 cm^{-1} (urea groups) as a function of time in a system composed of 100 parts reactive PUR resin 1 (Table 15.1) and 6 parts polyamine, it can be seen that the urethane groups are broken down in favor of the urea groups.

In Figure 15.8, the change in the concentration of the aromatic urethane groups in the two-component system (curve 1) as a function of time is compared with that of a three-component system (50 parts PUR resin 1 + 50 parts epoxy resin + 16.5 parts of polyamine). Most striking is the marked acceleration in the conversion of the capped isocyanate groups in the three-component system: 90% of

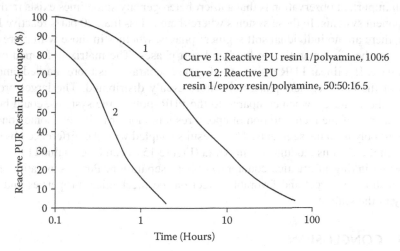

Curve 1: Reactive PU resin 1/polyamine, 100:6
Curve 2: Reactive PU resin 1/epoxy resin/polyamine, 50:50:16.5

FIGURE 15.8 Infrared spectroscopy results show a decrease in the concentration of blocked end groups in the reactive PUR resin as a function of time for a two-component system PUR–polyamine 100/6 system (curve 1) and a three-component PUR–epoxy–polyamine 50/50/16.5 system (curve 2). The curve indicates a faster reaction of the three-component system in relation to the catalyzed PUR system only.

the IR peak has disappeared after just 1.2 h, instead of the 30 h needed for the two-component system. A similar occurrence can be observed when comparing the same three-component system to a two-component system consisting of 100 parts epoxy resin and 33 parts polyamine [13]. In all cases, at least at the beginning of the curing process, the three-component system reacts much faster than a two-component one.

The results obtained by DTMA complement and support what is found by IR spectroscopy investigation of the kinetics of the curing process. A three-component epoxy–PUR–polyamine system was compared to two-component systems, namely, epoxy–polyamine and PUR–polyamine. The softest and slowest crosslinking system was found to be the PUR–polyamine system. The epoxy–polyamine system yields the hardest cured material but also eventually the most brittle one. It is also a characteristic of such systems that, at the very beginning, there is no mechanical evidence of curing, as is characteristic of many other polycondensation reactions. The three-component system has the fastest initial crosslinking reaction.

A few other points are of interest in these systems. At ambient temperature, curing is not complete. The crosslinking of reactive PUR–epoxy resin systems usually takes place between 20°C and 30°C. The lowest possible temperature of curing is between 5°C and 10°C. The crosslinking reaction can be considerably accelerated by increasing the reaction temperature, and this has no significant effect on the properties of the finished sealant. At both ambient and higher temperatures, it is possible to accelerate the curing considerably by adding a typical amine-type epoxy resin crosslinking catalyst, or a mixture of catalysts. In industrial practice, on bridge expansion joints application or repair, amine-catalyzed systems are generally applied at temperatures of about 70°C to be sure to reopen the road to traffic in about 3 h.

An important observation is that a microheterogeneity sometimes exists in three-component systems. In those systems where there is less than 20 parts reactive PUR resin, there are no individual soft segment phases, whereas in those with more than 30 parts, there is separation in hard and soft phases. The matrix is formed of the soft phase. If a linear PUR resin is used, phase separation is more marked, and the sizes of the hard segment phases are more evenly distributed. The larger proportion of these phases, when compared to the PUR–polyamine systems, could be the consequence of the incorporation of epoxy resin monomer units into some zones of the hard polyurethane segments. These results coupled with the difference observed between the viscous modulus G" maxima (Figure 15.6) can be explained by the fact that crosslinking in practice counteracts phase separation. Phase separation, however, also has a direct and favorable effect on the mechanical properties and flexibility of the sealant.

15.6 CONCLUSION

Sealants composed of epoxy/blocked polyurethane resins and crosslinked using a polyamine constitute the more suitable compositions for bridge sealants. They are able to adhere well to wet cement. They are strong enough to withstand the powerful stresses induced by the bridge expansion and contraction with the daily temperature variations. Their relative elasticity, elongation, toughness, and behavior under freezing winter conditions can be tailored according to needs by just varying the relative proportion of epoxy resin and polyurethane resin. Liquid tar is often added to reduce the cost of these systems without much drop in their effectiveness.

REFERENCES

1. P.D. Carter, *Concrete International*, *16(8)*, 60–62 (1994).
2. H. Lee and K. Neville, *Handbook of Epoxy Resins*, pp. 16–21 to 16–30, McGraw-Hill, New York (1967).
3. K.R. Cranker and A.J. Breslau, *Ind. Eng. Chem. 48(1)*, 98–103 (1956).
4. F. Setiabudi and J.-P. Wolf, U.S. Pat. No. 5143999, to Ciba-Geigy Corp. (1989).
5. J.S. Casemant, U.S. Pat. No. 3,090,793, to Minnesota Mining and Manufacturing Co. (1963).
6. J. Cameron, U.S. Pat. No. 3,352,810, to Dow Chemical Co. (1967).
7. S.N. Ephraim, U.S. Pat. No. 3,472,913 to Dow Chemical Co. (1969).
8. E.W. Garnish and R.G. Wilson, U.S. Pat. No. 4,126,505, to Ciba–Geigy Corp. (1976).
9. E.W. Garnish and C.G. Haskins, U.S. Pat. No. 3,914,288, to Ciba–Geigy Corp. (1975).
10. M. Ceintrey, Netherlands Pat. No. NL 8000718 A, to Britflex Resin Systems (1981).
11. K.-H. Hentschel, E. Jürgens, and W. Wellner, *Farbe Lacke*, *2*, 97–102 (1988).
12. T.G. Fox, *Bull. Am. Phys. Soc. 1*, 123 (1956).
13. G.M. Carlson, C.M. Neag, C. Kuo, and T. Provder, *Adv. Urethane Sci Technol.*, *9*, 47 (1984).

16 The Application of Sealants in Automotive Electronics Packaging

Robert E. Belke, Jr. and Kent Larson

CONTENTS

16.1   INTRODUCTION

Sealants prevent various external environmental agents from entering the interior of a protected enclosure. A successful sealant must withstand and provide protection against a number of location-specific external, mechanical, chemical, and electrical environmental influences over the life of a product. Resisting these influences in automotive electronics packaging is exceptionally challenging. The substrate materials, to which a sealant must form a joint, are equally diverse in both composition and configuration. Sealing electronic components from outside environmental agents in any one of a number of distinct automotive environments (schematized in Figure 16.1) poses many additional technical challenges. This is especially true considering high-volume/low-defect automotive electronic component manufacturing.

Automotive electronic sealants, in the broadest sense, are not limited to sealing box-type enclosures but can also be directly applied to components or groups of components in the form of liquids or molded forms that provide the required environmental protection.

In fact, with many electronic packaging applications, the function of a sealant is often difficult to separate from the other assorted roles provided by coatings and adhesive bonding materials. The distinctions between sealants and adhesives are not entirely obvious, as many sealing strategies rely on adhesion to one or more substrates to form a protective seal. There also exist a number of sealing/protection strategies ranging from hermetic glass-to-metal type sealing to metal crimp sealing and compression sealing. This chapter will focus on polymeric dispensed sealant materials that form a mechanical bond to at least one of the substrates to be sealed and result in a protective enclosure for an electronic interconnecting substrate. The unique requirements of sealants utilized in automotive electronics packaging will also be discussed along with a brief review of other common sealing methods.

16.2   ECONOMIC IMPACT OF AUTOMOTIVE SEALANT MATERIALS

The importance of the sealants in automotive electronics can be assessed by considering the global annual manufacture of 63 million vehicles, along with an average estimated electronics content of $1150 per car/minivan/light truck, with 4.0% projected annual growth, as shown in Figure 16.2 [1]. Automotive electronics content is usually subdivided into modules, based either on function or location. These modules incorporate a variety of diverse environmental protection strategies, ranging from reliance on the enclosing mechanical structure to using component-specific or entire module sealing, or simply fluid-shedding (drip-shielded) packaging.

Automotive electronics sealant usage is projected to outpace the overall sealant market-segment expansion. The rationalization for this includes the growth of new automotive markets, the tendency toward placing automotive electronic modules in

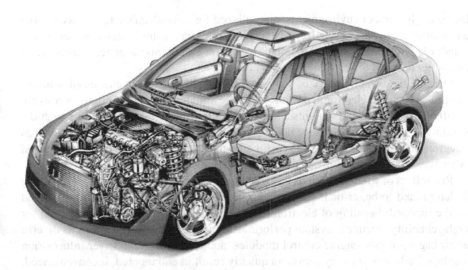

FIGURE 16.1   Automotive electronics can be found in nearly every location of a car, from harsh underhood and underbody environments to the relatively benign interior/passenger compartment as depicted in this figure. Electronic modules and sensors can even be present inside tires, fuel tanks, and transmission cases. (Drawing courtesy of Dow Corning Corp.)

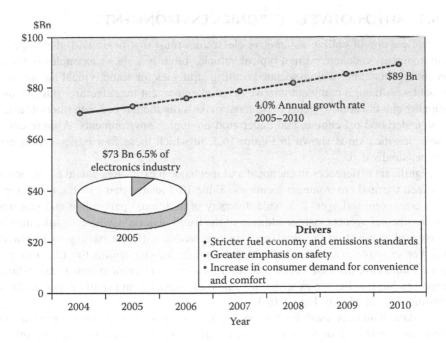

FIGURE 16.2   Projected growth of the global automotive electronics market. (Courtesy of Prismark Partners LLC, Cold Spring Harbor, NY.)

increasingly harsher environments, and reliance for critical safety functions. Global automotive electronics growth is further driven by increasingly strict fuel economy standards, greater emphasis on vehicle occupant safety, and a general increase in consumer demand for convenience and comfort.

Silicone (siloxane) materials comprise one of the dominant classes of sealants used in automotive electronic packaging. It is estimated that $70–$80 million worth of silicone adhesives and sealants alone were used in automotive electronics in 2005. A minority of this silicone usage is for interconnection substrate attachment, such as integrated circuit assembly, but it is estimated that about 70%–80% is for sealing and other types of electronics protection, as will be subsequently discussed.

Regardless of what the total usage of sealants in automotive electronics packaging is determined to be, critical automotive function and performance features depend on the successful sealing of electronic modules. There is an unquestioned need for high-reliability electronic system performance in automotive applications as diverse as air bag restraints, engine control modules, audio controls, and driver information displays. Failure of these systems can quickly result in endangered, inconvenienced, or dissatisfied consumers. Therefore, 5–7 g of dispensed sealant material on a module flange, if not selected, designed, or applied correctly, could potentially lead to severe quality issues with economic implications, drastically outweighing the quantity or cost of the original sealant material.

16.3   AUTOMOTIVE ELECTRONICS ENVIRONMENTS

The discussion of sealing automotive electronics must first begin with the range of environments encountered in a typical vehicle. Table 16.1 shows examples of these environments along with associated sealing strategies for some typical automotive modules. Although many automotive original equipment manufacturer (OEM) and supplier classification systems exist, most make a distinction between interior, exterior, underhood off-engine, and underhood on-engine environments. A more complete classification is shown in Figure 16.3, in which these four basic regions are further subdivided.

Significant differences in chemical and mechanical exposures would be expected between thermal environment locations. Table 16.2 relates many of the most common environmental agents. A wide diversity of additional particulate and gaseous environmental agents exist, in addition to the listed materials, whose significance is strongly dependent on the challenges encountered in different packaging locations [2]. For example, a transmission control module located within the transmission itself would have increased demands in terms of temperature and transmission-fluid exposure but less burden in terms of stone/rock impact than usually expected for a module in location G of Figure 16.3.

Sealant materials used for the packaging of automotive electronics must not only survive in one of these potentially demanding locations but must be inherently cost effective, both in terms of raw material and manufacturing expenditure in high-volume production. For instance, a fairly economic two-component sealant requiring precise metering and subsequent oven curing may end up costing more to produce than a 50% more expensive single-component UV/moisture cure sealant

TABLE 16.1
Some Typical Automotive Modules, Environments, and Associated Sealing Strategies

Module Type	Module Function	Typical Location (see Figure 16.3)	Typical Protection Strategy	Environmental/Sealing Issues
Powertrain control module	Engine/integrated powertrain electronics management	B, D, E, F	S, C/S/V, C	Temperature, automotive fluid resistance, sulfur contamination, dust/dirt protection, vibration
Audio system	Audio entertainment	A, B, C	Unsealed, uncoated	Relatively benign environment
Navigation system	Driver information	A, B	U/U	Relatively benign environment
Transmission control module	Transmission function management	D, G, inside-transmission	S, S/C	Harsh environment
Instrument cluster	Driver information	A, B	N, C	Relatively benign environment
ABS electronic module	Braking system control	C, D	S, C	Harsh environment, brake fluid, particulate
Suspension module	Suspension system	G	S, C	Harsh environment, dust/dirt/particulate protection
Fuel pump delivery module	Fuel system management	G	S, C	Harsh environment, fuel resistance requirements
Assorted sensors	Engine control, fuel ...	D, E, F, G	S, S/C, E	Harsh environment, location-specific requirements

Protection strategy: N = Not usually sealed—fluid shedding; S = sealed; C = conformal coated circuit; S/C = sealed/conformal coated; C/S/V = conformal coated/sealed/vented; E = encapsulated; U = uncoated.

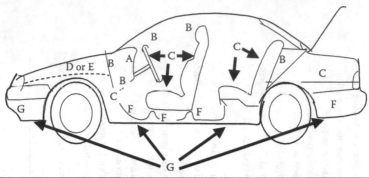

Designation	Vehicle Location	Typical Temperature Range	Some Chemical/Mechanical Agents
A	Interior, no sun exposure	−45°C ↔ +85°C	Water, cleaning agents, particulates
B	Interior, sun exposure	−45°C ↔ +105°C	Water, cleaning agents, particulates
C	Interior, moderate/low splash	−45°C ↔ +85°C	Water, cleaning agents, particulates
D	Underhood off-engine	−45°C ↔ +125°C	Water, fuel, oil, coolant, particulates
E	Underhood on-engine	−45°C ↔ +150°C	Water, fuel, oil, coolant, particulates
F	Interior, water/splash potential	−45°C ↔ +85°C	Water, salt, cleaning agents, particulates
G	Exterior, splash	−45°C ↔ +85°C	Water, salt, road debris, cleaning agents

FIGURE 16.3 Automotive electronics packaging environments diagram. (Figure courtesy of Visteon Corporation.)

TABLE 16.2
Common Automotive Chemical Exposure Agents

Gasoline (including alcohol additives)	Differential and transmission fluids and oils
Diesel fuel (including bio-diesel fuel)	Hydraulic fluid
Other fuels	Interior cleaners and protectants
Battery fluid (sulfuric acid or alternate electrolytes)	Refreshments—coffee/soda …
Brake fluids	Car wash chemicals
Coolants and antifreeze	Glass cleaner
Protective paints and lacquers	Wheel/tire cleaners
Paint and lacquer remover	Engine cleaners
Engine oil (including synthetic oils)	Kerosene
Engine-cleaning agents	Salt water
Methanol	Other location- and application-specific agents

after manufacturing cycle time, capital equipment cost, labor, and utilities are fully considered. The environments encountered in a car are rivaled by the diversity of electronic module types. Every automotive OEM has a methodology for packaging electronics. This strategy may change between car lines, or even from year to year in the same car model.

The chemical exposure agents listed in Table 16.2 can be encountered at elevated temperatures, depending on electronic component location and function. Fluids, such as automatic transmission fluid, oil, and antifreeze, can also chemically change over time, sometimes becoming acidic or caustic as they degrade. These changes can affect sealant properties and longevity. Fluid vapors can also present challenges to sealant integrity. For example, silicone sealants can absorb and swell in response to exposure to a variety of organic fluids, including many automotive fuels. The sealants are usually not degraded by this exposure, and the sealant joints return to original or near-original property values once the fluid evaporates. Some evaluations may incorrectly identify these temporary changes for permanent damage.

The list of automotive chemical exposure agents is very situational and dynamic. Emerging and growing technologies, such as electric, hybrid, and alternative fuel vehicles, require emphasis on different agents than a traditional gasoline-fueled vehicle. For instance, a sealant used on a gas/electric hybrid vehicle might need resistance to acidic vapor exposure and associated higher temperatures than encountered in the same location in a gas-powered vehicle. In all test situations, it is critically important to take the application into account when determining test conditions such as time, temperature duration, and whether to immerse, spray, drip, or temporarily wipe a specific chemical agent.

16.4　PROTECTION STRATEGIES FOR AUTOMOTIVE ELECTRONICS PACKAGING

The specialized area of automotive electronics packaging continues to evolve. Some of the major protection strategies are depicted in Figure 16.4. The selection of the appropriate strategy is a complex decision dependent on the previously mentioned factors of vehicle location, possible contaminants, and module function. In addition to the following general packaging schemes, there are many possible combinations. Novel packaging designs, materials, and approaches continue to advance the state of automotive electronics packaging art. Table 16.3 further defines and contrasts the risks and benefits of the general packaging approaches shown in Figure 16.4.

16.4.1　Unsealed Electronics

Many variables should be considered when developing a protection strategy for automotive electronic components. The crucial question is "Does this module really need to be sealed in the first place?" The intended mounting location provides the answer. For example, if a module is located above the beltline in an automotive interior with no adjacent or overhanging air ducts or other possible sites of condensation, the module probably does not need sealing. In this case, one might consider a

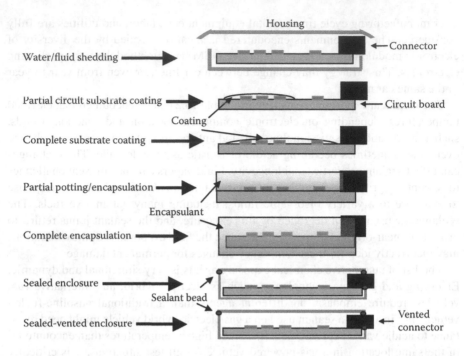

Water/fluid shedding ➡

Housing

← Connector

Partial circuit substrate coating ➡

Coating

← Circuit board

Complete substrate coating ➡

Partial potting/encapsulation ➡

Encapsulant

Complete encapsulation ➡

Sealed enclosure ➡

Sealant bead

Sealed-vented enclosure ➡

Vented
← connector

FIGURE 16.4 Schematic presenting the major electronic module environmental protection approaches, which may be used alone or in combination. Vehicle location and module orientation are critical determinants for an overall protection strategy.

water-resistant or fluid-shedding cover with no fluid entrapment recesses adjacent to fasteners and clip features. Figure 16.5 illustrates this point with two different modules, both packaged as "a box within a box" with reliance on the passenger compartment for protection from the most extreme exterior environmental factors. Control of module orientation is important with this strategy as well as the orientation of associated connectors. A wider systemic view of moisture management for this and

FIGURE 16.5 Not all automotive electronics assemblies are sealed due to their relatively protected locations. Automotive interior components such as instrument clusters (left) and audio components (right) are often unsealed, with encased electronics left uncoated. (Photos courtesy of Visteon Corporation.)

TABLE 16.3

Automotive Electronics Environmental Protection Strategies: Benefit/Risk Comparison

Environmental Protection Strategy	Benefit	Risk	Overall Cost	Comment
No protection	Lowest cost	Particulate, gas, liquid	No additional cost	Usually automotive interior. Protection afforded by interior location or surrounding structure
Water/drip resistant–fluid shedding construction	Low cost	Particulate, gas, liquid (indirect condensation)	Inherently low cost This approach must be considered in the initial module design	Usually automotive interior. Protection is provided by design. Orientation must be specified and predictable
Local conformal coating of circuit printed wiring board substrate	Protective coating material and location may be changed rapidly. Process usually incorporated into PWB process. Local protection of critical circuit regions	Repair/rework difficulty Compatibility with solder flux materials	Medium to high due to either selective application or part masking	Can be used in combination with sealing or other protection strategies
Overall conformal coating of circuit printed wiring board substrate	Protective coating/sealing material may be changed rapidly. Process usually incorporated into PWB process	Repair/rework difficulty Compatibility with solder flux and other process materials	Moderate cost	Can be used in combination with sealing or other protection strategies

—continued

TABLE 16.3 (continued)
Automotive Electronics Environmental Protection Strategies: Benefit/Risk Comparison

Environmental Protection Strategy	Benefit	Risk	Overall Cost	Comment
Module potting (complete encapsulation of module with polymeric material)	Overall robustness. Some protection from competitive evaluation	CTE mismatch repair/ remanufacture. Device thermal arrangement must be considered	Medium to high	Extent of environmental robustness depends on potting material and process as well as on management of internal stresses
Overmolding/ encapsulation	Overall robustness. Some protection from competitive evaluation	CTE mismatch does not allow easy repair/remanufacture of enclosed electronics. Device thermal management must be considered	Medium	Extent of environmental robustness depends on potting material and process as well as on management of internal stresses. Can be used in combination with other protection strategies
Formed-in-place-gasket (FIPG)	Robust protection approach. Proven in high-volume manufacturing. No additional mechanical fastening needed in most applications	Does not allow easy accessibility to enclosed electronics. Leak test verification recommended	Medium	Can be used in combination with sealing or other protection strategies. Enclosure may be vented
Cured-in-place-gasket (CIPG)	Allows disassembly for repair/ rework of electronics	Requires mechanical fastening to compress gasket	Medium to high	Can be used in combination with sealing or other protection strategies. Enclosure may be vented
Elastomeric gasket/O-ring	Allows disassembly for repair/ rework of enclosed electronic device	Requires mechanical fastening to compress gasket	Medium to high	Gasket cost may be offset by lower equipment cost

other approaches considers the connected wire harness location with recommended drip loops. The function of the module might dictate more protection. One example would be critical safety items such as air bag electronics, which can stand-by, nearly dormant, for years but must be absolutely reliable when needed.

Vigilance should be used with the general approach of designing unsealed vehicle electronics. There are certainly many successful applications of unsealed and uncoated automotive radio and instrument cluster (combimeter) circuit boards. However, environmental conditions may produce microclimates allowing infiltration of saturated, humid air, and subsequent moisture condensation on a cold printed wiring board surface.

16.4.2   Unsealed-Coated Electronics

A somewhat different example would be encountered in an underseat interior location (location "B" in Figure 16.3), in which there is a greater splash and moisture infiltration threat. A stronger argument for a sealed or sealed/vented module can be made for this location. Figure 16.6 shows two views of an underseat-mounted powertrain control module. In this example, the circuit board electronic components are coated/sealed with a silicone material. The lid provides mechanical protection as well as a measure of drip and liquid protection; however, flooding or submersion of the module would result in eventual failure.

16.4.3   Coating as an Environmental Seal

Coating an electronic interconnection substrate, regardless of the underlying circuit composition (organic circuit board, ceramic thick-film hybrid, or other electrical interconnection technology), is a common method of protecting and sealing individual electronic components from environmental agents. This method, often known as conformal coating, can be used alone in relatively benign environments, such as car interiors, or used in combination with an additional sealed, vented enclosure as part of an overall protection strategy. In automotive applications, this coating may consist of a number of different materials. The most commonly used include acrylics, epoxies, polyurethanes, or blends of these generic polymer classes and, most

FIGURE 16.6   External (left) and internal (right) views of underseat-mounted powertrain control module. (Photos courtesy of Visteon Corporation.)

often, silicone resins. Coatings with thicknesses ranging from 50 to 200 μm afford temporary protection against condensed moisture droplets [3].

Several general trends are found in the area of conformal coating of automotive electronic components. The use of volatile organic solvents have been reduced and sometimes eliminated during the past decade. This is usually accomplished by selection of 100% solids systems, as found with many silicone and polyurethane coatings, or by use of newer waterborne systems as exemplified by some recent polyurethane formulations.

The selective deposition of conformal coating only on substrate areas that need protection (usually by spraying, as shown in Figure 16.7) has also seen rapid growth. These substrate regions include condensed moisture-sensitive circuitry and/or components. Such selective application methods leave connectors, displays, thermal conduction pathways, and other similar regions free from coating. One more important trend is the consideration of any circuit coating as a component in a multiple material system. For automotive electronics, the proliferation of "no-clean flux"-containing solder materials leaves residues on the substrate surfaces, which may contribute to long-term corrosion as well as interaction with a particular coating. These interactions

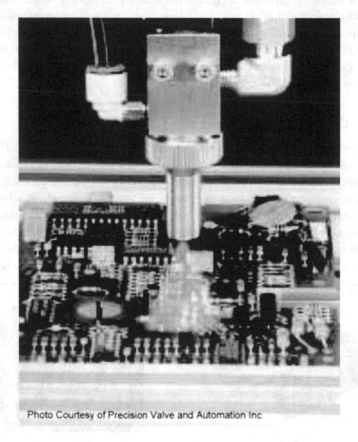

Photo Courtesy of Precision Valve and Automation Inc

FIGURE 16.7   Conformal coating being selectively sprayed onto a component-populated printed wiring board to effectively seal the surface from condensed moisture and other contaminants.

must be understood to ensure long-term device reliability. Recently, there has been strong interest in eliminating conformal coating in many module designs, particularly those in which another level of protection, such as location within a formed in-place gasket (FIPG)-sealed enclosure, exists for environmental protection.

16.4.4 ENCAPSULATED ELECTRONICS

Sealing electronic modules and their components can be accomplished in many different ways, depending on the exact feature needing protection and the environmental agents it needs protection against. Sealing at the circuit board component level is achieved with various coatings and with localized potting, sometimes called "glob-top" potting. When sensitive components or wires extend upward beyond the reach of these local coating methods, it is common to partially or completely encapsulate or "pot" the whole module. Boards are mounted into an outer housing, and all electrical interconnection is completed. An encapsulant is then dispensed into the module to displace air and to cover all sensitive components, wires, and wire bonds.

Such encapsulation provides the single best environmental protection, although at the penalty of considerable weight and material expense. One solution is to encapsulate only moisture-sensitive devices or regions of an electronic module. Figure 16.8 shows one such example in which leads of two neighboring integrated circuits are coated with a thixotropic silicone sealant.

The two most commonly used encapsulants are epoxy and silicone, although polyurethanes (PU) and polyisobutyrals (PIBs) are also occasionally utilized. Epoxies offer the dual advantages of lower cost and extreme toughness. If the entire module is filled, no top cover may be required, as is often the case with ignition coil modules. However, if there is significant coefficient of thermal expansion (CTE) mismatch between constituents, the unyielding epoxy may cause wire

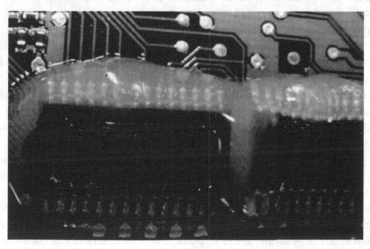

FIGURE 16.8 Two adjacent integrated circuit packages locally protected with a UV–moisture combination cure thixotropic sealant. (Photo courtesy of Visteon Corporation.)

FIGURE 16.9 A thick film-interconnection substrate-based automotive electronic module encapsulated with a rigid epoxy resin. The stresses between the relatively low coefficient of thermal expansion ceramic substrate and higher CTE epoxy encapsulant must be accommodated in the design and validated through testing. (Photo courtesy of Visteon Corporation.)

bond and solder joint damage during thermal cycling. Figure 16.9 shows wire-harness signal-conditioning circuitry on a ceramic substrate, completely encased in an epoxy encapsulant. The preferred lower-elastic modulus, softer encapsulants normally require a covered module to prevent impact damage. Polyurethanes and PIBs do not meet the temperature range requirements for some automotive applications and may not have a low-enough modulus for especially sensitive components. Silicone encapsulants are available as moderately hard to extremely soft elastomers, with the lowest modulus versions comprising a category called "gels" that begin to bridge property characteristics between liquids and solids. One application of the use of a fuel-resistant fluorinated silicone gel is shown in Figure 16.10. In general, soft and compliant silicone gels act to reduce and spread out thermal movement-induced stresses while simultaneously electrically isolating boards, wires and components, thereby providing protection from moisture condensation and ionic transfer.

Encapsulants work by displacing all, or nearly all, air from within a module, leaving no place for moisture condensation against components or conductors. Condensed moisture becomes the charge-carrying medium necessary for corrosion, dendrite growth, and leakage current. With this function in mind, it is important that the encapsulant fully wets all surfaces with no entrapped air under components or at the board or wire surface [4]. In some cases, cleaning may be required to remove oily contaminants that could prevent the desired wet-out and air displacement.

Low-pressure injection overmolding of automotive electronics has recently been developed. This type of encapsulation involves the use of hot-melt-type adhesive materials, which are injected around an electronic substrate to provide protection [5]. An automotive application of this technology is shown in Figure 16.11. It is vital that the lower elastic modulus/compliance and inherently high CTE of the encapsulant be reconciled with the generally lower in-plane CTE characteristics of the interconnection substrate. The substrate-to-encapsulant bond integrity over many environmental thermal cycles is necessary to prevent a pathway for the infiltration and stagnation of water and other liquid contaminants. For these reasons, the first applications of this

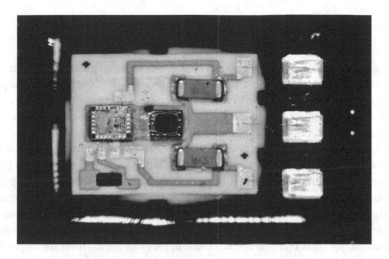

FIGURE 16.10 Fuel system sensor electronics coated with a transparent silicone gel encapsulant. (Photo courtesy of Visteon Corporation.)

FIGURE 16.11 An example of a low-pressure injection-molded automotive electronics module. This prototype module uses a nonpigmented translucent amorphous polyamide resin to reveal underlying circuitry. Production modules are generally opaque. This design incorporates an aluminum heat sink (hidden from view on opposite side), separate connector (with encapsulated flange), and overall molded resin thickness sculpted to the profile of the underlying electronic components. (Photos courtesy of Visteon Corporation.)

technology have been small geometry parts to minimize stress. More recent applications involve encapsulation of larger substrates by sculpting of the molding tool to yield a uniform resin thickness over components to minimize resin usage, mold cycle times, and thermal trauma during the molding operation [6].

16.4.5 SEALED ENCLOSURE AND SEALED-VENTED ELECTRONIC MODULE STRATEGIES UTILIZING DISPENSED SEALANT

The use of a formed-in-place-gasket (FIPG), or, closely related, cured-in-place-gasket (CIPG), remain the most important methods with which to form a protective enclosure. In automotive electronics packaging, the basic sealed-box design is complicated, in most instances, by the need to remove heat from the enclosed electronics. The successful design must also allow for connectors, often containing over 100 individual signal and power conductors, to enable communication with the outside world. This situation is further complicated with the persistent need for cost-effective, high-volume production, often meaning hundreds of thousands of parts per year (300 or more parts per hour in a typical manufacturing setting). The insulation of a large number (sometimes approaching 200) of individual electrical conductors often means the utilization of a polymeric connector or polymer housing/connector combination. Module thermal management drives the use of high thermal conductivity materials such as die-cast metals and metal stampings for at least one surface of the enclosure. These two basic design drivers require the sealing of materials with drastically different thermal and mechanical properties, presenting a significant challenge for high-reliability sealing.

An FIPG is usually defined as a joint that results from placing a wet adhesive/sealant on one surface of an eventual joint, mating the two parts to contact and partially displace the sealant, followed by subsequent curing. This is the type of sealing usually associated with underhood automotive electronics packaging. Therefore, a successful module protection strategy must rely on a sealant forming strong, durable bonds to a variety of metallic and polymeric materials. Bonding dissimilar materials, such as cast metal housings (relatively high thermal conductivity) to polymeric connectors (low cost, electrically insulating) can present sealing surfaces with different linear coefficients of thermal expansion as well as different abilities to form chemical bonds to a sealant material.

A sealant having adhesive properties is commonly used to bond two surfaces together. The adhesive/sealant is applied to one side, and the other mating surface is brought into contact prior to cure. Adhesion is normally very good to both surfaces. This type of seal does not require compression but also is rather permanent; thus, it is very difficult to reopen lids sealed in this way without deforming or destroying the enclosure. Figure 16.12 shows two views of a completely sealed underhood module requiring strong and compliant bonds between aluminum and polyester resin surfaces.

16.4.6 ELECTRONIC ENCLOSURE JOINT DESIGN USING DISPENSED SEALANTS

One successful FIPG sealant strategy is to incorporate a "tongue-in-groove" joint design. An example is a stamped metal cover with a curved edge interposed with a groove in a cast metal or injection-molded plastic housing in which an adhesive/sealant has been dispensed. The edge of the metal cover displaces the sealant during assembly, thereby increasing both leak path length and wetted area. Mechanical fastening or clamping, or built-in clips, may be used to locate and secure the lid to the housing, but are not absolutely necessary. This type of sealing requires clean and easily wettable surfaces, along with management of the potential effects of

FIGURE 16.12 Top (left) and bottom (right) views of a completely sealed underhood-mounted automotive electronic control module enclosure. The white sticker is used to seal the vent hole designed to prevent internal pressure buildup during thermal curing of the FIPG sealant. This sealant joint requires durable adhesion between both the stamped metal cover and injection-molded polyester housing. (Photos courtesy of Visteon Corporation.)

any elevated curing temperature on the expanding internal air volume within the package. In addition to tongue-in-groove designs, other common designs include flat-to-flat flange designs, crimped flat flange, and overcoating the sealant onto pre-mated substrate surfaces. Dissimilar substrates and resulting thermal movements often impact bond strain dynamics. The use of low-modulus sealants can mitigate the resulting shear, tensile, and peel stresses.

There are a multitude of FIPG sealant chemistries available. Thermal cured systems remain the most prevalent FIPG material types. The pressure buildup due to entrapped air during heating can result in leakage path formation unless effectively managed. Temporary or permanent vents are designed to permit pressure equalization during cover placement and subsequent curing to allow an undisturbed bond-line. A module vent may be filled following sealant cure with an adhesive, label, tape, or mechanical plug. Other single- and dual-component adhesives may be used for this application but must be adaptable to high-volume manufacturing.

A vent may be a permanently designed feature that lets the module "breathe" during temperature and pressure transients. This is an added benefit since the elimination of a pressure differential with the exterior environment will prevent water and other contaminants from being drawn through defects into the module interior. There have been several successful polymer membrane-type vents that allow free passage of air (and water vapor) but restrict passage of liquid water at worst-case automotive environment pressures. These breather vents can be placed in either metal or polymeric module components. Figure 16.13 shows one such vent in a connector, to be sealed into an underhood electronics module. This feature can also be incorporated into the connector of the production underhood-located engine control module as shown in Figure 16.14.

CIPGs are typically dispensed onto one mating surface and then cured, forming an adhesive bond to that surface. A lid is then mechanically attached with screws or bolts to compress the gasket and form a mechanical seal. Generally, a compression of 30%–50% is desired to obtain a robust seal.

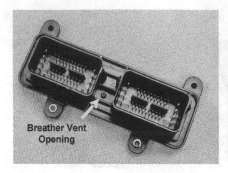

FIGURE 16.13 External (left) and internal (right) close-up views of a connector showing breather vent. This type of vent allows free exchange of gases but prevents condensed liquid infiltration. (Photos courtesy of Visteon Corporation.)

FIGURE 16.14 External (left) and internal (right) views of a powertrain control module, incorporating a vented connector, which has been sealed with an FIPG for survivability in the challenging underhood environment. The vent location is shown at the end of the white arrow in the right-hand photograph. (Photos courtesy of Visteon Corporation.)

Closely related to CIPGs are dispensed foam gaskets (DFGs; usually silicones) that foam after they are dispensed, creating a lightweight closed cell structure and a highly compressible seal. A benefit of using the DFG approach is that it can work with looser part fit and tolerances than a free-standing elastomeric or O-ring seal. A compression of 50%–80% is commonly used to obtain a robust seal.

For both CIPG and DFG beads, proper joint designs are crucial to obtaining a good seal. As shown in Figure 16.15, the best gasket bead design is normally a rounded "hump" with a height-to-width aspect ratio of 0.75.

Higher aspect ratios leave a relatively small area for adhesion of the total gasket bead size, and can also experience problems with "rollover," where the bead is pushed and deformed to one side when compressed, thereby, compromising good sealing. There have been attempts at application of a second layer of gasket bead over the top of the first layer. This is not recommended because of the poor aspect ratio and because it is very easy to entrap air in the knitline area where dispensing began for the first bead layer.

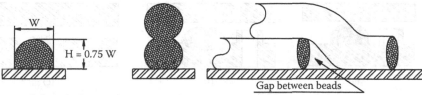

Recommended single dispensed sealant bead cross section

Double-dispensed sealant bead–not generally recommended

FIGURE 16.15 CIPG and DFG bead and joint design considerations. Ideal designs include height-to-width aspect ratios above 0.75 and do not utilize double-dispensed sealant beads.

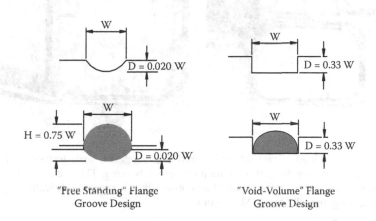

"Free Standing" Flange
Groove Design

"Void-Volume" Flange
Groove Design

FIGURE 16.16 Two common sealant joint variations to improve sealing viability.

There are several seal joint designs that can improve sealing efficacy. One common method is to design a flange groove. This is a peripheral recess in the substrate into which the gasket material is dispensed. Figure 16.16 depicts a shallow radius "free-standing" flange groove in a substrate along with a "void-volume" flange groove. This second variation is often designed as a square-cornered trench in the substrate with the best seals obtained when the volume of the CIPG bead is 60%–75% of the groove volume.

In designing these groove seals, the dispensed bead should still maintain the shape and aspect ratio described. The overall bead height from the bottom of the groove to the top should be 3/4th of the maximum bead width.

Additional design features include compression limiters. These are raised areas on the substrate next to a groove that are designed to allow the proper compression of the gasket needed to obtain a good seal but provide protection against overcompression, which could damage the seal material and/or deflect out of its proper alignment. The main design styles are shown in Figure 16.17.

In some cases a very tall gasket is required with a height much larger than its width. This type of gasket design is most effective in a void-volume groove arrangement. No-slump CIPG materials must be used with precise dispensing to achieve a well-

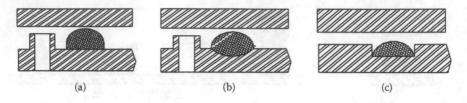

FIGURE 16.17 CIPG or no-slump DFG bead using (a) flat surface with a compression limiter, (b) grooved surface with a compression limiter, and (c) void-volume groove.

FIGURE 16.18 Views of two different cured-in-place-gasket seals formed by precision robotic dispensing on a die-cast aluminum electronic housing. This method allows nearly perfect and repeatable seam-free gasket fabrication. (Photos courtesy of Sealant Equipment and Engineering, Inc., Plymouth, Michigan.)

formed seal. Two views of a typical precision valve dispensing a DFG sealant bead are shown in Figure 16.18.

Acceptable sealant bead process control can only be achieved with reliable part fixturing and robotic dispensing. Care must also be taken to properly mix the gasket materials in the case of two-part systems, and to fully cure the gaskets prior to compression.

16.4.7 GASKETS AND O-RING SEALS

The use of discrete stand-alone gaskets and O-rings present two alternatives to CIPG or DFG joints. These freestanding gaskets may be composed of a variety of elastomers or foams and, as with the related CIPG and DFG joints, require mechanical compression to produce an effective seal. Preformed gaskets are often more expensive than dispensed material, based on comparative polymer cost and associated mechanical fastening. However, this may still be a cost-effective option with low production volume or in low-labor-cost manufacturing sites requiring less automation to obtain reproducible, high-integrity sealing. Common gasket materials suitable for automotive applications are shown in Table 16.4. Compressive loads for CIPG, DFG, and gasket joints may be applied with a large variety of methods ranging from the use of stand-alone fasteners and built-in clip features to more involved crimp designs [7].

TABLE 16.4
General Properties of Common Elastomers Associated with Automotive Electronics Gaskets and O-Rings

ASTM D1418 Designation	Common Names	Temperature Range	Attributes	Deficiencies
CR	Polychloroprene	−50°C to 120°C	Good weathering, oil and petroleum resistance	Widely variable properties depending on exact specification
EPDM	Ethylene propylene diene monomer	−50°C to 125°C	Low-temperature stability	Temperature stability, oil resistance
FPM	Fluorocarbon elastomer	−30°C to 200°C	High-temperature stability Oil, fuel, and solvent resistance	Higher cost, low tear resistance in some compounds
NBR	Nitrile	−50°C to 135°C	Good oil, solvent, and fuel resistance	Poor ozone, sunlight resistance
HNBR	Hydrogenated nitrile butadiene rubber	−50°C to 150C	Good oil, solvent, and fuel resistance	Higher cost, poor acid resistance
Q	Silicone	−45°C to 230°C	High-temperature stability Low-temperature flexibility Good mechanical strength retention at elevated temperature	Only fair mechanical strength Oil/fuel swelling
FVMQ	Fluorosilicone elastomer	−60°C to 170°C	Good fuel and solvent resistance Good thermal stability	Higher cost, only fair mechanical strength, poor tear resistance, low brake fluid resistance

Numerous commercial sources exist for stock and custom-made elastomeric seals, from which more detailed design and usage information can be obtained. The ultimate properties of the final gasket strongly depend on the exact elastomer constituents and precise specification of required final properties. Additional key properties for gaskets and O-rings include durometer hardness/elastic modulus, compression set, tear strength, thermal resistance, and solvent/fuel resistance.

The choice between FIPG, CIPG, DFG, and freestanding gaskets is a complex decision based on many factors including overall design, cost, frequency and need for disassembly/repair/service, usage environment, and the overall manufacturing process. This decision is further complicated by the diversity of available materials. Some discussion of dispensed materials will follow in Section 16.6.

16.4.8 PROTECTION STRATEGIES USING MULTIPLE APPROACHES

The various protection approaches may be combined to offer multiple levels of environmental protection and, in some instances, to allow the module to serve another function in the assembly. The speed control amplifier module shown in Figure 16.19 is one such example. The circuit board substrate is encapsulated by a transparent silicone gel encapsulant, and the entire amplifier assembly is used as one side of the mechanical underhood gear box for the speed control system.

Another commonly encountered example is the utilization of a conformally coated circuit board within a sealed or sealed/vented enclosure. The growing trend of integrating control electronics in proximity to, or integral to, the associated mechanical system provides many opportunities for creative electronics packaging, yet places

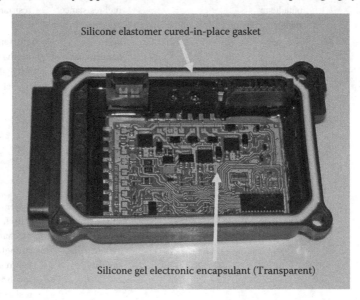

FIGURE 16.19   Speed control amplifier showing both a CIPG seal and a silicone gel encapsulant. (Photo courtesy of Visteon Corporation.)

many demands on module environmental protection strategies. One example is the placement of fuel pump system electronics within a fuel tank. The trend toward locating sealed control electronics within transmission cases, partially or completely submersed in hot transmission fluid, is arguably one of the best examples of a challenging packaging environment.

In contrast to the use of multiple protection strategies is the concept of not using polymeric materials to establish a seal. There has been interest in crimped metal-to-metal closure of modules. Examples of reliable leak-resistant crimped cans can be seen in food stores and pharmacies. Some of this technology may be adaptable to automotive underhood module sealing with appropriate modification to meet the stringent requirements of an underhood automotive environment.

16.5   SOME CONSEQUENCES OF POOR ELECTRONICS PROTECTION

A major consequence of condensed moisture on an unsealed, uncoated, or otherwise unprotected powered circuit board is metal electromigration leading to dendritic corrosion. A module containing an energized printed wiring board assembly contains a multitude of electrical conductors at different potentials in close proximity. Printed wiring board processing necessitates the attachment of components to provide electrical interconnection, usually with any of a number of solder attach processes. Fluxing agents are usually necessary to activate oxidized and/or contaminated metal surfaces to allow reliable soldering. Corrosive, ionic flux residues may not be entirely cleaned from the board surface after the component soldering operation. Additionally, the use of "no-clean" flux systems has found increased usage since the early 1990s. Although not solely responsible for the onset of metal electromigration, these fluxes and other contaminants have constituents that can either partially or completely dissociate with the addition of water, thereby hastening the onset of the reaction. The resulting dendrites form low-resistance pathways between conductors intended to be isolated, thereby compromising electronic device functionality and reliability.

Figure 16.20 depicts the overall metal electromigration reaction. The resulting corrosion is shown in the circuit trace photographs in Figure 16.21 taken from circuit boards that failed humidity testing.

In addition to dendritic corrosion, there are a variety of additional environmentally induced failure mechanisms, including, but not limited to, other corrosion processes (especially near electrical contact surfaces), and conductive or partially conductive pathways caused by infiltration of external contaminants such as dust, dirt, metallic particles, whiskers, or shavings.

These corrosion and contaminating agents, along with the resulting failure modes, are not exclusive to automotive electronics. However, the unique aspect in this situation is the consumer expectation that vehicle electronics will survive for long periods of time without replacement or defects in severe mounting locations. One of the most notable locations is underhood, where a constant barrage of contaminants is encountered at varying temperature and vibration levels. Depending on the electronic module function, failure can also compromise passenger comfort and safety.

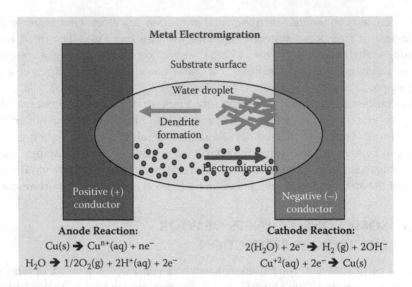

FIGURE 16.20 If water, whether distilled or contaminated, is placed across unprotected circuit conductors, the critical elements needed for electromigration are present, namely, potential difference, ionic mobilizing medium (water), ionic contamination (from substrate, liquid, or even from initial dissociation of the water), and pathway between the conductors.

FIGURE 16.21 Two pictures of dendrites formed between adjacent printed circuit conductors. The dendrites pictured in both the left and right photos formed under a solder mask coating on an uncoated printed wiring board substrate following moisture exposure. (Photos courtesy of Visteon Corporation.)

Of course, water and other liquids are not the only contaminants from which automotive electronics need protection. Polymeric sealant, coating, and encapsulation materials also allow diffusion and permeation of gaseous species at markedly different rates, depending on the exact polymer sealant composition, temperature, and concentration of the diffusing species, as well as other factors.

Sulfur-containing molecular species, such as hydrogen sulfide (H_2S), may diffuse through some silicone protective coatings to react with underlying conductor

FIGURE 16.22 View of black sulfur corrosion on component termination metallization beneath a silicone coating. (Photo courtesy of Visteon Corporation.)

and component metallization, with the potential for causing component failure. The sulfur corrosion shown in Figure 16.22 was caused by a partially vulcanized rubber component adjacent to the lower instrument panel-mounted silicone-coated device.

The infiltration of H_2S and other small molecules into an underhood silicone FIPG-sealed module may be remedied by the substitution of a flexible epoxy FIPG material. The significantly slower diffusion of sulfur species through the epoxy sealants is due, in part, to tighter molecular packing [8]. It should be noted that the use of a low-permeation-rate material will slow down but will not entirely prevent the diffusion of small deleterious molecules, and therefore the effectiveness of any proposed solution should be verified.

16.6   COMMON AUTOMOTIVE ELECTRONICS SEALANTS

16.6.1   SILICONE SEALANTS

Silicones have a set of characteristics that make them particularly well suited for use as sealants for underhood automotive electronics. The term *silicone* generally refers to elastomers that contain alternating silicon and oxygen atoms in long chains. Several varieties exist, each with its own set of unique attributes.

By far the most common type of silicone is poly(dimethylsiloxane), or PDMS (Figure 16.23). The basic monomer repeat unit in this polymer has two methyl groups attached to the silicon atom on the polymer backbone.

Other types of silicones are made by exchanging one or more of the methyl (CH_3) groups on each silicon atom in PDMS with an alternative group such as phenyl or trifluoropropyl, as shown in Figure 16.24. Adding phenyl groups to PDMS imparts

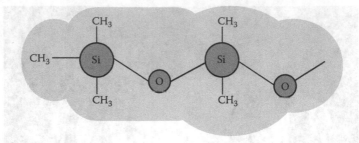

FIGURE 16.23 The chemical structure of poly(dimethylsiloxane), or PDMS. Generally when one hears the term *silicone*, it refers to a PDMS-based material (although there are other readily available siloxane polymers).

very low temperature flexibility, while adding fluoro groups provides for significant resistance to swelling in most solvents and fluids. These groups also add considerable cost to the manufacturing process and, therefore, are only utilized in applications requiring these uniquely special attributes.

As shown from their structures, the polymeric backbone of silicones is inorganic. The Si-O bond in silicones is both very stable and quite flexible, which gives rise to many of its distinctive characteristics.

Silicones are one of the most thermally stable elastomers known. Many silicone sealants are rated for continuous use at 150°C, and some can remain elastomeric for years at temperatures approaching 200°C. Weight loss of a typical PDMS sealant during four temperature exposures is shown in Figure 16.25 to illustrate this point. Short-duration exposures can, in many cases, be tolerated up to 260°C or even 300°C. Silicone elastomers are thermosetting materials and, therefore, do not melt or soften at these high temperatures.

Figure 16.26 illustrates the robustness of a typical PDMS-based sealant to various temperatures for extended periods (265 days). Although there can be some mechanical property change with such high-temperature exposures, the material usually retains some level of elasticity, thereby allowing functionality in the intended application. For applications requiring only little change in properties over very long durations, an upper temperature use limit of 150°C to 175°C is often recommended.

The mechanical properties of silicone elastomers are also quite stable over a wide range of temperature. Figure 16.27 shows high degrees of both tensile strength and elastic modulus for a common PDMS-based sealant used in automotive electronic packaging. Many organic elastomers, such as butyl rubber, will soften considerably and/or degrade at temperatures that may be considered rather "mild" to silicones.

Silicone polymers are highly resistant to crystallization. Freezing is observed at a very low temperature. Silicones are also quite prone to supercooling, where crystallization can be delayed both in time and temperature. The glass transition point of silicones is generally about −115°C. They will, however, undergo an often misunderstood transition in the −40°C to −85°C temperature range. This transition is often called a *melting* or *freezing point*, although these terms are misleading [9].

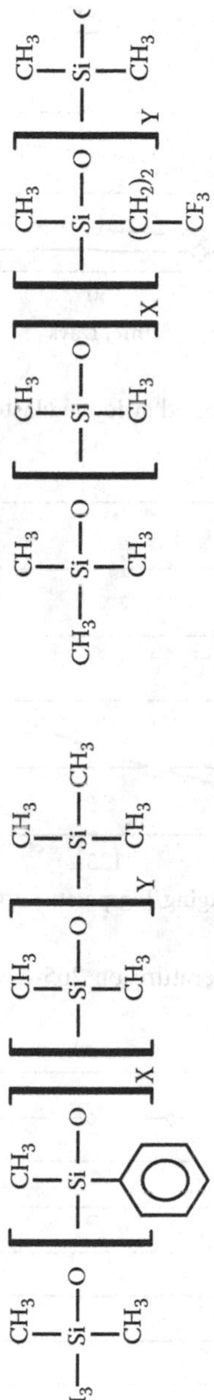

FIGURE 16.24 Structural representations of methylphenyl (left) and methyl trifluoropropyl (right) siloxane polymers.

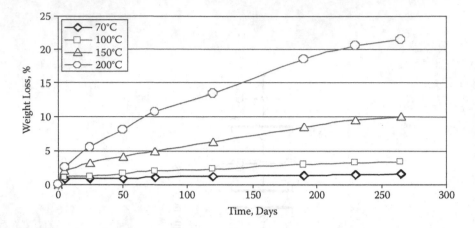

FIGURE 16.25 Weight loss of silica-filled silicone elastomer as a function of time at four exposure temperatures.

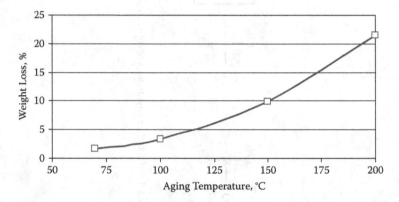

FIGURE 16.26 Effect of aging temperature on 265-day weight loss for a typical PDMS sealant.

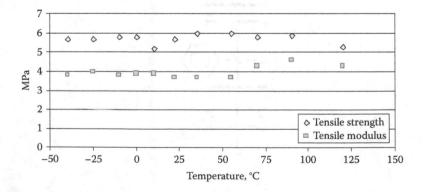

FIGURE 16.27 Typical silicone elastomer tensile strength and modulus stability versus temperature.

In thermal excursions past this transition, cured elastomeric silicones have been found to shrink by about 1.5%, and Shore A durometer hardness values are found to increase by 15 to 30 points. At temperatures below the transition, silicones typically remain elastomeric, though with a higher modulus attributed to the tighter molecular packing through the formation of semicrystalline domains. Misunderstanding can come from the perception the terminology evokes—when a material like water freezes, it undergoes a tremendous change to a hard and brittle solid. Likewise, when a semicrystalline linear thermoplastic is taken above its melting point, it will typically soften and lose its dimensional stability. The elastomeric properties of cross-linked silicones also change during this thermal transition but to a much lesser degree than most materials. In many regards, it can be considered that a silicone transitions from a soft rubbery elastomer to a harder rubber/elastomer typically around −45°C.

The low-temperature properties of silicones allow them to be used in many applications where most organic elastomers would become hard and brittle. Indeed, the mechanical properties of silicone elastomers change relatively little as the temperature is lowered.

The molecular flexibility of silicones also gives rise to their typical low-durometer hardness, low modulus, and high elongation over wide temperature ranges as shown in Figure 16.28. Applied stresses are quickly distributed throughout the elastomer so that properties are not generally sensitive to the rate of applied strain. Therefore, the elongation or modulus measured at two different rates would be nearly equivalent.

Since silicones at most operating conditions have essentially no crystallinity, and the polymer chains have high mobility and movement, the elastomers they form tend to be quite soft over a relatively large range of temperatures and crosslink densities.

The molecular flexibility discussed earlier allows silicone elastomers to be used down to quite low temperatures. The thermodynamically stable molecular bonds in silicones endow them with great utility over a wide range of environmental conditions.

Many elastomers will harden, embrittle, and crack after long-term exposure to outdoor conditions. There are many factors that can influence this behavior. Some elastomers contain additives or plasticizers that will slowly leach or evaporate out of the cured matrix. Plasticizers act to soften the material from its otherwise natural

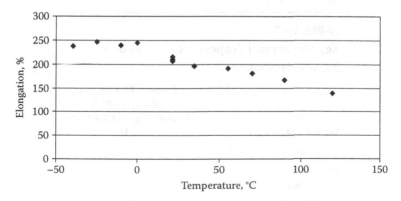

FIGURE 16.28 Elongation versus temperature behavior for a typical silicone elastomer.

hard and rather brittle state. Additives are often incorporated in many elastomers to provide resistance to oxidation caused by long exposures to warm temperatures, high humidity, sunlight and ozone, or other airborne contaminants.

Plasticizers are not generally added to silicone elastomer formulations since they are normally quite soft and flexible. Silicones also do not require additives to provide resistance to typical environmental conditions. Since they are already in a highly oxidized state, further oxidation requires rather extreme conditions. The extreme stability of the Si-O bond that makes up the polymer chain backbone of silicones gives these elastomers unparalleled longevity.

Silicones are very hydrolytically stable. Exposures to liquid water and water vapor have only a few negative effects on these elastomers at ambient temperatures. For this reason, silane coupling agents are a preferred route to attaining good adhesion performance for very many elastomers, both silicone and nonsilicone. Excerpts from hydrolytic stability testing follow:

- Silicone adhesives have been aged in 80°C saturated salt water for several weeks with no loss of adhesion strength, and in some cases even in boiling water for a week or longer.
- Silicone sealants have been aged for over 12 weeks by immersing in 60°C water without showing any loss of adhesion.
- Silicone sealants aged for 28 days at 90°C/95% relative humidity (R.H.) showed no significant change in durometer hardness.

Silicones demonstrate high stability and mechanical property retention in relatively high-temperature/high-humidity environments. This is exemplified by the lack of change in several key sealant mechanical property attributes during extended elevated temperature and humidity exposure (see Table 16.5).

Partly because of their high molecular flexibility, silicones expand a great deal when they are heated, and likewise contract a great deal when they are cooled. This property of expansion and contraction is typically characterized by the coefficient of thermal expansion (CTE), measured as the slope of the linear or volumetric change

TABLE 16.5
Key Mechanical Property Attributes of a
Silicone Sealant

Mechanical Property	Property Change (%) following 500 h Conditioning at 85°C/85% R.H.
Tensile strength	0 to +5
Elongation	−5 to +12
Tensile modulus	0 to +2
Durometer hardness	0 to +2

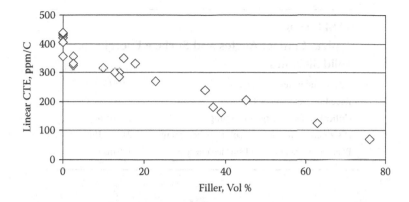

FIGURE 16.29   Linear coefficient of thermal expansion versus filler loading for a PDMS-based silicone.

in dimension over a given temperature range. For silicones, the dimension change is very linear and reproducible from about +200°C down to about −40°C. For most silicone sealants, the linear CTE is between 300 and 350 ppm per degree centigrade (micrometer per meter per degree centigrade). The CTE of a filled silicone elastomer is directly related to the volume faction of filler as shown in Figure 16.29. Particulate fillers, particularly metal oxides, usually have CTE values less than 10 ppm per degee centigrade and therefore can have a significant effect on the resulting CTE of compounded material. Since unfilled and particulate-filled silicones exhibit isotropic behavior, the volmetric CTE is mathematically approximately three times larger than the linear CTE.

Silicones have much larger CTE values than the electronics module sealing surfaces they contact. This difference is sometimes over an order of magnitude larger. Thermal transients, therefore, create a great deal of stress in parts. The inherently low modulus and high elongation of silicone elastomers allows sealing to be maintained over significant temperature excursions. Therefore, silicones not only accommodate the stresses induced by their own expansion and contraction but can also accommodate the stresses from the expansions and contractions of adjacent components.

Silicones have extremely low surface energies, as summarized in Table 16.6. This means that liquid silicones will not bead up but instead will wet out and flow out over nearly all surfaces they come into contact with. This characteristic allows silicones to not only flow over and fill in very finely detailed substrate topography but also to lift and displace air bubbles and particulates on surfaces.

Silicones are known for their ability to form hydrolytically stable adhesive bonds to a wide variety of surfaces. Their surface activity is, in part, due to the low surface energy they possess, allowing them to wet out nearly all substrate surfaces.

With the wide variety of silane coupling agents available, silicone elastomers can bond to many surfaces. These bonds are typically quite strong, but the joint strength is, of course, limited to the strength of the silicone elastomer matrix. Given the generally low modulus and high elongation of silicone elastomers, adhesive joints must be stressed over a fairly long distance before they will break.

TABLE 16.6

Typical Contact Angles and Surface Energies of Solid Silicones

Advancing water contact angle	110°C–115°C
Receding water contact angle	57°C–64°C
Critical surface-energy of wetting	24 mJ/m^2
Dispersion component of solid surface energy	21.7 mJ/m^2
Polar component of solid surface energy	1.1 mJ/m^2

TABLE 16.7

Comparison of Adhesion Test Results for a Typical Silicone Sealant Cured at 150°C (Test Temperature = 25°C, Bondline Thickness = 0.75 mm)

Lap shear adhesion, MPa	5.10
Tensile adhesion, MPa	3.76
180° Peel strength, kg/cm	11

Adhesion testing for elastomers is commonly carried out in three modes: lap shear, tensile, and peel. Typical values for silicone sealants are shown in Table 16.7.

Lap shear adhesion testing typically involves applying sealant to a 25 mm wide × 7–10 cm long flat substrate panel. A second identical panel is placed over the sealant for an overlap bond area of 6.25 cm^2, and the sealant is cured. The panels are then pulled apart at 180° from each other. Bondline thickness is often held constant at 0.5 to 1.5 mm. This is thinner than what is often used with higher-rigidity sealant and adhesive materials, where fracture mechanics plays a much more important role in joint failures. Crack propagation can play a role in the failure mode, and this can be significant since the cracks or tears can travel up to 2.5 cm of joint distance. For a soft adhesive/sealant with high elongation, much thinner bondlines can be used with good reproducibility. Lap shear testing is probably the most common adhesion test for silicone elastomers. Lap shear strength values will commonly be less than the tensile strength of the product being tested at bondlines above nearly 0.5 mm thick.

Tensile adhesion testing also utilizes a 6.25 cm^2 bonding area. The two substrates are pulled apart at 90° from each other. In this mode, crack propagation is less of an issue since such a crack can only travel along the bondline thickness.

Peel strength testing is most commonly used with one-part condensation (moisture) cure products. The sealant is applied to a flat substrate, and a flexible strip is placed over it. Additional sealant is then applied over the flexible strip to fully cover and encapsulate it. The strip may be a solid narrow thin metal, a metal screen, or a nonmetal mesh. A common feature of this flexible member is that it allows a large surface area of the sealant to be exposed to air to allow for a reasonably short cure time. The flexible strip is pulled away from the substrate at either a 90° or 180° angle from the solid base substrate.

Adhesion performance can be affected by a variety of factors:

- Surface—reactivity and cleanliness
- Cure—time and temperature
- Joint design—area, movement, and bondline thickness
- Environmental—temperature, humidity, and accelerated aging

Silicones can be made to bond to most surfaces. Most chemical bonding occurs through hydroxyl sites on the surface, though other chemical species can also have sufficient reactivity to achieve bonding. Most metal substrates have a large number of hydroxyl sites, enabling very strong bond formation with relative ease. In contrast, many industrial polymers have only a few hydroxyl sites, though when compounded into useful plastics they often contain fillers for improved stiffness and strength. It has been found that a glass or mineral (usually calcium carbonate) filler content of 20% or higher significantly improves bonding capability for most plastics. For some plastics, bonding is very difficult in the absence of sufficient concentrations of such fillers.

Surface modifications are commonplace to improve sealant bonding. Plasma, corona, UV/ozone, and flame treatments chemically alter surface chemistries, often by oxidation and the formation of reactive free radicals [17]. Some of the reactive species that are formed are somewhat transient, and so sealants must be applied within a certain amount of time to obtain the benefits of the treatment.

Primers are also used to chemically modify surfaces. These are typically dilute solutions of silane coupling agents in a solvent. The silanes cure via a condensation reaction and are very similar (and in some cases identical) to the adhesion promoters used in silicone sealants. For a sealant to strongly bond to a given substrate, these species must migrate to the bonding surface to give any benefit, and so the majority of the silane molecules remain in the bulk of the sealant, thereby not participating in bonding. In primers, all of the silanes are applied to a surface to take part in bonding.

Silicones will wet out on nearly all surfaces. Because of this, they can actually lift many contaminants off surfaces and achieve good bonding in spite of the contamination. However, some contaminants can remain. Obviously, anything that interferes with good contact between the sealant and the actual substrate would be expected to deter strong bonding. Surface contamination can come from release agents, lubricants, plasticizers, oils, and processing aids and/or airborne species. Washing parts is occasionally required to obtain the desired adhesion. It should be noted that some contamination can reside in the porosity of the substrate and can come out during exposure to heat or in use, and can disrupt bonds.

For heat-cure silicones, the time required to cure is directly related to the cure temperature. For most of the self-priming heat-cure silicone sealants, temperatures above 80°C are essential for adequate adhesion, and for applications requiring cure times of 1 h or less, cure temperatures of above 100°C are typically required. For these products, the higher the temperature, the faster the cure will be completed. This is shown in Figure 16.30 for two typical heat-cure (addition-cure) PDMS-based automotive electronics packaging sealants.

For most heat-cure silicone sealants, bond strength develops more slowly than mechanical properties such as tensile strength, elongation, modulus, or durometer hardness. This translates to cure times that extend beyond the point where the material appears to have reached full property development. Generally, the time to reach full adhesion is 1.5 to 2 times longer than the time required to reach complete or nearly complete development of the mechanical properties, such as commonly expressed in terms of the t_{90} the time it takes to form 90% of the total possible cross-links in the compound (t_{90}), determined via rheometry.

Bond strength is directly proportional to the bond area. Maximum sealant joint strength is obtained by ensuring the largest contact area between the sealant and the surfaces to be bonded. While large overlaps of flat surfaces can achieve this objective, it is often preferable to utilize a tongue-in-groove design. This maximizes bonding area while minimizing total packaging space. It can also limit the total distance the joint can be moved in at least one direction. For effective sealing, a long, circuitous potential leak path afforded by a tongue-in-groove joint is preferred over a straightline design.

One of the benefits that silicone elastomers bring to sealing applications is their high elongation. Though this can cause difficulties, successful sealing results if bonded surfaces are allowed to have considerable relative movement. This compli-

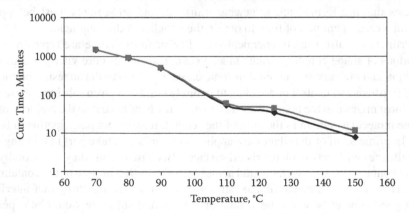

FIGURE 16.30 Cure time to obtain maximum aluminum lap shear strength to aluminum versus temperature for two different PDMS-based heat-cure silicone sealants.

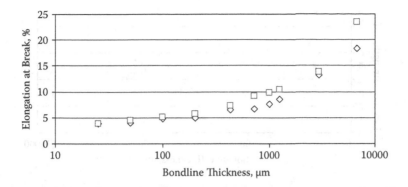

FIGURE 16.31 Elongation of lap shear joints versus bondline thickness for two different PDMS-based silicone adhesive/sealants commonly used in sealing of automotive electronics.

ance can relieve stresses due to differing thermal expansions and contractions and other application-induced movements.

As previously discussed, optimizing joint designs, such as tongue-in-groove architectures, is critical to successful sealing. Another design aspect that needs careful attention is bondline thickness. Joints must be large enough to accommodate the expected thermal contraction and expansion for sealing applications where large substrates are attached. Figure 16.31 shows typical lap shear joint elongation at break as a function of bondline thickness for two PDMS-based silicone sealants.

Determining the optimal bondline thickness may be a bit more complex. Besides the thermally induced movement of substrates, dimensional tolerance stack-ups and substrate placement accuracy need suitable consideration. In some cases, automated placement can misalign or press the substrates together to the point of actual direct contact so that the sealant is nearly completely squeezed out. Such joints will have low robustness to environmental stresses. Even in a lap shear joint configuration, bonds can be stressed over a considerable distance before failure.

To avoid some of these issues, spacer beads can be added to the sealants. These beads are typically made of silicate glass in nominal sizes of 75 to 250 μm diameter. The slight degradation in overall sealant strength resulting from the addition of glass beads to a sealant formulation is compensated by the assurance of a minimum bondline thickness.

While a certain minimum bondline thickness is required, it is not advantageous to use any more sealant than is required to perform the job at hand for both economic as well as aesthetic reasons. There are also performance benefits provided by *thinner* bondlines. Generally, adhesion strength of a silicone sealant rises as the bondline depth decreases. This general trend holds true to a limiting thickness of about 100 μm, as the data in Figure 16.32 show for a typical PDMS-based sealant.

In automotive applications, individual electronic components may see actual use temperatures as low as −40°C and as high as +200°C. Stresses arise mainly from unavoidable thermal expansion mismatch between the various components of a final part or module. These movements, although primarily focused on the in-plane (x-y) orientation, can also be significant in the vertical (z) axis direction. Figure 16.33 shows

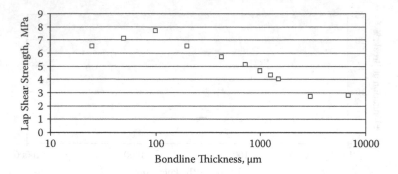

FIGURE 16.32 Adhesion strength (lap shear strength) versus bondline thickness for a commonly used PDMS-based silicone automotive electronics sealant.

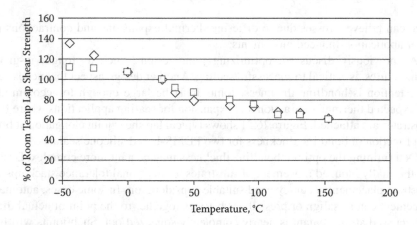

FIGURE 16.33 Adhesion strength versus application temperature for two different PDMS-based silicone sealants used in sealing automotive electronics. Lap shear strength is shown with respect to the value at room temperature.

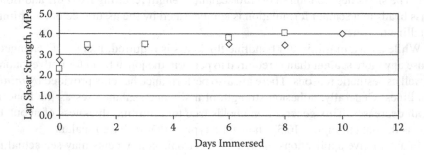

FIGURE 16.34 Effects of immersion in 85°C saturated salt water solution on adhesion of two different automotive silicone sealants to 30% glass-filled poly(butylene terephthalate), a common automotive connector material.

that, with decreasing temperature, the lap shear strength of silicone sealants typically increases slightly, while elevated temperatures tend to decrease adhesion strength.

Accelerating the effects of environmental aging can take on a variety of forms. The main techniques used to evaluate the long-term durability of silicone elastomers often involve subjecting the materials to cycling from very low to very high temperatures, or aging at set temperatures coupled with exposure to salt water (immersion, fog, or spray) or high humidity. The stable adhesion strength performance of two typical automotive silicone sealants as a function of time of immersion in saturated salt water at 85°C is shown in Figure 16.34. For electronics, some of these tests may also be conducted with electrically energized components or modules to evaluate the potential for galvanic-induced corrosion.

16.6.1.1 Some Types of Silicone Sealants

Silicone sealants come in several different forms, commonly differentiated by their cure chemistries.

16.6.1.1.1 One-Part Moisture-Cure Silicones

This type of silicone cures by reacting with moisture in the air. Probably the most widely recognized variety of this type gives off acetic acid (acetoxy-cure) and is not generally suitable for electronics due to metal corrosion concerns. Those most commonly used in automotive electronics applications give off an alcohol such as methanol or ethanol (alkoxy-cure), although there are other available types. Once applied, these sealants will cure as long as they are exposed to moist air. The cure proceeds from the outside inward, first forming a "skin" at the surface and then solidifying in the bulk. Since moisture from the air must diffuse into the sealant and react, the time to fully cure can take days. During this cure time, the sealant cannot be exposed to high temperatures without the risk of boiling the alcohol cure by-product, resulting in a seal containing bubbles and voids.

This cure chemistry is very robust and typically provides adhesion to the widest set of substrates. It is one of the least expensive types of silicones, but the long cure time can be detrimental to high-volume manufacturing due, in part, to the handling of parts with uncured material. For lid sealing, however, some electronic module manufacturers commonly use a moisture-cure silicone sealant. The sealant can survive a low-pressure leak test almost immediately after dispensing and will continue to cure during subsequent manufacturing steps, and even while sitting in inventory or being shipped to an end user. There are other related cure chemistries suitable for automotive electronics sealing, including those that evolve oxime and ketone species during cure.

One-part moisture-cure silicones are typically cured at room temperature. They require a relative humidity of only 20% in the air to cure properly and in a reasonable time. Cure begins at the outer surface and proceeds inward. Full cure can take several days—the longer the seal path length, the longer the time it will take to achieve full cure. In many cases, sealant can be applied and parts can be handled within a few minutes, allowing the material to complete its cure during subsequent manufacturing steps or even while packaged and shipped to final end users. Note that exposing these materials to temperatures above nearly 65°C before they are fully cured can create bubbles from the cure reaction by-product, which could compromise the seal. After they are fully cured, the seals can be exposed to much higher temperatures.

16.6.1.1.2 Heat-Cure Silicone Resins

A very common type of silicone sealant used in automotive electronics is supplied in one- or two-part formulations, and requires heat to cure and activate the adhesion. The advantages of this cure system are that it provides a material that can have a long shelf life and is manufacturing-friendly in that it remains in an as-supplied condition until it is heated. Once brought to its cure temperature (usually between 125°C and 150°C), it solidifies and fully cures within 30–60 min. All curing takes place on the manufacturing floor under controlled conditions. Figure 16.35 shows a single-component silicone sealant being dispensed onto a die-cast module housing for subsequent thermal curing.

Besides achieving full cure rapidly, this type of silicone sealant is generally stronger than most of the moisture-cure variety and has a wider range of flow characteristics, but must often be formulated with adhesion-promoting materials.

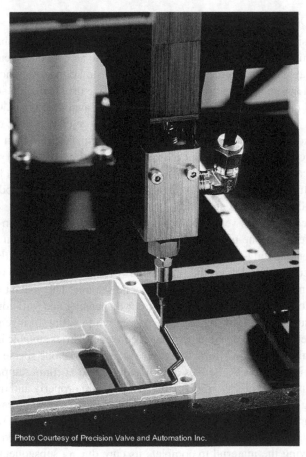

Photo Courtesy of Precision Valve and Automation Inc.

FIGURE 16.35 A single-component thermal-cure silicone CIPG gasket being dispensed onto a metal module housing.

Heat-cure silicones (addition-cure type) are somewhat sensitive to certain contaminants, which can cause the cure to be retarded or poisoned. Cure poisoning (also called "inhibition") can be evidenced by a tacky or wet feeling contact surface and/or outer surface, by loss of adhesion, or, in some cases, with total noncuring of the sealant. The most common sources of contamination that can cause this are from amine and sulfur species. Contact of these species with the silicone in dispensing equipment (butyl, natural, polychloroisoprene, and some other types of rubbers used in hoses and gaskets), on substrates or in the cure oven (use of amine-catalyzed epoxies in the oven, for instance), can retard or halt the cure. Precautions should be taken to avoid such cross-contamination.

Typical oven temperatures used with heat-cure silicone sealants are 125°C to 150°C, with cure times ranging from 5 to 60 min. Since adhesion is often maximized in the last stages of the material cure cycle, it is important to ensure that full cure is achieved. The best way to accomplish this is to run an adhesion test on the parts at several cure times and/or temperatures to determine when optimal adhesion is obtained.

16.6.1.2　Silicone Mechanical Property Considerations

In choosing a material for a seal, one must take into consideration the expected movement of the substrates. Large movements from mismatches in coefficients of thermal expansion and mechanical shock may require the seal to stretch and compress considerably. In some designs, this seal movement could create considerable stress to the joint. This can be managed, in part, by choosing a lower durometer hardness and/or low-modulus material that will deform rather easily to provide stress relief. Silicone sealants and gaskets typically range from a Shore A durometer hardness of 20 up to 70, providing a range from a very soft to a rather hard rubber. Soft seals are commonly chosen when using a thin lid to provide the greatest stress relief, while more firm seals are desired when joint movement should be limited. Material strength is also generally higher with the harder, higher elastic modulus, sealants.

Extremely soft closed-cell foam seals are chosen when part tolerance and movement are rather high, as they can be compressed much further than a more solid seal and they usually provide the lowest possible hardness and modulus. Silicones generally have very low compression set values compared to other seal materials. Fully cured silicones often have compression sets between 20% and 30%. While this is not an issue with adhesive seals since they are chemically bonded to both substrate surfaces, however, when designing a CIPG or DFG seal, compression set must be taken into consideration. Sealants must be compressed more than their compression set so that the seal will be maintained over long periods even after high-temperature exposures.

16.6.2　Polyurethane Sealants and Encapsulants

Polyurethane (PU) resins play an increasingly important role in sealing and protecting automotive electronics. The ability of these materials to form strong stable bonds to a variety of substrates and yet allow compliance following cure makes them most useful for module perimeter sealing (CIPG and FIPG) as well as potting/encapsulation resins. The highly versatile chemistry variations allow the specific formulations to achieve desired properties for a wide variety of automotive sealant and encapsulation applications. PU resins can covalently bond to active substrate surface hydrogen and can also bond to fairly low surface energy substrates.

In the most general terms, PU materials depend on the reaction of an isocyanate group (either aromatic or aliphatic) with any number of materials having an active hydrogen. The most common and useful reaction for sealant materials is the reaction between a diisocyanate material such as methylene diphenyl diisocyanate (MDI) or toluene diisocyanate (TDI) and a polyol. As with all sealant chemical classes, PU resins may be modified not only with chemical reactant variants but with numerous fillers, thixotropes, and other additives, resulting in desired properties. The polyol material is usually varied to change end characteristics. Elastomeric chain segments, such as polybutadiene, are incorporated to improve flexibility to seal dissimilar surfaces. For sealant materials, side reactions such as the reaction between an isocyanate and water with the formation of undesirable products and, more importantly, the liberation of CO_2, should be avoided.

The critical characteristics of a successful PU sealant would include:

Thermal stability	125°C (Underhood off-engine applications)
Viscosity	800–1600 Poise with shear-thinning characteristics
Cure conditions	125°C for 30 min (sealant bondline temperature)
Durometer hardness/elastic modulus	40–80 Shore A scale
Compression set	Low
Adhesion to substrate	Seal dependent, can range from no bond to over 3.4 MPa lap shear strength

As with epoxy and silicones, PUs represent a large chemical class of materials, so it is very difficult to make general statements concerning their curing and use. Much activity and interest in PU sealants continue in the case of automotive electronics.

There are a few general drawbacks to using PUs as with all sealant material classes. Generally, there is moisture sensitivity during application and cure. Side reactions leading to CO_2 gas and undesired urea-based side reactions are possible from moisture exposure. Metal adhesion, especially to lightly contaminated metal castings and stampings, where oil is present, may need special cleaning, priming, or precoating to obtain high adhesion for applications such as FIPGS. Some two-component sealant formulations require precise metering and mix ratio monitoring. Ratios ranging from 100:4 to 100:10 are not uncommon, and result in dispensed sealant property variations unless controlled. As with all two-component systems, PU sealant formulations require complete mixing for homogeneity of the dispensed material. This would necessitate dynamic mixing, as shown in Figure 16.36, rather than lower-cost static mixing equipment in a high-volume manufacturing environment.

FIGURE 16.36 Dynamic mixing dispense valve dispensing a 4:100 mix ratio PU sealant material used in sealing an underhood-packaged engine control module. (Photo courtesy of Bartec, Inc.)

PU materials have also fallen under scrutiny from the standpoint of worker health and toxicity. Common, cost-effective diisocyanate materials such as TDI and MDI are often prereacted into larger adducts, or even replaced entirely in formulations [10]. Appropriate health and safety precautions are mandatory, and the Material Safety Data Sheet (MSDS) should be consulted at a minimum with all sealant usage.

The largest concern with using PU encapsulants and sealant materials in underhood environments is the upper continuous usage temperature of the application in comparison to the claimed upper continuous usage temperature of the PU sealant. The trend in the underhood automotive environment is to compress increased functionality in an ever-decreasing volume. The applications that locate individual electronic components closer to the engine, as well as associated exhaust, climate control, and other heat-generating components, are increasing [11]. The situation is worsened by the associated trend toward packing increasing numbers of individual heat-generating electrical components, such as power drivers, diodes, and ICs into these smaller, higher-temperature locations. One of the demands this places on the automotive packaging engineer is that the sealant or encapsulant of choice must withstand peak temperature over 125°C and 150°C in off- and on-engine applications, respectively, plus the internal heat contribution of enclosed or encapsulated devices [12]. The upper usage temperature of PUs, though greatly dependent on the specific formulation and additives, can range from 115°C to 150°C. This greatly affects general utilization of PUs for underhood sealants. With this being said, there are current production modules, such as the example shown in Figure 16.37, designed with PU sealants for the underhood automotive environment. As with all sealant materials,

Photo Courtesy of Lord Corporation

FIGURE 16.37 Automotive electronic module potted with a two-component PU sealant. The printed wiring board interconnection substrate is thermally coupled in critical regions to the die-cast aluminum housing. The rubber-modified PU material is characterized by low initial viscosity, high adhesion to all surfaces, and room temperature cure can be accelerated with the application of heat.

a complete understanding of the usage environment and OEM qualification testing is needed before selection or application. Significant variations also exist in defining thermal stability and continuous-use temperatures. Testing following 2000 to as much as 3000 h at the maximum usage temperature may be required. A "safety margin" is also useful for a sealant in terms of thermal stability since it is difficult to anticipate all usage profiles.

16.6.3 EPOXY SEALANTS AND ENCAPSULANTS

Epoxy-based materials are widely used for sealing automotive electronic enclosures, most commonly in off-engine underhood or exterior automotive module locations. Epoxy resins are considered hard, rigid materials with generally good resistance to water, most automotive solvents, heat, and cold. The success of this class of sealants and encapsulants depends not only on inherent properties but also on both the design and usage environment.

16.6.3.1 Epoxy Chemistry

Epoxy chemistry is well known and diverse. In the most basic terms, a resin usually containing two or more epoxide functional reactive groups (such as the commonly used epichlorohydrin adduct of bisphenol A) is reacted with a multifunctional reactive species such as a polyamine, to yield a thermosetting matrix. Aside from the basic resin and curing agent constituents, an epoxy sealant will contain fillers, thixotropes, reinforcements, pigments, antioxidants, and a variety of other components suited to a particular application [13].

It must be noted that since one of the most economic synthesis routes for epoxy resins is to react epichlorohydrin with a hydroxyl-containing material such as bisphenol A, many epoxy resins of this type contain inherently high chloride content as a reaction by-product. This chloride content can contribute to subsequent corrosion in proximal electronic circuitry. This reaction by-product may be substantially reduced to yield "electronic-grade" epoxy resins at an increased cost.

Epoxy sealants are generally characterized by varying degrees of rigidity that make them most suitable for the bonding of dissimilar linear CTE materials. Epoxy sealants may wet a variety of different surface energy substrates. Low surface energy or contaminated substrates must be treated and/or primed to yield acceptable bond strengths.

The bonding of a glass-filled PBT polyester module lid to a housing of the same resin for an off-engine automotive module or sensor would be a typical example of an automotive use for a rigid epoxy sealant. In this example, a high-volume manufacturing environment would favor a single-component room-temperature-stable resin/curing agent mixture that would cure at about 125°C. Other desirable attributes would include a suitable glass transition temperature (T_g), a viscosity of about 900–1500 Poise, a sufficiently shear-thinning rheology to permit flow during dispense and allow bead formation, suitable cure time and temperature for both the manufacturing line rate and survival of internal components, and sufficient adhesion to both substrates (in the case of an FIPG). The ideal location of the glass transition temperature in a lid sealing epoxy resin is a point of some debate. One viewpoint is that the ideal resin would have a Tg above the upper use temperature of the sealed device, particularly since the resin matrix would have a lower linear coefficient of thermal expansion. Another perspective is that the location of the resin T_g is of secondary importance since although the CTE of the matrix substantially increases, it is offset by the general decrease in elastic modulus. The correct decision is strongly dependent on the specific materials being sealed, along with the geometry, dimensions, and other factors. It is recommended that the proposed seal joint be modeled and its integrity verified through appropriate thermal and mechanical testing, including thermal cycling, vibration, humidity, and mechanical shock testing.

16.6.3.2   Flexible Epoxy Resins

Epoxy resins are generally considered to be rigid materials not appropriate for materials with dissimilar thermal and mechanical properties. The applicability of the general class of epoxy-based materials to sealing automotive electronic enclosures is greatly increased with the incorporation of flexible modifiers [14]. Flexible epoxies and related materials have useful elongations of a few percentage points in contrast to elastomers, which can exhibit elongations well over 100%.

Additionally, there are potential nonreactive additives to enhance epoxy flexibility. The use of flexibilizers in epoxy resins is often at the expense of other desirable properties, including chemical, moisture, and thermal resistance, as well as some physical properties.

16.6.4 OTHER AUTOMOTIVE ELECTRONICS SEALANTS

There are many other sealants used in selected automotive applications in addition to the major classes mentioned. Acrylic-based sealants may be used for lower-temperature sealing of similar coefficients of thermal expansion materials. Foamed materials, other than the previously mentioned silicone foam, are also used in niche applications. Butyl mastic-type materials have been used in an FIPG-type construction in conjunction with a crimped-metal mechanical closure. Polysulfide [15] and fluorosilicone elastomeric sealants have also found utility in selected applications. However, the odor and outgasing constituents of polysulfide sealants must be managed. The expense of fluorosilicone and related materials must be balanced against the low degree of swelling and retention of flexibility in proximity to fuel.

Sealing compounds based on polyacrylate rubber represent one of the latest developments in FIPG sealing automotive powertrain modules against harsh hot engine oil and transmission fluid. This recent electronics sealing option is available with a compatible noncorrosive condensation-cure chemistry and exhibits mechanical properties similar to those of PDMS-based silicone FIPG materials. Retention of adhesion, tensile strength, and elongation is exceptional during hundreds of hours of immersion in various commercial engine oils and transmission fluids at 150°C [16].

16.7   THE SEALING SUBSTRATE

The condition of the surface to which a sealant will adhere is important with CIPG, DFG and, especially, with FIPG designs. Both curing and adhesion of the resulting sealant must be verified on both material test plaques and production representative parts. Stamped metal parts must be free from stamping oils, shavings, scale, corrosion, burrs, or oxide buildup. Die-cast metal parts must be free from die lubricant and other processing oils. Silicone die lubricants should be used with great caution since bond compromising traces following any aqueous cleaning may often persist. Mechanical adhesion should also be maximized by use of an appropriate surface topography that has been verified through testing [17].

Die-cast metal housings using silicone oil-based die lubricants present significant issues for eventual sealant adhesion due to their persistence after deflashing and deburring operations. Tumbling media do not properly impinge on bonding troughs and recesses due to geometrical constraints. Water-based surfactant systems have limited effectiveness in removing these materials—especially in grooves destined to be the bonding/sealing surface. Painting, electrolytic, electrostatic, epoxy, or other coating technologies resulting in another layer of material over a stamped or die-cast metal housing can be a benefit, provided the underlying metal cleaning and priming results in superior coating adhesion. Painting usually results in a high-surface-energy surface that offers good bonding with most sealants (although this needs to be checked and controlled). The risk is for poorly adhering coating to be pulled from the metal surface by the sealant, with subsequent leaking. The subject of priming will be discussed in the next section.

Consistent bondable metal surfaces are needed for robust FIPG-type sealing. Adequate process controls should be established with the metal foundry to guarantee clean surfaces, with a special emphasis on monitoring of any batch-cleaning processes. Periodic part analysis or, more practical, periodic bond tests with the adhesive sealant to be used, is preferable. Even with the best measures in place at the die-cast or stamping facility, long supply pipelines from low-cost global manufacturers along with contamination from packaging material lend unpredictability to the potential of the metal substrate to accept a sealant when received at the electronics assembly plant. Some automotive electronics manufacturers have resorted to oxidative and combustive methods of cleaning, such as the use of gas flame (Figure 16.38) or flume plasma cleaning (Figure 16.39) with robotic or conveyorized equipment close to the time and place of sealant dispense to minimize unpredictable opportunities for contamination [18].

These dry methods are highly preferable over wiping, cleaning, or rinsing methods. Solvent wiping, although low cost, is labor intensive, raises obvious health and safety concerns, and is of questionable effectiveness. Attempting to remove a hydrocarbon processing oil with an isopropyl alcohol cloth, for example, would only serve to incompletely remove the contaminant and with potential redeposition of contaminants on the part upon alcohol evaporation. Another glaring deficiency is that the chosen solvent might be ineffective in solubilizing the mix of contaminating materials present on a part at a particular point in time.

FIGURE 16.38 Gas flame cleaning of die-cast aluminum castings prior to FIPG bonding to insure adequate bond strength. The picture shows a row of metal housings traveling toward the vertically and horizontally oriented flame burners. (Photo courtesy of Visteon Corporation.)

FIGURE 16.39 Open-air flume plasma treatment of a housing prior to sealant dispensing and bonding, to both clean and increase the polarity of a low-surface-energy polymer material. (Photos courtesy of Visteon Corporation.)

The verification of cleaning process effectiveness can be performed with relatively sophisticated means such as solvent rinsing of the sealant region of the part, followed by a number of relevant analytical techniques, including IR spectroscopy or gas chromatography. Process monitoring on a frequent basis is best performed by actually taking the sealant to be tested and performing a "spot adhesion test." This fairly empirical test simply involves curing a spot (a few grams) of sealant on the cleaned metal part and pushing the cured material with a metal spatula. A poorly cleaned part would result in the sealant being easily removed from the part with a clean interfacial release, while a clean part is defined by a high degree of sealant adhesion [19]. Cleaning of metallic stampings and castings should be done immediately prior to sealant dispense since there is less chance then for recontamination from plant environments, shipment, packaging material, and handling. This is especially valid for die-cast parts, which might have entrapped contaminants in surface porosity that could exude to the surface over time.

Plastic parts, typically injection molded, should likewise be free from contaminants. Externally applied mold lubricants, especially silicone mold release agents, should not be permitted for parts requiring strong adhesion. The effect of internal mold lubricants and other compounding materials such as antioxidants and flame retardants on sealant adhesion and cure should be investigated prior to molding tool manufacture. As with metal parts, the adhesion characteristics of plastic materials can be tested using the "spot adhesion test," contact angle measurements, or wetting tests.

The inherent surface lubricity and/or low surface polarity of some plastic molding compounds, especially aliphatic and aromatic polyamide and poly(phenylene) sulfide) materials, to name a few, may need modification, either chemically or, more commonly, using dry oxidative methods such as flaming, corona discharge, UV, UV-ozone, or open air (flume) plasma etching (shown in Figure 16.39). There has been an increased use of lower cost polyolefin materials, such as polypropylene, in some sealed housing applications, which most probably needs surface treatment for reliable sealing. The longevity and handling characteristics of the surface treatment need to be understood. The time dependency of a treatment or the sensitivity to handling or packaging contamination often necessitates on-site part treatment rather than treatment at a subsupplier.

The surface topography of the plastic housing or connector to be sealed must also be optimized to provide adequate sealing. Molding process can have a significant effect on sealant adhesion since parameters such as injection molding tool surface temperature and resin packing pressure, among others, can alter the extent of surface crystallinity (with semicrystalline polymers), increase or decrease surface filler or reinforcement exposure, or cause increased surface concentrations of internal lubricants, antioxidants, flame retardants, and other compounding materials.

The long-term integrity of a seal depends on the selection of the substrate as well as the sealant. For example, the corrosion of an aluminum module cover in a saltwater environment might cause the failure of a silicone FIPG seal from erosion of the aluminum metal adjacent to and under the seal. A similar situational example would be the failure of a poly(butylene terephthalate) (PBT) connector to aluminum die-cast housing silicone FIPG seal due to the hydrolysis and erosion of the polyester resin underlying the seal. There are many similar examples, and much care needs to be directed at all the materials of an electronic enclosure as an environmental protection system, not solely focusing on the sealant.

16.7.1 PRIMING

Another method to ensure adequate FIPG sealant adhesion to both metal and plastic substrates involves the use of a separate primer material. This method is often a last resort since it involves the dispensing of another material system with associated process controls for consistency. Additionally, many primer materials contain solvents, thereby presenting accompanying health and safety concerns.

Primers are typically solvent dilutions of moisture- or heat-cure adhesion promoting substances. The purpose of the primer is to create a more reactive surface that a sealant can bond to. In some cases, primers are also used to "seal" the surface, that is, to prevent mobile species in the substrate from blooming to the surface, which could interfere with bonding.

There are several benefits to using a primer. First, the solvents used in the formulation can act to further clean a substrate. Second, the concentration of adhesion promoter chemicals in the primer is often much higher than can be practically incorporated into adhesive or sealant formulations. In a sealant, the only adhesion promoter molecules that actively take part in the adhesion process are those that are in contact with the surface when the sealant is applied and cured. An excess of such adhesion promoters has to be included in the formulation since only a small fraction will be present at the substrate interface. A balance must often be achieved between obtaining sufficient quantity of the adhesion promoter at the interface and the negative impact the excess will have on properties such as tensile strength, elongation, and shelf life.

With a primer, a much larger concentration of the adhesion promoters can be used since 100% is deposited at the interface between the substrate and the sealant. In addition, a much wider selection of adhesion promoters can be used since they will not have an impact on other properties of the applied adhesive or sealant.

A primer works by chemically bonding to the substrate while creating a multitude of new bonding sites specifically targeted for reactivity of the adhesive or sealant. The primer must be applied and then cured, usually by exposure to moisture in the air. Some primers will additionally penetrate into the surface porosity to create new bonding sites deep (on a molecular level) into the surface. On some plastic and rubber surfaces, the primer may also begin to plasticize the outer layer of the surface, allowing adhesion promoters to penetrate deeper still where they react to form what is called an interpenetrating network (IPN). IPN formation provides an additional anchor for the primer to the substrate surface to further enhance bond strength.

In order for a primer to wet out, penetrate into, and react with a given substrate, it must be chemically tailored to be a good fit for that substrate. When dealing with plastic surfaces, solvents and other ingredients ideally should mildly plasticize the surface without creating stress cracks or other forms of degradation. Relatively polar substrates such as nylon require polar solvents, surfactants, and/or adhesion promoters to achieve good bonding, while nonpolar substrates require relatively nonpolar additives. For extremely nonpolar and/or nonreactive substrates such as polyethylene, polypropylene, gold, platinum, fluorinated surfaces, and others, it may be necessary to use a more aggressive surface treatment such as corona, flame, plasma, acid etch, or others to create enough chemical bonding sites for adhesion.

Surface contamination can impede or completely prevent a primer from wetting out and reacting to a substrate. While the solvents in the primers can sometimes act to remove or redistribute such contamination, good adhesion requires surfaces to be relatively clean before primer application. Cleaning techniques vary considerably, based on the substrate and the types of expected contaminations that may be present. Metal surfaces are commonly washed with a hot-water-based detergent to remove cutting oils, lubricants, waxes, and debris, and then thoroughly rinsed. Alternatively, some metal surfaces are exposed to a propane flame for a few seconds to burn off contamination. Plastic parts are more commonly treated with UV, UV-ozone or plasma to both oxidize away contamination and create a more reactive oxidized surface [20].

Primers are typically formulated with moisture-reactive adhesion promoters (often silane coupling agents) and cure catalysts. These must be diluted with a solvent so that they can be applied in layers often no more than 0.1 to 1 μm thick. Primers are usually applied by spraying, dipping, wiping, or brushing. To maintain a uniform and very thin coating, it is common to further dilute primers with additional solvent when spraying or dipping. In nearly all cases, the best performance will be obtained by applying as thin a coating to the surface as possible. Overapplying is commonly evidenced by the formation of a whitish chalky surface that may not provide optimum bonding. A common creed used with priming is "less is better" when referring to their application.

Since the active ingredients in primers generally cure on exposure to moisture, time must be allowed for the primer to fully cure before applying the adhesive or sealant. A somewhat common mistake is to assume that, as soon as the solvent from the primer has evaporated and dried, the surface is ready for sealant application. Many primers require cure times from 20 min to 2 h at room temperature, and can be somewhat accelerated with low heat (but bear in mind that a hot oven usually has a very low moisture content in the air). When designing a process, it is a good protocol to evaluate several primer cure schedules to see what meets adhesion performance targets.

In general, once a primer has been sufficiently cured, the surface will remain in peak condition for bonding for quite some time. Various time lengths have been published, ranging from 24 h to a week or more. In most cases, the surface should remain active for bonding nearly indefinitely, as long as it is protected from contamination or abrasion.

16.8 SEALANT DISPENSING

Nearly all automotive electronic modules and enclosures are manufactured in volumes over 50,000 units per year, and many in volumes exceeding hundreds of thousands per year. For every automobile, there may be twelve or more individual major or minor electronic control modules, many of these being located in harsh environments. In addition to proper material selection and design, a leakproof, reliable enclosure depends on a highly precise and reproducible seal. These factors usually require utilization of automated robotic sealant and encapsulant dispensing.

The first major element in automated dispensing is the automated work cell as shown in Figures 16.40 and 16.41. This work cell can be conveyorized, and can use a robot, walking beam, or a number of other automatic or even manual methods to introduce a part to the sealant dispense valve. The sealant valve is moved by use of either a multiaxis robot or, more commonly, gantry-type automation that can move the dispense valve or spray valve in X, Y, and Z directions (with other possible articulation) as shown in Figure 16.41. Alternatively, the workpiece can move under a stationary valve.

There are numerous feed system options focusing on transporting the sealant material from the shipping container to the dispense valve at the necessary flow rate and pressure. These feed systems range from rather simple gravity systems to more common pressure tanks and ram-type push-plate transfer pumps shown in Figure 16.42. The use of specialty and small sealant volumes might require use of refrigerated storage or heated delivery (specifically in the case of hot melt sealants) or smaller syringe plunger-type containers, all with an endless variety of construction materials and coating variations. Feed systems may also include container agitation to prevent separation, or additional fluid pumps to facilitate material movement. It is important to match the size of the feed system with the expected material usage to minimize the potential for exceeding the recommended floor life of the material or, conversely, to require unmanageably frequent feed container replacement.

Depending on the sealant type and specific part requirements, the sealant may have an additional inline pump or even a metering device or combination. Two com-

FIGURE 16.40 Dual-station production dispense station showing dispense of a two-component dispensed silicone foam sealant on a plastic housing. Note the use of replaceable static mixing element and replaceable dispenser tip. Parallel dispensers double the overall cell throughput. The foam sealant was chosen because of large Z-axis tolerances between the cover and adjacent housing.

mon options are shown in Figure 16.43. One- and two-component metering may also be accomplished at the dispense valve.

There are many dispense valve variations in the market with features enabling one- and two-component dispensing, the latter with any mix ratio dictated by the sealant formulation. Valves can range from simple air-actuated designs to more involved positive displacement, dynamic mixing, and meter/mixing designs. Additionally, valves may be sized to handle most sealant volumes and sealant bead geometries. The best approach is to first select a viable sealant for an application and then work with the material supplier to find a viable and robust manufacturing solution. Most sealant suppliers have extensive experience with the packaging and dispensing of the sealants they provide. The chances are overwhelmingly great that either a sealant supplier or a sealant equipment manufacturer has set up dispense installations similar to a planned application. Therefore, it is beneficial to leverage this collective industrial experience prior to trying to "go it alone."

There are many types of dispense valve head design variations. The variety ranges from relatively simple single- and double-component air-actuated valves, possibly requiring upstream metering or pumping, to the more accurate piston and rod positive displacement pump valves that combine metering and dispense valve functions.

Two-component valves often have static, or motionless, mixing nozzles attached, which contain a series of alternating right- and left-hand helical elements at a 90° orientation to one another. These disposable, low-cost, mixers have no moving parts, and are available in a wide variety of sizes and materials. Some nozzles combine material metering with both ratio control and dispense.

FIGURE 16.41 Silicone CIPG being dispensed on a module cover with X-Y gantry-type automated cell.

Figures 16.44 and 16.45 show views of two different two-component positive displacement sealant dispense valves complete with disposable static mixing nozzles. These valves may be constructed with specialty coatings and components to resist specific physical or chemical interactions with the dispensed materials. The portion of the valve in contact with the sealant or coating is usually kept relatively simple for rapid disassembly for cleaning and maintenance.

The manufacture of modules containing high-integrity/high-reliability seals requires several key elements, including precise controls over the quantity of dispensed sealant, consistent and accurate placement of the sealant bead, and a capable curing process. Most automated dispensing machines, including the machine shown in Figure 16.46, allow control over the quantity of dispensed sealant by several means, such as exclusion of air or other contaminants into the fluid-handling system, monitoring of line pressure in several locations to identify line blockage or bubbles, incorporation of air bleeder valves, and laser or vision verification of

 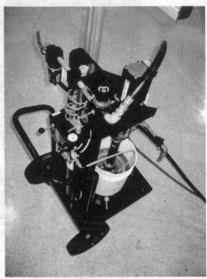

FIGURE 16.42 Two views of piston-type pail pumps used to move the sealant from shipping container to the sealant fluid-handling system. The right picture shows a view of the piston that exerts high pressure on the material to transport it to the adjacent dispense cell. (Photos courtesy of Visteon Corporation.)

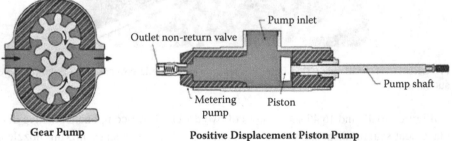

Gear Pump   **Positive Displacement Piston Pump**

Reprinted with permission of Graco Ohio Inc.

FIGURE 16.43 Schematic diagrams of sealant flow through two types of metering pumps.

dispensed sealant bead continuity, along with other methods. High-volume manufacturing also requires precise equipment setup routines that record and verify critical process characteristics at regular intervals during a production run. Sealant container replenishment and changeover requires detailed procedures aimed at reducing the opportunities for contamination, and in the case of two-component materials, eliminating the chance for any cross-contamination of individual components. Two-component systems require additional control to guard against off-ratio dispensing stemming from line clogs, pump and valve malfunctions, or numerous other possible causes. Sealant bead integrity can also be machine or operator vision inspected with

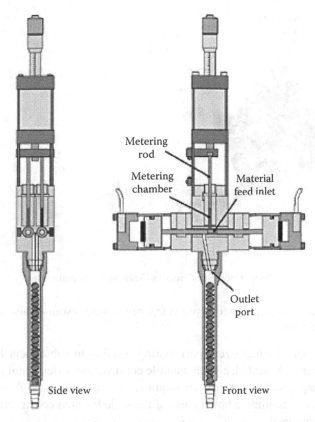

Reprinted with permission of Graco Ohio Inc.

FIGURE 16.44 Side and front views of two-component positive displacement metering/ mixing valve complete with static mixing element.

a variety of different lighting sources, an operation that may be enhanced by the incorporation of special pigments or dyes (including UV fluorescent pigments and dyes) into the sealant formulation.

Compatibility of all fluid-handling components contacting the sealant should be considered. Ignoring this point can lead to substantial system expense. For example, the use of silicone tubing and seals with silicone sealants will lead to long-term swelling as sealant components are absorbed. Lubricants in dispense equipment and valves can also lead to contamination issues. Fillers and other additives can also lead to rapid or longer-term wear of fluid-handling system components, amplifying the need for appropriate components selection, such as ceramic valves and metering components, along with regular preventive maintenance.

High-integrity seal formation is strongly dependent on how substrates are brought together after an FIPG sealant bead is dispensed. Manual module cover and/or board/connector assembly processes without at least mechanical fixturing to assure precise relative part location are to be avoided. Wet sealant beads can be wiped from

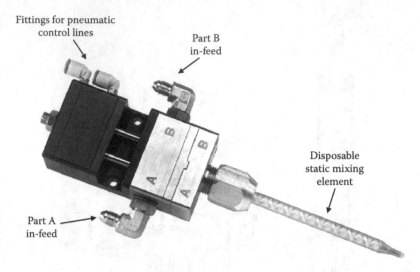

Fittings for pneumatic
control lines

Part B
in-feed

Disposable
static mixing
element

Part A
in-feed

Photo Courtesy of Precision Valve and Automation Inc.

FIGURE 16.45 Close-up view of a two-component sealant dispense valve complete with static mixing element.

intended locations by haphazard part mating, leading to subsequent leaking. The closure of completely sealed, airtight module constructions offers a unique challenge since any compression of internal air might cause "blowout" void formation of the seal. This issue is minimized by the use of the sealed/vented construction approach previously mentioned.

16.9 SEALANT CURING

Some common sealant cure technology options are outlined in Table 16.8. Thermal curing in high-volume manufacturing is accomplished by continuous vertical or horizontal conveyor ovens similar to the large conveyor oven shown in Figure 16.47. Batch curing is sometimes used for lower volume or very small modules, but precise control over the exact temperature is diminished due to the issues of oven recovery time, full versus partial oven loading, and general work flow. There has been some notable activity in variable frequency microwave curing of sealant materials, which effectively increases the kinetic energy of reactive groups without heating the entire module or component mass. The major heating modes in continuous conveyor ovens are electric forced-air convection and infrared heating. Infrared ovens generally provide less uniform heating than convection ovens since module surfaces often have wide variations in energy absorption, heat capacity, and thermal conductivity.

Successful thermal curing of a sealant material requires accurate temperature profiling of the bonded joint as the part moves through the oven to insure adequate temperature to provide cure. Oven profiling at frequent time intervals is critical for acceptable results. Oven loading often influences temperature; therefore, it is

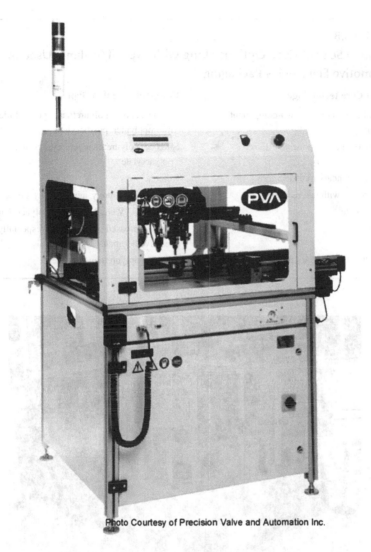

Photo Courtesy of Precision Valve and Automation Inc.

FIGURE 16.46 Automated sealant dispense production cell capable of precise sealant dispensing on hundreds of parts per hour.

critical to understand the oven temperature profiles for both low- and high-volume production scenarios.

Ultraviolet radiation curing of coatings and sealants is an option in a high-volume manufacturing environment, which is often used with specially formulated silicone, PU, epoxy, and acrylic coatings and sealants. This option is easier to use with potting, encapsulation, coating, and CIPG sealing since the entire bulk of the material is exposed. FIPG seals are possible if the seal joint is designed with an exposed sealant fillet to provide initial substrate adhesion, and a secondary cure mechanism such as thermal or moisture is enabled by the

TABLE 16.8

Common Sealant Cure Options along with Typical Sealants Used in Automotive Electronics Packaging

Sealant Cure Technology	Compatible Sealant Type
Thermal cure of one- or two-component systems	Silicone, epoxy, polyurethane, polysulfide, fluoropolymer, polyacrylate
UV cure of one-component system	Silicone, polyurethane, epoxy, acrylic, polyacrylate
UV cure/secondary moisture cure	Silicone
Two-part mix with fast reaction	Polyurethane, epoxy, acrylic, polysulfide
Moisture cure	Silicone (RTV condensation), polyacrylate
Hot melt	Polyamide-based polyurethane (especially reactive hot melt)
Hot melt with secondary moisture cure	Reactive polyurethane hot melt

FIGURE 16.47   Large conveyor oven used to simultaneously cure a silicone FIPG sealant and a thermally conductive silicone adhesive on a large engine control module. The oven holds over 100 parts at any given time for an elapsed time of 1 h. (Photo courtesy of Visteon Corporation.)

material. UV-cure silicones with a secondary moisture-cure mechanism are in common use in automotive underhood electronics packaging due to shadowing. The UV-curing equipment is usually conveyorized with either mercury vapor lamp or fusion lamp sources.

Simultaneously curing two or more materials with the same method, usually thermally, is often used in automotive module manufacture. Conformal coatings, encapsulants, and/or perimeter bead sealants can be cured during the same oven cycle. Additional module materials, such as thermally conductive adhesives or staking adhesives, may also be cocured. This practice has an obvious economic advantage from the standpoints of facilities and energy usage, and overall process efficiency. The realization of such efficiency requires selection of all materials, including the sealant, in the initial stages of module design. For example, if an addition-cure thermal transfer adhesive/sealant, selective conformal coating, and an FIPG gasket material are to be cocured, it makes sense for all materials to be chemically compatible in the uncured state as well as during cure. Worst-case contact and intermixing scenarios should be anticipated, characterized and, if needed, prevented.

The thermal curing of a sealant on FIPG modules presents the special challenge of accounting for the volume expansion of entrapped air. Venting, either permanent or temporary, must be adequate to allow pressure equalization in order to prevent leak formation in the curing sealant.

The curing process, whether thermal or another method, should not compromise the function of either the electronics substrate or components, including the connector. Thermal limitations of all module components need to be known before a sealant material and cure mode can be specified.

16.10 ELECTRONICS RELIABILITY CONSIDERATIONS

The sealing and packaging of electronics has implications on the function of the enclosed circuitry from both perspectives of sealant materials and associated processes. Simple factors such as the impact of curing temperature on the function of individual electronic components have been discussed. Material and process interactions must also be considered. Unintentional dripping of sealant onto an interconnect substrate must either be completely prevented, or determined to be a negligible factor to solder joint life, for example.

Some precautions are less obvious. Sealing and coating materials generally have significantly higher dielectric constants than dry or humid air. The presence of unintentional coating over high-speed traces can have consequences to the overall circuit performance.

Automotive electronics sealants generally outgas materials over their life, especially when heated. In the case of silicone sealants and coatings, the small linear and cyclic molecular species evolved may interfere with nonsealed relays, nearby motor brushes, and other selected components. For this reason, silicones are not commonly used directly near relays with open contacts or brush motors due to volatiles that may condense, oxidize, and cause abrasion or loss of electrical contact [21,22]. Dedicated ovens are typically used to cure silicones to avoid any potential for cross-contamination from volatiles.

Sealants may also affect the electromagnetic performance of an electronic enclosure. For example, if a module metal cover is bonded to a die-cast aluminum housing, electrical continuity, or at least a low electrical resistance pathway, might be required between these components to reduce electromagnetic susceptibility to and/

or emissions from the enclosed device. The sealant strategy must meet this requirement either by allowing use of an interposing electrically conductive spring/tab or fasteners, or with the use of an electrically conductive sealant. The sealant in the latter case might contain a conductive filler or reinforcement. The sealant type and matrix might be rigid or elastomeric, with the selection dependent on the property and design guidelines previously discussed.

16.11   SPECIFYING A SEALANT

The specification of a sealant material requires detailed knowledge of the environment in which the electronic device will be located. The selection of a sealant also necessitates the understanding of customer specifications and, sometimes, even undocumented expectations for durability and performance.

Figure 16.48 shows a successful under-chassis fuel system module designed to survive many years in a harsh environment. Water ingress, salt corrosion, stone impact damage, and effective elimination of heat were just some of the considerations during design.

Automotive OEM manufacturers have sets of additional design and process validation tests in which sealant integrity plays a large role, as the failure of a seal during this testing will, most often, lead to component malfunction.

A sealed electronic module would be expected to survive the following:

- Thermal cycling
- Thermal shock
- Humidity exposure
- Vibration testing

FIGURE 16.48   Overall view of a completely sealed under-chassis fuel system module incorporating a silicone FIPG. The sealant is dispensed into a groove in the polyester lid, which is then mated to the die-cast aluminum bottom and engaged with integral clips. (Photo courtesy of Visteon Corporation.)

- Car-wash spray
- Salt spray
- Mechanical drop/shock (3-axis)
- Fluid/chemical exposure
- EMI shielding
- A variety other tests specific to both function and vehicle location

These customer-driven requirements, along with the module construction, must be used to select the sealant system, together with the other packaging enclosure materials. Often, the exact sealant selection is a complex decision based on numerous factors, including:

- Worker health and safety
- Cost-effectiveness of sealant versus alternatives
- Sealant properties that will meet requirements
- Sealant availability, including use of sealant with known performance or already inventoried
- Commonality of sealant to other products made in the same production line/lean cell, or manufacturing plant
- Rate at which the part must be manufactured/sealed
- Degree of automation available
- Facilities assumptions and budget (Do unused legacy facilities exist?)

If one is approaching the sealant selection with a "clean sheet of paper," which is to say that both the design (not carryover) and manufacturing line are new, then the sealant choice should be decided on the overall location, requirements, and customer expectations. Table 16.9 outlines a checklist for combining sealant and manufacturing attributes. Many of these have been previously discussed and should be incorporated into a decision analysis for sealant selection. The other key decision point is the overall cost of the proposed sealant system.

The importance of thermal and environmental resistance of the sealant cannot be overemphasized. The location of a glass transition temperature or a crystallization transition within the storage or usage temperature range, although to be discouraged, must be considered and evaluated if encountered. Some parameters would be most important in specific situations or designs. For example, a compression set would be critical in compression gaskets such as CIPGs but not for an FIPG seal formed with a sealant. Conversely, substrate adhesion would be more important for an FIPG seal.

16.12   SEAL AND SEALANT TESTING

The best method to test the integrity of a sealed module in a manufacturing environment is through inline leak testing. This testing may be performed on either an audit basis or on every unit. There are several commercially available leak test units that can function either as vacuum decay or pressure decay test modes. In general terms, a predetermined vacuum or pressure is applied to the sealed module through either a temporary or permanent vent and monitored during an established time period

TABLE 16.9

Checklist of Key Sealant Parameters

Heat resistance/aging/continuous use temperature	Fatigue resistance
Thermal transitions encountered during process or application lifetime (including T_g)	Resilience
Compression set	Durometer hardness
Weathering/ozone/oxidation resistance	Organic/acid/base resistance
Low temperature flexibility	Staining resistance
UV resistance	Solvent/fuel resistance or swelling
Flame resistance/limited oxygen index	Abrasiveness to equipment and abrasion resistance
Moisture/hydrolysis resistance (including sealing surfaces)	Adhesion to sealing substrates
Catalyst poisoning	Carbon tracking resistance
Gas permeation	Shelf life and floor life
Tear resistance	Viscosity/thixotropy
Impact resistance	Cure characteristics
Elastic modulus/load deflection behavior	Volatile content/solvent content
Dynamic mechanical properties	Other location-specific design/functional characteristics

to provide a leak rate. The leak testing apparatus is calibrated by using a standard orifice with a known leak rate. Such leak test units can be programmed to set off an alarm at a predetermined leak rate level. False failures may result from causes unrelated to seal integrity. Faulty coupling of the module to the leak testing apparatus, caused by dents, debris, or misalignment, may show failure of an otherwise acceptable part. One way to determine the exact location of the leak is to temporarily pressurize the module at a low level (several kilopascals) and immerse in a water bath to identify the defect location and mark for possible repair.

There are many possible leak repair strategies, strongly dependent on the type of seal and seal/part geometry. For high-volume manufacturing, the best approach is to "make the seal right the first time" through strict automated process setup and control.

Production leak test failures may stem from a variety of causes, the most common being substrate contamination, bonding surface mechanical defects, inadequate sealant cure, contaminated sealant, inadequately mixed sealant, mating part misalignment, and sealant dispense defects such as bead gaps, bubbles, voids, and skips. The prevention of many of these has been previously discussed. Figure 16.49 shows bubbles formed in uncured sealant during the dispensing process in a prototype module. These bubbles were caused by the interaction of a surface moisture contaminant with the sealant adhesion promoter.

The first symptom of substrate contamination is usually the clean interfacial release of a sealant from that surface (in the case of FIPG and CIPG seals). Parts

FIGURE 16.49 Bubbles in an uncured FIPG silicone sealant could lead to leak paths in the assembled module. (Photo courtesy of Visteon Corporation.)

that have not had sealant dispensed should first be visually inspected for any obvious signs of contamination. Surface energy readings using one of the commercially available "dyne" pens or solutions should be determined to ensure proper wetting of the specific sealant used. More involved solvent-rinsing procedures can also be used with subsequent quantification and identification of rinsed contaminants with analytical tools such as infrared spectroscopy or GC/MS. Once a contaminant is identified, the history of the part should be retraced to identify the source. If the source of the contamination cannot be identified (and eliminated), the process is modified to be robust to the contaminated condition.

Sealant cure issues are easier to identify and correct. Improperly cured sealant may be identified by tackiness, high compression set, or even a semifluid condition. Recycling the part through an oven, in the case of a thermally cured sealant, with subsequent sealant cure will point to the time and temperature as potential issues. Alternatively, a sampling of the partially cured material followed by differential scanning calorimetry (DSC) analysis may indicate undercuring by the presence of a residual cure exotherm. This result would demonstrate that the sealant has the potential for further reaction. If these methods do not point to a sealant in which the cure can be advanced, then sealant contamination, or in the case of two-component systems, an off-ratio mix situation, might be the culprit. Several different sealant chemistries have the potential for contamination and subsequent cure inhibition.

Sealant defects and resulting leaks also stem from production issues such as incomplete or off-ratio mixing, part misalignment, sealant path misregistration from bent dispense needles, or part fixture issues. These concerns can be minimized with the establishment of process setup, control, and maintenance procedures. Setup procedures should address three basic points, namely, (1) quantity of dispensed sealant, (2) location of dispensed sealant, and (3) condition of the sealant bead. Scrap parts, or surrogate parts composed of cardboard, polyester transparency sheets, or other material may be used at regular time intervals to answer these questions. An outline of the dispensed pattern, or even a test pattern, may be transferred to the sheet to provide dispensed pattern location information. The sheet may be tared and weighed following dispensing to track the weight of sealant. The sheet can be cured to assess the quality of the cured sealant bead.

Sealant bead skips, bubbles, and other defects in the uncured material may stem from leaks in the fluid-handling system, or be a reaction by-product with two-component systems. Small bubbles may form as a reaction by-product of the sealant with the substrate, or fluid-handling system. Alternatively, larger bubbles may form by the interaction of the dispensed liquid bead with the geometry of the sealant groove in which it is being dispensed. For example, a moderate viscosity sealant might be dispensed as a cylindrical bead in a "V-shape" cross-section sealant groove, thereby forming an air pocket that will form bubbles over time as the sealant settles. This condition can usually be identified by dispensing and observing the sealant without placing the mating cover or part over the groove.

16.13   REPAIRING, REMOVING, AND REWORKING SEALS

There is a strong emphasis on sealing automotive electronics at very low first-pass defect rates. This is due to the high cost of repair and the fact that sealing is the last step in the construction of a high-value part. The latter point makes it expensive to discard or even inventory a large number of leaking modules. Modules are often functionally tested prior to sealing to afford an opportunity to make minor repair possible without the added time and expense of seal removal.

The practicality of seal repair in automotive electronics depends on the seal type. CIPG, stand-alone gaskets/O-rings, and DFGs, being mechanically compressed and not bonding to at least one of the sealing surfaces, may be exposed for repair and replacement by removing the associated fastening mechanism. The exposed seal material may then be removed and replaced following any other needed module repairs. CIPG sealants may sometimes need to be mechanically removed from one of the two sealing substrate surfaces. Care must be taken with any abrasion, machining, or scraping operations since residual sealant, or sometimes conductive substrate material, may find its way to electrically shorting adjacent uncoated circuitry.

FIPGs present a more complicated repair and replacement situation. Many form relatively strong bonds to both adjacent sealing surfaces, and might not be adequately exposed for mechanical removal; module opening is often not practical without damaging one or both sides of the sealing surfaces. For example, if one side of the seal is composed of a relatively fragile stamping, this can be removed and discarded to expose the sealant material and other sealing surface. The residual sealant may be mechanically removed by taking appropriate precautions as mentioned earlier.

Although there is a variety of chemical means to dissolve specific sealants, these are usually not practical for use with electronics destined to be repaired and used in the field. The precision of chemical dispensing and migration is usually difficult to control, thereby entrapping potentially corrosive agents inside module crevices, and under components and other mechanical features.

Automotive electronic modules that fail either vacuum decay or pressure decay type leak tests may often be set aside for leak location identification and rework. A water immersion test or soap bubble test can identify the location of a minor leak along the sealing surface, and this can often be repaired using the same or similar sealant as used on the original seal. Leak test failures may have causes other than defective seals.

Mechanical interface defects with the leak tester, cracked or porous die-castings, and leaking knit lines in injection-molded plastic parts are just a few of the many defects that can place unwarranted blame on the sealing materials and processes.

16.14   CONCLUSION

Only a few sealant application areas can match automotive electronics with the simultaneous need for rapid, precise sealing at high manufacturing rates to yield an end product that can withstand increasingly severe environments. The overall sealing strategy, seal type, specific sealant material, and associated manufacturing process offer many options based on complex performance requirements. For many automotive component suppliers, electronic device sealing and protection represents one of the most challenging aspects of the assembly process due to the many variations in sealing materials, sealing surfaces, and process conditions. Seldom does such a small amount of material have so much impact on the long-term performance of a complex system.

ACKNOWLEDGMENTS

The authors wish to thank Visteon Corporation and Dow Corning Corporation for assistance in the preparation of this chapter. In addition, they would like to thank all colleagues and suppliers who contributed photos, valuable insights, and other material to this document.

REFERENCES

1. Economic Data, courtesy of Prismark Partners LLC, Cold Spring Harbor, NY (2007).
2. C. Larner, Proc. of SMTA Harsh Environment Electronics Workshop, Dearborn, MI (2003).
3. R Pound (Ed.), *Electronic Materials Handbook,* Vol. 1, ASM International, Materials Park, OH (1989).
4. K. Larson, Proc. of the Surface Mount Technology Association International Conference, Orlando, FL (2007).
5. K. Carlson, R. Hanson, and W. Thilo, U.S. Patent 6,821,110 (2004).
6. S. Brandenburg, M. Koors, and J. Daanen, U.S. Patent 6,285,551 (2001).
7. A.N. Satullo, R.E Belke, H. Zhou, D. M Jett, B.R. Mohr, J. Trublowski, M.S. Schulke, and A.W.Schubring, U.S. Patent 6,977,377 (2005).
8. D.J. Thompson, R.E. Belke and E.P. McLeskey, U.S. Patent 6,752,015 (2004).
9. K. Larson, Proc. of the IPC Midwest Conference, Schaumburg, IL (2007).
10. O. Figovsky and L. Shapovalov, Proc. First International IEEE Conference on Polymers and Adhesives in Microelectronics and Photonics, Potsdam, Germany (2001).
11. R.W. Johnson, J.L. Evans, J.R, Thompson, and P. Jacobsen, *IEEE Trans. J. Electronics Manufacturing Packaging, 27(3),* 77–87 (2004).
12. R. Thompson, J. Freytag, W. Senske, and W. Wondrak, Proc. of SMTA (Surface Mount Technology Association) Conference, Chicago, IL (2002).

13. H. Lee and K. Neville, *Handbook of Epoxy Resins*, McGraw-Hill, New York (1967).
14. E. Maurice (Ed.), *Electronic Materials Handbook* Vol. 1, ASM International, Materials Park, OH (1989).
15. N. Akmal and A. Usmani, in *Handbook of Adhesives Technology*, 2nd edition, A. Pizzi, and K. L. Mittal (Eds.), Marcel Dekker, New York (2003).
16. C.S. Lin, D.M. Headley, and M. Neuenschwander, Proc. SAE World Congress, Detroit, MI (2008).
17. R.C. Snogren, *Handbook of Surface Preparation*, Palmerton Publishing, New York (1975).
18. C.W. Granville, U.S. Patent 4,549,866 (1985).
19. ASTM D3808-01 "Standard test method for qualitative determination of adhesion of adhesives to substrates by spot adhesion."
20. K.L. Mittal (Ed.) *Polymer Surface Modification: Relevance to Adhesion*, Vol. 4, VSP/Brill, Leiden (2007).
21. F. Gubbels, *Global SMT and Packaging*, No. 4, 10–14 (2004).
22. R. Reinders, F. Gubbels, and R. Dandois, *Global SMT and Packaging*, No. 5, 20–24 (2005).

17 Fibrin Sealants

Ronit Bitton and Havazelet Bianco-Peled

CONTENTS

17.1 INTRODUCTION

Fibrin sealants (FS), also termed *fibrin glues* or *fibrin adhesives*, are the most successful of the tissue sealants in terms of tissue compatibility, biodegradation, and clinical utility [1,2]. FS consists of naturally occurring molecules and proteins, the main ingredients being fibrinogen and thrombin, but additional components such as calcium, factor XIII, or antifibrinolytic agents being occasionally added. As the thrombin and fibrinogen solutions combine, a clot develops in the same way as it would form during normal blood clotting through a series of chemical reactions known as the *coagulation cascade*. At the end of the cascade, the thrombin breaks up the fibrinogen molecules into smaller segments that arrange themselves into strands termed *fibrin*. The fibrin is then crosslinked by a blood factor known as *Factor XIII* to form a net-like pattern that stabilizes the clot.

Fibrin sealants have several advantages over older methods of stopping bleeding (hemostasis). They speed up the formation of a stable clot, can be applied to very small blood vessels and to areas that are difficult to reach with conventional sutures, reduce the amount of blood lost during surgery, lower the risk of postoperative inflammation or infection, and are conveniently absorbed by the body during the healing process. They are particularly useful for minimally invasive procedures and for treating patients with blood-clotting disorders.

This chapter describes the current status of FS. Following a brief historical review, we will detail the chemical composition of FS and describe their structural, mechanical, and adhesion properties. Finally, the wealth and extensiveness of the published literature describing the clinical application of FS will be demonstrated.

17.2 HISTORY

The use of fibrin for surgical purposes goes back to 1909 when Bergel used fibrin emulsion as a hemostatic agent [3]. A few years later, fibrin was used as a hemostat in cerebral surgery and as parenchymal tissue dressing [1]. In 1940, fibrinogen was first used as a tissue adhesive to attach the peripheral nerve [4]. Purified thrombin became available in 1938, and the combination of thrombin and fibrinogen was first used in 1944 to enhance the adhesion of skin grafts [1].

Despite these early applications, poor adhesion strength, premature repair failures, and viral infections transmitted by human fibrinogen caused a dramatic decrease in the use of FS. The revival of FS occurred only during the early 1970s, when techniques for isolating and concentrating clotting factors were improved. Cryoprecipitate, a plasma product used in Factor VIII deficiency therapy, in combination with bovine thrombin solution, was used as glue in peripheral nerve repair [5]. The improved adhesion strength of this product was attributed to the elevated concentrations of fibrinogen, Factor XIII, fibronectin, and other factors in the cryoprecipitate. After further refinement of the cryoprecipitate components and production methods, the first commercially available FS, Tissucol® (Immuno A.G., Vienna) and Beriplast HS® (Centeon Pharma GmbH, Germany), were launched in Europe in 1982. These FS have also been used in Japan and Canada ever since. In the United States, however, the perceived risk of blood-borne diseases caused the Food and Drug Administration (FDA) to approve the Tisseel VH (Baxter Health Corp.) only in 1998 [6]. Tisseel VH is indicated for use as an adjunct to hemostasis in surgeries involving cardiopulmonary bypass and treatment of splenic injuries, when control of bleeding by conventional surgical techniques is ineffective or impractical. The thrombin and fibrinogen components in the Tisseel VH sealant are made from screened pooled human plasma; a two-step vapor heating and freeze drying is used to eliminate the risk of disease transmission.

Deriving FS components from pooled human or bovine blood inherently involves some risk of virus transmission as well as the potential for immunological reactions to foreign proteins. Autologous fibrin sealants were developed to avoid these hazards by extracting the sealant components from the patient's own blood [7,8]. The Vitagel™ surgical hemostat (marketed by Orthovita, USA) was approved by the FDA in 2000 [9]. It is indicated for use in cardiovascular, orthopedic, urologic, and general surgery to control bleeding. Vitagel™ is approved for use only in conjunction with the Cellpaker® plasma collection system (Orthovita, USA). Upon application, it works by combining a bovine thrombin/collagen suspension with the patient's own plasma. A similar product is marketed in Europe as Vivostat® (Vivostat A/S, Denmark) [10]. The Vivostat system is an automated system for the on-site preparation and application of patient-derived fibrin sealant that produces an autologous sealant from 120 mL of the patient's own blood in 23 min.

Crosseal™, developed by Omrix Biopharmaceuticals Ltd. (Israel), and now exclusively marketed and distributed by Ethicon, Inc. (USA), a Johnson & Johnson Company. It was approved by the FDA in 2003, and is an all-human protein, bovine component-free fibrin sealant approved for as an adjunct to hemostasis in liver surgery.

17.3 CHEMISTRY

Fibrinogen is the main structural component in FS. Another major component is thrombin, which acts proteolytically to cleave fibrinopeptides A and B from fibrinogen (Figure 17.1), leaving soluble fibrin monomers, which then polymerize by hydrogen bonding and electrostatic interactions to form an unstable, soft clot. Thrombin also activates Factor XIII to XIIIa in the presence of calcium ions. In the presence of Factor XIIIa, fibrin undergoes covalent crosslinking to form a crosslinked network of fibrils and fibers. The formation of this network can be influenced by factors such as pH, temperature, and presence of fibronectin. The degradation of FS, occurring by plasmin, can be controlled by adding antifibrinolytic agents [11]. A detailed description of the main components in FS is given in the following text.

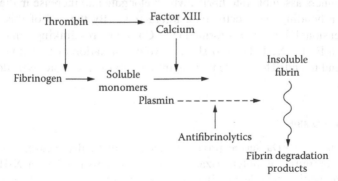

FIGURE 17.1 Fibrin sealant activation mechanism, including fibrinolytic degradation [12]. (Reprinted from W. D. Spotnitz and R. Prabhu, *J. Long Term Effects Med. Implants 15*, 245, 2005. With permission from Begell House.)

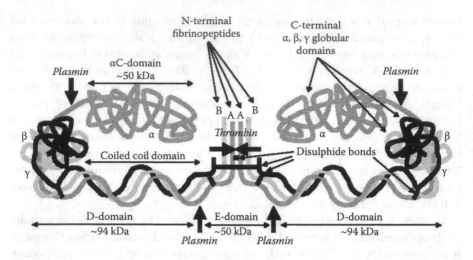

FIGURE 17.2 Schematic representation of a fibrinogen molecule [15]. (Reprinted from A. M. Afify, M. Stern, M. Guntenhoner, and R. Stern, *Arch. Biochem. Biophys. 305*, 434, 1993. With permission from Elsevier.)

17.3.1 Fibrinogen

Human fibrinogen is an elongated 340 kDa glycoprotein. The primary structure of fibrinogen includes three pairs of polypeptide chains termed α, β, and γ (Figure 17.2), composed of 610, 461, and 411 amino acids, respectively [13]. The three formative chains and the two halves of fibrinogen molecule are held together by a series of 29 disulphide bonds with 58 cysteine residues [14].

Fibrinogen is converted to its monomeric form by α-thrombin, which releases the fibrinopeptides A and B from the amino-terminal ends of α and β chains. The fibrin monomers assemble into fibrils, which elongate and increase in diameter due to hydrogen bonding and electrostatic interactions. By the end of this process, a three-dimensional fibrin clot is formed [14]. Covalent crosslinking can be achieved by activated Factor XIII. The crosslinks involve formation of γ–γ dimers, linking the fibers end to end. The covalent crosslinking retards the proteolytic degradation of the clot.

17.3.2 Thrombin

Thrombin is a 39 kDa serine protease that enzymatically cleaves fibrinogen to fibrin, thereby initiating polymerization [2]. It also activates Factor XIII by cleavage of the activation peptide and, in physiological clots, promotes platelet activation by activating protease-activated receptors on the platelet. Thrombin is produced by the enzymatic cleavage of prothrombin (Factor II) by Factor Xa in the presence of Factor V [1].

17.3.3 FACTOR XIII

Zymogene Factor XIII is a 320 kDa glycoprotein consisting of four subunits, two α2-subunits and two β2-subunits. It is transformed into its active form, termed Factor XIIIa, when Ca^{2+} ions bind to the α2-subunits and thrombin cleaves the β2-subunits.

Factor XIIIa is a transglutaminase that catalyzes the formation of covalent bonds between glutamyl and lysyl residues of adjacent fibrin molecules. Factor XIIIa crosslinks both γ- and α-chains of fibrin, with the crosslinking of γ-chains occurring at a faster rate than that of α-chains [16].

17.3.4 CALCIUM CHLORIDE

Ca^{2+} ions are required for the two major steps involving thrombin in the coagulation cascade: conversion of fibrinogen to fibrin and activation of Factor XIII to Factor XIIIa [1].

17.3.5 ANTIFIBRINOLYTIC AGENTS

The fibrin matrix is degraded primarily through the action of enzyme plasmin [13,14]. Plasmin also degrades fibrinogen, thereby preventing its conversion to fibrin [1]. Adding antifibrinolytic agents to the FS is useful in some clinical applications in which a reduced rate of degradation is required.

A commonly used antifibrinolytic agent is aprotinin, a protein that inhibits plasmin activity [1]. Other agents are transexamic acid and ε-aminocaproic acid, which are synthetic derivatives of lysine. They hinder fibrinolysis by blocking lysine-binding sites on plasminogen and inhibiting its interaction with fibrin [2].

17.4 PHYSICAL PROPERTIES OF FS

17.4.1 STRUCTURE OF FIBRIN CLOTS

When fibrinogen is cleaved by thrombin, the monomers further assemble into fibrils that form a three-dimensional network of branched fibers. Scanning electron microscopy observations [17–19] showed that the individual fibrin fibers can vary in size from a small protofibril, containing only two molecules in the fiber cross-section, to very large fibers of about 200 nm in diameter [20,21]. It was suggested that fiber growth was limited by the amount of energy necessary to stretch a protofibril [22]. An alternative explanation was that fibers were prevented from associating laterally by being tied down in a three-dimensional network [23]. Network formation was also characterized using light-scattering techniques [24–32] and scanning confocal 3D microscopy of gel matrices [22,33,34]. Despite the limited resolution of the latter technique, it could detect ordered networks in which straight fiber strands, of slightly different widths, crossed each other in different directions in space. Such ordered structures were not observed with electron microscopy, possibly due to the dehydration of the clots during sample preparation [22].

Turbidity [33] and light-scattering measurements [25] indicated that the gel structure could vary between two limiting classes: a coarse gel having large pores and thick fibers, and a fine, narrow-pore gel made of thin fibers [25]. There are a number of factors affecting the gel structure, such as the concentrations of the reagents, pH, and ionic strength. As the $CaCl_2$ concentration is increased, both the average fiber diameter and their length increase [17]. Increasing the thrombin concentration diminished the clotting time and caused smaller fiber length and diameter [17,20,35]. Both fiber and branch-point densities increase manyfold as the thrombin concentration is increased, whereas a decrease in these parameters occurs when the $CaCl_2$ concentration is increased [17]. These findings have led to the suggestion that fiber size was determined kinetically [23], and a kinetic model that predicted the effects of the various parameters on gel structure has been developed [36].

Network features were also characterized using scanning electron microscopy at cryogenic temperature (cryo-SEM) [37]. This technique allows examination of the hydrated gel without drying. It was shown that by gradually increasing the calcium ion concentration at constant thrombin and fibrinogen concentrations, the network appearance changed dramatically, from an almost isotropic structure at low concentration to highly oriented fibers at higher concentration. In addition, it was found that further increase in the calcium ion concentration induced fiber aggregation and formation of large clusters coexisting with fine fibers. Changes in thrombin concentration at constant calcium and fibrinogen concentrations induced only slight changes in the gel appearance.

The time course of morphological changes observed in fibrin sealants was found to be similar to natural wound-healing processes [38,39]. A dense and thick fibrin network was formed on day 1 after surgery, followed by fibroblasts formation and degradation of the clot within 7 days.

17.4.2 MECHANICAL PROPERTIES

The mechanical and viscoelastic properties of FS are essential for its functions and, therefore, have been investigated extensively [40–44]. Fibrin gels ranging from "coarse" to "fine," prepared by the action of thrombin on purified fibrinogen under various conditions of pH and ionic strength, with and without calcium ions, Factor XIII, and fibrinoligase, were investigated. The shear modulus G at a time scale of seconds was found to be proportional to fibrinogen concentration and followed the power-law relationship of $G\sim[\text{fibrin}]^{1.5}$ in uncrosslinked clots, and $G\sim[\text{fibrin}]^{2.3}$ in coarse crosslinked gels.

These early experiments established the basic features of fibrin's viscoelastic behavior [45]: (1) For small strains, that is, over a short time scale, fibrin exhibits nearly perfect elastic behavior; (2) slow creep was observed for uncrosslinked gels under constant stress, while maintaining a nearly constant initial modulus; (3) incomplete recovery of strain was detected for uncrosslinked gels following the release of stress; (4) for both fine and coarse gels prepared in the presence of Factor XIIIa and Ca^{2+}, creep was essentially eliminated.

The creep of coarse fibrin gels was proposed to be due to slippage of protofibrils past each other within fibrin bundles. The small creep of uncrosslinked fine gels was attributed to sliding of rod-like protofibrils through a network of interpenetrating rods, lacking interactions other than steric hindrance. The elastic modulus for both fine and coarse gels was explained in terms of bending of fibril bundles or protofibrils.

The storage modulus (G′) increases with time due to the buildup of structure within the clot. Increased fibrinogen concentration elevates clot rigidity through the establishment of greater fiber and branchpoint densities [17,46]. The effect of other FS components on its viscoelastic behavior was also investigated. A decrease in calcium ion concentration below a value of 10 mM leads to a decrease in the shear modulus [47–49]. In the presence of Factor XIIIa, stiffness of the clot is increased substantially, and the creep or irreversible deformation is decreased by stabilizing the interactions between preassembled protofibrils, thereby increasing fiber flexural stiffness [17,34,43].

Although the mechanical properties of fibrin clots are well documented, the origin of these properties is not well established. Collet and coworkers [34] measured the elastic modulus of individual fibrin fibers in fibrin clots with and without crosslinking. The elastic modulus of the crosslinked fibers was higher, supporting the idea that the mechanical properties of any branched network depend on the mechanical properties of an individual fiber. The mechanical properties of single fibrin fibers in the presence and absence of Factor XIIIa were also investigated by Liu et al [50] who found increased extensibility and elasticity in crosslinked clots, indicating a directional alignment of the crosslinks along the fiber axis.

Recently, the viscoelastic behavior of FS was used to compare the kinetics of polymerization between 20 and 300 s post application [51]. Variation in G′ over time was found to be associated with the degree of crosslink density of the clot, thus providing a measure of the kinetics of crosslinking and polymerization.

The tensile mechanical properties of Tissucol® were also evaluated [52]. Tensile strength and Young's modulus were found to be proportional to both thrombin and fibrinogen concentrations. Increasing the concentration of Factor XIII did not alter the clot rigidity. Contradictory results were found when Beriplast P®, a commercial FS containing 40–80 U/mL Factor XIII, was compared with FS having different concentrations of Factor XIII. Increasing the concentration of Factor XIII up to 70 U/mL enhanced the tensile strength of the resulting fibrin clot.

17.4.3 ADHESION CHARACTERISTICS

One of the requirements for a tissue adhesive is the ability to hold the tissue together with adequate strength for sufficient time. The adhesion properties of fibrin sealants have been investigated extensively by several groups using various techniques such as tensile, blister, and shear tests. As would be expected, the type of tissue and the test method used will vary, depending on the particular surgical interest of the investigator. Different methods were developed in an effort to achieve reproducible adhesion strength results in these tests [53,54]. Yet, a large range of measured values, from 2.45 kPa [51] to 36 kPa in tensile [55] and up to 38 kPa in shear tests [56], have been reported.

Investigators were especially interested in exploring the influence of the glue's components on its adhesion strength. Additionally, several studies compared the adhesion characteristics of autologous FS to those of other commercially available FS.

The adhesion strength of an autologous fibrin tissue adhesive (AFTA), developed to fulfill the need for a fibrin tissue adhesive for use in ear surgery, was examined by an adhesion test involving two pieces of human dura attached to a silk string and weights [56]. The adhesion strength of AFTA was shown to be directly related to the concentration of fibrinogen precipitated from the patient's blood. No correlation was found between the fibrinogen level in the blood and the adhesion strength. Other commercially available fibrin sealants exhibited higher adhesion than AFTA when the measurement was performed 10 minutes after application; however, when the bond strength was tested after 30 min, the AFTA gave better results.

Human dura was also one of the substrates used by Laitakari and Luotonen [57], who compared the adhesion strength of Beriplast® to that of two different AFTAs. One formulation was prepared with ammonium sulfate and the other with poly(ethylene glycol). Pieces of cotton cloth, rat skin, or human dura were glued and incubated for 30 min at 20% and 100% relative humidity, and subsequently the tensile strength was measured. As in the previously described research, the bond strength was directly related to the fibrinogen concentration. Relative humidity, however, did not have a significant influence on adhesion strength.

The suitability of FS for use in large-diameter vascular anastomoses was investigated using the porcine aorta model [55]. Three clinical methods of attachment, end-to-end anastomosis, vessel overlapping, and sutures, were compared by performing tensile and burst tests at high and low fibrinogen concentrations, at two different setting times (5 min and 45 min). Higher fibrinogen concentration and longer setting time resulted in a moderately higher adhesion strength. The overlap technique resulted in a significantly weaker anastomosis than the end-to-end technique, and suturing was much stronger than both, leading to the conclusion that FS are only suitable for small-diameter vascular anastomoses.

In an attempt to overcome the difficulties of controlling the test conditions affecting the measured adhesion values, a method based on an ASTM test procedure for evaluation of shear adhesion strength was developed [58]. In this method, split thickness rabbit dorsum was attached to two aluminum jigs. The jigs were then attached to each other using FS. The assemblies were incubated for 90 min, and then strained to failure by tensile loading. In a comparison between Tissucol® and cryoprecipitated AFTA, the former had better adhesion strength. In an attempt to fine-tune the test method to serve as a model for a specific clinical in-situ action, porcine skin grafts were used [59]. The adhesion strengths of Tissucol® and those of six cryoprecipitated AFTAs having different fibrinogen concentrations were measured. As in previous studies [55], a higher shear strength was obtained for a higher fibrinogen concentration. There appeared to be a difference between Tissucol® and the AFTAs. While the AFTAs showed no significant increase in adhesion strength after 5 min of bonding, Tissucol® exhibited a time-dependent behavior and attained the adhesion strength equivalent to that of AFTA only after 90 min. It should be noted that the adhesion strengths reported in this study were an order of magnitude lower than the values reported for other FS having similar concentrations.

Another method for measuring the adhesion strength of FS was developed by Kjaergard et al. [54]. In this method, veins leftover from coronary artery bypass grafting were fastened to a tensiometer linked to a computer. FS was applied to the tissue, and the surfaces of the samples were held together and then automatically pulled apart by the tensiometer. This method was used to investigate the adhesion characteristics of Vivostat®, a patient-derived FS, compared with Tissucol® and Beriplast HS®. Vivostat® provided enhanced instant adhesion strength and elasticity in comparison to the cryoprecipitated AFTA investigated by Sierra et al. [59]. The adhesion strengths of Tissucol® and Beriplast HS® at 20 s were about 30% of the values attained after 300 s of polymerization. The Vivostat® polymerized faster, and its strength after 20 s was 68% of the final value.

Over the years, various synthetic substrates have been used in in vitro adhesion strength tests instead of tissues. The effect of calcium chloride and sodium chloride on the gel time and shear bond strength of fibrin glue to collagen films was studied by Wang et al. [60]. The results showed that, in the absence of NaCl, clotting time was reduced and shear strength was enhanced. As for $CaCl_2$, an optimal concentration of 20 mM gave the highest bond strength and the shortest gel time. Shehter-Harkavyk and Bianco-Peled [37] studied the influence of fibrinogen, thrombin, and $CaCl_2$ concentrations on the adhesion performance of FS in a lap-shear test using Mylar surfaces. An optimal $CaCl_2$ concentration was again detected. Cryo-SEM images showed that increasing the $CaCl_2$ concentration at constant thrombin and fibrinogen concentrations changed the network appearance from an isotropic structure to a highly oriented one. The formation of an ordered network was correlated with the maximal adhesion strength. The adhesion strength increased with an increase in fibrinogen concentration, whereas it was almost unaffected by the thrombin concentration. On the contrary, an optimal thrombin concentration of 30 U/mL was found for cryoprecipitated AFTA, while the $CaCl_2$ concentration had no effect on adhesion strength [61]. The adhesion strength was well correlated with the total concentration of fibrinogen.

As the use of FS became more abundant, new application methods were developed, and their impact on adhesion strength was investigated. Sierra et al. [62] investigated the effect of mixing on the mechanical properties of a two-component tissue adhesive. Ex vivo tensile adhesion strength measurements using porcine skin grafts showed that more thoroughly mixed sealants demonstrated higher adhesion strength. Kaetsu et al. [63] explored the adhesion performance of FS using a spraying technique in which the fibrinogen solution was mixed with a smaller volume of thrombin solution. The spray applicator allows various mixing ratios of fibrin/thrombin as well as higher fibrinogen concentrations. The adhesion strength was found to be directly proportional to the final concentration of fibrin.

The adhesion performance of fibrin glue applied by a dual-channel gun onto rat skin was also investigated [64]. Increasing fibrinogen levels increased adhesion strength, while increasing the thrombin level had no significant effect. Heparin reduced the adhesion strength by about 20%. Endogenous Factor XIII did not contribute significantly to skin bonding, indicating that the tissue supplied all the necessary transglutaminase activity required to crosslink the fibrin and to covalently link it to the tissue.

A comprehensive study performed by Dickneite et al. [65] compared the adhesion strength of 12 different combinations of commercially available FS and applicators. A bottle-shaped mold with an open bottom was pressed against the inner side of a fixed porcine skin sample, and FS was applied *via* a cannula into the mold (thus bonding the tissue by forming a sealant clot). After 3 min, a tensiometer was used to measure the adhesion strength of the FS–tissue bond. As in other cases, the concentration of fibrinogen as well as the amount of FS applied were critical for optimal adhesion strength. Although all FS were based on combining fibrinogen and thrombin, different formulations and varying concentrations of key components such as Factor XIII caused variations in the properties of the clots. No conclusions were made regarding the reasons for these differences.

17.5 CLINICAL AND IN VIVO APPLICATIONS

Fibrin sealants have been gaining popularity as an adjunct for many surgical procedures since the FDA approval in 1998. Even though FS were approved only for a limited number of medical indications, they have been successfully employed in a countless number of off-label surgical applications. The majority of studies confirm the effectiveness and safety of FS as a hemostatic agent, sealant, and adhesive. A recent review of the data from controlled randomized trials that studied the influence of FS on blood lost in adult elective surgeries [66] showed that applying FS resulted in a relative reduction of blood loss of about 150 mL per patient and in about 60% reduction in the number of patients who needed blood transfusion.

This section intends to demonstrate the wealth and extensiveness of the published literature describing clinical application of FS in different surgical specialties. The reader is also referred to several recent reviews [12,67,68].

17.5.1 CARDIOVASCULAR AND VASCULAR SURGERY

The application of FS in cardiovascular surgery was pioneered by Spangler et al. in 1976 [69]. They have been widely used since, mainly as a supplement to hemostatic and sealing suture holes [70,71].

FS are primarily used in combination with vascular sutures since sutureless application increases the risk of intravasal thrombemboli due to an accidental deposit of the sealant [72]. Dickneite et al. [70] used pig vascular model to investigate the efficacy of FS in reducing suture hole bleeding. This study has shown that FS are indeed effective in improving homeostasis and reducing blood loss. A comparison between Beriplast P® that contains Factor XIII, and the same product from which this factor was removed, showed that the presence of Factor XIII significantly reduced suture hole bleeding and bleeding time. In a later work, the properties of several commercially available fibrin sealants were compared [65]. Low content of factor XIII was found to be associated with the formation of a soft clot, leading to low adhesion strength, and increased risk of late hemostatic and premature clot lysis. Jackson et al. [73] performed a randomized clinical trial that compared the efficacy of Tisseel VH human FS to that of a thrombin-soaked gelatin sponge. Both products were found to be equivalent in their ability to control suture hole bleeding from expanded poly(tetrafluoroethylene) (ePTFE) patch angioplasty during vascular surgery. Another randomized trial showed that applying FS significantly decreased the time to hemostasis during vascular surgery with poly(tetrafluoroethylene) (PTFE) patch closure [74]. Blood loss was also reduced, but the change in this parameter was not statistically significant. A recent animal study [72] suggested a new sutureless application of FS in combination with an autologous muscle pad. This combination shows high potential for hemostasis in microvascular surgery.

Kheirabadi and coworkers [75] investigated both the in vivo and the in vitro performance of Crosseal® (Ethicon, Inc., USA) human fibrin sealant in a rabbit model, attempting to reliably assess the functional strength of fibrin sealant required to prevent arterial bleeding during the operation (short-term hemostasis) and over the few weeks required for normal healing of vascular tissue (long-term hemostasis). All rabbits that were treated with a placebo bled to death after the vessel was unclamped. Treatment of suture line with FS at standard fibrinogen concentration (120 mg/mL) sealed the anastomosis and prevented blood loss. The hemostasis achieved by the sealant was sustained during the time required for the injured vessel to heal properly. On the contrary, dilution of the fibrin sealants resulted in a significant blood loss, failure in long-term hemostasis, and only limited wound healing. These results were well correlated with the in vitro experiments, leading to the conclusion that both the strength of the fibrin clot and its ability to secure hemostasis ultimately depend on the fibrinogen concentration. A later publication from the same group [76] compared the efficacy of fibrin sealant to that of other commonly used hemostatic agents. End-to-end anastomosis of transected abdominal aorta was performed in moderately anticoagulated rabbits using four or six interrupted sutures. Bleeding from nontreated sutures, and from sutures covered with either absorbable gelatin sponges, bovine-derived microfibrillar collagen, oxidized regenerated cellulose, or human FS was investigated. The main finding was that application of FS immediately achieved hemostasis in all the animals. Moreover, the FS was superior to all other agents in terms of blood loss and time to hemostasis.

17.5.2 OCCLUSION OF FISTULA

A fistula is an abnormal connection between an organ, vessel, or intestine and another organ. Fistulas are usually the result of injury or surgery, but can also result from infection or inflammation. Sealing a fistula often requires surgical intervention; however, the procedure in debilitated patients might be exceptionally risky. Fibrin sealants can be applied using endoscopic procedure and, therefore, offer a safe alternative to surgery [77].

Many of the published reports on surgical applications of FS discuss their use for closure of fistula [67]. Successful repair of postoperative upper gastrointestinal fistulae [78], esophagobronchial fistula [79], and tracheoesophageal fistula [80] was reported. Recently, a nonhealing gastrocutaneous fistula was successfully treated using a combination of Tisseel VH and hemostatic clips [77]. A large bronchopleural fistula, a complication associated with pulmonary resection, was permanently sealed in a critically ill patient with a combination of FS and a small section of demineralized human spongiosa using flexible bronchoscope [81].

FS also offer a simple and safe alternative to surgical correction of other types of fistula, such as low-input pancreas fistula [82] and anal fistula [83]. Despite reported healing of up to 77% of patients suffering from an anal fistula [83], delayed recurrence was recognized [84]. A study in porcine model [85] showed that Tisseel VH FS therapy diminished formation of granulation tissue and decreased the fistula volume compared to the control group. Incomplete healing in the fibrin-treated group was attributed to large volumes of granulation tissue that were still observed, possibly due to fast degradation of the sealant.

17.5.3 UROLOGIC SURGERY

FS are finding increasing roles in urology, although most applications are off-label. Successful management of Fournier's gangrene sequelae [86], prostatectomy [87], fluid collection after renal transplantation [88], and treatment of injuries and surgical complications [89] has been reported. A case study described an unhealing urinary leak after gunshot that was successfully treated with FS [83]. Hick and Allen [90] reported on their initial experience with FS in urethral reconstruction. Although the patient groups were not randomized, the impression was that sealants improved wound healing and allowed early catheter removal. Diner et al. [91] treated 32 patients with prostate cancer with radical retropubic prostatectomy. In half of the patients the suture line was covered with Tisseel VH FS. Close bulb suction drains were installed and left in place until the drainage was minimal. The follow-up suggested that the application of FS decreased postoperative drain output and allowed earlier drain removal with less discomfort to the patients. On the contrary, insertion of the drain could not be avoided after percutaneous nephrolithotomy procedure, performed as a treatment for large renal calculi, even when FS were applied [92]. Pruthi et al. [93] used Tisseel VH in 15 patients with renal tumors as a haemostatic agent and collecting-system sealant during hand-assisted laparoscopic partial nephrectomy. None of the patients suffered from delayed urine leakage or other delayed complications. FS was also useful in open partial nephrectomy in a porcine model, where immediate hemostasis was achieved [94]. A model of kidney stab injury was studied in a porcine model [95]. The wounds were treated in a sutureless procedure consisting of application of Hemaseel APR FS (Haemacure Corporation, USA), augmented by an external gelatin sponge or with an external microfibrillar collagen sheet. The treatment significantly decreased blood loss as well as time to hemostasis. CT scans performed on postoperative day 8 revealed no significant perirenal hematoma or urinoma in any group.

A variety of other urologic applications of FS for hemostasis, tissue adhesion, and urinary tract sealing were recently reviewed by Evans and Allen [96]. The authors conclude that fibrin sealants should not be viewed as a replacement for conventional sound surgical judgment or technique, but rather as a complementary adjunct to improve surgical outcome and enhance wound healing.

17.5.4 NEUROSURGERY

The use of a fibrin sealant in nerve repair was first introduced by Young and Medawar in 1940 [4]. Despite a great deal of experimental research and clinical practice, the quality of result is generally no different when compared with nerve repair using conventional stitches [4,97].

A challenging postoperation complication after transsphenoidal surgery is cerebrospinal fluid leak that results from inadequate repair of a fistula. To prevent such complication, different materials have been used to fill and close the sellar cavity during surgery [99]. Floris and coworkers partially succeeded in treating eight patients with CT-guided introduction of FS into the sphenoidal sinus [98] . In a later study, 56 patients were treated with Tissucol and/or collagen fleece [99]. The occurrence of postoperation complications was compared with a group of 34 patients that were treated with other conventional sellar repair methods. A synergistic effect of the FS and the collagen fleece was found, with fewer complications when both materials were used.

17.5.5 PLASTIC AND RECONSTRUCTIVE SURGERY

Experiments in mice model showed that the Hemaseel™ FS improved the graft fixation of cultured human epidermal sheets used for the treatment of extensive burns [100]. These improved results were attributed to the rapid and better immobilization of the graft on the wound bed. A recent study used Tisseel VH in facial augmentation with cartilage grafts in a rabbit model [101]. Significant improvements in both the gross and histological features of implanted cartilage using fibroblast growth factor and FS, or even FS alone, were demonstrated. Grant et al. [102] suspended cultured cells in a spray of fibrin sealant. A study in a porcine wound model detected formation of a multilayered undulating epidermis two weeks after the treatment. On the contrary, application of FloSeal™ (Baxter Health Corp.) and Tisseel VH did not improve the survival of transferred tissue in a rat model [103]. A recent study explored the optimal conditions for skin graft fixation on full-thickness wounds with FS in a porcine model [104]. The study protocol included a slow-clotting FS as a spray to fix the autologous split thickness skin grafts. An excellent outcome was notable on day 21 in the FS group as well as in the control suture group (Figure 17.3), with no statistical difference in the graft uptake between the groups.

17.5.6 ONCOLOGY SURGERY

The treatment for early breast cancer is excision of the primary tumor. The most common postoperative complication is fluid collection in the wound (seroma). Two randomized controlled studies were recently performed to evaluate the efficacy of fibrin glue over drain placement in preventing seroma after breast surgery [105,106]. Both studies showed a lower rate of seroma formation; however, this difference was statistically significant only in specific cases. It should be noted that the study by Johnson et al. [106] was terminated early because of withdrawal of two principal surgeons who believed that the procedure for the application of the fibrin glue was

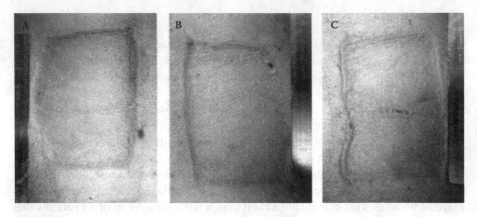

FIGURE 17.3 Full-thickness defects covered with split-thickness autografts 21 days after grafting (A-fibrin sealant 0.15 mL/cm²; B- fibrin sealant 0.05 ml/cm²; C-sutures). (Reprinted from R. Mittermayr, E. Wassermann, M. Thurnher, M. Simunek, and H. Redl, *Burns 32*, 305, 2006. with permission from Elsevier.)

cumbersome and was hampered by the additional surgical personnel required for mixing the glue. Also, applying fibrin sealants involved additional cost of a few hundred dollars per procedure.

17.5.7 ORTHOPEDIC SURGERY

Jones et al. [107] used Vivostat® fibrin sealant to prevent adhesion formation following flexor tendon repair in a rabbit model. While surgical injury alone resulted in the development of adhesion with acute and chronic inflammatory reactions, in the Vivostat® group the injury was less apparent, with evidence of repair, less or no adhesion, and less inflammation within the adjacent tendon. A study in minipigs evaluated the efficacy of Tissucol® in the healing of bone defects [108]. The main findings were that FS did not increase formation of new bone but did not suppress it either. Addition of growth factors had no effect on bone formation. A clinical study examined the use of allografts transplantation in the management of disorders of the knee joints [109]. Applying FS did not improve the outcome.

17.5.8 THORACIC SURGERY

Vivostat FS® was found to be superior to human albumin in controlling air leakage after lung resection in a pig model [110]. In a clinical study, Vivostat® was used during pulmonary surgery [111] and was found to significantly reduce air leakage and bleeding. O'Neill et al. [112] reported on a patient who suffered from persistent air leakage. Application of FS under continuous computed tomography fluoroscopy allowed real-time catheter manipulation for precise placement of glue, which prevented the leak.

17.5.9 A SELECTION OF OTHER SURGICAL APPLICATIONS

Splenic injuries, a clinical situation in which traditional hemostatic agents are often unsuccessful, were treated with topical application of FS [113,114]. Rapid, reliable, and lasting hemostasis was achieved, decreasing the need for blood transfusion and splenectomy.

A randomized study on 121 patients showed that Crosseal FS significantly reduced the time to hemostasis following liver resection, compared to standard hemostasis topical agents [115]. TachoSil® (Nycomed, Denmark), a collagen fleece carrying fibrin glue components, was found to be more efficient than Argon Beamer for the same application [116].

A pilot study demonstrated that donor cornea could be grafted to the patient's eye in a sutureless procedure [117]. After one month, all recipients' eyes remained clear and inflammation free. The authors pointed out drawbacks of the Tisseel VH FS in eye surgery, and in particular the high cost, the bulky application kit, and the relatively long preparation time.

The use of FS during the embryo transfer stage of in vitro fertilization (IVF) was found to be beneficial for older women and for women with previous recurrent IVF failures [118].

Fibrin sealants were applied in 69 patients on oral anticoagulant therapy after surgical teeth extraction [119]. Anticoagulant therapy prevents thromboembolism and therefore increases the risk of bleeding. Due to the ability of FS to induce clot formation at the site of the wound without affecting the therapeutic level of anticoagulant, no major complications were observed.

A study in a rat model of abdominal hernia showed that macroporous mesh, used for the repair of the hernia, could be fixed with FS instead of staple [120]. Immediately after implantation, fibrin glued meshes were evenly attached to the underlying tissue, as the sealant dispersed very well between the mesh interstices. Stapled meshes were more mobile centrally since they were only fixed at the four corners. Nevertheless, all implanted meshes displayed satisfactory burst and tensile strengths. Histology investigation indicated that FS was equivalent to staples in terms of neovascularization, foreign body reaction, and tissue integration.

17.6 SUMMARY

FS are a surgical tissue adhesive comprised of blood-derived proteins, mainly fibrinogen, thrombin, and Factor XIII. Upon application of the FS, fibrinogen is converted to soluble fibrin monomers, which then polymerize to form a clot—a three-dimensional crosslinked network of fibrin fibrils and fibers. The structure of the network, the mechanical properties of the clots, and the adhesion strength are strongly influenced by selection of the glue's components and their concentrations.

FS have been gaining popularity as an adjunct for many surgical procedures since their FDA approval in 1998. Even though they were approved only for a limited number of medical indications, they have been successfully employed in countless number of off-label surgical applications. The majority of studies confirm the effectiveness and safety of FS as a hemostatic agent, sealant, and adhesive.

REFERENCES

1. D. H. Sierra, *J. Biomater. Appl. 7*, 309 (1993).
2. G. Wozniak, *Cardiovasc. Surg. 11 Suppl.* 1, 17 (2003).
3. S. Bergel, *Dtsch. Med. Wochenschr. 35*, 663 (1909).
4. I. M. Jou, W. C. Chen, C. I. Shen, and H. Matsuda, *J. Hand Surg. 24*, 707 (1999).
5. H. Matras, *J. Oral Maxillofacial Surg. 43*, 605 (1985).
6. http://www.advancingbiosurgery.com/us/products/tisseel/.
7. B. R. Davidson, S. Burnett, M. S. Javed, A. Seifalian, D. Moore, and N. Doctor, *British J. Surg. 87*, 790 (2000).
8. H. K. Kjaergard and H. R. Trumbull, *Ann. Thorac. Surg. 66*, 482 (1998).
9. http://www.angiotech.com/?seek=133.
10. http://www.vivostat.com/.
11. I. Webster and P. J. West, in *Polymeric Biomaterials*, S. Dumitriu (Ed.), 2nd edition, p. 703, Marcel Dekker, New York (2002).
12. W. D. Spotnitz and R. Prabhu, *J. Long Term Effects Med. Implants 15*, 245 (2005).
13. C. Fuss, J. C. Palmaz, and E. A. Sprague, *J. Vasc. Interven. Radiol. (JVIR) 12*, 677 (2001).
14. S. Herrick, O. Blanc-Brude, A. Gray, and G. Laurent, *Int. J. Biochem. Cell Biol. 31*, 741 (1999).
15. A. M. Afify, M. Stern, M. Guntenhoner, and R. Stern, *Arch. Biochem. Biophys. 305*, 434 (1993).
16. G. Dickneite, H. J. Metzner, M. Kroez, B. Hein, and U. Nicolay, *J. Surg. Res. 107*, 186 (2002).
17. E. A. Ryan, L. F. Mockros, J. W. Weisel, and L. Lorand, *Biophys. J. 77*, 2813 (1999).
18. O. V. Gorkun, Y. I. Veklich, J. W. Weisel, and S. T. Lord, *Blood 89*, 4407 (1997).
19. C. V. Dang, C. K. Shin, W. R. Bell, C. Nagaswami, and J. W. Weisel, *J. Biol. Chem. 264*, 15104 (1989).
20. M. E. Carr, Jr. and J. Hermans, *Macromolecules 11*, 46 (1978).
21. M. F. Muller, H. Ris, and J. D. Ferry, *J. Mol. Biol. 174*, 369 (1984).
22. B. Blomback, *Thromb. Res. 83*, 1 (1996).
23. R. R. Hantgan and J. Hermans, *J. Biol. Chem. 254*, 11272 (1979).
24. M. De Spirito, G. Arcovito, F. A. Bassi, M. Rocco, E. Paganini, M. Greco, and F. Ferri, *Nuovo Cimento Soc. Ital. Fis. 20D*, 2409 (1998).
25. M. De Spirito, G. Arcovito, F. A. Bassi, and F. Ferri, *Macromol. Symposia 162*, 263 (2000).
26. F. Ferri, M. Greco, G. Arcovito, F. A. Bassi, M. De Spirito, E. Paganini, and M. Rocco, *Phys. Review E. 63*, 031401/1 (2001).
27. R. W. Greene, *J. Clin. Invest. 31*, 969 (1952).
28. G. Arcovito, F. A. Bassi, M. De Spirito, E. Di Stasio, and M. Sabetta, *Biophys. Chem. 67*, 287 (1997).
29. S. Bernocco, F. Ferri, A. Profumo, C. Cuniberti, and M. Rocco, *Biophys. J. 79*, 561 (2000).
30. F. Ferri, M. Greco, and M. Rocco, *Macromol. Symposia 162*, 23 (2000).
31. A. Haeberli, P. W. Straub, G. Dietler, and W. Kaenzig, *Biopolymers 26*, 27 (1987).
32. R. Kita, A. Takahashi, M. Kaibara, and K. Kubota, *Biomacromolecules 3*, 1013 (2002).
33. B. Blomback, K. Carlsson, B. Hessel, A. Liljeborg, R. Procyk, and N. Aslund, *Biochim. Biophys. Acta 997*, 96 (1989).
34. J.-P. Collet, H. Shuman, R. E. Ledger, S. Lee, and J. W. Weisel, *Proc. Natl. Acad. Sci. U. S. A. 102*, 9133 (2005).
35. J. D. Ferry and P. R. Morrison, *J. Am. Chem. Soc. 69*, 388 (1947).
36. J. W. Weisel and C. Nagaswami, *Biophys. J. 63*, 111 (1992).
37. I. Shehter-Harkavyk and H. Bianco-Peled, *J. Adhesion Sci. Technol. 18*, 1415 (2004).
38. R. Hattori, H. Otani, H. Omiya, S. Tabata, Y. Nakao, T. Yamamura, M. Osako, Y. Saito, and H. Imamura, *Annals Thoracic Surg. 70*, 2132 (2000).

39. M. Kroez, W. Lang, and G. Dickneite, *Wound Repair Regen. 13*, 318 (2005).
40. R. W. Rosser, W. W. Roberts, and J. D. Ferry, *Biophys. Chem. 7*, 153 (1977).
41. G. W. Nelb, C. Gerth, J. D. Ferry, and L. Lorand, *Biophys. Chem. 5*, 377 (1976).
42. C. Gerth, W. W. Robert, and J. D. Ferry, *Biophys. Chem. 2*, 208 (1974).
43. W. W. Roberts, O. Kramer, R. W. Rosser, F. H. M. Nestler, and J. D. Ferry, *Biophys. Chem. 1*, 152 (1974).
44. G. W. Nelb, G. W. Kamykowski, and J. D. Ferry, *Biophys. Chem. 13*, 15 (1981).
45. M. D. Bale Oenick, *Biophys. Chem. 112*, 187 (2004).
46. J. L. Velada, D. A. Hollingsbee, A. R. Menzies, R. Cornwell, and R. A. Dodd, *Biomaterials 23*, 2249 (2002).
47. L. L. Shen, R. P. McDonagh, J. McDonagh, and J. Hermans, Jr., *Biochem. Biophys. Res. Commun. 56*, 793 (1974).
48. L. L. Shen, J. Hermans, J. McDonagh, R. P. McDonagh, and M. Carr, *Thromb. Res. 6*, 255 (1975).
49. M. E. Carr, L. L. Shen, and J. Hermans, *Anal. Biochem. 72*, 202 (1976).
50. W. Liu, L. M. Jawerth, E. A. Sparks, M. R. Falvo, R. R. Hantgan, R. Superfine, S. T. Lord, and M. Guthold, *Science 313*, 634 (2006).
51. H. K. Kjaergard, J. L. Velada, J. H. Pedersen, H. Fleron, and D. A. Hollingsbee, *Thromb. Res. 98*, 221 (2000).
52. R. Nowotny, A. Chalupka, C. Nowotny, and P. Boesch, *Adv. Biomaterials 3*, 677 (1982).
53. D. H. Sierra, A. W. Eberhardt, and J. E. Lemons, *J. Biomed. Mater. Res. 59*, 1 (2002).
54. H. K. Kjaergard, J. L. Velada, T. Pulawska, V. S. Ellensen, S. S. Larsen, and D. A. Hollingsbee, *Europ. Surg. Res. 31*, 491 (1999).
55. C. Flahiff, D. Feldman, R. Saltz, and S. Huang, *J. Biomed. Mater. Res. 26*, 481 (1992).
56. K. H. Siedentop, D. M. Harris, and B. Sanchez, *Laryngoscope 98*, 731 (1988).
57. K. Laitakari and J. Luotonen, *Laryngoscope 99*, 974 (1989).
58. D. H. Sierra, A. J. Nissen, and J. Welch, *Laryngoscope 100*, 360 (1990).
59. D. H. Sierra, D. S. Feldman, R. Saltz, and S. Huang, *J. Appl. Biomaterials 3*, 147 (1992).
60. M.-C. Wang, G. D. Pins, and F. H. Silver, *Mater. Sci. Eng. C C3*, 131 (1995).
61. H. Yoshida, K. Hirozane, and A. Kamiya, *Biol. Pharm. Bull. 23*, 313 (2000).
62. D. H. Sierra, K. O'Grady, D. M. Toriumi, P. A. Foresman, G. T. Rodeheaver, A. Eberhardt, D. S. Feldman, and J. E. Lemons, *J. Biomed. Mater. Res. 52*, 534 (2000).
63. H. Kaetsu, T. Uchida, and N. Shinya, *Intl. J. Adhesion Adhesives 20*, 27 (2000).
64. G. Marx, *Transfus. Med. Rev. 17*, 287 (2003).
65. G. Dickneite, H. Metzner, T. Pfeifer, M. Kroez, and G. Witzke, *Thromb. Res. 112*, 73 (2003).
66. P. A. Carless, D. M. Anthony, and D. A. Henry, *British J. Surg. 89*, 695 (2002).
67. D. M. Albala and H. Lawson Jeffrey, *J. Am. College Surg. 202*, 685 (2006).
68. K. I. Schexneider, *Current Opin. Hematol. 11*, 323 (2004).
69. H. P. Spangler, F. Braun, J. Holle, E. Moritz, and E. Wolner, *Wiener medizinische Wochenschrift 126*, 86 (1976).
70. G. Dickneite, H. Metzner, and U. Nicolay, *J. Surg. Res. 93*, 201 (2000).
71. T. E. MacGillivray, *J. Cardiovasc. Surg. 18*, 480 (2003).
72. N. P. Fehm, B. Vatankhah, S. Dittmar Michael, Y. Tevetoglu, G. Retzl, and M. Horn, *Microsurgery 25*, 570 (2005).
73. M. R. Jackson, D. L. Gillespie, E. G. Longenecker, J. M. Goff, L. A. Fiala, S. D. O'Donnell, E. D. Gomperts, L. A. Navalta, T. Hestlow, and B. M. Alving, *J. Vascular Surg. 30*, 461 (1999).
74. A. A. Milne, W. G. Murphy, S. J. Reading, and C. V. Ruckley, *European J. Vascular Endovascular Surg. 10*, 91 (1995).
75. B. S. Kheirabadi, R. Pearson, K. Rudnicka, L. Somwaru, M. MacPhee, W. Drohan, and D. Tuthill, *J. Surg. Res. 100*, 84 (2001).

76. B. S. Kheirabadi, A. Field-Ridley, R. Pearson, M. MacPhee, W. Drohan, and D. D. V. M. Tuthill, *J. Surg. Res. 106*, 99 (2002).
77. J. Akhras, M. Tobi, and A. Zagnoon, *Digestive Diseases Sci. 50*, 1872 (2005).
78. C. Cellier, B. Landi, A. Faye, P. Wind, P. Frileux, P. H. Cugnenc, and J. P. Barbier, *Gastrointestinal Endoscopy 44*, 731 (1996).
79. K. Harries, A. Masoud, T. II. Brown, and D. G. Richards, *Diseases Esophagus 17*, 348 (2004).
80. N. E. Wiseman, *J. Pediatric Surg. 30*, 1236 (1995).
81. W. R. Baumann, J. L. Ulmer, P. G. Ambrose, M. J. Garvey, and D. T. Jones, *Annals Thoracic Surg. 64*, 230 (1997).
82. C. C. Cothren, C. McIntyre Robert, Jr., S. Johnson, and V. Stiegmann Gregory, *Am. J. Surg. 188*, 89 (2004).
83. S. M. Baughman, F. Morey Allen, H. Van Geertruyden Peter, G. Radvany Martin, E. Benson Amy, and P. Foley John, *J. Urology 170*, 522 (2003).
84. J. R. Cintron, J. J. Park, C. P. Orsay, R. K. Pearl, R. L. Nelson, J. H. Sone, R. Song, and H. Abcarian, *Dis. Colon Rectum 43*, 944 (2000).
85. G. N. Buchanan, P. Sibbons, M. Osborn, I. Bartram Clive, T. Ansari, S. Halligan, and C. R. G. Cohen, *Dis. Colon Rectum 48*, 532 (2005).
86. B. J. DeCastro and F. Morey Allen, *J. Urology 167*, 1774 (2002).
87. A. F. Morey, R. C. McDonough, 3rd, S. Kizer William, and P. Foley John, *J. Urology 168*, 627 (2002).
88. A. I. Chin, N. Ragavendra, L. Hilborne, and H. A. Gritsch, *J. Urology 170*, 380 (2003).
89. L. A. Evans, K. H. Ferguson, J. P. Foley, T. A. Rozanski, and A. F. Morey, *J. Urology 169*, 1360 (2003).
90. E. J. Hick and F. M. Allen, *J. Urology 171*, 1547 (2004).
91. E. K. Diner, S. V. Patel, and A. M. Kwart, *J. Urology 173*, 1147 (2005).
92. M. W. Noller, S. M. Baughman, A. F. Morey, and B. K. Auge, *J. Urology 172*, 166 (2004).
93. R. S. Pruthi, J. Chun, and M. Richman, *British J. Urology Int. 93*, 813 (2004).
94. E. Kouba, C. Tornehl, J. Lavelle, E. Wallen, and S. Pruthi Raj, *J. Urology 172*, 326 (2004).
95. B. C. Griffith, A. F. Morey, T. A. Rozanski, R. Harris, R. Dalton Scott, J. Torgerson Sigurd, and R. Partyka Scott, *J. Urology 171*, 445 (2004).
96. L. A. Evans and F. Morey Allen, *Intl. Brazilian J. Urology 32*, 131 (2006).
97. M. Jubran and J. Widenfalk, *Expl. Neurology 181*, 204 (2003).
98. R. Floris, C. Salvatore, B. Fraioli, F. S. Pastore, R. Vagnozzi, and G. Simonetti, *Neuroradiology 40*, 690 (1998).
99. P. Cappabianca, M. Cavallo Luigi, V. Valente, I. Romano, I. D'Enza Alfonso, F. Esposito, and E. de Divitiis, *Surg. Neurol. 62*, 227 (2004).
100. W. Xu, H. Li, T. Brodniewicz, F. A. Auger, and L. Germain, *Burns 22*, 191 (1996).
101. M. R. Kaufman, R. Westreich, M. Ammar Sherif, A. Amirali, A. Iskander, and W. Lawson, *Archives Facial Plastic Surg. 6*, 94 (2004).
102. I. Grant, K. Warwick, J. Marshall, C. Green, and R. Martin, *Br. J. Plastic Surg. 55*, 219 (2002).
103. S. Jorgensen, A. Bascom Daphne, A. Partsafas, and K. Wax Mark, *Archives Facial Plastic Surg. 5*, 399 (2003).
104. R. Mittermayr, E. Wassermann, M. Thurnher, M. Simunek, and H. Redl, *Burns 32*, 305 (2006).
105. P. K. Jain, R. Sowdi, A. D. G. Anderson, and J. MacFie, *British J. Surg. 91*, 54 (2004).
106. L. Johnson, T. E. Cusick, S. D. Helmer, and J. S. Osland, *Am. J. Surg. 189*, 319 (2005).
107. M. E. Jones, S. Burnett, A. Southgate, P. Sibbons, A. O. Grobbelaar, and C. J. Green, *J. Hand Surg. 27*, 278 (2002).
108. G. Fuerst, R. Gruber, S. Tangl, F. Sanroman, and G. Watzek, *Clin. Oral Implants Res. 15*, 301 (2004).

109. A. Bakay, L. Csonge, G. Papp, and L. Fekete, *Int. Orthop. 22*, 277 (1998).
110. H. K. Kjaergard, J. H. Pedersen, M. Krasnik, U. S. Weis-Fogh, H. Fleron, and H. E. Griffin, *Chest 117*, 1124 (2000).
111. A. Belboul, L. Dernevik, O. Aljassim, B. Skrbic, G. Radberg, and D. Roberts, *Europ. J. Cardio-Thoracic Surg. 26*, 1187 (2004).
112. P. J. O'Neill, H. L. Flanagan, M. C. Mauney, W. D. Spotnitz, and T. M. Daniel, *Ann. Thoracic Surg. 70*, 301 (2000).
113. E. D. Canby-Hagino, A. F. Morey, I. Jatoi, B. Perahia, and J. T. Bishoff, *J. Urology 164*, 2004 (2000).
114. P. Modi and J. Rahamim, *Europ. J. Cardio-Thoracic Surg. 28*, 167 (2005).
115. M. Schwartz, J. Madariaga, R. Hirose, T. R. Shaver, L. Sher, R. Chari, J. O. Colonna, II, N. Heaton, D. Mirza, R. Adams, M. Rees, and D. Lloyd, *Archives Surg. 139*, 1148 (2004).
116. A. Frilling, A. Stavrou Gregor, H.-J. Mischinger, B. de Hemptinne, M. Rokkjaer, J. Klempnauer, A. Thorne, B. Gloor, S. Beckebaum, F. A. Ghaffar Mohamed, and E. Broelsch Christoph, *Langenbeck's Archives Surg./Deutsche Gesellschaft fur Chirurgie 390*, 114 (2005).
117. H. E. Kaufman, S. Insler Michael, A. Ibrahim-Elzembely Hosan, and C. Kaufman Stephen, *Ophthalmology 110*, 2168 (2003).
118. I. Bar-Hava, H. Krissi, J. Ashkenazi, R. Orvieto, M. Shelef, and Z. Ben-Rafael, *Fertil. Steril. 71*, 821 (1999).
119. L. Bodner, J. M. Weinstein, and A. K. Baumgarten, *Oral Surg. 86*, 421 (1998).
120. A. H. Petter-Puchner, R. Fortelny, R. Mittermayr, W. Ohlinger, and H. Redl, *Hernia 9*, 322 (2005).

Index

Printed in the United States
by Baker & Taylor Publisher Services